Neurophysiology and Standards
of Spinal Cord Monitoring

T.B. Ducker   R.H. Brown
Editors

Neurophysiology and Standards of Spinal Cord Monitoring

With 245 Illustrations

Springer-Verlag
New York Berlin Heidelberg
London Paris Tokyo

Thomas B. Ducker, M.D., F.A.C.S.
Clinical Professor, University of Maryland, and Associate Professor,
Johns Hopkins University, Baltimore, MD, USA

Richard H. Brown, Ph.D.
Director, Division of Surgical Research, Research Director,
St. Luke's Spine Center, Cleveland, OH, USA

Library of Congress Cataloging-in-Publication Data
Neurophysiology and standards of spinal cord monitoring.
   Based on a conference held in Annapolis, Md., Oct.
21-24, 1986.
   Includes bibliographies.
   1. Evoked potentials (Electrophysiology)—Congresses.
2. Spinal cord—Diseases—Diagnosis—Congresses.
I. Ducker, Thomas B., 1937–   . II. Brown, Richard H.
[DNLM: 1. Evoked Potentials—congresses. 2. Monitoring,
Physiologic—methods—congresses. 3. Spinal Cord—
physiology—congresses. 4. Spinal Cord Diseases—
physiopathology—congresses. WL 400 N494 1986]
RC402.2.E94N48 1988   612'.813   88-20182

Camera ready copy provided by the editors.
Printed and bound by Arcata Graphics/Halliday, West Hanover, Massachusetts.
Printed in the United States of America.

9 8 7 6 5 4 3 2 1

ISBN 0-387-96634-X Springer-Verlag New York Berlin Heidelberg
ISBN 3-540-96634-X Springer-Verlag Berlin Heidelberg New York

Table of Contents

Standards

Ascending Recordings

Operative Data

Continuous, Chronic Changes in Spinal Cord Evoked Potentials

Conclusions

Standardization of Evoked Potential Recording

R. H. Brown;[*] C. L. Nash, Jr.

Introduction

There exists in the application of evoked potential monitoring today such a variety of techniques that it is virtually impossible for any one group to directly compare their results with those obtained from any other group. This is due in large part to the number of different systems, both in-house designs and commercial, that are in use today. Even among the commercial systems so many different settings are available to the end user that generally no two groups utilize exactly the same set-up procedures on their instrumentation. Additionally, the specific choices allowed often differ considerably from one manufacturer to another with the use of non-standard in-house designs further compounding this problem.

Even when the system hardware parameters are set in the same or even approximately the same manner, variations in method of instrumenting the subject under evaluation can also create problems in comparing obtained data. These variations involve such factors as type and point of application of stimulus, type and locus of recording sites, type of electrodes used and whether the recording systems are differential or single ended. When taken individually any of these factors can create problems in comparing data, taken together they create an almost endless variation on the one common theme we call evoked potential monitoring.

If evoked potential monitoring is ever to come into its own, such as ECG and EEG have, there must be an overriding cry for "STANDARDIZATION". While recognizing that our knowledge of many of the modalities employed may at this point in time be insufficient to agree on the best technique to apply for a given situation, there must be agreement on at least standardization in reporting of one's results. This will not necessarily allow direct comparison of data obtained from one institution with that obtained elsewhere, but it will at least give the reader a thorough understanding of the techniques employed in obtaining the presented data. This standardization must involve all aspects of evoked potential monitoring, including both hardware and patient interfacing.

Therefore, any report involving the use of evoked potential data should include a complete description of the system used in terms of all its hardware settings and a comprehensive description of the patient interfacing relative to standard anatomical sites.

Standardizing your System

Before any given system can be put into routine clinical use, normative data must be acquired for that particular system. This is primarily due to the fact that many of the systems available on the market today are configured with subtle but significant dif-

* Surgical Research Division, Saint Luke's Hospital, 11311 Shaker Boulevard, Cleveland, OH 44104

ferences which can affect the actual waveshape obtained. Even when two systems are identical in manufacture, different methods of applying the technique from institution to institution can also have a significant effect on the obtained evoked potential. Also, for intraoperative use the anesthetic considerations involved must be standardized and normalized for a given application. This requires that for each intended application implemented by a user, a strict protocol be developed and adhered to in the process of acquiring this normative data.

In the case of diagnostic applications where manufacturer supplied procedures and data are not utilized, a series of normal volunteers or patients with no apparent or known neurologic compromise must be evaluated with the system to document the variability and reproducibility of the system. This is best done by scheduling each volunteer for a minimum of two test sessions separated by at least two weeks. The data from the first test session should then be compared to the data from the second test session and any variances noted. For a good system, with rigid protocol application, the variances should be slight or non existent. If reproducible data cannot be obtained from the normal volunteers with as little as a two week interval, then the reliability of any such system applied to diagnostic evaluations would obviously have to be suspect. Once the system has been shown to be reproducible on an inter and intrasubject population, then the entire series of normals needs to be compared in terms of amplitude and peak latency. These data then become the normative data base from which all other measurements can be made.

In the case of intraoperative application of the system, a similar standardization needs to be undertaken. In this instance it is obviously not possible to use normal volunteers for intraoperative procedures. Instead, those patients in whom monitoring is being used prophylactically and who have no known neurologic or other mediating central nervous system dysfunction should be chosen for normalizing the system. All of these patients should receive standard preoperative medications and be subjected to a complete preinduction evaluation in Surgery. A full set of data, including backgrounds, should be obtained. This same data set needs also to be obtained as shortly as possible after induction, just prior to and subsequent to any significant surgical manipulation and at the time of closure. Then retrospectively, all those individuals in whom no significant monitoring changes were noted as the result of the procedure should be culled and used to construct the normative intraoperative data base. This approach is utilized to exclude any significant input from the actual surgical procedure itself and to account for the predictable changes occurring from the effects of preoperative medication and the routine anesthetic agents utilized for the procedure. A particular attempt should be made to include in this group individuals undergoing two stage procedures who are otherwise neurologically intact. This will serve to further demonstrate the intrasubject variability resulting from the standardized operative protocol. These data then, as in the case of the diagnostic data base, serve as the basis from which to evaluate changes which may occur intraoperatively. The expected values for a given system under a specified protocol should be included in any serious presentation.

General Guidelines

Regardless of the type of evoked potential monitoring being undertaken, be it somatosensory, spinal, visual, motor, auditory or any other modality, certain basic components are involved in obtaining the desired evoked potential; a system for generating the appropriate stimuli to the CNS pathway being provoked, a method for recording the associated evoked response from the appropriate anatomical site and some overall piece of hardware to synchronize the two and average the responses to increase the signal to noise ratio. This system needs to be described in terms of a unique descriptor of the stimulus employed, a unique description of the type of recording montage used, all of the engineering specifications relative to the hardware system utilized to record and average the evoked potential and the analysis conducted once the averaged evoked potential has been obtained. Any other salient specifications relative to this system which would help to clarify any variants or obtained data must by necessity also be included in this description. Only when this is done can the evoked potential community begin to compare and learn from the many and various approaches that are being conducted throughout the world.

Standardization of Stimulus Parameters

The generic type of stimulus utilized needs be specified such as constant current or voltage, tactile or vibratory, visual etc. In addition to the type of stimulus being used, the parameters relative to that particular stimulus need also to be precisely defined. For instance, in the case of constant current stimulus whether it is bipolar or monopolar, whether percutaneous or transcutaneous electrodes are employed, the stimulus duration and intensity, the stimulus frequency and whether it is periodic or random. In addition to describing the technical aspects of the actual stimulus, it is also necessary to accurately describe the point or region of application to the central nervous system. In the case of the above example of a constant current stimulus simply saying it was applied to the lower extremity is inadequate in describing its point of application. Specifically, a more precise description would include application to the posterior tibial nerve just posterior to the medial malleolus or the median nerve just proximal to the wrist.

A good way to record and/or present these parameters is by use of a table including at least electrode type, stimulus modality, intensity, duration, frequency and location (see appendix).

Standardization of Recording Parameters

Since the method utilized to record an "evoked" potential from a given individual can significantly alter the waveshape obtained it is equally imperative that any report involving recording of evoked potentials accurately and completely define the methodology employed in acquiring these signals. The first and most obvious descriptor needs to be the anatomical location from which the recordings were made. These are generically classified as cortical, brainstem, spinal, and/or peripheral. In order to understand the actual recording procedure it is necessary to specify the type of recording employed and whether single-ended or true differential recording is being utilized at the input amplifier to one's system. Also, the specific placement of each electrode utilized in acquiring a given evoked potential needs to be precisely specified. This can be either in terms of the international 10-20 system or any other recognized system or a modification thereof. Regardless of the system used for placing the recording electrodes, the anatomical location of the active, indifferent and reference electrodes needs to be stated in unequivocal terms. In those systems that do not float their input, the active, reference and ground electrode locations need to be specified. The type of electrodes utilized also needs to be specified in terms of either needle or surface electrodes. Where surface electrodes are utilized the type of surface electrode, skin preparation employed and method of fixation to the skin also needs to be described.

Once the electrode interface scheme has been adequately described the system's input specifications need to be listed. These involve specifying the system's input impedance, the input amplifier's characteristics in terms of its common mode rejection ratio in db, its input filter's -3 db points (half-power points) and the fall off slope of the filters utilized. This latter is as important as specifying the -3 db points in that a system utilizing 6 or more pole filters will obviously fall off and attenuate the input signal much more rapidly than a system using a 2 or 4 pole input filter. These filters are generally specified in terms of db per decade or db per octave. Additionally, the inclusion of any type of 60 Hz notch filter needs also to be documented.

The input amplifier specifications also need to be documented in terms of the overall system gain from the recording electrodes to the averager. At this point in most systems the input amplifier signal is digitized for storage and post processing by the system's computer. This portion of the system needs to be defined in terms of the resolution of the system's analog to digital converters. A system with an eight bit analog to digital converter can only resolve the total range of the input signal into one in 256 while an analog to digital converter with a 12 bit resolution can resolve the input signal to one part in 4,096. As you can see, the addition of 4 more bits to a system's input A-to-D significantly improves the system's resolution by a factor of 16.

Once the analog to digital system's resolution has been specified, it is also necessary to specify the rate at which the samples were acquired, i.e. how many samples per second were utilized to digitize the analog waveform into the system. This has two very important characteristics associated with it. The first is obviously the temporal resolution one can claim on measuring the latencies associated with the obtained response. A system which is sampling at 1,000 Hz can only resolve a peak latency to within one msec. However, a system running at 10,000 Hz can resolve a peak latency to 0.1 msec or 100μsec. Depending on the type of potential being analyzed, this can be significant. The second has to do with a little understood theory in digital signal processing known as the Nyquist criteria. This theorem states that to faithfully bring data into a digital system, one must sample at a frequency at least twice that of the highest frequency component being allowed through the system. In the case of most evoked potential systems, the required sample frequency would be at least twice the system's upper filter's - 3 db set point. If this rule is not adhered to, any complex post processing attempted on the data stored in one's system will be fraught with many errors.

Once the manner of digitization has been specified, including whether any automatic artifact rejection has been incorporated into the system, the number of averages or other type of signal enhancement actually utilized to obtain the final response needs to be specified. This information is not just part of the standard specification for one's system but also gives the observer an appreciation of a given system's signal to noise figure of merit. That is, a system that can obtain highly reproducible responses with fewer averages would be the system with the higher inherent signal to noise ratio.

Once the final averaged response is obtained, additional information needs to be conveyed to the observer. The most important of these are polarity and time base. The first involves the age old debate is positive up, or is positive down? This author has a strong bias that in the natural sciences and in all engineering and mathematical calculations, positive has assumed an upward direction and time directed towards the right I strongly support this convention for the displaying of neurophysiologic data. However, it is honorable among learned men to agree to disagree and in this vein it is therefore acceptable to display one's data with positive down, provided one specifies that positive is down. In addition to polarity, a calibrated amplitude scale needs also to be displayed concomitant with any evoked potential data. This way polarity and magnitude are easily discernible by the reader. In terms of time base, again an overall time base needs to be specified and the scale of the time base needs to be included with the presented data. Two successive averaged evoked potentials should be presented to show system reproducibility as well as a no stimulus background record of an equal number of averages to demonstrate the overall systems inherent signal to noise merit. The "overall" system includes not only the hardware but the electrode interface and the patient.

As with the stimulus parameters, the recording parameters might be summarized in tabular form (again see appendix).

Environmental Considerations

Ideally, evoked potential monitoring of any type should be conducted in a quiet comfortable atmosphere conducive to achieving maximal relaxation and cooperation in the subject undergoing evaluation. This would be a totally shielded, quiet, pleasantly appointed room maintained at a comfortable temperature and humidity, furnished with comfortable furniture, including an examination table for supine testing. This is what we all tend to wish for when we use the term "controlled laboratory conditions". This is because even the best system in use today will have difficulty in obtaining clean responses in the presence of the resultant significant myoelectric iatrogenic contamination if the subject is in a highly anxious state. It is in this environment that we can achieve the optimally clean signals from which we can compile our normative data base. Also, it should be under these idealized conditions that any diagnostic evaluations are conducted whenever possible, since these conditions tend to put the individual at their ease, thereby facilitating maximum subject cooperation. It also allows the expeditious taking of any pertinent history as well as the conduction of any other requisite pretesting or evaluations.

Conversely, because data taken in other than ideal conditions can be fraught with problems such as competing diagnostic monitors both attached to the individual as well as in any adjoining areas, data taken in such areas as the Emergency Room, Recovery Room, ICU and/or the Operating Room should be clearly labeled as such. In these instances it is also important that a clear yet succinct summary of the patient's condition dictating the use of evoked potential monitoring be listed along with the data. While recognizing how nebulous the concept of normative data relative to evoked potential testing in these areas might be, any deviations from these standards of performance for a given system functioning in these areas should also be noted when reporting the data. Perhaps the one area where a reasonable degree of control can be exercised outside ideal laboratory conditions is the use of monitoring as an adjunct to surgery. Since these are usually scheduled procedures, a protocol can and should be developed for the use of monitoring as an adjunct to surgical procedures. Such a protocol might be configured as follows:

l. Preoperative medication

Use of a standardized preoperative medication whose effects are either minimal or well defined relative to the changes they cause in the monitoring records from accepted standardized values. One such medication that we have used routinely is Secobarbital 2 mg/kg I.M. on call to Surgery. Atropine 0.4 is usually added by Anesthesia to control secretions. In those instances where a patient is exhibiting above normal anxiety, morphine sulfate 5-10 mg may be added to the preoperative medication to help control this anxiety. The effect of this latter is to slightly increase the later components of cortically obtained evoked potentials while essentially having little or no effect on the onset of the primary peaks of the complex.

2. Preoperative monitoring

Preoperative monitoring should be conducted whenever possible. The most ideal time to do this is just prior to the procedure, following the administration of any preoperative medication but with the patient awake prior to the onset of anesthesia. This is because given individuals react differently to the effects of anesthesia and this will not be appreciated unless preanesthesia records have been obtained. Secondarily, this also allows the conduction of what essentially amounts to a diagnostic evaluation immediately prior to surgery. This evaluation can often identify individuals who tend to fall outside established limits thereby indicating an increased at risk status during the procedure. This information can be extremely beneficial to the surgeon in planning his procedure. It also has the tertiary benefit of making available control records on each individual should postoperative monitoring be required as a follow-up in the Recovery Room, ICU or laboratory.

Finally, it is far easier to trouble-shoot any problems which may have arisen in implementing the surgical monitoring with the patient awake, not paralyzed and especially not out of sight underneath many layers of surgical drapes in the middle of a sterile field. High quality preoperative records gives one a sense of confidence that any changes observed during surgery are the result of either the surgical intervention or changes in anesthetic protocol and not due to a problem with the implementation of the monitoring.

3. Anesthetic factors

Once quality preoperative records are obtained, it is necessary that anesthesia be used that is consistent with the continued acquisition of quality evoked potentials. While a fair amount of literature reports that the halogenated agents do not seriously affect the recording of direct spinal responses, these agents do indeed seriously affect the recording of brainstem and cortical responses because they are highly lipid soluble agents and readily pass through the blood-brain barrier directly affecting the cortex. While low-dose Ethrane (0.5% or less) can sometimes be used without seriously effecting the cortical responses, this is not the anesthesia of choice for cortical monitoring. Unfortunately, many of the problems with cortical monitoring reported in the literature imply the use of halogenated agents but fail to adequately describe their technique. This must raise the question - were these failures of monitoring or were they failures to monitor?

Somatosensory-Cortical evoked potential monitoring is perhaps the most sensitive of all the monitoring techniques to the effects of anesthesia. The following is an example of how these effects are controlled by STANDARDIZING the anesthetic protocol. Good results have been reported in the use of a balanced nitrous-narcotic and muscle relaxant technique for intraoperative monitoring. With this type of anesthesia the muscle relaxants have been shown to have little if any effect on the transmitted evoked potential. Also, by utilizing standard infusion techniques, the analgesic can be infused at a constant rate, eliminating the variations in the response oft times seen with the administration of significant boluses of these agents. That is, the alternately depressing effect on cortical amplitudes and latencies immediately following the administration of a bolus and the concomitant increases in these parameters as the bolus dose wears off. Prior to adopting the infusion technique it had been our experience that just when the monitoring became critical, anesthesia would succumb to the urge to administer a healthy bolus of analgesic. The use of the infusion technique not only eliminated this variant but also allowed the level of analgesic usage in a given procedure to be standardized.

When Fentanyl is the analgesic of choice, good results can be obtained by giving a reasonable loading dose at the time of induction and then maintaining the subject at a continuous infusion of nominally 2-4 mcg/kg/hr. This rate can then be individualized to each individual's stress response as surgery progresses without rapidly changing blood levels.

4. Physiologic factors

The one drawback to this technique is that it is not in and of itself conducive to controlling hypertension. However, if this is recognized it poses no significant problem. At the time of induction the patient is routinely started on a controlled infusion of nitroglycerine prophylactically to control blood pressure. If this fails to be effective, sodium nitroprusside is used to control the pressure. The advantage of this technique is that it facilitates not only rapid but repeated wake-up tests should the monitoring prove to be equivocal. Simply by placing the individuals on 100% oxygen and reversing muscle relaxants as required, the wake-up test can generally be accomplished in ten minutes or less with deepening of anesthesia being accomplished in most instances by just readministering the nitrous. Occasionally Ethrane may be briefly required to expedite the deepening of anesthesia.

5. Biochemical factors

Since individuals respond differently to even the most standard of medications, during intraoperative monitoring a continuous log needs to be kept by the monitorist which lists all anesthetic interventions. Specifically, the time, amount and nature of the agent administered needs to be recorded. Even such benign interventions as repeated doses of muscle relaxants and the administration of blood and blood products should be recorded. Of course, exact ratios of nitrous to oxygen, levels of any inhalation agents and the infusion rate of any continuous medications need to be recorded throughout the procedure. Frequent blood gases need also to be obtained to insure that the patient is not being maintained too hypocarbic since this tends to increase the permeability of the blood-brain barrier and increases the anesthetic effect on the cortex beyond that which would normally be expected. In general the patient should be maintained normocarbic, normotensive, normothermic and normovolemic. The importance of maintaining blood pressure near normotensive values has been repeatedly demonstrated by the deterioration of monitoring during periods of hypotension sufficient to call for a wake-up test only to have the monitoring return toward baseline as the pressure increased as a result of initiating the wake up test. We now routinely raise the blood pressure prior to actually initiating a wake up and have eliminated the need for many a wake-up test.

As an adjunct to this list, all supplemental medications and the time of last administration should also be recorded separately on the intraoperative log sheet. These are medications separate from those anesthetically administered and generally consist of routine medications that may have been discontinued prior to surgery. Occasionally supplemental agents will be discontinued prior to surgery but will have a half-life which extends well beyond the time of last administration and as a result can have a distinct affect on the type of evoked potentials obtained. Many of the medications routinely used to control seizures or to treat depression or other psychotic disorders act directly or indirectly on the central nervous system and can have half lives measured in days. This is another reason to utilize awake preoperative monitoring as part of a standard protocol.

All of these data need to be recorded prior to any evoked potential session and periodically throughout any surgical monitoring. This is most easily done using tables and/or flow sheets which includes at least the location of the procedure, any routine medications on board, diagnosis if known, type of anesthesia if a surgical procedure and the patient's biochemical status if available. A more detailed listing of these factors is given in the appendix.

Terminology

Of all the problems inherent in reporting results of evoked potential monitoring, perhaps the most significant problem being faced by the evoked potential community is that of the standardization of a reporting format. This is due primarily to the fact that the human central nervous system can be perturbated in an almost endless variety of ways any one of which can result in an evoked response being measurable at some other point in the system.

In reviewing 46 manuscripts for these proceedings, nine did not use any mnemonic to describe their technique. Of the remaining 37, 13 different mnemonics were employed. In rank order the terms used among the 46 manuscripts were;

SEP	17	EEP	1
SER	1	SSEP	3
NONE	9	BSEP	1
ESCAP	1	CAP	2
SCP	3	SCM	1
CSEP	1	ESEP	1
EP	3	SCEP	1

Unfortunately, attempts to define a unique list of descriptors has resulted in what amounts to a geometrically expanding, unending list of abbreviations. Recognizing the futility of this approach, the following modified system is postulated for consideration as a standardized reporting format.

Since all "evoked" potential monitoring results from stimulating the patient at one site and recording the evoked potential at a separate site, these basic descriptors need to be included with each record. To facilitate this, the following list is postulated for use by the evoked potential community. It will be noted that each abbreviation consists of the first TWO letters of each word abbreviated. The use of the two letter notation hopefully will obviate the confusion that has been rampant with the use of the letter S. In current literature one finds the letter S used to represent spinal, somatosensory, sacral and subdural. Reference to the list given will show that these all can be specified simply by using the first two letters of the desired term. With this introduction the list is given as follows;

Mnemonic Terminology

Au - Auditory	Ep - Epidural	So - Somatosensory
Br - Brainstem	Er - Erb's Point	Sp - Spinal
Ce - Cervical	Lu - Lumbar	Su - Subdural
Co - Cortical	Pe - Peripheral	Th - Thoracic
De - Dermatomal	Sa - Sacral	Vi - Visual

From this chart one can pick the appropriate descriptors of both the stimulus site & recording site and follow them with the capital letters E.P. to denote that this is an evoked potential record. For instance, if one was doing spinal-spinal recording the denotation would be Sp-Sp E.P. which would clearly represent a spinal-spinal evoked potential recording. Conversely if somebody was doing somatosensory cervical, they would denote that it was So-Ce E.P. recording. The specific placement of their stimulus electrodes and recording electrodes, method of recording, band widths, etc. would then be clearly defined in their methodology section.

Methods of Reporting

In summarizing, any serious work on evoked potential monitoring should meet basic criteria for both hard copy presentation and information to be contained in the methodology section.

1. Hard copy records

For those papers presenting hard copy records, at least two successive records should be presented to show that the data being acquired is indeed reproducible. Additionally, a no stimulus background record should be included somewhere in the paper or presentation such that the signal to noise merit of the system being utilized can also be appreciated. On the hard copies themselves there needs to be a calibrated amplitude scale including polarity, a calibrated time base in msec, the peak latencies if desired of the primary complex and sufficient other information to render the stimulus and recording sites to be unambiguous.

2. Methodology

The specifics of what should be included in the methodology section are the parameters associated with application of stimulus, specifically the type of stimulus, the type of electrodes utilized in applying the stimulus, the stimulus intensity, the stimulus frequency and whether it is fixed or aperiodic, and the number of stimuli delivered. Relative to the recording, the type of recording, whether it is single ended or differential, the type of electrodes utilized, the actual electrode montage referenced to unambiguous anatomical landmarks or conventions, the bandwidth, both upper and lower, of the amplifiers, the overall gain of the amplifiers, the sample frequency of the analog to digital converters, the number of averages summed to get the final hard copy, any artifact rejection schemes that are employed as well as any post processing or smoothing of the data prior to presentation. For the papers reviewed for this proceedings, only 28 of 46 or 60% included even a basic system description.

In the case of intraoperative monitoring, the use of any preoperative medications and time of administration should be noted, the type of anesthesia utilized, listing not only the agent but the ratios in terms of oxygen to nitrous oxide or percent of inhalation agent and/or whatever else may have been used, the infusion rate or frequency and magnitude of bolus administration, and the use of any other adjuncts to the anesthetic regime. Of particular interest in cortical monitoring is also frequent evaluations of blood gases to show that the patients were not being kept too hypocarbic which as noted above, effects the blood-brain barrier. Other notations such as deliberate use of hypotensive anesthesia, whether the individuals were allowed to become hypovolemic, hypothermic or any other physiologic parameter that was purposely allowed to deviate from normative baseline conditions for that individual need to be recorded. As with nonoperative diagnostic evaluations, any ongoing medications that the individual has been on should also be noted. Again, from the reviewed manuscripts, only 10 of the 46 included sufficient anesthesia/medication information.

Summary

The foregoing is presented as an approach to the standardization of the methods of reporting of evoked potential data. If the evoked potential community can come to terms with this problem, the dissemination of information and resultant increase in the speed of our mutual learning and understanding of the full potential of our techniques will be greatly enhanced. It will also insure that some of the more deleterious reports that have appeared in the literature appear so with the proper documentation and substantiation. Any report that lacks a detailed description of the system utilized will seriously hinder the interpretation of its results. This matter is of critical importance in studies where there is a question of "false negative" results of monitoring. The ultimate effective utilization and determination of the role and place of spinal cord monitoring can only be built on a base of clearly defined system design and application.

APPENDIX

Example Evoked Potential Worksheets

Stimulus Parameters

Modality:	constant current, voltage, strobe, clicks etc.
Electrodes:	type - percutaneous, transcutaneous, transcranial, eye goggles, ear phones etc
Intensity:	volts, milliamps, db, lumens etc.
Duration:	effective length of each stimulus delivered
Frequency:	Interstimulus interval - periodic/aperiodic
Location:	specific anatomic description of electrode site
Other:	any other information required for the reader to replicate the set-up

Recording Parameters

Location:	specific anatomic description of electrode site
Modality:	differential or single-ended input to amplifier
Electrodes:	type - transcutaneous, percutaneous, epidural, bone etc.
Gain:	overall system gain from electrodes to display
Bandwidth:	-3db bandwidth and filter type utilized, plus notation if a 60 Hz. notch filter was employed
Sample rate:	the rate at which the digitizer samples each incoming response for enhancement and display
Enhancement:	the technique used to enhance the desired response - generally straight averaging
Polarity:	which way is up - positive or negative?

Anesthetic/medication Parameters

Routine meds:	any routine meds being administered or taken by the subject need to be recorded
Pre-op meds:	dosage and time of administration of any preoperative medications given
Induction:	the agent used for induction of anesthesia together with time of induction
Maintenance:	the agents and level of administration employed for maintenance on a chronological basis
Analgesic:	agent and method of administration (infusion or bolus) on a dosage/time basis
Normos:	usually blood gas parameters but core temperature, volemic status and blood pressure should be recorded on a periodic basis
Other:	any other factor that might affect the response being recorded should be recorded for retrospective analysis

Clinical Neurophysiology of Neural Stimulation

M. R. Dimitrijevic[*]

Advances in biomedical engineering (equipment, recordings), and in the neurophysiological analysis of the peripheral and central nervous system, combined with the need to monitor functions of the nervous system during general anesthesia and surgical interventions on the spine and spinal cord, have led to the recent development of clinical neurophysiology for spinal cord monitoring.

Long before spinal cord structures were stimulated and monitored, brain cortex stimulation became a clinical neurophysiological procedure during the surgical treatment of epilepsy, the removal of brain tumors adjacent to the motor cortex, and stereotaxic procedures for the treatment of movement disorders. This is probably due to the fact that the brain's gray matter is on the surface like all other parts of the central nervous system, unlike the spinal cord where the gray matter is situated only inside white matter pathways.

Clinical diagnostic or monitoring routines include today the stimulation of sensory peripheral nerve fibers and recordings of cortical somatosensory responses or spinal cord somatosensory potentials. It is possible to also stimulate the posterior structures of the spinal cord from epidural or subdural sites, record from the sensory cortex, or stimulate the motor cortex transcranially, and record from the descending spinal motor tracts, motor neurons, or muscles. Epidural or subdural stimulation of any of the spinal cord tracts will also elicit orthodromic and antidromic volleys along caudal and rostral portions of the spinal cord, and electrical events of conducting or tract waves. These procedures are not invasive when surface electrodes and the averaging technique are used. They are invasive when stimulating and recording electrodes are placed transcutaneously through the needle inserted into the epidural or subdural space. The risks of infection and destruction are minimal. By means of radiological techniques, the electrode can be guided toward the desired site in the spinal cord. The accuracy of the position can be verified through neurophysiological stimulation and recording techniques. Thus, it is possible to monitor the physiological condition of particular ascending and descending conducting white matter structures by recording conducting, traveling and segmental stationary waves.

Each of the techniques has a particular technical characteristic and specific neurophysiological significance. However, so far none can fulfill all the expectations required of a fully reliable technique to recognize and predict the immediate changes in spinal cord functions during manipulation of the spine or intervention on the spinal cord.

---

* Section of Restorative Neurology and Clinical Neurophysiology, Department of Rehabilitation, Baylor College of Medicine, Houston, TX 77030

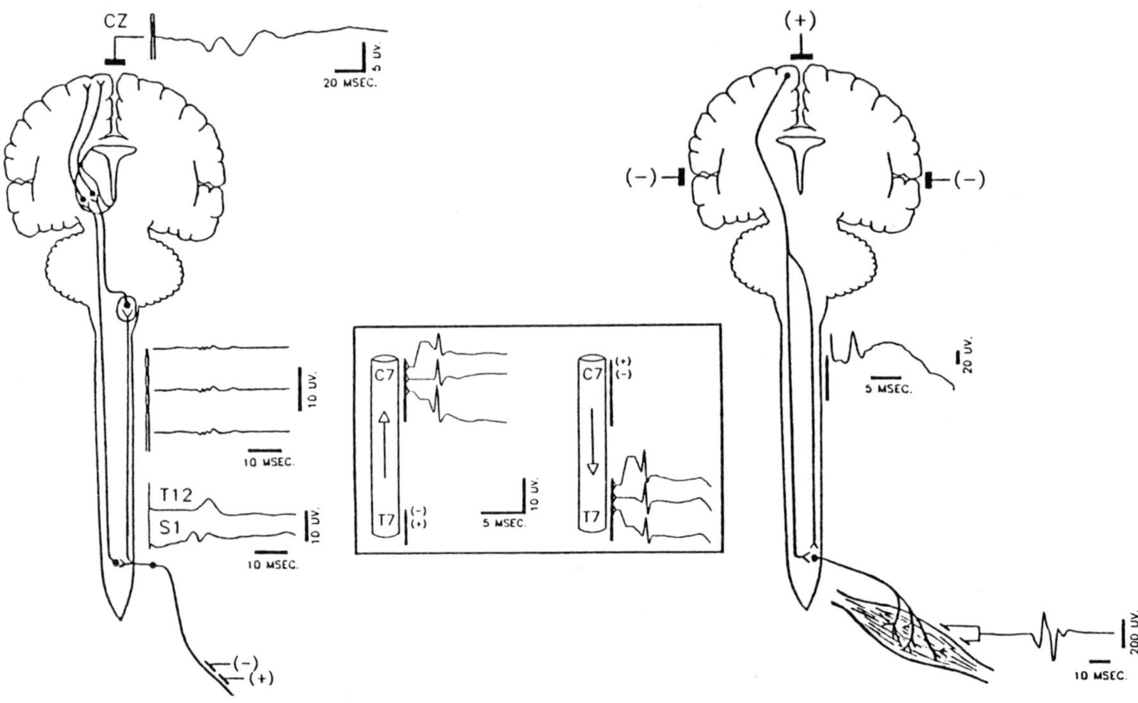

Fig. 1. Schematic illustration of presently used clinical monitoring techniques of spinal cord functions during surgical interventions on the spine or spinal cord: Cortical somatosensory evoked responses, spinal somatosensory evoked responses and responses elicited through motor cortex stimulation.

In the past, the orthopedic surgeon assessed spinal cord functions by awakening the patient during surgery. At that time, clinical neurophysiology of spinal cord functions was not available to monitor the activities of certain ascending and descending pathways during general anesthesia, and detect alarming transitory changes in spinal cord functions. In contemporary neurophysiological monitoring approach there are still different practical problems. The neurophysiological approach to monitor spinal cord functions is rooted in three different methodologies. The first, the method of somatosensory cortical evoked potentials, is based on recording activity from specific cortical regions of the brain cortex after stimulation of peripheral nerve structures and mediation of generated volleys through posterior spinal cord structures. Changes in electrophysiological characteristics, which occur while monitoring spinal cord functions, indicate that the functions of large diameter fibers of the posterior columns are impaired. No definite information is obtained about the anterolateral spinal cord ascending systems, or the functional condition of the spinal cord gray matter. Thus, this method is restricted to the pathways which are under examination and represents only the aspect of the sensory system which is involved in the intact modalities for touch perception, vibratory sensibility and position sense. Nevertheless, provided that general anesthesia does not interfere with the conducting and processing functions within the lemniscal system, somatosensory cortical evoked potentials are valuable indicators of the dysfunctions of the fast conducting system of the spinal cord.

The second method, more recently developed, is based on transcranial cerebral motor cortex stimulation to elicit conducting potentials of spinal lateral corticospinal volleys from subdurally or epidurally placed electrodes. Descending volleys depolarize the lower motor neuron and elicit postsynaptic conducting waves within the motor nerve fibers in the peripheral nerves and, if neuromuscular conduction is not blocked, the motor unit activity can be recorded by surface or needle electromyography. The value of this procedure is very similar to that of cortical somatosensory evoked responses. It is restricted to the fast conducting anatomical system of the spinal cord, the lateral and anterior corticospinal system. Therefore, when monitoring spinal cord functions by stimulating the motor cortex and recording presynaptic or postsynaptic electrical events, we can follow changes occurring within the lateral portion of the spinal cord. However, this method of transcranial stimulation does not test or monitor the bulbospinal portion of the motor system, known as the "extrapyramidal" motor system, which is functionally highly integrated with the "pyramidal" corticospinal tract within the spinal interneuron system.

During transcranial cortical stimulation and recordings of volleys elicited postsynaptically from spinal motor nerves or EMG responses, the presence of these postsynaptic events will indicate the integrity of the corticospinal tract and the successful transmission of corticospinal volleys to the lower motor neuron, which depend upon the functional condition of the interneurons and of the anterior horn of the gray matter. Thus, transcranial motor cortex stimulation tests not only the integrity of the functional condition of the fast corticospinal lateral system but other in line systems. This postsynaptic property of corticospinal tracts, when used to monitor spinal cord functions, sig-

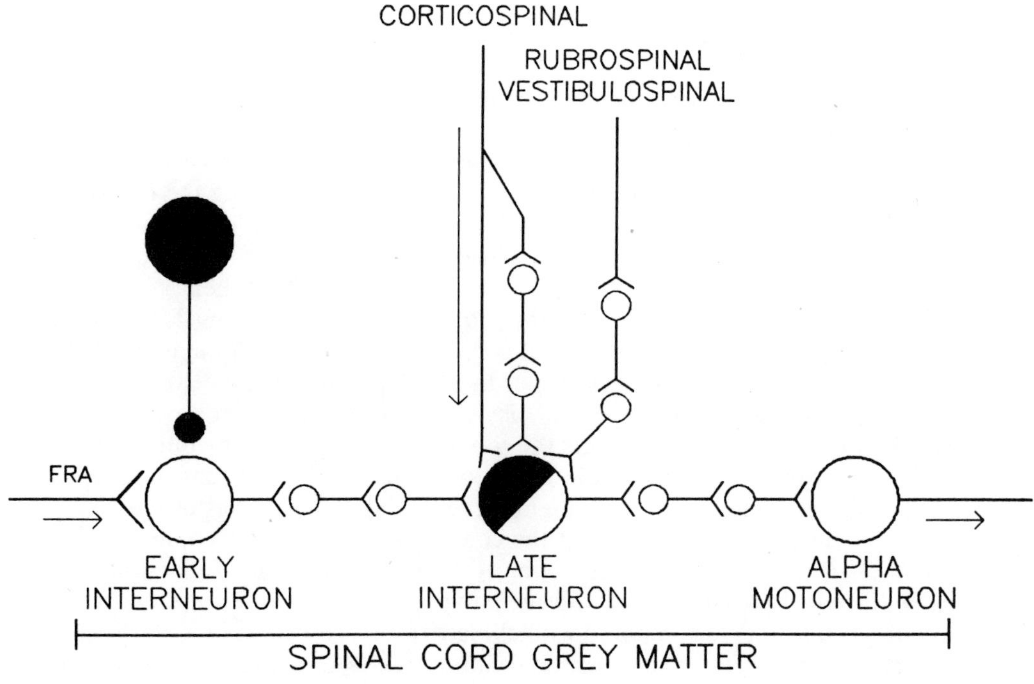

Fig. 2. Spinal motor cell presynaptic and postsynaptic aspect is situated in the gray matter of the spinal cord contrary to the white matter of the ascending spinal cord system.

nificantly enhances the value of this method as compared with spinal cord evoked potentials, which merely reflect the activity of the white matter within the spinal cord. This difference explains why there are more and more reports on experimental animals about corticospinal tract tests in induced spinal cord injury. This technique is more sensitive than tests carried out to measure responses of somatosensory cortical potentials (8).

The third clinical method for the monitoring of spinal cord functions is to record conducting waves from the spinal cord and segmental stationary waves from the posterior horns of the spinal cord's gray matter with multisite electrodes placed epidurally, caudally, and rostrally around the spinal cord on the segments which are under observation (6, 1). With this approach it is possible to record conducting waves by stimulating the peripheral nerve and to use the rostral and caudal electrodes either to stimulate or record, and to record conducting waves and spinal cord segmental waves through antidromic stimulation of the ascending tracts of the spinal cord. This methodology has not yet been fully developed, and mainly uses electrodes placed over the posterior structures of the spinal cord to record conducting waves and to elicit stationary segmental waves. It is thus possible to acquire more information about the functional condition of the posterior horn of the spinal cord. In the near future, we expect to benefit from the epidural placement of electrodes to record conducting and stationary waves after transcranial electrical stimulation. Moreover, it will also be feasible to examine that portion of the spinal cord under examination by placing it between two multisite ring electrodes. This approach is under development and is known as "SESCE" (System for Epidural Spinal Cord Evaluation) (2). We expect that SESCE will monitor more efficiently the functions of the spinal cord in neurological conditions (3). On the basis of the findings thus obtained, different restorative neurological procedures will be selected to upgrade motor and sensory functions in patients with acute or chronic neurological disorders (4, 5).

As a conclusion to all these remarks on clinical neurophysiology of neural stimulation to monitor spinal cord functions, let us not forget that patients during surgical spinal manipulation are not awake. Our goal, when monitoring spinal cord functions, is to demonstrate that these functions are fully preserved without having the opportunity to examine the patient's sensation and volitional activity. The difficulty lies in the fact that intact sensation and motor activity result from the integrated activity of descending and ascending tracts, their fast and slow conducting fibers, as well as their integration in the spinal interneuron system. When we monitor spinal cord functions during general anesthesia we aim to detect transitory spinal cord dysfunctions which might be short or long lasting. We are limited in our endeavor by the drawback that we are monitoring only part of the spinal cord and the electrophysiological events which are relevant for neurological functions. We do not assess the functions which are highly integrated with the ascending, descending, and spinal interneuron system but, by means of a selected methodology, we artificially disconnect them by recording the activities of the fast conducting fibers of the posterior columns or the corticospinal tracts. This basic difference between clinical neurophysiological and clinical neurological evaluation of spinal cord functions will help us understand the negative as well as the positive clinical neurophysiological results as compared with clinical results (7).

The capacity to monitor the functional integrity of the spinal cord, different posterior and lateral white matter conducting structures, and the posterior and anterior horns of the gray matter of the spinal cord represents a significant achievement in the neurology of subclinical events. The degree to which subclinical neurology is pertinent to clinical neurology depends upon the extent to which we can integrate monitored and recorded subclinical events.

References

1. Beric, A.; Dimitrijevic, M.R.; Prevec, T.S.; Sherwood, A.M.: Epidurally recorded cervical somatosensory evoked potentials in humans. Electroencephal. Clin. Neurophysiol., 65: 94-101, 1986.
2. Dimitrijevic, M.R.; Sherwood, A.M.: A system for epidural spinal cord evaluation. Appl. Neurophysiol., 45: 465-467, 1982.
3. Dimitrijevic, M.R.; Faganel, J.; Sherwood, A.M.: Spinal cord stimulation as a tool for physiological research. Appl. Neurophysiol., 46: 245-253, 1983.
4. Dimitrijevic, M.R.: Restorative neurology: Introductory remarks. In: J. Eccles; M.R. Dimitrijevic (eds.): Recent Achievements in Restorative Neurology: Upper Motor Neuron Functions and Dysfunctions. Vol. 1. Karger: Basel, pp. 1-9, 1985.
5. Dimitrijevic, M.R.: Residual motor functions in spinal cord injury. In: S.G. Waxman (ed): Physiologic Basis for Functional Recovery in Neurological Disease. Raven Press: New York (in press), 1987.
6. Halter, J.; Dolenc, V.; Dimitrijevic, M.R.; Sharkey, P.C.: Neurophysiological assessment of electrode placement in the spinal cord. Appl. Neurophysiol., 46: 124-128, 1983.
7. Lesser, R.P.; Raudzens, P.; Luders, H., et al.: Postoperative neurological deficits may occur despite unchanged intraoperative somatosensory evoked potentials. Ann. Neurol., 19: 22-25, 1986.
8. Simpson, R.K.; Baskin, D.S.: Corticomotor evoked potentials in acute and chronic blunt spinal cord injury on the rat: Correlation with neurological outcome and histological damage. Neurosurg., Vol. 20, 1:131-137, 1987.

Ascending Recordings

Human Spinal Cord Potentials (SCPs): Ascending Recording Variations -- an Update

K. Shimoji[*]

Summary

The basic physiological features of spinal cord potentials (SCPs) must be well-documented for accurate monitoring of human SCPs during surgery and also to improve our understanding of the neurophysiological background involved in neurological diseases. This paper surveys the fundamental patterns of human SCPs, recorded from the epidural space (ES) in response to various peripheral nerve stimulations, based on the data in our laboratory.

I. Human SCPs evoked by stimulation of the segmental nerves (segmental SCPs)

1. The fundamental pattern

Stimulation of the peripheral nerve produces a potential change in the posterior ES of the same or adjacent segment innervating the nerve (1). Examples of the segmental SCPs recorded in 4 different persons by stimulation of various peripheral nerves at 3 x threshold are shown in Fig. 1. The potential (segmental SCP) consists of the initially positive spikes (P1), which are sometimes not recordable, and the subsequent slow negative (N1) positive (P2) waves (2-7). The P2 wave is sometimes followed by a slower negative (N2) wave when the stimulus strength is increased (3). This fundamental pattern is very similar to that of the cord dorsum potential in animals (8-13). The N1 wave has a relatively short duration (5.5-8ms). The P2 wave has a longer duration of 27-90ms, depending on its decay (6). The waveforms of the segmental SCPs recorded in the cervical and lumbar enlargements are very similar to each other. The central latencies of the N1 components, however, calculated from the peak of the initially positive spike (P1), are shorter for the SCPs evoked in the cervical enlargement than for those in the lumbar enlargement (14).

On the decaying phase of the P2 wave, the secondary positive peak (P2s), preceded by the secondary negative dip (N1s), sometimes appears in the wakeful subject when the stimulus intensity is increased or when the subject is excited (5, 6). Both the N1s and P2s potentials are very vulnerable to anesthetic drugs (3, 4). Similar secondary waves can be demonstrated in the lightly anesthetized rat and are abolished by transection of the cord at C1 through C2 level (15). Therefore, these secondary waves, N1s and P2s, are considered to be produced by a feedback loop via supraspinal structures

* Department of Anesthesiology, Niigata University School of Medicine, 1-757 Asahi-Machi, Niigata 951, Japan

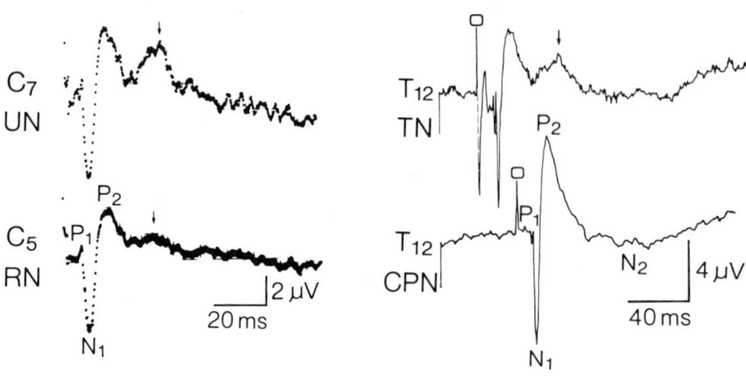

Fig. 1. Fundamental patterns of the SCPs evoked by the segmental nerves in four different subjects (6). All SCPs were recorded from the posterior ES. The vertebral level of the recording electrode and the peripheral nerve stimulated are shown at the left of each sweep. UN = ulnar nerve stimulation at the wrist. RN = radial nerve stimulation at the wrist. TN = tibial nerve stimulation at the popliteal fossa. CPN = common peroneal nerve stimulation at the popliteal fossa. Arrows indicate the secondary component of the P2 wave. Symbol O denotes the stimulus artifact.

(3-7). Several lines of evidence indicate that the P1, N1 and P2 waves reflect the afferent volley of the roots, synchronized activity of dorsal horn neurons and primary afferent depolarization (PAD), respectively.

2. Peripheral origins

Although a relatively weak stimulation of the large nerve (CPN) produces the clear configuration of the segmental SCP, even an intense stimulation of a more peripheral site, such as the fifth toe, evokes no response in the corresponding posterior ES (6). Fig. 2 shows the segmental SCP recorded from the posterior ES at the T12 vertebral level (lower traces) and the simultaneously recorded SERs from the scalp (upper traces) in response to the CPN (left) stimulation at two times threshold and painful stimulation of the fifth toe (right) (6). The typical segmental SCP is produced by weak stimulation of CPN, but not by even an intense stimulation of the fifth toe. On the other hand, the SERs from the scalp are clearly demonstrated by both stimulation. It is therefore postulated that stimulation of the large nerve fibers at a rostral site is more effective for producing the segmental SCP than that of the small skin nerves at a distal site.

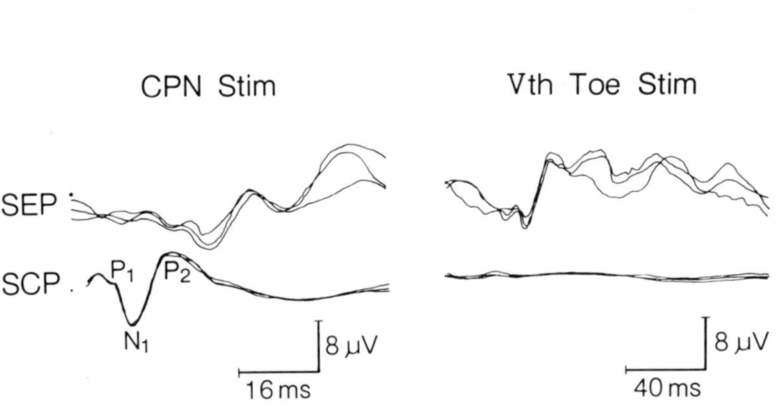

Fig. 2. The peripheral origin of the segmentally evoked SCP (6). Note that common peroneal nerve stimulation (CPN Stim.) at the popliteal space (two times the threshold strength) produces a clear configuration of the SCP in the posterior ES at T12 vertebral level, whereas an intense stimulation (painful) of the fifth toe (Vth Toe Stim.) does not produce the SCP but elicits the scalp recorded SEP.

3. Effects of graded stimulation

Fig. 3 shows the segmental SCPs produced by graded stimulation of the tibial nerve during the quiet resting state of a subject (5). The SCPs were recorded from the posterior ES of the T12 vertebral level. Stimulus strength is expressed at the left of each trace in multiples of the threshold of the initial spike potentials (P1).

At near threshold strength for the P1, both the N1 and P2 potentials are hardly noticed, although they increase in height with the amplitude of the incoming afferent volley. The threshold to evoke the P2 wave is slightly higher than that for the N1 wave. With a strong stimulation, the P2 wave changes its configuration with prolongation of the peak latency and sharpening of the decay, and a second component appears on its decaying phase, as demonstrated by an arrow.

4. Effects of repetitive stimulation

By brief trains of repetitive pulses (4 at 300Hz), the N1 wave

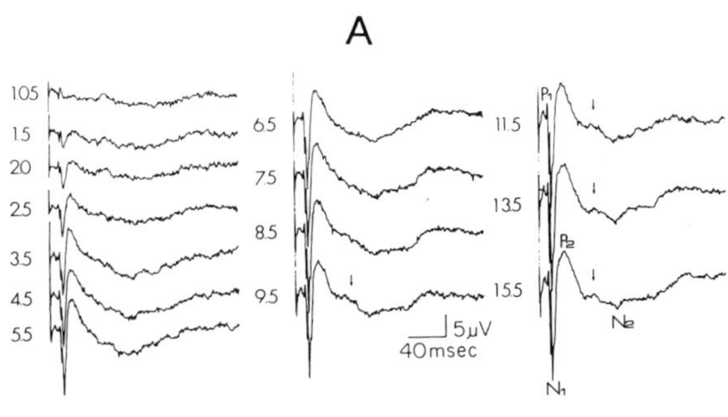

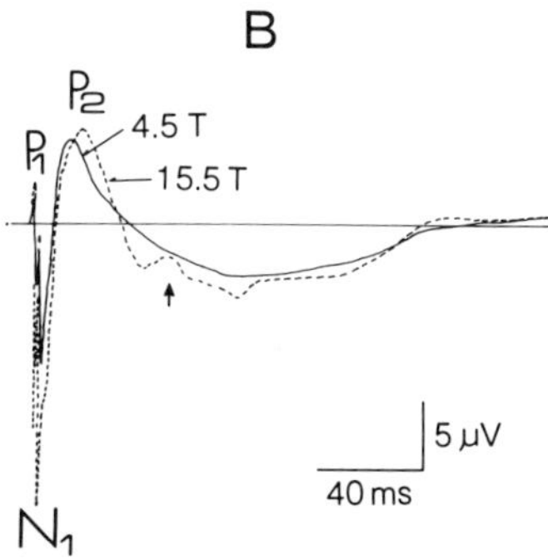

Fig. 3 (A). The effect of graded stimulation on the segmental SCP (5). See the text for explanation. (B). Superimposition of two segmental SCPs evoked by moderate (4.5 times threshold) and strong (15.5 times threshold) stimulus strengths. Note that the amplitude and peak latency of the P2 wave are increased with appearance of its secondary component by a stronger stimulation.

response to the first and third pulses with attenuation but not to the second and fourth pulses as shown in Fig. 4 (A). With higher frequency (9 at 800Hz) stimulation, the attenuated N1 wave responding to the fifth pulse is observed [bottom trace in Fig. 4 (A)]. These results may indicate that the absolute refractory period for the N1 wave lies within 5-7ms. By more prolonged stimulations (the right traces in B), the P2 wave increases in duration, making several summits during its time course, and is preceded by a large negative wave [Fig. 4 (B)]. Several pieces of evidence are now accumulating to indicate that these late components of the P2 potentials are produced by a feedback loop via supraspinal structures (3-6, 15).

5. Interactions between the segmental SCPs

When two stimuli, applied to the same segmental nerve, are separated by more than 100ms, both the N1 and P2 waves produced by the second volley are hardly affected by the preceding volley. However, when the interval between the two volleys is reduced to 10-20 and 20-25ms, amplitudes of the N1 and P2 waves following the second volley start to decrease, respectively, as shown in Fig. 5. The area of the P2 potential responding to

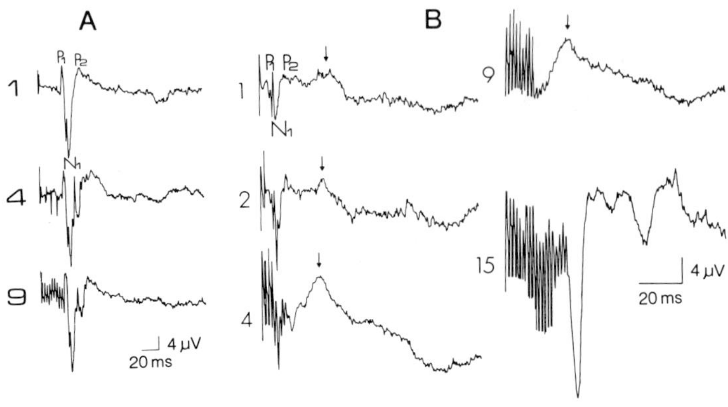

Fig. 4. Effects of repetitive stimuli on the segmental SCP (5). Note that the second negative wave appears, responding only to the third pulse at 300Hz and the fifth pulse at 900Hz, respectively (A). Prolonged repetitive pulses evoke a large negative wave and several late components of the P2 wave (B). An arrow indicates the second component of the P2 wave.

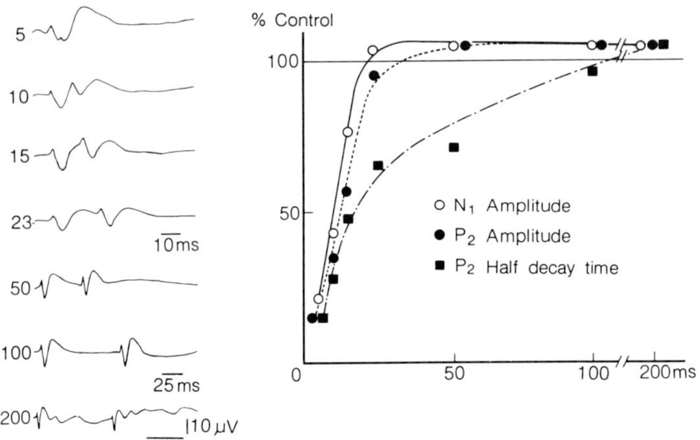

Fig. 5. Effects of double shocks on the segmental SCP (5). Specimen records are shown on the left. The number at the left of each trace represents the interval between the two stimulus volleys. The recovery curve of each component of the segmental SCP is shown on the right. Abscissa indicates the interval between the two volleys, and the ordinate, per cent of the control (produced by the initial volley) in each component.

the second stimuli decreases up to an interval of 100ms, due to the decrease in the half-decay.

When two stimuli are delivered to the contra- and ipsilateral tibial nerves at short intervals, a slight but definite occlusion is demonstrated in the P2 wave, whereas there is a linear addition of the N1 waves without facilitation or occlusion. An example is demonstrated in Fig. 6. Electrical stimuli (4T) are delivered to the left tibial nerve (upper traces) or both sides of the nerve (middle traces) at several intervals in the popliteal spaces. For easier calculation a simple summation of the two potentials evoked independently by left and right tibial nerve stimulations is shown in the bottom traces. Occlusion of the P2 wave amounts to about 87 and 92% of the control at intervals of 7 and 12ms, respectively, when the amplitude of the P2 wave produced by testing stimuli (left tibial nerve) to that produced by a simple summation is calculated. At intervals of more than about 25ms interaction is hardly noticed. There is a linear addition of the N1 waves without facilitation or occlusion. Therefore, it may be postulated that ipsi- and contralateral stimulations evoke separate interneurons, respectively, and that the PAD's produced by ipsi- and contralateral stimulations have in part common afferent fibers receiving the depolarizing impulses (5).

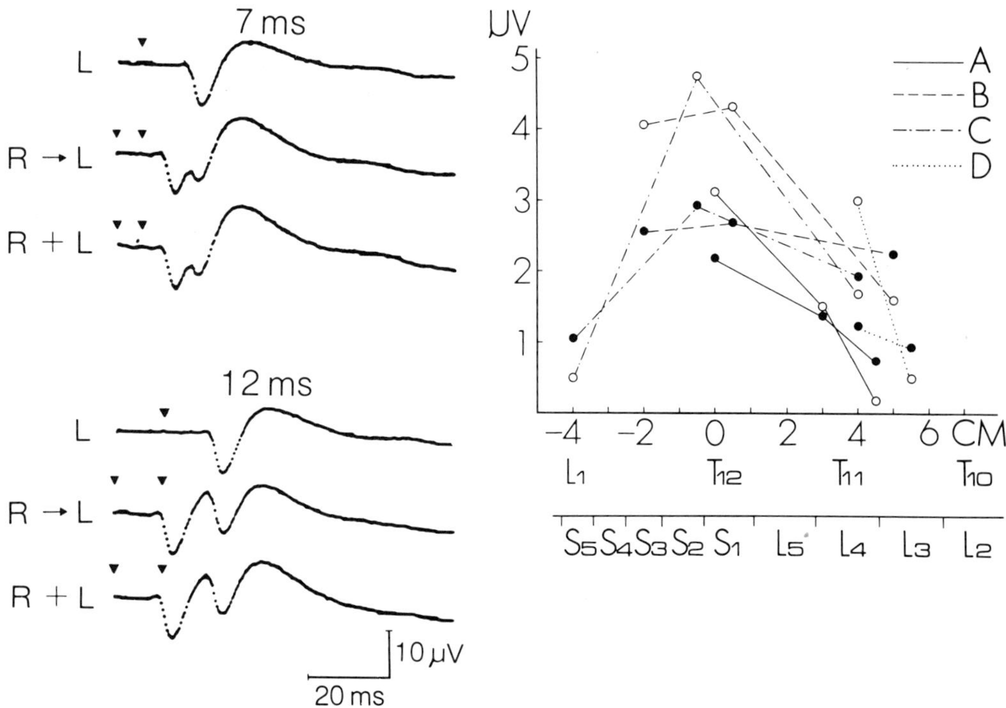

Fig. 6. Effects of contralateral nerve stimulations on the segmental SCP (5). L = left tibial nerve (TN) stimulation. R L = right TN stimulation is followed by left TN stimulation. R + L = simple summation of two SCPs evoked independently by right and left tibial nerve stimulations. The interval between the two stimulus volleys is shown on the top of each setting.

Fig. 7. Amplitude changes (ordinate) of the N1 and P2 components in relation to the recording level along the cord axis according to the x-ray finding (abscissa) in 4 subjects (A to D) (6). Note that the N1 potential (open circle) shows steeper decreases in amplitude than the P2 wave (closed circle) when the recording electrodes are situated rostral or caudal to the T12 vertebral level. Bottom scale is inserted in reference to the Pernkoph's atlas.

6. Longitudinal distribution of the segmental SCP

In the region of lumbar enlargement, the SCP by stimulation of the tibial nerve is recordable between the T10 and L1 vertebral levels, approximately corresponding to L2 through S5 time spinal segments (6). Below the L1 vertebral level only the spike potentials are recorded. When the tibial nerve is stimulated, the maximal amplitude of both the N1 and P2 waves is achieved in the region of the T12 vertebral level where the spinal segments S1 and S2 are situated. The decrement of the N1 wave from maximum is steeper than that of the P2 wave along the cord, as shown in the graph in Fig. 7. This may indicate that the interneurons which activate PAD, as demonstrated in the P2 wave, have their fibers spread to the primary afferents beyond the segment along the longitudinal axis of the spinal cord (4).

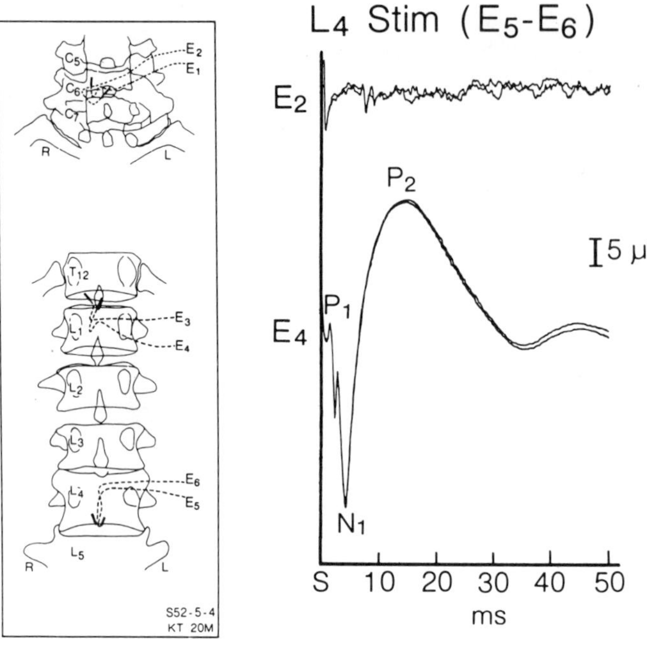

Fig. 8. The SCPs, recorded from C6 (ascending SCP) and T12 (segmental SCP) vertebral levels, in response to epidural stimulation of the cauda equina at L4 vertebral level. Note the difference in amplitude between the ascending and segmental SCPs. The drawing inside the frame shows the x-ray photograph, showing the positions of the epidural electrodes.

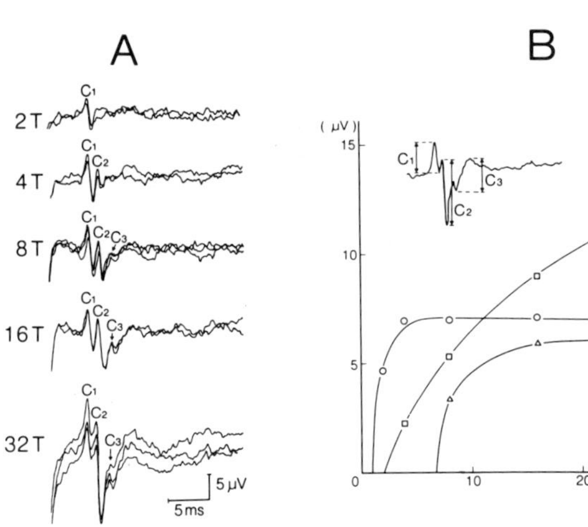

Fig. 9. The effect of graded stimulation on the ascending SCP (14). (A). Specimen records of the ascending SCPs. T represents the threshold strength (2T = electrical stimulation of 2 times the threshold strength). (B). Amplitude change in each component of the ascending SCP (ordinate) in relation to stimulus strength (abscissa).

II. Human SCPs evoked by ascending volleys along the cord

1. Waveform characteristics and effects of graded stimulation

The segmental SCPs are recordable over several segments. Recording electrodes situated more rostral, however, do not detect the slow negative and positive waves, but only small spikelike potentials. The spikelike potentials reflect compound action potentials in ascending tracts within the cord (14). These spikelike potentials are more consistently evoked by epidural stimulation of the cauda equina than by peripheral nerve stimulation (14) (Fig. 8).

Epidural stimulation of the cauda equina at 2 times the threshold strength (2T) of the initial spike produces only a single small spike similar to that evoked by peripheral nerve stimulation (Fig. 9). With increasing stimulus strength, a second and then a third spike appears [Fig. 9 (A)]. Strong stimulation often produces a slow positive-negative wave of 20-25ms duration after the polyphasic spikes [Fig. 9 (A), 32T]. The first and third components reach maximum amplitudes at 4-6 and 12-14 times the threshold strength, respectively, while the second component is not maximal even at 30 times threshold [Fig. 9 (B)]. These facts may indicate that these three components are different from each other in their synaptic origins (14).

2. Effects of two ascending volleys along the cord

When double shocks are applied within 5ms intervals, the first and second components evoked by the test stimuli show facilitation as shown in the recovery curve [Fig. 10 (C)]. On the other hand, the third component has a different pattern and is reduced for more than 20ms. Therefore, both the first and second components probably reflect action potentials conducted through certain tracts within the spinal cord with at least one synapse in their pathways. The third component appears to be an action potential of a different spinal tract also having at least one synapse in its course with different characteristics (14). Monitoring of spinal cord function during spine or spinal surgery is carried out mainly by this type of the SCP in several institutes (17-20).

3. Conduction velocities of ascending volleys along the cord

The mean maximum conduction velocity of an ascending volley along the human spinal cord is 82.0 ± 5.4 (mean ± SE) ms calculated by measuring the distances between the cervical and lumbar electrodes and the latencies of the initial dip of the first

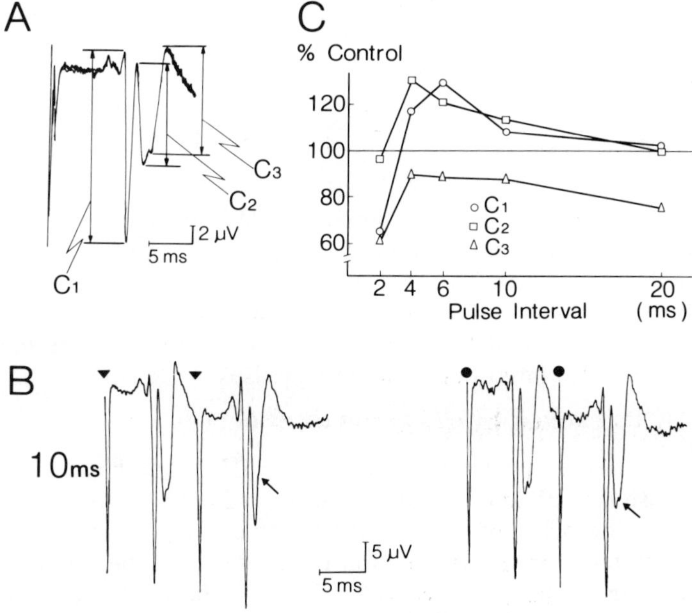

Fig. 10. Effects of double shocks on the ascending SCP (14). (A). Amplitude measurement of each component. (B). Examples of the ascending SCPs produced by the two ascending volleys at an interval of 10ms (left) and those produced independently by two single volleys separated by 10ms (right). Arrows indicate the third component. Note the inhibition of the third component by the double shocks. (C). The recovery curve of each component in response to double shocks. The ordinate represents per cent of the control value (the amplitude evoked by the initial volley), and the abscissa, the interval between the double shocks.

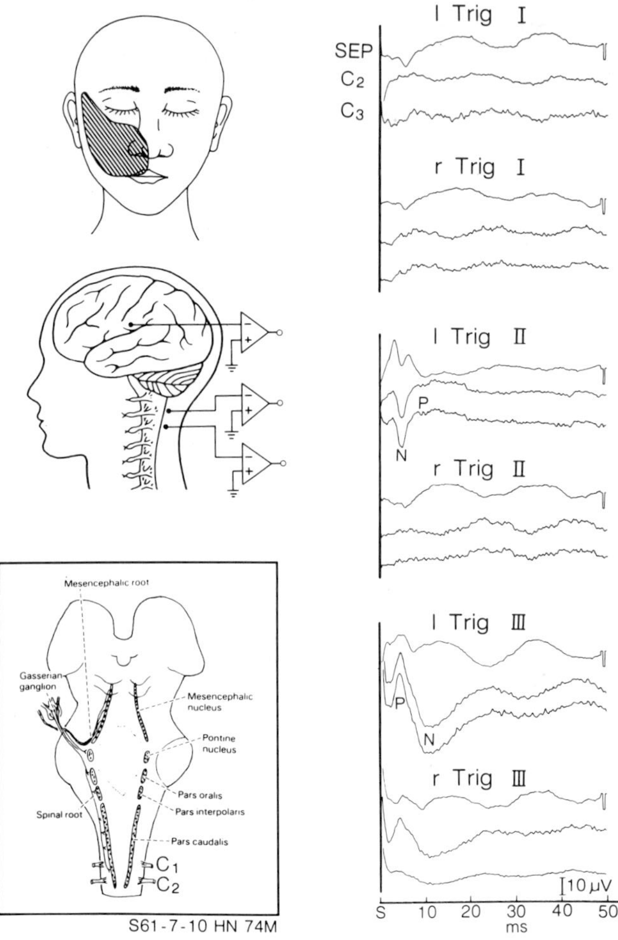

Fig. 11. The SCPs produced in the upper cervical posterior epidural space (C2 and C3) in response to stimulation of the three branches of the trigeminal nerve, shown on the right. Drawings on the left show the involved area of herpes zoster (top), electrode positions (middle) and anatomy of cervical nuclei of the trigeminal nerve in man (17). See the text for explanation.

spike of the ascending SCPs. There is a great deal of individual variation in conduction velocities along the cord evoked by the ascending volley among the subjects tested, as shown in Table 1. Sex and age differences have a minimal effect on the conduction velocities between the ages of 10 and 62. The great variability of maximum conduction velocities among the subjects is thought to be caused mainly by differences in the current distributions of the submaximal stimulating pulses in the ES (14).

III. Human SCPs evoked by the trigeminal nerves

Traces in Fig. 11 depict specimen records of the SEP from the scalp (top traces), the SCPs from the posterior ES at C2 (middle traces) and at C3 (bottom traces) evoked by stimulation of three branches of the trigeminal nerve in a patient with herpes zoster in the right maxillary nerve area. Stimulation of the supraorbital nerve does not evoke any potential change in the upper cervical ES. On the other hand, left infraorbital nerve stimulation produces the initial spike and a subsequent slow negative-positive complex. Both the slow negative and positive complexes evoked by infraorbital nerve stimulation show the time courses similar to those of slow negative waves of the segmental SCP, to those of slow negative and positive waves of the segmental SCP, respectively. Right infraorbital nerve stimulation does not evoke any potential changes in either the scalp or the ES probably due to herpes zoster and/or permanent chemical block of the nerve. Interestingly enough, mental nerve stimulation now produces a characteristic potential in the posterior ES consisting of an initially positive spike followed by slow positive and negative waves. This finding was just the same in all five subjects tested. Thus, the two evoked potentials elicited by infraorbital and mental nerve stimulations constitute a mirror image of each other. The waveform characteristics of the negative and positive complexes of the SCP evoked by infraorbital nerve stimulation are very similar to those of the N1 and P2 waves in the segmental SCP, respectively. Therefore, the negative and positive complexes of the SCP evoked by infraorbital nerve stimulation might reflect the synchronized activity of the neurons in the spinal nucleus of trigeminus and the PAD of trigeminal nerve terminals, respectively. We have no idea why the SCP evoked by mental nerve stimulation shows a reversed polarity. Absence of potential change by stimulation of the supraorbital nerve is probably due to an onion-like topographical representation of the face in the nucleus of trigeminal nerve (16).

Summary

Fundamental patterns of human SCPs, recorded from the epidural space in response to various afferent volleys, were surveyed, and their several physiological characteristics were discussed based on the data from our laboratory..

The SCPs (segmental SCPs) evoked by segmental peripheral nerve stimulation consist of the initially positive spike (P1) followed by sharp negative (N1) and slow positive (P2) waves.

Several electrophysiological studies in this laboratory indicate that the P1, N1 and P2 waves reflect the action potentials of the segmental roots, synchronized activities of dorsal horn neurons and primary afferent depolarization, respectively.

In wakeful subjects, the secondary positive peak preceded by the secondary negative dip can be produced. These secondary potentials are suggested to be evoked by a feedback loop via supraspinal structures.

Synchronized activities of large nerve fibers are needed to evoke the segmental SCP.

The SCPs recorded from the rostral epidural space beyond the segment (ascending SCPs) consist of the spike potentials made up of three components, followed by slow waves.

The three components of the ascending SCP are considered to be action potentials of afferent spinal tracts having at least one synapse each in their conducting courses.

The maximum conduction velocity of an ascending volley through the human spinal cord is about 80ms.

The SCP in response to stimulation of infraorbital and mental nerves can be evoked in the upper cervical epidural space.

Fundamental patterns of the SCP evoked by trigeminal nerve stimulation are similar to those of the segmental SCP evoked by spinal nerves. The SCP produced by mental nerve stimulation, however, shows a reversed polarity in comparison to that evoked by the intraorbital nerve.

SCPs evoked by stimulation of the trigeminal nerves probably reflect the evoked potentials in the spinal nucleus of the trigeminal nerve.

References

1. Shimoji, K.; Higashi, H.; Kano, T.: Epidural recording of spinal electrogram in man. Electroenceph. Clin. Neurophysiol., 30: 236-239, 1971.
2. Shimoji, K.; Kano, T.; Higashi, H.; Morioka, T.; Henschel, E.O.: Evoked spinal electrograms recorded from epidural space in man. J. Appl. Physiol., 33: 468-471, 1972.
3. Shimoji, K.; Ito, Y.; Ohama, K.; Sawa, T.; Ikezono, E.: Presynaptic inhibition in man during anesthesia and sleep. Anesthesiology, 43: 388-391, 1975.
4. Shimoji, K.; Kano, T.: Evoked electrospinogram: Interpretation origin and effects of anesthetics. In: K. Mori (ed), Effect of Anesthesia on the Central Nervous System. Little Brown, Boston, 1975, pp. 171-189.

5. Shimoji, K.; Matsuki, M.; Ito, Y.; Masuko, K.; Maruyama, M.; Iwane, T.; Aida, S.: Interactions of human cord dorsum potential. J. Appl. Physiol., 40: 79-84, 1976.
6. Shimoji, K.; Matsuki, M.; Shimizu, H.: Wave-form characteristics and spatial distribution of evoked spinal electrogram in man. J. Neurosurg., 46: 304-313, 1977.
7. Shimoji, K.; Shimizu, H.; Maruyama, Y..; Matsuki, M.: Dorsal column stimulation in man: Facilitation of primary afferent depolarization. Anesth. Analg., 61: 410-413, 1982.
8. Gasser, H.S.; Graham, H.T.: Potentials produced in the spinal cord by stimulation of dorsal roots. Am. J. Physiol., 103: 303-320, 1933.
9. Barron, D.H.; Matthews, B.H.C.: The interpretation of potential changes in the spinal cord. J. Physiol. (Lond.), 92: 276-321, 1938.
10. Beall, J.E.; Applebaum, A.E.; Foreman, R.D.; Willis, W.D.: Spinal cord potentials evoked by cutaneous afferents in the monkey. J. Neurophysiol., 40: 199-211, 1977.
11. Bernhard, C.G.: The spinal cord potentials in leads from the cord dorsum in relation to peripheral source of afferent stimulation. Acta. Physiol. Scand. 29, Suppl. 106: 1-29, 1953.
12. Bernhard, C.G.; Widen, L.: On the origin of the negative and positive cord potentials evoked by stimulation of low threshold cutaneous fibres. Acta. Physiol. Scand. 29, Suppl. 106: 42-54, 1953.
13. Carpenter, D.O.; Rudomin, P.: The organization of primary afferent depolarization in the isolated spinal cord of the frog. J. Physiol. (Lond.), 229: 471-493, 1973.
14. Maruyama, Y.; Shimoji, K.; Shimizu, H.; Kuribayashi, H.; Fujioka, H.: Human spinal cord potentials evoked by different sources of stimulation and conduction velocities along the cord. J. Neurophysiol., 48: 1098-1107, 1982.
15. Shimoji, K.; Maruyama, Y.; Shimizu, H.; Fujioka, H.; Taga, K.: Spinal cord monitoring -- A review of current techniques and knowledge. In: J. Schramm; S.J. Jones (eds), Spinal Cord Monitoring. Springer-Verlag, Berlin, 1985.
16. FitzGerald, M.J.T.: Neuroanatomy - Basic and Applied, Bailliere Tindal, London, 1985, pp. 238-241.
17. Kurokawa, T.: Functional spinal cord monitoring in spinal surgery through evoked spinal cord action potential measurement (in Japanese). Clinical Electroenceph., 22: 464-470, 1980.
18. Jones, S.J.; Carter, L.; Edgar, M.A.; Morley, T.; Ransford, A.O.; Webb, P.J.: Experience of epidural spinal cord monitoring in 410 cases. In: J. Schramm; S.J. Jones (eds), Spinal Cord Monitoring. Springer-Verlag, Berlin, 1985, pp. 215-220.
19. Tamaki, T.; Takano, H.; Takakuwa, K.; Tsuji, H.; Nakagawa, T.; Imai, K.; Inoue, S.: An assessment of the use of spinal cord evoked potentials in prognosis estimation of injured spinal cord. In: J. Schramm; S.J. Jones (eds), Spinal Cord Monitoring. Springer-Verlag, Berlin, 1985.
20. Kotani, H.; Hattori, S.; Senzoku, F.; Kawai, S.; Saiki, K.; Yamasaki, H.; Omote, K.: Evaluation of cord function in cervical spondylosis by a combined method using segmental and conductive spinal evoked potentials (SEP). In: J. Schramm; S.J. Jones (eds), Spinal Cord Monitoring, Springer-Verlag, Berlin, 1985.

Evaluation of Segmental Spinal Evoked Potential with Topographic Computer Display and Dipole

A. Dezawa;[*] T. Tamaki; S. Homma; Y. Nakajima; Y. Okamato; T. Musha

Introduction

Since Gabor (2) first described the relationship between potential distribution recording from the body surface and epicardial electric surface, many authors such as Brazier (1) have applied the theory to locating an electric dipole in the brain. Homma et al., (3) have lately developed a new computerized procedure to locate the origin of a current dipole in the human brain. We call this work Dipole Tracing.

On the other hand, in 1933 Gasser and Graham reported a cord dorsum potential elicited by stimulating the dorsal root, and since that time many researchers have studied the location of the sources of the early components of the segmental spinal evoked potential. This study aimed to analyze dipoles which evaluate the location and vector components of an equivalent electrical source in segmental spinal evoked potential (SSEP) elicited by electrical stimulation of peripheral nerves.

We recorded segmental spinal evoked potentials with 21-42 surface electrodes surrounding the lumbar spinal cord. This system is a better one in high diagnostic sensitivity and specificity for qualitative assessments. The vector moment and location of the electrical source were evaluated as a dipole using a new computer aided dipole tracing method. The location and vector components of an equivalent electric dipole which simulates electric activity of the spinal cord, were evaluated from its surface distribution. We applied this dipole tracing to the cat's

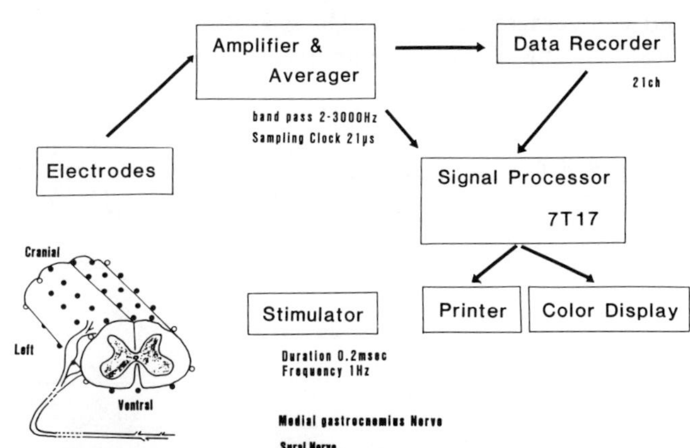

Fig. 1. Block diagrammatic representation of experimental arrangement.

* Department of Orthopedics and Physiology, Chiba University School of Medicine, 1-8-1 Inohana, Chiba 280, Japan

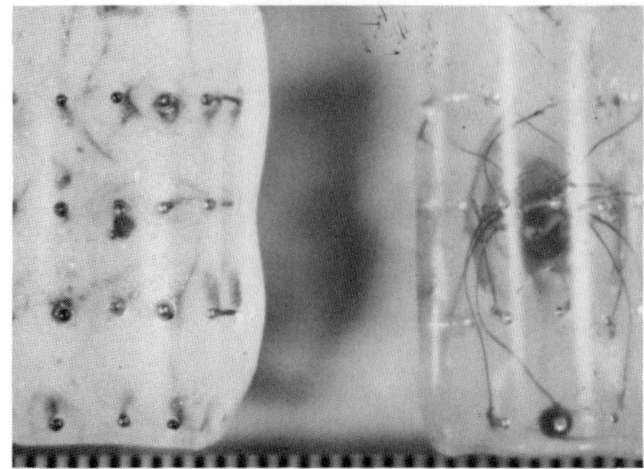

Fig. 2. Silver ball electrodes mounted on the transparent silicone plate, are placed craniocaudally at 5mm intervals and cross sectionally 2.5mm.

spinal cord surrounded with homogeneous volume conductors to record a SSEP elicited by peripheral nerve stimulation.

Materials and Methods

The experiments were performed on 11 adult cats ranging from 3.1 to 5.0 kg. Anesthesia was induced with 1% alphachlolarose and 10% urethan and paralyzed with pancuronium bromide and ventilated via a tracheostomy with a respirator.

The sural nerve, or the medial gastrocnemius nerve was exposed for electrical stimulation. A pair of silver hook electrodes were applied to the nerve in the popliteal fossa. To prepare for compression and recording, a wide laminectomy at L4 through S1 was carried out. 21 silver ball electrodes were positioned epidurally over cord dorsum and ventrum respectively. The SSEPs were amplified and averaged 32 times. 21 differential amplifiers with bandpass from 2 to 3,000Hz were used in conjunction with a

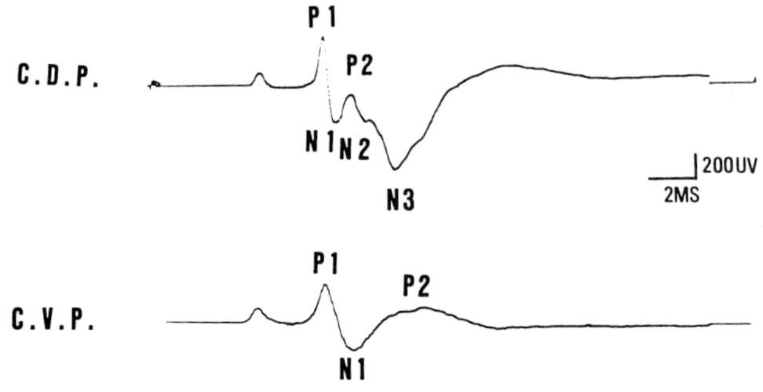

Fig. 3. CDP in segmental SSEP elicited in peripheral nerve stimulation consists with P1 through N1 spike potential and slow negative N2 through N3. Upward deflections are positive.

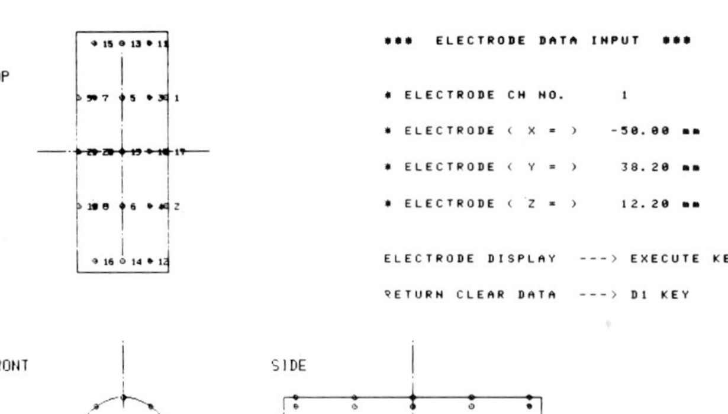

Fig. 4. The cat spinal cord model in this study. Potentials at the time indicated by arrow were analyzed.

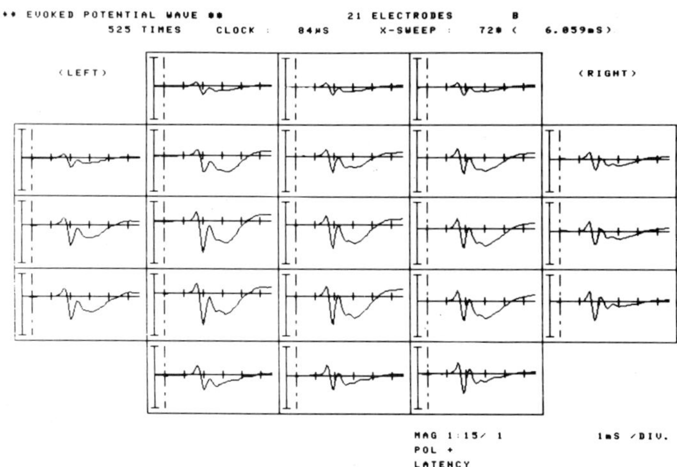

Fig. 5. One case of spinal cord surface distribution from dorsal side evoked by stimulation of the left side medial gastrocnemius nerve.

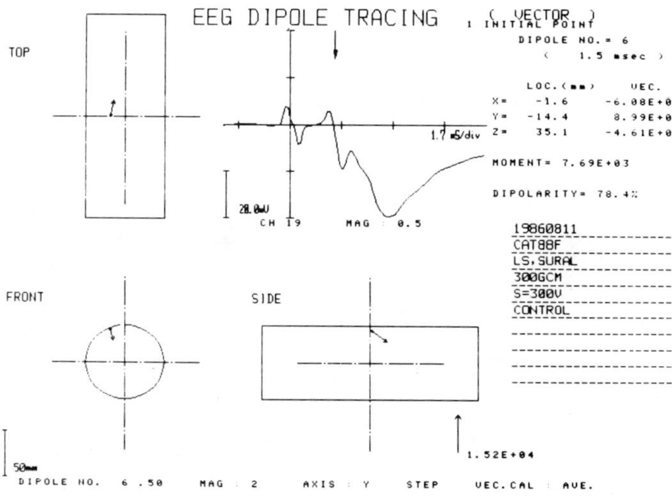

Fig. 6. Potential at the time (P1 through N1) indicated by arrows were calculated.

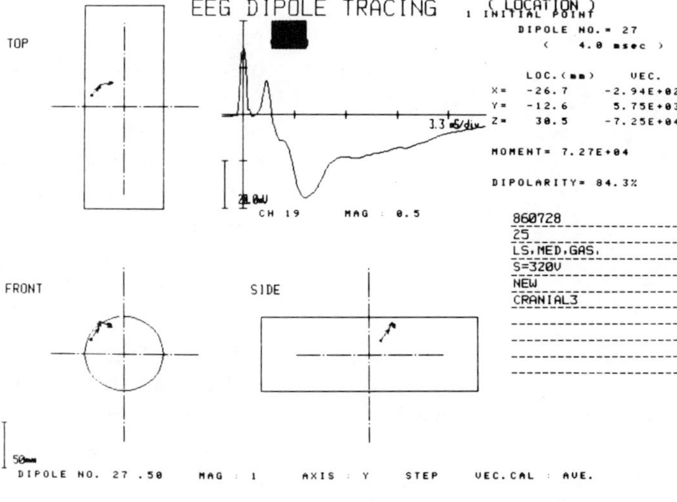

Fig. 7. Trajectories of source of dipole obtained from P1 onset.

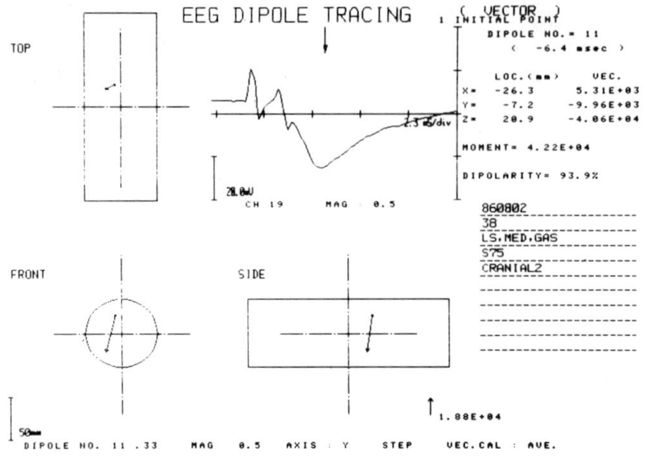

Fig. 8. Locations and vector moments corresponding to N3 wave in CDP
elicited by left side medial gastrocnemius nerve.

NEC 7T17 signal processor (San-Ei Instrument Co., Japan). All data were recorded on a 21-channel data recorder (TEAK SR71) and played back at one-quarter speed to improve sampling time. 7T17 signal processor has new computer program to estimate a current dipole of SSEP.

Fig. 2 shows 21 electrodes that were used for recording simultaneously. Each of 21 electrodes was placed at 5mm distance cephalocaudally, and 2.5mm cross sectionally. The electrodes were covered with a transparent silicone tube.

Results and Discussion

Generally, cord dorsum potentials (CDP) in SSEP elicited by peripheral nerve stimulation were generated as P1 through N1 spike potential and slow negative N2, N3, while cord ventrum potentials (CVP) appeared as triphasic wave with P1, N1, and P2.

P2 deflexion in CVP showed the mirror image of N3 in CDP. There was a slight delay in peak latency in P2 in CVP from that of N3 in CDP (Fig. 3).

The model is projected on three planes indicated as cross sectional view, side view, and top view. The system is formed in which the YZ plane is cross sectional, XY plane is top, and XZ plane is side. The Y axis lies horizontally in the frontal plane, positive toward the right side. The X axis lies axially in top and side view that is sagittal plane, positive toward the caudal side. Surface recording electrode positions were measured from the actual positions surrounding the spinal cord surface and dotted on the simulation model. The distance between the electrodes and respective axis was measured (Fig. 4).

Fig. 5 shows one case of CDP from top elicited by electrical stimulation to the left side medial gastrocnemius nerve. No. 19

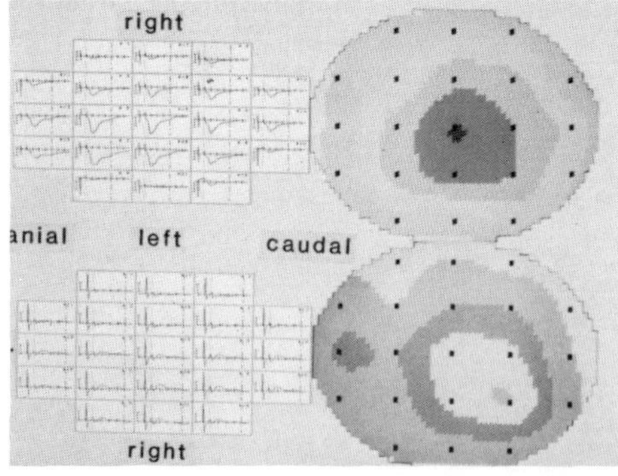

Fig. 9. Contour map of the distribution of SSEP corresponding to the N3 wave recorded from cord dorsum (upper) and ventrum (down) following stimulation of left side medial gastrocnemius nerve. The signal averaged SSEP shown in CVP was recorded near the site of maximum peak positivity at the time corresponding to the maximum N1 wave of No. 19 electrode in CDP.

electrode, located in the center of 21 electrodes and 0, 0, and 40mm along X, Y, and Z axis respectively, gained the highest amplitude of P1 through N1.

The source of the potential at P1 through N1 1.35ms latency following stimulation of left side sural nerve, was situated at -1.6, -14.4, and 35.1mm on respective X, Y, and Z axis. The dipolarity was 78.4%. The dipolarity shows the validity of an equivalent current dipole to normalize the nondipolar contribution. It seems possible that this potential corresponds to the conduction in the dorsal column (Fig. 6).

P1 through N3 electrical events can be best understood by considering the tracing of the sources of many dipoles representing an activation of SSEP. Fig. 7 shows the tracing of a source which was conducted after P1 onset. Dipole conduction was from primary afferent to dorsal horn.

Fig. 8, displays the location of the N3 dipole by electrical stimulations to the left side medial gastrocnemius nerve. This source is -26.3, -7.2, and 20.9mm on X, Y, and Z axis respectively, and its dipolarity was 93.9%.

N3 was considered as an interneuronal activity projecting from dorsal horn to anterior horn.

To estimate the location and vector moment of the dipole, distribution of spinal cord surface potentials were analyzed using an isopentential mapping technique (Fig. 9).

Negativity in CDP represented the sink of N3 current dipole, while positivity in CVP corresponded to the source of N3 dipole.

Summary

Segmental spinal evoked potential (SSEP) elicited by electrical stimulation on the medial gastrocnemius and sural nerves were recorded simultaneously with 21 electrodes over the cord dorsum and ventrum. To analyze the location and vector moment of the source of the early components of SSEP as an equivalent dipole, we used a new computer aided method called dipole tracing (DT).

The source of the N1 wave seems to be situated in the entry zone ipsilateral to stimulation, and P1 through N1 potential corresponded to the conduction in dorsal column. The dipole of N3 wave was suspected as being located in dorsal horn and projected to anterior horn. It seems likely that this method will prove useful in the experimental study on certain aspects of spinal cord function.

References

1. Brazier, M.A.B.: A study of the electrical fields at the surface of the head. Electroenceph. Clin. Neurophysiol., 2: 38-52, 1949.
2. Gabor, D.; Nelson, C.V.: Determination of the resultant dipole of the heart from measurements on the body surface. J. Appl. Phys., 25: 413-416, 1954.
3. He, B.; Musha, T.; Okamoto, Y.; Homma, S.; Nakajima, Y.: Evaluation of errors in estimating an electric dipole in the brain. The Japan Society of Medical Electronics and Biological Engineering, 24: 23-28, 1986.
4. Homma, S.; Hayashi, K.; Nakajima, Y.: Unidimensional latency topography of the spinal evoked potentials in the cat. S. Homma; T. Tamaki (eds.): Fundamental and Clinical Application of Spinal Cord Monitoring. Saikon Publishing, Tokyo, 23-32, 1984.
5. Lorent, D.E.: A study of nerve physiology. Stud. Rockefeller Inst. Med. Res., 132, 1947.
6. Okamoto, Y.; Teramachi, Y.; Musha, T.: Motion of equivalent electric dipole to cardiac activity estimated from body surface potentials. Jpn. Heart J., 21: 761, 1980.
7. Okamoto, Y.; Musha, T.: Moving multiple dipole model for cardiac activity. Jpn. Heart J., 21: 761, 1980.
8. Shaefer, H.; Trautwein, W.: Quantitative zur theorie lokaler potential abgriffe beim elektrencephalogram. Archiv fur Psychiatrie und Zeitschrift Neurologie, 183: 175-188, 1949.
9. Willis, W.D.: Evoked Spinal Potentials in the Cat and Monkey: Use in the Analysis of Spinal Cord Function. Fundamental and Clinical Application of Spinal Cord Monitoring. Saikon Publishing, Tokyo, 23-23, 1984.

Intraspinal SEPs Recorded from the Vicinity of the Dorsal Root Entry Zone

J.A. Campbell;[*] J.B. Miles

Introduction

Intraspinal recordings of somatosensory evoked potentials have been performed at this centre for several years in patients undergoing radio-frequency percutaneous cervical cordotomy for the relief of malignant pain (Campbell and Lipton, 1983, 1984, 1985; Campbell and Bowsher, 1985; Campbell, 1985). These recordings have been used both as a research tool in the investigation of sensory transmission in the human spinal cord and as an extra aid for accurate positioning of the lesion electrode within the spinal cord. Some open neurosurgical procedures also require accurate localization of spinal tracts and it was felt that similar techniques may prove valuable in these cases. As a result, responses evoked by peripheral nerve stimulation have been recorded from various positions within the spinal cord during ablative neurosurgical procedures. Somatosensory evoked potentials have been recorded from the cords of 6 patients undergoing destruction of the dorsal root entry zone (DREZ), 2 patients having a midline myelotomy and 1 patient on whom an open thoracic cordotomy was performed.

Methods

The procedures were performed under general anaesthesia with a muscle relaxant. A hand-held concentric bipolar electrode (external diameter 0.55mm, central conductor protruding 1mm) was used to record the intraspinal potentials. The stimulating electrodes were securely attached to the skin over a peripheral nerve and the threshold for motor activity established prior to the induction of anaesthesia. The responses were evoked by peripheral stimulation well above motor threshold (usually at 50mA) using rectangular electrical pulses with a width of 0.1ms at a frequency of 3 per second from a constant current stimulator (Medelec ISC).

This electrode was positioned at depths of 1-3mm within the cord in the dorsal columns, dorsal root entry zone, dorsal spinocerebellar tract and/or anterolateral quadrant after a minute puncture had been made in the pia mater using an arachnoid knife. The spinal level at which these recordings were made varied from T6-7 to C1 depending on the surgical procedure.

The signals were amplified using a Medelec MS6 modular EMG system and recorded onto FM tape (Racal Store 4) whilst being simultaneously monitored and averaged. The amplifier filters were set at 8Hz to 3.2kHz (3dB points) and the tape had a frequency response of DC to 2.5kHz. The trigger pulse for the stimulator was also recorded onto tape and was used as a trigger for the averager during play-back at a later

* Pain Relief Foundation, Rice Lane, Liverpool L9 1AE, U.K.

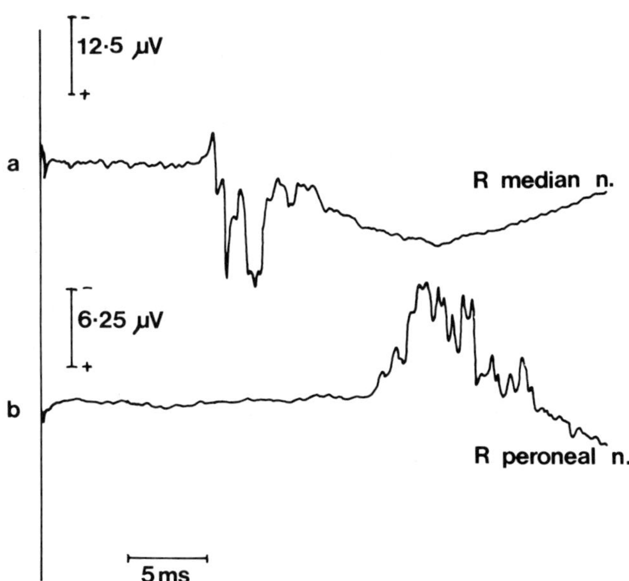

Fig. 1. This figure illustrates the response recorded from within the right dorsal column at the C1 root level after stimulation of the ipsilateral median nerve at the wrist (a) and ipsilateral peroneal nerve at the knee (b) in a patient undergoing a high cervical myelotomy. The electrode position was estimated to be near the boundary between the gracile and cuneate funiculi.

date. The average responses were then transferred onto computer disc via an interface between the Medelec MS6 and a DEC MINC-11 computer where subsequent analysis could take place.

Results

A. Dorsal column recordings

Recordings of somatosensory evoked potentials were made after positioning the electrode within the dorsal column, located by eye using an operating microscope. The electrode was inserted through the pia (after a minute puncture was made using an arachnoid knife to allow easy penetration) and into the cord to a depth of approximately 1mm from the outer conductor. A cuff around the electrode at this level allowed the electrode depth to be kept constant throughout the recording. A peripheral nerve whose spinal roots were caudal to the recording site was then electrically stimulated. The resulting responses were clear and well-defined with little residual noise after averaging 128-256 responses. Most could be seen on the unaveraged trace. Fig. 1(a) illustrates a recording from the dorsal column at the level of the C1 root after stimulation of the ipsilateral median nerve at the wrist. The onset of this response occurred 4.2ms after the onset of the potential recorded over Erb's point in this lady and the response lasted a typical 6.3ms. No discernible response was recorded from this electrode position after stimulation of the contralateral median nerve. Fig. 1(b) shows the response recorded from the same recording position after stimulation of the ipsilateral peroneal nerve at the knee. The multiphasic morphology of the dorsal column potential is particularly apparent here and has been found to be a consistent feature of all the potentials arising from the dorsal columns that we have recorded from sites within the spinal cord.

The difference in the onset latencies of the responses evoked by stimulation at the wrist and knee in this subject was 11ms. This is considerably longer than the mean difference of 6.6ms seen in intraspinal recordings during percutaneous cordotomy reported elsewhere (Campbell, 1985) in which the recordings were made at a similar vertebral level. This phenomenon was further investigated by recording responses at more than one level in the same patient. When recording from the dorsal column at the level of the C8 root in one patient after stimulation of the peroneal nerve the onset latency was 0.6ms earlier than the response recorded at the level of C6. This corresponds to a conduction velocity of 50ms-1. However, when similar recordings were made from the dorsal column of another patient the conduction delay over 2 segments (T1 to C7) was 1.2ms. The increase in latency was linear over the three segments investigated (T1, C8, C7) and corresponded to a cord conduction velocity of only 25ms-1. This slowing of conduction velocity during open surgery is thought to be a result of cord cooling, as is seen frequently in animal studies. Other cases of reduced conduction velocity in man during open spinal surgery have recently been reported (Jones and Thomas, 1985).

The position of the electrode used for the recordings shown in Fig. 1 was judged to be near the boundary between the gracile and cuneate funiculi. Other recordings have been made in which the electrode was positioned both laterally and medially within the dorsal column. A response evoked by peroneal nerve stimulation was easily recorded from the medial position (gracile funiculus) but no response could be recorded from the lateral position (cuneate funiculus). This showed the specificity of recordings made using this electrode configuration.

B. Spinocerebellar tract recordings

Recordings of somatosensory evoked activity were also made from positions just lateral to the entrance of the dorsal roots at a depth of approximately 1mm from the outer conductor. This was judged to be within the dorsal spinocerebellar tract and Fig. 2(a) illustrates a typical response obtained by stimulation of the ipsilateral peroneal nerve and recorded from between sixth and seventh thoracic vertebrae. It was possible in this case also to record from a position just anterior to the dentate ligament at a similar depth and this is shown in Fig. 2(b). As the electrode position was quite superficial it as presumed to have been within the ventral spinocerebellar tract.

It can be seen that the first part of the ventral response is very similar to the dorsal spinocerebellar recording. It is well known that both spinocerebellar tracts carry ac-

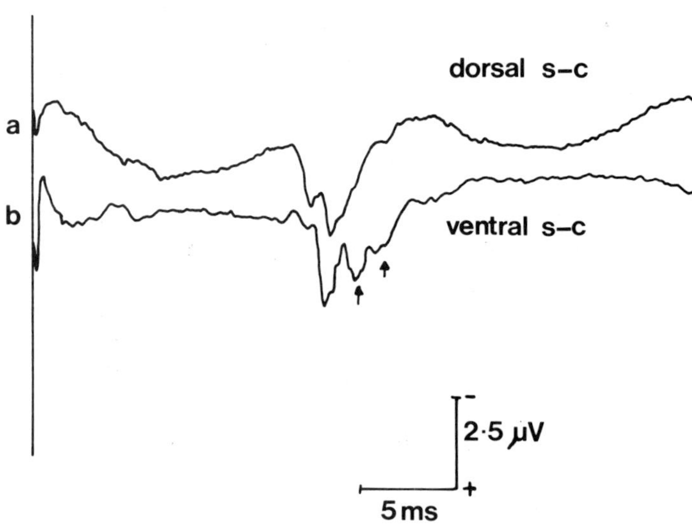

Fig. 2. This figure compares the responses obtained after stimulation of the ipsilateral peroneal nerve at the knee when the electrode was positioned so that the outer conductor penetrated the cord to a depth of approximately 1mm within the dorsal spinocerebellar tract (dorsal s-c, (a)) and just anterior to the dentate ligament in the superficial anterolateral quadrant of the cord (ventral s-c, (b)) between vertebrae T6 and T7 in a patient undergoing an open thoracic cordotomy. The later peaks in the ventral response that do not have correlates in the dorsal recording (see text) are arrowed.

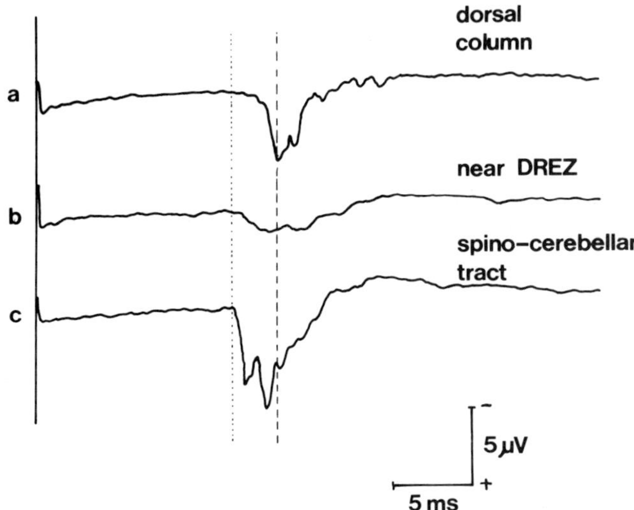

Fig. 3. This figure shows recordings made from within the cord at the T6 vertebral level of a patient with postherpetic neuralgia who had apparently normal cord anatomy. The responses illustrated were recorded after ipsilateral peroneal nerve stimulation at the knee when the recording electrode was at a depth of 1mm medial to the dorsal roots (dorsal column, (a)), in line with the roots (near DREZ, (b)) and 1mm lateral to the roots (spinocerebellar tract (c)). The dotted line indicates the onset of the response recorded from the spinocerebellar tract. The dashed line denotes the position of the first peak of the dorsal column response.

tivity from similar primary afferents and both would be activated on lower limb stimulation. The ventral spinocerebellar tract contains both crossed and uncrossed fibres whereas the dorsal tract only contains uncrossed fibres (Brodal, 1981) but this would affect these recordings only in that the activity from the ventral tract would be smaller in amplitude due to the reduced number of uncrossed fibres. However, as the amplitude of these responses is also a function of electrode position this would not necessarily be apparent.

The later peaks of the ventral response (arrowed in Fig. 2(b) and occurring 3.5ms and 5ms after the onset of the response) do not appear in the dorsal spinocerebellar recording, although there is a very small 'notch' in the ascending limb corresponding to the first arrowed peak.

C. Recordings made in the vicinity of the dorsal root entry zone

Fig. 3 shows recordings made from the cord at the T6 level of a patient with postherpetic neuralgia who had apparently normal cord anatomy. The responses illustrated were recorded after peroneal nerve stimulation at the knee when the electrode was at a depth of 1mm medial to the dorsal roots (within the dorsal columns, Fig. 3(a)), in line with the roots (near DREZ, Fig. 3(b)) and lateral to the roots (within the dorsal spinocerebellar tract, Fig. 3(c)). The last position was only 1mm lateral to the DREZ position. It can be seen that the onset of the response recorded from the spinocerebellar tract (indicated by the dotted line on Fig. 3) occurs 1.2ms before that recorded from the dorsal column. The peak of the dorsal column response (denoted by the dashed line) is later than the peak of the spinocerebellar response, as would be expected due to the peripheral input to the spinocerebellar tract containing fibres with higher conduction velocities than that mediated centrally by the dorsal columns. Both of these recordings were clear and well-defined and could be seen on the unaveraged recordings. In contrast, the response recorded when the electrode was in line with the spinal roots and was judged to be within the dorsal root entry zone (between the other two recording positions) was very small and ill-defined. This could also be predicted as the peroneal nerve has no segmental input at the mid-thoracic level.

Similar recordings have successfully been used to assist in the identification of the dorsal root entry zone in cases where the cord is deformed or where the roots have been avulsed.

Recordings of the spontaneous activity from the dorsal root entry zone may also assist in the positioning of an electrode prior to making a radio-frequency DREZ lesion. The amplitude of the spontaneous activity is significantly greater when the electrode is advanced into the cord. This amplitude difference was particularly obvious when the signals were fed to a loudspeaker during acquisition. There was a marked increase in volume as the electrode was advanced into the cord which then diminished as the electrode was again withdrawn into a superficial position. It is presumed that the deeper position was within the grey matter of the dorsal horn where the ongoing cell activity could be recorded. This change in amplitude proved to be a useful indicator of depth and assisted in the accurate placement of the radio-frequency lesions.

Summary

Responses evoked by peripheral nerve stimulation were recorded from various positions within the spinal cord during open surgery using a concentric bipolar electrode.

When the electrode was within the dorsal columns, a complex waveform could be recorded after ipsilateral peripheral nerve stimulation. This response contained many small peaks which could not be attributed to noise in the recordings. This response was long-lasting (6.5ms at T6-7 after peroneal stimulation) and easily recorded, being seen on the unaveraged responses. No response could be recorded when the electrode was within the lateral part of the dorsal column (cuneate funiculus) after leg stimulation although moving the electrode more medially (into the gracile funiculus) produced a clear evoked response. No response was recorded bipolarly from the dorsal column after stimulation of the contralateral leg. Responses could be obtained from the cervical dorsal column after stimulation of both the peroneal nerve at the knee and the median nerve at the wrist. The difference in onset latencies could then be used as an estimate of the conduction velocity within the dorsal columns. Cord conduction velocities could also be measured by recording at more than one level in the cord, keeping the stimulation site constant. Such measurements from one subject gave a conduction velocity of approximately 50ms-1. However, in two subjects, the conduction velocity within the dorsal columns was estimated as only 25ms-1. This slowing of conduction within the cord may be due to cooling of the cord during open surgery.

Responses recorded from superficial electrode positions lateral to the dorsal root entry zone were also clear and well-defined and similar to the early responses recorded from the ventral spinocerebellar tract. The onset of this response was earlier than that recorded from the dorsal column at the same level and this suggests that the electrode was within the dorsal spinocerebellar tract which is activated by peripheral A-alpha fibres.

Recordings made from superficial positions between the dorsal columns and the dorsal spinocerebellar tract (in line with the dorsal roots when present) were small and ill-defined when a peripheral nerve having no segmental input was stimulated. This lack of activity marked the dorsal root entry zone. The entry zone could be localized in a deformed or abnormal cord by its electrical inactivity to remote peripheral nerve stimulation between two regions of clear evoked responses with the onset of the lateral response being earlier than that of the response from the medial position. Recordings were also made at deeper positions (2-3mm) within the entry zone. The spontaneous activity was markedly larger than that seen from the corresponding superficial position and was thought to represent cell activity within the dorsal horn. This increase in amplitude of the spontaneous activity proved to be a useful indicator of the depth of the electrode prior to a radio-frequency lesion being made in the dorsal root entry zone.

References

1. Brodal, A.: Neurological Anatomy in Relation to Clinical Medicine. Oxford Univ. Press, N.Y., 1981.
2. Campbell, J.A.: Observations on somatosensory evoked potentials recorded from within the human spinal cord. Ph.D. Thesis, University of Liverpool, 1985.
3. Campbell, J.A. and Bowsher, D.: Electrical activity recorded from within the anterolateral funiculus during percutaneous cordotomy in man. Phil. Trans. Roy. Soc. London, B, 1985, 308, 422.
4. Campbell, J.A. and Lipton, S.: Somatosensory evoked potentials recorded from within the anterolateral quadrant of the human spinal cord, In: J.J. Bonica et al. (Eds.). Advances in Pain Research and Therapy, Vol. 5, 1983, pp. 193-196.
5. Campbell, J.A. and Lipton, S.: Intraspinal evoked potentials in man during cervical cordotomy, In: S. Homma and T. Tamaki (Eds.). Fundamentals and Clinical Application of Spinal Cord Monitoring, Saikon Pub. Co. Tokyo, 1984, pp. 245-252.
6. Campbell, J.A. and Lipton, S.: Intraspinal evoked potentials in man, In: C. Morocutti and P.A. Rizzo (Eds.). Evoked Potentials. Neurophysiological and Clinical Aspects, Elsevier Science Publishers B.V., 37-43, 1985.
7. Jones, S.J. and Thomas, D.G.: Assessment of long sensory tract conduction in patients undergoing dorsal root entry zone coagulation for pain relief, In: J. Schramm and S.J. Jones (Eds.). Spinal Cord Monitoring, Springer Verlag, Berlin, 1985.

Interpretation of Anterior and Posterior Spinal SEPs During Scoliosis Surgery

J.P. Halonen;[*] M.A. Edgar; S.J. Jones; A.O. Ransford

Of the various techniques for monitoring spinal cord function during surgery, that of recording spinal somatosensory evoked potentials (SEPs) from the epidural space following electrical stimulation of peripheral nerves is often preferred to less invasive methods (Jones et al., 1982, 1983; Macon and Poletti, 1982; Macon et al., 1982; Bradshaw et al., 1984; Beric et al., 1986; Nainzadeh et al., 1986; Schramm et al., 1986; Whittle et al., 1986; see Dinner et al., 1986 for review). In order fully to appreciate the significance of changes occurring during surgery we need to understand in detail the physiological characteristics of the potentials recorded, including which sensory modalities and spinal cord tracts are represented.

When potentials are recorded simultaneously or sequentially from multiple levels of the spine it is possible to estimate the conduction velocities (CVs) of the various components. With stimulation of the posterior tibial (peroneal) nerve at the knee, at least 3 components can be recognised with CVs ranging from about 40 to more than 80m/sec over midthoracic to cervical segments (Fig. 1). The fastest component (which also has the lowest activation threshold) is

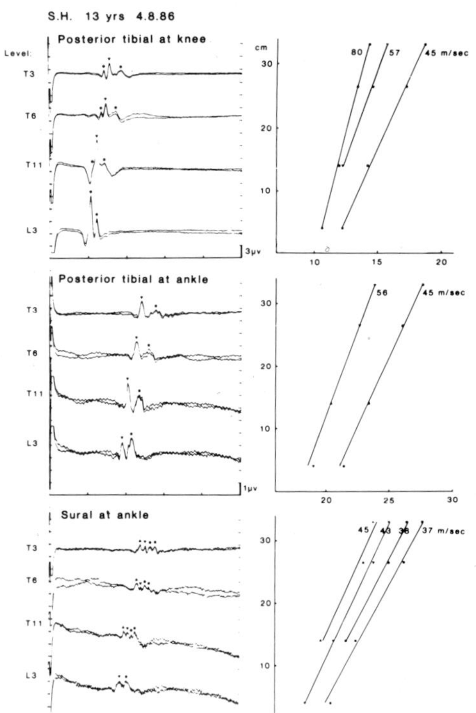

Fig. 1. Epidurally recorded spinal SEPs at 4 levels following stimulation at 3 sites in the lower limb, plus graphs of latency against distance along the cord for the components presumed to be equivalent at each level.

* Medical Research Council and Royal National Orthopaedic Hospital, Queen Square, London WC1N 3BG, England

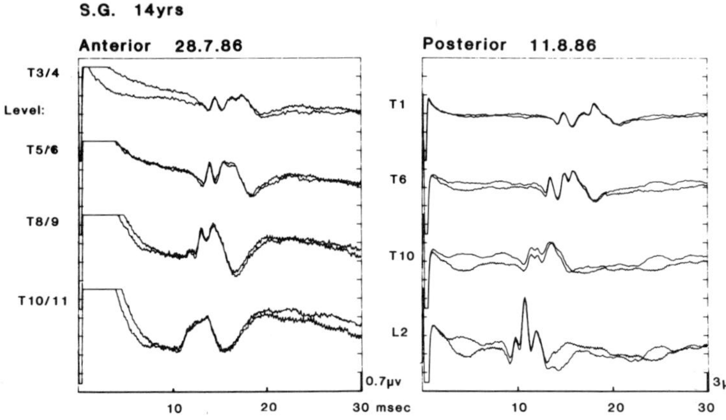

Fig. 2. Spinal SEPs recorded at 4 levels with needle electrodes inserted in the vertebral discs during anterior spinal surgery, and with dorsal epidural electrodes in the same patient during posterior surgery 2 weeks later.

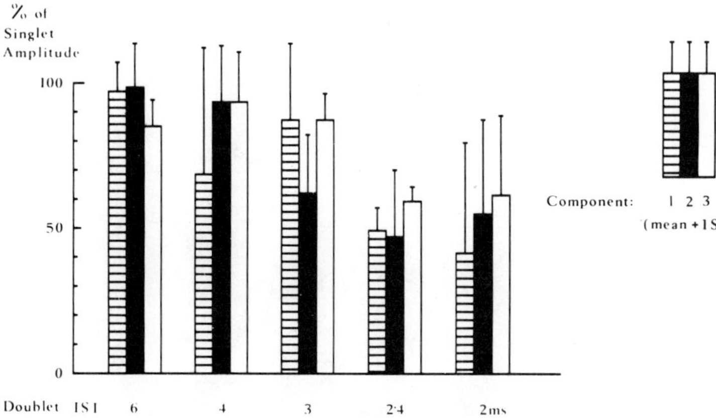

Fig. 3. Amplitude decrement of components I, II and III at upper thoracic level, following doublet stimulation with ISIs of 6 to 2msec.

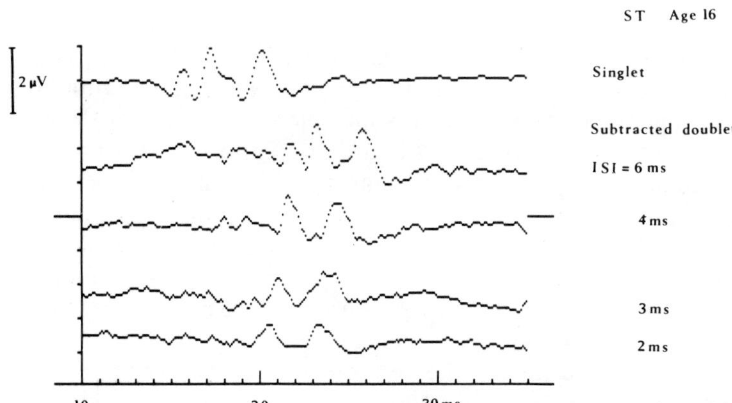

Fig. 4. Spinal SEPs following singlet stimuli (10/sec) and doublet responses with ISIs of 6-2msec after subtraction of the initial response, showing abolition of component I at ISIs of 4, 3 and 2msec.

greatly diminished or absent when the stimulus is delivered to the posterior tibial nerve at the ankle, and the 2 components remaining have CVs which are similar to those of components II and III following stimulation at the knee. Sensory fibres deriving from cutaneous receptors, small muscles in the foot and joint capsules will be present in the nerve at both levels, but a large proportion of fast-conducting Group I afferents innervating the calf muscles (primary muscle spindle and tendon afferents) as well as the Group II afferents from secondary spindle endings will not be present in the tibial nerve at the ankle. It can therefore be concluded that component I following stimulation at the knee is most likely due to Group I muscle afferents, which (from the lower limbs) are mainly conducted in the dorsal spinocerebellar tracts.

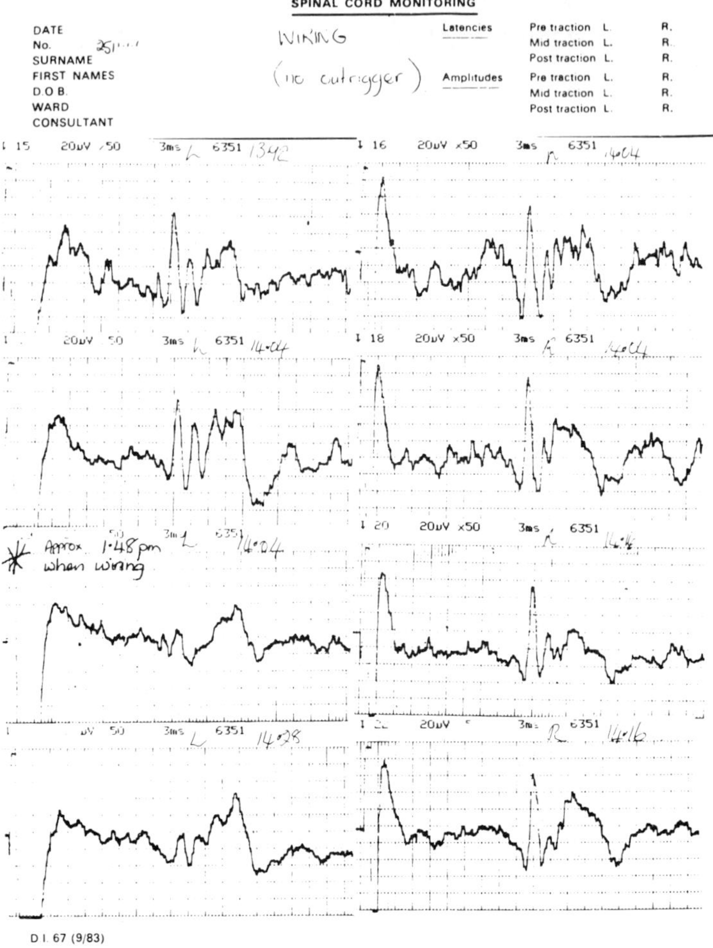

Fig. 5. Sudden onset of decrement confined to components I and II with stimulation of the left leg during sublaminar wiring.

When the stimulus is delivered to a purely cutaneous nerve, the sural, all components are conducted at around 40m/sec from mid-thoracic to cervical levels, corresponding approximately to the CV of component III following tibial nerve stimulation at the knee (Fig. 1). This suggests that the fibres responsible for component III are likely to derive from cutaneous receptors in the foot and that component II, which is present in the response to stimulation of the mixed tibial nerve but absent from the cutaneous sural nerve response, may be a second muscle afferent potential originating in small muscles of the foot. Another possibility, however, is that component II may be due to fibres arising in a class of cutaneous receptor which is prominent in the glabrous skin regions innervated by the posterior tibial nerve but is not present in the hairy skin served by the sural nerve.

When the recording electrode is located to the left or right of the cord in the epidural space it can be seen that all 3 major components are generated on the side ipsilateral to the stimulus and that component I is the most strongly lateralized. This is clearly compatible with activity in the dorsal spinocerebellar tract, the most lateral of afferent cord pathways. The dorsal columns also relay information from the ipsilateral side of the body, so may be responsible for some if not all of the later components. Other afferent pathways, for example the spinothalamic tracts, are mainly contralateral

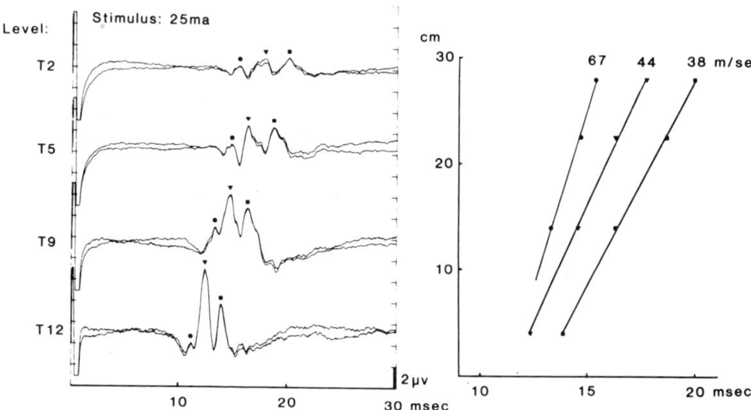

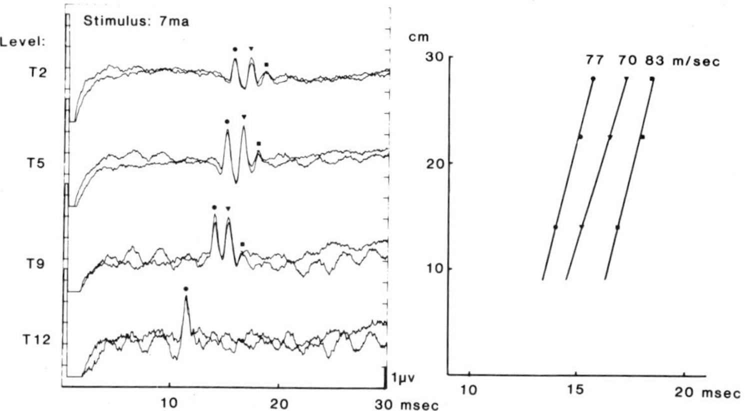

Fig. 6. Spinal SEPs recorded at 4 levels plus graphs of latency against distance, with stimulus intensities of 25ma (upper) and 7ma (lower). Stimulus duration was 0.2msec.

after decussation of postsynaptic fibres within a few segments of root entry level, and so can be virtually eliminated as important generators of the spinal SEP.

During anterior procedures it is possible to record spinal SEPs with needle electrodes inserted into the exposed discs on the ventral side of the cord. The potentials are of similar morphology to those recorded from the dorsal epidural space but smaller in amplitude, and there is a suggestion that component III may be relatively more attenuated than I or II (Fig. 2). This is clearly compatible with an origin in the dorsal columns.

The spinocerebellar tracts are postsynaptic pathways, and may therefore be refractory at longer inter-stimulus intervals (ISI) than projections which are essentially asynaptic such as the dorsal columns. Increasing the frequency of stimulation from 20Hz to 50Hz sometimes resulted in a slight decrement in the amplitude of the second component (Jones et al., 1982), while components I and III were virtually unaltered. Another method of determining the absolute and relative refractory period of neuronal potentials is to deliver double stimulus pulses separated by a variable interval. Since at upper thoracic level the duration of the spinal SEP is in the order of 7-8msec, with ISIs of less than 8msec, overlap will occur between late potentials evoked by the first stimulus and early potentials evoked by the second. To circumvent this problem it is possible to record responses to a single stimulus and subtract this waveform from the doublet response, thereby revealing the second response of the doublet undistorted by the first.

Although data obtained from 6 patients demonstrated only a general trend for amplitudes to be smaller at shorter ISIs (Fig. 3), without significant dissociation between the 3 components, in 2 patients component I was absent at intervals which left components II and III fully or partially preserved. Fig. 4 illustrates such a case, where the response to the second stimulus of the doublet (the first response having already been subtracted) is very similar to the singlet waveform at ISI = 6msec, whereas at 4, 3 and 2msec component I is absent while II and III undergo a more gradual decrement in amplitude.

In support of the hypothesis that components I and II are postsynaptic in origin, these potentials are often found to be more vulnerable than component III to events occurring during surgery. On tightening of sublaminar wires, transient and sustained amplitude decrements of components I and II have been observed while component III was not significantly altered (Fig. 5), and a transient loss of component I was also seen when a wire briefly came into contact with the conus medullaris (Jones et al., 1983). In the former case the precipitating factor may have been temporary, local anoxia of the cord in the region of synaptic transmission, while in the latter the mechanism may have been related to that of spinal "shock".

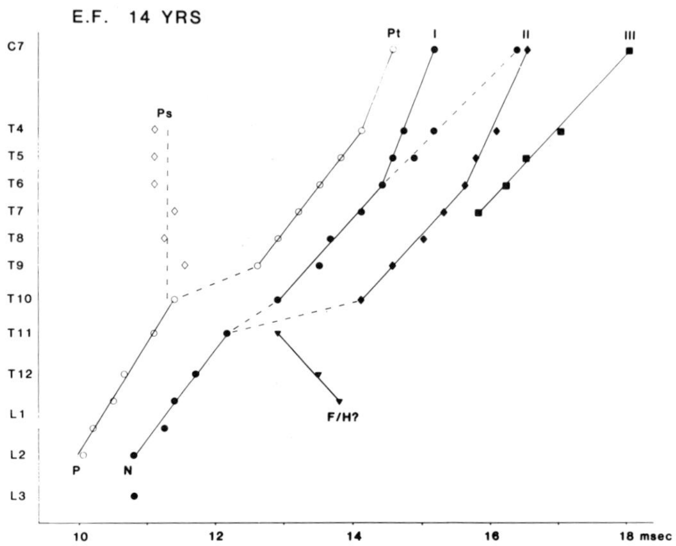

Fig. 7. Graph of SEP latency against vertebral level. Filled symbols represent negative peaks and open symbols positive. The initial positivity recorded below T11 splits into travelling (Pt) and stationary (Ps) components. From T11 to L1 an efferent potential is recorded, possibly related to the F or H reflex.

At high stimulus intensities, cord conduction velocities frequently appear to be fairly uniform from lumbar to cervical levels (Fig. 1), although at caudal sites the waveform is markedly different from that recorded more rostrally and it is difficult to be sure which peaks (if any) correspond to components I, II and III. In the high-intensity responses illustrated in Fig. 6 (upper) components II and III have fairly uniform CV but the initial component is apparently conducted more slowly between T12 and T9 than over mid-thoracic segments. At just supra-threshold stimulus intensity (Fig. 6, lower), sufficient to evoke only a single component at T12 and therefore presumed to activate only the largest (Group I) afferent fibres, there is a more marked delay between the potentials recorded at T12 and T9 and the 3 components present at T9, T5 and T2. All have CVs equal to or faster than the fastest component following high intensity stimulation. Their combined amplitudes are also considerably greater than that of the potential recorded at T12, and the fact that the 3 peaks do not diverge in latency from the single peak at T12 suggests that they reflect repetitive volleys in a postsynaptic tract. The marked hiatus between the T12 potential and initial potential recorded at T9 may also be due in part to tapering or branching of the presynaptic axons over the lumbar enlargement. Very similar findings were reported in the cat (Lloyd and McIntyre, 1950), where potentials recorded from the dorsum of the cord in response to stimuli just above threshold underwent a marked slowing of conduction and decrease in amplitude over upper segments of the lumbar enlargement, and were followed after a short interval (less than 0.75msec) by a fast-conducted, multiphasic volley in ipsilateral dorsolateral tracts.

Recordings made at 14 levels in one neurologically normal patient illustrate the evolution of the spinal SEP from lumbar to cervical regions (Fig. 7). It should be emphasized, however, that the pattern varies considerably according to the intensity of stimulation. At lumbar (cauda equina) level the waveform is relatively simple, essentially biphasic (positive/negative) with rostrally increasing latency. Between L1 and T11 an additional peak may be seen, decreasing in latency towards rostral sites. This is believed to reflex discharge in the ventral roots, perhaps responsible for the F or H reflex. Between T11 and T10 in this case there is a local region of conduction slowing and/or synaptic delay and above T10 at least 2 negative peaks are recorded, both with rostrally increasing latency. From T9 to T4, in addition to the "travelling" positive wave preceding the 2 negative peaks there is a "stationary" positivity, recorded with fairly constant latency but rostrally decreasing amplitude. "Stationary" or "far-field" potentials are frequently encountered in the context of propagating nerve potentials and tend to occur wherever there is a discontinuity in the characteristics of the volume conductor (in this case associated with entry of the volley into the spinal cord). Rostral to T8, 3 or even 4 travelling negative peaks can be distinguished. It now appears certain that the first is generated in the dorsal spinocerebellar tract and the last probably in the dorsal columns, but whether or not other tracts are involved in generation of the intermediate potentials awaits further evidence.

References

1. Beric, A.; Dimitrijevic, M.R.; Prevec, T.S. and Sherwood, A.M.: Epidurally recorded cervical somatosensory evoked potential in humans. Electroenceph. Clin. Neurophysiol. 1986, 65: 94-102.
2. Bradshaw, K.; Webb, J.K. and Fraser, A.M.: Clinical evaluation of spinal cord monitoring in scoliosis surgery. Spine 1984, 9: 636-643.
3. Dinner, D.S.; Lueders, H.; Lesser, R.P. and Morris, H.H.: Invasive methods of somatosensory evoked potential monitoring. J. Clin. Neurophysiol. 1986, 3: 113-130.
4. Jones, S.J.; Edgar, M.A. and Ransford, A.O.: Sensory nerve conduction in the human spinal cord: Epidural recordings made during scoliosis surgery. J. Neurol. Neurosurg. Physiat. 1982, 45: 446-451.
5. Jones, S.J.; Edgar, M.A.; Ransford, A.O. and Thomas, N.P.: A system for the electrophysiological monitoring of the spinal cord during operations for scoliosis. J. Bone Joint Surg. 1983, 65B, 134-139.
6. Lloyd, D.P.C. and McIntyre, A.K.: Dorsal column conduction of Group I muscle afferent impulses and their relay through Clarke's column. J. Neurophysiol. 1950, 13: 39-54.
7. Macon, J.B. and Poletti, C.E.: Conducted somatosensory evoked potentials during spinal surgery. Part 1: Control conduction velocity measurements. J. Neurosurg. 1982, 57: 354-359.
8. Macon, J.B.; Poletti, C.E.; Sweet, W.H.; Ojemann, R.G. and Zervas, N.: Conducted somatosensory evoked potentials during spinal surgery. Part 2: Clinical applications. J. Neurosurg. 1983, 57: 354-359.
9. Nainzadeh, N.K.; Neuwirth, M. and Bernstein, R.: Direct recording of spinal evoked potentials (SEP) to peripheral nerve stimulation, by a specially modified electrode. Proceedings of Third International Symposium on Spinal Cord Monitoring, Annapolis, October 1986.
10. Schramm, J.; Watanabe, E. and Romstöck, J.: Cortical and spinal intraoperative recordings in uneventful monitoring and in cases with neurologic changes. Proceedings of Third International Symposium on Spinal Cord Monitoring, Annapolis, October 1986.
11. Whittle, I.R.; Johnson, I.H. and Besser, M.: Recording of spinal somatosensory evoked potentials for intraoperative spinal cord monitoring. J. Neurosurg. 1986, 64: 601-612.

Experimental Ascending Evoked Potentials in Spinal Cord Injury

H. Baba;[*] K. Tomita; S. Umeda; N. Kawahara; S. Nagata; S. Nomura; H. Yugami

Summary

The basic waveform of the conducted epidural SEP of cat in ascending volleys consisted of two major components, namely N1 and N2, with mean conduction velocity of 74ms and 55ms, respectively. Significant amplitude attenuation of N1 and N2, and considerable latency delay of N1 rather than N2 were demonstrated in ventral epidural recordings. Latency delay of the spinal cord field potentials recorded from the dorsal part of the spinal cord occurred even if there was no latency delay in the ventral part, in 100g/cm contusion injury. There was no distinct relationship between severity of the injury and appearance of positive-going killed end potentials in acute contusion injury of the spinal cord.

Introduction

Modern technological advancement in computerized measuring equipment has accelerated physiological testing of spinal cord function (2, 7). In many papers recently published on the physiological meanings of spinal cord evoked potentials in spinal cord injury, positively-reversed evoked response has been described to be one of the major parameters of neurological deterioration (3, 4, 11, 12). The authors reported that these positive-going potentials frequently appeared at the injured vertebral levels in cervical and/or thoracic myelopathy (3). In order to investigate these positive-going potentials of the ascending SEP in various degrees of spinal cord injury, an experimental study using a modified Allen's method was carried out. This report describes (a) characteristics of the basic waveform of the ascending SEP, (b) influences of spinal cord contusion injury on SEP, and (c) changes of the intramedullary field potentials in spinal cord contusion injury.

Materials and methods

Experiments were carried out in 40 adult mongrel cats weighing 2.0kg to 5.8kg, average of 3.6kg. Anesthesia was induced by intraperitoneal injection of pentobarbital sodium (30mg/kg) and maintained using halothane (0.5% to 1.0%), nitrous oxide, and oxygen (1:1) by artificial ventilation. Animals were positioned prone fixed with a small animal stereotaxic spine instrument (Narishige Model ST-7, SM-15) for strict restriction of hazardous spinal cord movement during SEP recording. A bipolar pick-up electrode was placed at the dorsal midline epidural surface through a laminectomy of T6 through T8 and a bipolar stimulating electrode was wrapped around the exposed left

---

* Department of Orthopaedic Surgery, School of Medicine, Kanazawa University, Kanazawa 920, Japan

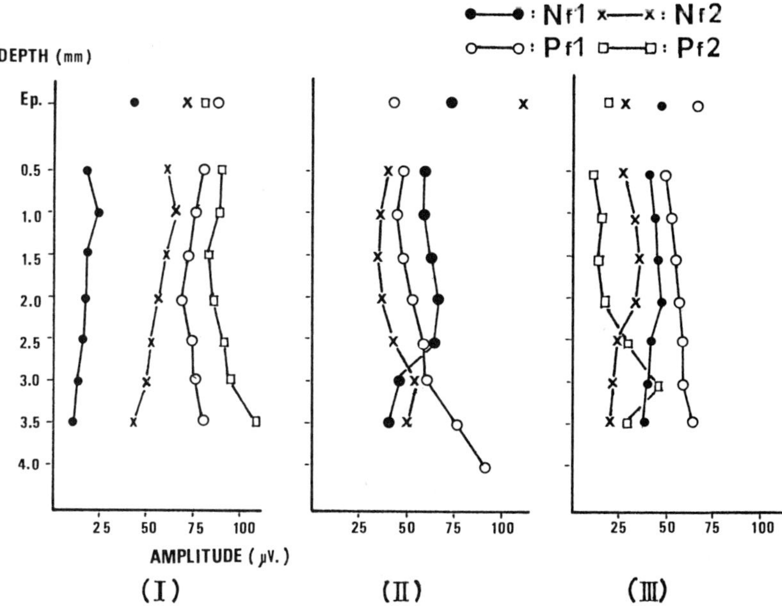

Fig. 1. Amplitude changes in the field potentials from various sites of the spinal cord. (I) microelectrode truck of 1.0mm left to midline, (II) midline, (III) 1.0mm right to midline. Stimulus was applied to the left sciatic nerve (n = 4, Ep. :Dorsal epidural surface).

sciatic nerve at hip. In the elicitation of evoked signals, care was taken to place the pick-up electrode close to the the midline surface of the dura. Using a Medelec NT-6 stimulator, rectangular pulses of 0.2ms duration were applied at 10Hz or 20Hz with intensities well above the threshold required to obtain a minimal motor response in the hind limb (mean 36 V). Two hundred fifty-five repetitive evoked signals in the ascending volleys were averaged by Medelec DAV6, which was preamplified by Medelec AA6, with a bandwidth of 10Hz to 10KHz.

The spinal contusion injury was produced by the standardized weight-drop technique of Allen (1) modified by the authors (4). A copper weight of 1.0g was lowered within a plastic cylinder onto the dorsal surface of the dura mater, which produced contusion injury. The ascending SEP was recorded stepwise at and near the injury site.

The spinal cord field potentials (SFP) were recorded from a monopolar tungsten microelectrode (A-M Systems Inc., Everett, WA 98201, U.S.A., Cat. No. 5735) which was insulated by Paralene-C, from the midline microelectrode truck. The exposed microelectrode tip was 10μm in diameter and had an impedance of 2M ohms. The input impedance of the Medelec AA6 preamplifier was 20M ohms and it fully received the respective signal from the microelectrode. Both SEP and SFP were recorded at the injury site. An intermediate electrode was placed in the paravertebral muscle between the recording and the stimulating electrode.

Results - Basic waveform of the spinal cord evoked potentials

The basic waveform of the ascending SEP consists of two major negative waves, namely N1 and N2, and whose durations are nearly 1ms and several ms, respectively. Relatively higher amplitude of N1 was followed by a short positive deflection (P1), and N2, by a very small positive potential (P2) with longer latency. The peak-to-peak amplitude of N1 through P1 and N2 through P2 ranged 40 to 120μV and 22 to 56μV, respectively. The mean conduction velocity of the stable components of N1, P1, and N2 was 73.65 ± 2.23ms, 63.79 ± 2.54ms, and 55.43 ± 1.37ms, respectively (n = 10). In the measurement of latencies from the various recording sites around the spinal cord, it revealed that there was significant difference of the latencies of N1 among the recording sites. In recording SEP at T6 through T8 level with the left sciatic nerve stimulation, the latency of N1 in the ventral epidural space recording was 0.3ms longer than that from the dorsal recording. There were also significant amplitude reduction of N1 and N2 in the ventral recording compared with those in the dorsal epidural recording. The mean amplitude of N1 in dorsal, lateral, ventro-lateral, and ventral recording were 23.47 $\pm$ 5.71mV, 19.53 $\pm$ 3.18μV, 12.44 $\pm$ 2.57μV (n = 5, p < 0.005), and 11.62 $\pm$ 2.83μV (n = 5, p < 0.005), respectively. Also in these five animals, the mean amplitude of N2 was 32.19 $\pm$ 6.41μV in the dorsal epidural recording, 9.64 $\pm$ 1.16μV in the lateral (p < 0.005), 9.37 $\pm$ 0.37μV in the ventro-lateral (p < 0.005) and 4.72 $\pm$ 0.53μV in the ventral recording (p < 0.005). Amplitude attenuation of N2 component in the ventral recording was more significant than N1.

Results - Spinal cord field potentials without spinal cord injury

In order to investigate physiological alterations in the electrical activities within the injured spinal cord, the intramedullar SFP without any spinal cord injury was studied.

Although it does not seem appropriate to correlate monopolar SFP with bipolar ascending SEP, it is reasonable to assume that epidurally recorded SEP well reflects electrophysiological activity within the cord. In this study, SFP in the midline microelectrode truck was investigated.

The basic SFP consists of the first wave (Nf1, Pf1) followed by the second component (Nf2, Pf2). In the majority of the SFP data, only the first waves were stably recorded. Fig. 1 shows amplitude changes in the SFP from various depths of the midline microelectrode trunk. There is amplitude increase in the negative component (Nf1) in the midline dorsal column, and relative amplitude increase in the positive component (Pf1) in the central and ventral areas of the spinal cord. In four animals, the largest amplitude of the Nf1 was elicited at 2.0 and 2.5mm in depth from the dorsal surface of the spinal cord, and those of Pf1 were at 2.5 to 3.5mm in depth.

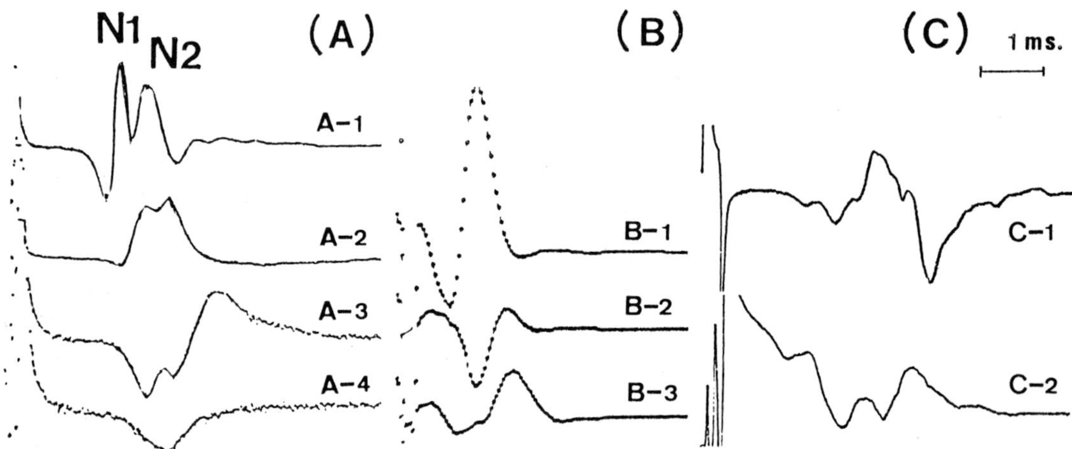

Fig. 2. Spinal cord evoked potentials recorded from the dorsal epidural surface in acute spinal cord contusion injury. Positive-going potentials are recorded in all types of injury. (A) 80g.cm injury; Trace of A-1 was recorded 10mm rostral from the injury site, A-2, 5.0mm rostral, A-3, injury site, and A-4, 5.0mm caudal. (B) 100g.cm injury; B-1 was recorded 5.0mm rostral from the injury site, B-2, injury site, and B-3, 5.0mm caudal. (C) 400g.cm injury; Trace C-1 was recorded from 10mm rostral from the injury site, and C-2, from the injury site.

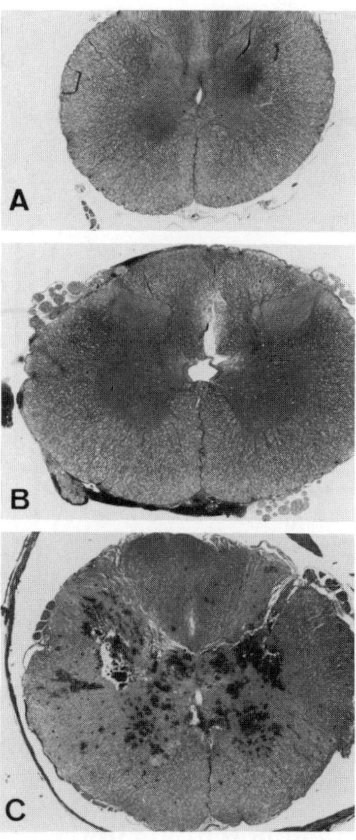

Fig. 3. Histological changes of the spinal cord in the acute spinal cord contusion injury. Slightly edematous changes were demonstrated in the 80g.cm injury (A) and those became significant with hemorrhage within the gray matter and the lateral column in the 100g.cm injury (B). Marked hemorrhagic necrosis with cavity formation within the gray matter and hemorrhage within lateral, dorso-lateral column associated with structural destruction were observed in the 400g.cm injury (C).

Results - Spinal cord evoked potentials in the spinal cord contusion injury

The main purpose of the present study to observe the positive-going killed end potentials was carried out in 15 animals; in 4 with 80g.cm injury, in 6 with 100g.cm injury, and in 5 with 400g.cm injury. Stepwise recordings of the ascending SEP were carried out in craniocaudal direction across the lesion with 5mm intervals.

In 80g.cm contusion injury, N1 component reversed to be positive-going only at the injury site, whereas, N2 revealed no marked amplitude change nor phase alteration. It was noteworthy, however, that the latency of N2 at the injury site markedly delayed (Fig. 2-A).

In 100g.cm contusion injury, all of the negative components (N1, N2) reversed to be positive-going at the injury site, and, additionally, delayed in latencies (Fig. 2-B). In rostral recording site, the amplitude of N1 and P1 increased almost up to twice than those before injury (Trace B-1 in Fig. 2-B).

In the heavier injury with 400g.cm contusion, N1 and N2 components markedly delayed and the amplitude attenuation of N2 occurred at the injury site (Fig. 2-C). SEP recorded from 10mm rostral from the injury site showed relatively increased amplitude of N1 and positively reversed N2 with delayed latency.

Typical histology is shown in Fig. 3, contrasting extension of the hemorrhagic necrosis between 100g.cm and 400g.cm contusion injury. There was less evident histological changes in 80g.cm and 100g.cm injury, however, extensive hemorrhagic necrosis within the central gray matter and moderate hemorrhage in dorsal and lateral column occurred in 400g.cm contusion injury.

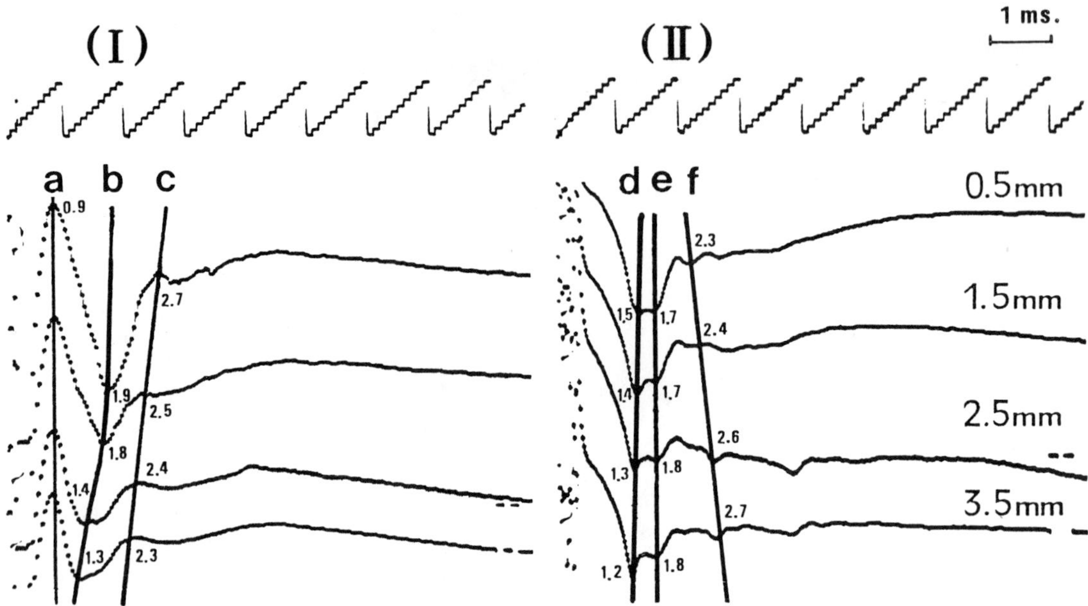

Fig. 4. Difference of the spinal cord field potentials in the 100g.cm injury (I) and 400g.cm injury (II). Latency delay is significant only in the dorsal part of the spinal cord in the 100g.cm injury but is marked in all parts of the spinal cord in the 400g.cm injury. (a) Nf1 (b) Pf1 (c) Nf2 (d) Pf1 (e) Pf2 (f) Pf3

Results - Spinal cord field potentials in the acute spinal cord contusion injury

In 100g.cm injury, there was remarkable latency delay in SFP recorded at 0.5mm and 1.5mm in depth (Pf1, Nf2), but those latencies recorded at 2.5mm and 3.5mm in depth were only slightly delayed (Fig. 4-I, Animal No. 23). In this animal, P1 component of the epidural SEP had latency of 1.55ms, which was somewhere between the two points of SFP recorded at 0.5mm in depth (1.9ms) and 3.5mm in depth (1.3ms).

In 400g.cm contusion injury, contrasting to 100g.cm injury, mostly positive SFP became markedly polyphasic without any significant latency difference among the various recording sites (Fig. 3-II, Animal No. 22).

Discussion

Electrodiagnostic testing and intraoperative monitoring has become more useful part in the assessment of the spinal cord function (10). The present study was designed to investigate characteristics of N1 and N2 and to observe positive-going killed end potentials in the widespread method of the ascending SEP recording.

In consideration of the bipolar recording technique, it must be taken into account that there exists phase cancellation when the evoked response is elicited between two different points. The subtracted output between two points within the lesion may be eliminated in the amplitude, which makes identification of each component of SEP very difficult. The common reference technique or monopolar recording technique may be one of the alternatives to improve this technical problem. In the meantime, the present study examines ascending SEP in the bipolar recording mode.

The latency of SEP recorded from the dorsal epidural space was shorter than that from the ventral epidural space, which implies methodological superiority of ascending

SEP in the dorsal epidural surface. The recording electrode must be placed just in the midline dorsal surface because the amplitude attenuation is so significant elsewhere.

It has been frequently reported that the positive-going killed end potential is one of the major electrodiagnostic signs manifesting lesion within the cord. In experimental works, however, Deecke (9) reported that these killed end potentials had no distinct correlation with the severity of the spinal cord injury in the acute phase. D'Angelo (8) reported that there was slightly edematous and hemorrhagic changes in 100g.cm injury and significant hemorrhage in the central gray matter in 300g.cm. In 500g.cm injury, markedly hemorrhagic necrosis in the lateral column was also demonstrated. D'Angelo concluded that the spinal cord was irreversibly damaged at least in 300g.cm injury. The similar histological findings were demonstrated in the author's work in the acute 400g.cm contusion injury. Our results revealed that positive-going killed end potentials were elicitable even though the amount of the injury was as little as 80g.cm or 100g.cm contusion, which demonstrated minimal histopathological changes. Thus, it is concluded that there is no distinct correlation between frequency of appearance of the positive-going killed end potentials and histopathological changes in the acute phase of the spinal cord contusion injury.

In the assessment of the latency of SEP, it must be noteworthy that the retardation of the ascending conductivity in the dorsal part of the spinal cord, which was clearly demonstrated by the monopolar SFP, could be concealed by the least damaged ascending volleys in the ventral part of the spinal cord (5, 6). Whittle (13) reported that the insult to the anterolateral cord might not be reflected in the changes of SEP, and that furthermore, the minimal damage to the dorsal cord does not necessarily bring any remarkable changes in SEP. The usefulness of the ascending SEP recording in the physiological evaluation of experimental spinal cord injury is valid to some extent, however, with these limitations in mind, other studies concerning reference mode and/or experimental models must be carried out.

References

1. Allen, A.R.: Surgery of experimental lesion of spinal cord equivalent to crush injury of fracture dislocation of spinal column. A preliminary report. JAMA, 57: 878-880, 1911.
2. Anderson, T.E.: Spinal cord contusion injury. Experimental dissociation of hemorrhagic necrosis and subacute loss of axonal conduction. J. Neurosurg., 62: 115-119, 1985.
3. Baba, H.; Shima, I.; Tomita, K. et al.: Clinical usefulness of spinal cord evoked potentials. In: J. Schramm; S.J. Jones (eds), Spinal Cord Monitoring. Berlin Heidelberg: Springer-Verlag, 1985, pp. 245-249.
4. Baba, H.: Experimental study of the spinal cord evoked potentials in cats. J. Jpn. Orthop. Ass., 60: 623-636, 1986.
5. Campbell, B.: The distribution of potential fields within the spinal cord. Anat. Record, 91: 77-88, 1945.
6. Coombs, J.S.; Curtis, D.R.: Spinal cord potentials generated by impulses in muscle and cutaneous afferent fibers. J. Neurophysiol., 19: 452-467, 1956.
7. Cracco, R.Q.; Evans, B.: Spinal evoked potential in the cat. Effect of asphyxia, strychnine, cord section and compression. Electroencephalogr. Clin. Neurophysiol., 44: 187-201, 1978.
8. D'Angelo, C.M.; Vangilder, J.C.; Taub, A.: Evoked cortical potentials in spinal cord trauma. J. Neurosurg., 38: 332-336, 1973.
9. Deecke, L.; Tator, C.H.: Neurophysiological assessment and afferent conduction in the injured spinal cord of monkeys. J. Neurosurg., 39: 65-74, 1973.
10. Fernandez de Molina, A.; Gray, J.A.B.: Activity in the dorsal spinal gray matter after stimulation of cutaneous nerves. J. Physiol., 137: 126-140, 1957.
11. Greenberg, R.P.; Ducker, T.B.: Review article. Evoked potentials in the clinical neurosciences. J. Neurosurg., 56: 1-18, 1982.
12. Schramm, J.; Krause, R.; Shigeno, T. et al.: Experimental investigation on the spinal cord evoked injury potential. J. Neurosurg., 59: 485-492, 1983.
13. Whittle, I.R.; Johnston, I.H.; Besser, M.: Recording of spinal somatosensory evoked potentials for intraoperative spinal cord monitoring. J. Neurosurg., 64: 601-612, 1986.

Effects of Spinal Cord Compression on Repetitive Impulse Conduction of Ascending Fibers in the Dorsal Column

K. Sakatani;[*] T. Ohta; Y. Yamagata; M. Shimo-Oku

Abstract

The effects of spinal cord compression on dorsal column conduction potentials at various stimulus frequencies were analyzed in pentobarbital anesthetized cats. The L6 dorsal root was given 1 to 500Hz stimuli, and the CAPs were recorded from the L2 dorsal column. The L4 cord segment was compressed by stepped increments (0.5mm/5min.) until the responses at 1Hz disappeared. Before compression, each CAP at all frequencies showed almost uniform amplitudes and latencies. During compression, the CAPs were not altered significantly at any frequency until the spinal cord was compressed 2.5-4.0mm. Further compression produced a progressive decrease in amplitude and increase in latency of the CAPs. At 500Hz, however, the conduction block was much severer than that produced at lower frequencies. After decompression, the amplitude decreased and the latency increased progressively at 500Hz. At 100Hz, however, the amplitude increased and the latency

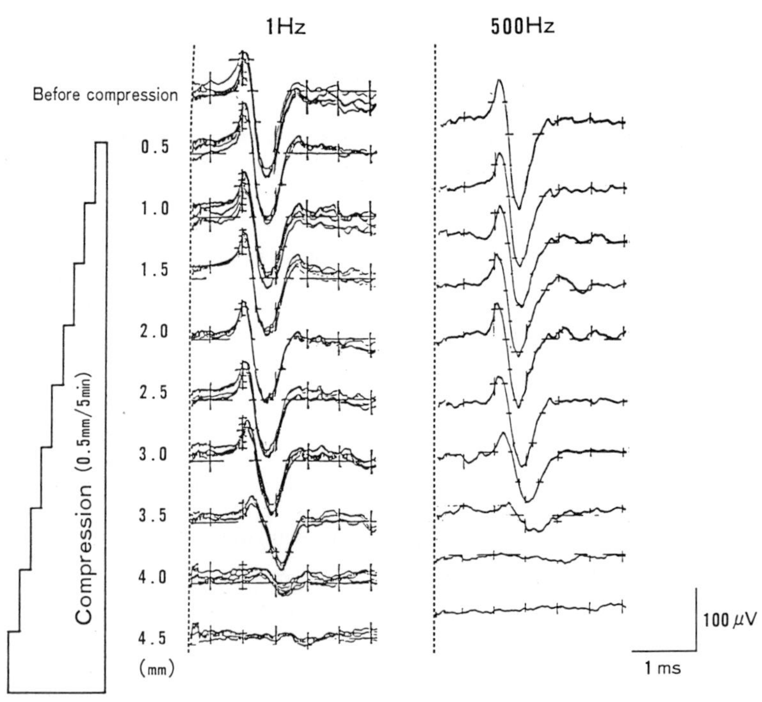

Fig. 1. Representative recordings of dorsal column potentials before compression and during compression. Positivity is upward. Amount of compression is given in mm to the left of the tracing.

* Department of Neurosurgery, Osaka Medical College, 2-7 Daigakucho Takatsuki, Osaka 569, Japan

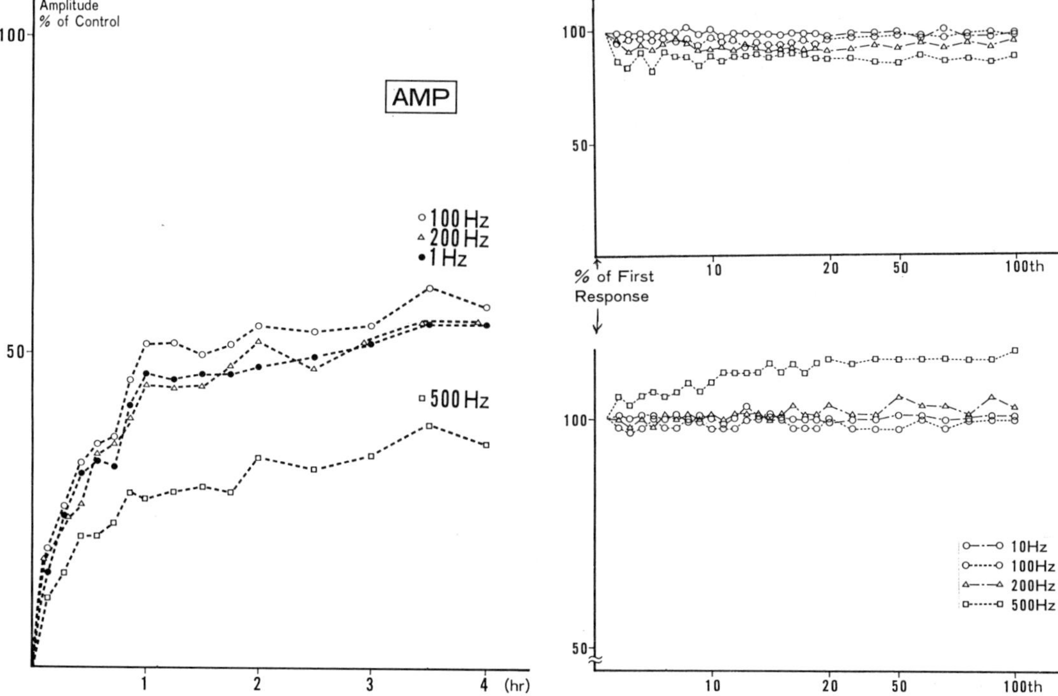

Fig. 2. Recovery curves of amplitude after decompression. The ordinate indicates amplitude as a percentage of the potentials before compression.

Fig. 3. Response curves of amplitudes (upper section) and latencies (lower section) of dorsal column potentials before compression. The ordinates indicate amplitude and latency as a percentage of the first potentials. The abcissa indicates the number of stimuli.

decreased. And they reached to almost steady level. In conclusion, the inability to transmit impulses at high stimulus frequencies indicates incomplete impairment of spike generation in the CNS fibers injured by compression. And it may be that the impulse conductivity of CNS fibers injured by compression which were less excitatory at low stimulus frequency are facilitated by a train of moderately high frequency stimuli.

Introduction

Normally, excitation of the ascending fiber in the spinal cord through the peripheral afferent axon and its receptor results in a train of impulses at various frequencies. In reported studies of spinal injury, however, the conductivity of ascending fibers in the spinal cord has been examined only at low stimulus frequencies, namely 1 to 10Hz. Therefore, it is important to examine the effects of spinal cord compression on the conduction of ascending fibers in the spinal cord at various stimulus frequencies.

In this study, the difference between the conductivity of dorsal column fibers at low and high stimulus frequencies was examined during compression and after decompression.

Materials and Methods

Twelve adult cats were anesthetized with sodium pentobarbital; and tracheal, arterial and venous catheters were placed. The cats were fixed in a stereotactic spinal unit after exposing the lumber spinal cord with an L1 - L7 laminectomy. Respirations were controlled with a volume-limit respirator after muscle paralysis with pancuronium bromide.

The central cut ends of the L6 dorsal roots were placed on bipolar electrodes and stimulated supramaximally for myelinated fibers (2-5 V, pulse duration 0.1ms). Spinal evoked potentials were recorded at the L2 dorsal column by a monopolar recording electrode, and L6 dorsal root potentials were monitored throughout the experiment to

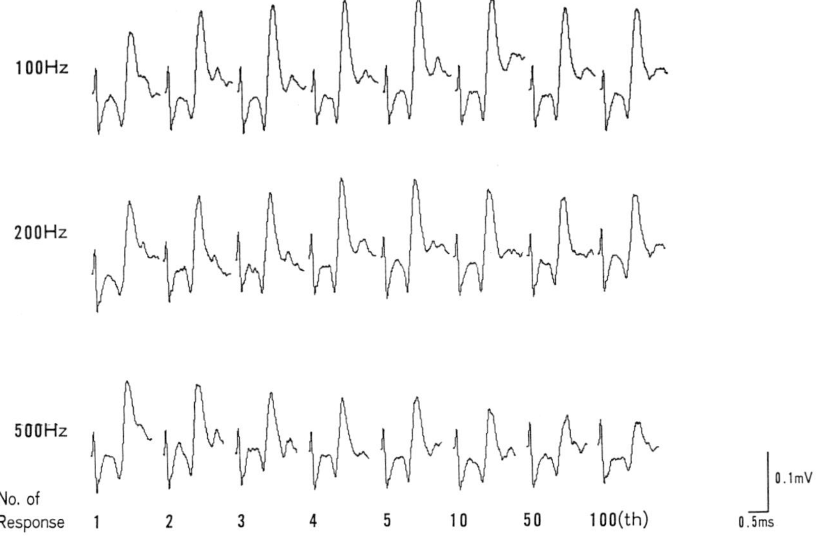

Fig. 4. Dorsal column potentials at 100Hz, 200Hz and 500Hz one hour after decompression. The 1st-5th, 10th, 50th and 100th response are shown.

confirm that stimulus conditions were constant. The antidromic root potentials were recorded in three cats. The L2 dorsal column was stimulated (20-50 V, pulse duration 0.1ms) and antidromic responses were recorded at the L6 dorsal root by a monopolar electrode.

A stimulus frequency of 1Hz was used as a low stimulus frequency and that of 100Hz, 200Hz and 500Hz as high stimulus frequencies. At high stimulus frequencies the response to the 100th stimulus was analyzed. In five cats all the responses at high stimulus frequencies, namely, from the 1st response to the 100th response, were analyzed using a personal computer system (PC-9801, NEC, Tokyo) with an original program.

The L4 cord segment was compressed by the plastic plate (4mm x 10mm) attached to the micromanipulator for glass electrodes that was fixed on the stereotactic frame. Compression was started from the point of contact with the dorsal surface of spinal cord and increased every 5 min. in increments of 0.5mm. The compression was carried on until the evoked potentials at 1Hz were flat; the spinal cord was released completely 10 min. after completion of compression.

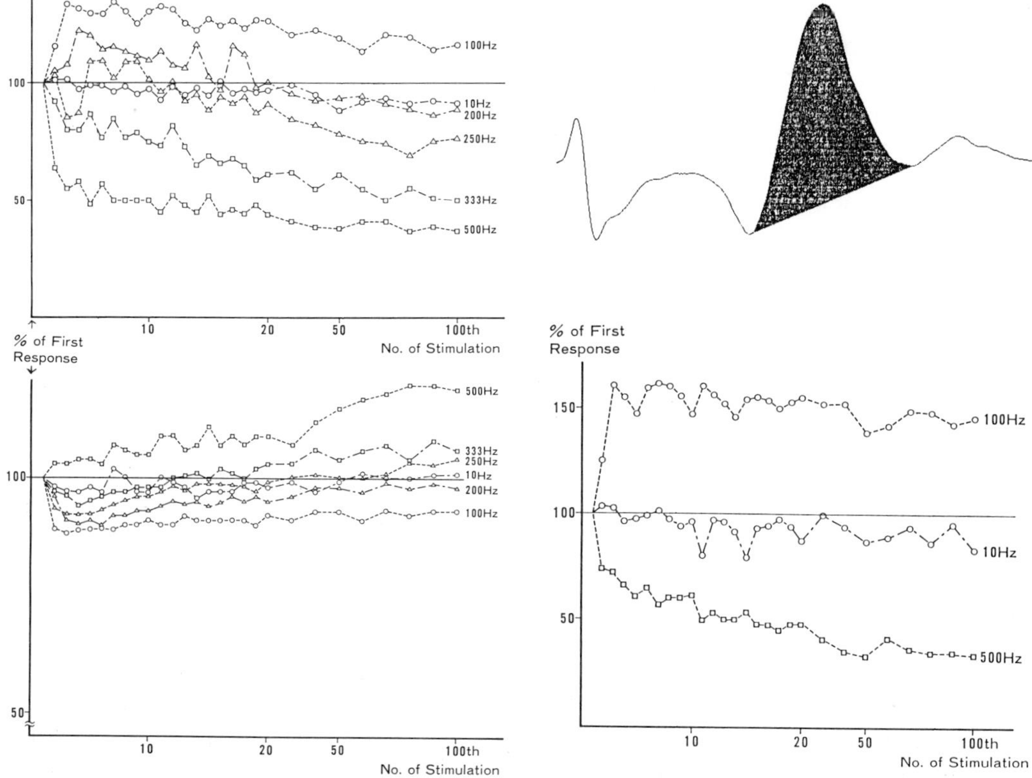

Fig. 5. Response curves of amplitudes (upper section) and latencies (lower section) of dorsal column potentials one hour after decompression.

Fig. 6. Measurement of the area of potentials (black area, upper section) and the response curves of area are shown. Positivity is downward.

Results

Before compression, triphasic waves were recorded at all frequencies (Fig. 1, uppermost tracing). They showed almost uniform amplitudes (initial positive peak to negative peak) and latencies (initial positive peak). The latencies were 0.7-0.8ms. The distance between recording and stimulating electrodes ranged from 5cm to 7cm. The latencies of the antidromic dorsal root potentials were the same as those of the antidromic dorsal column potentials. During compression, the potentials at all frequencies were not altered significantly until 2.5mm (Fig. 1). Further compression produced progressive conduction failure (latency increase, amplitude decrease). Especially at 500Hz, the conduction failure was much more severe than that produced at lower frequencies.

Although the changes in potentials showed a similar time course in all cats, the amount of compression that produced no significant changes in potentials at all frequencies varied from 2.5mm to 4.0mm because of individual differences in the volume of the spinal cord. For the same reason, the amount of compression that produced a complete conduction block at 1Hz varied from 4.5mm to 6.5mm.

After decompression, the amplitudes of the potentials recovered progressively for one hour, but only slowly thereafter in all cats. However, the recovery rates of amplitudes and latencies differed greatly according to the stimulus frequency (Fig. 2). Four hours after decompression, the amplitude at 1Hz was 50% of that before compression. At 500Hz, the recovery rate of amplitude was markedly less (30%) than that at 1Hz, and a greater increase in latency was exhibited for up to four hours. The amplitude at 200Hz was similar to that at 1Hz, but the recovery rate of amplitude at 100Hz (65%) exceeded that at 1Hz.

To estimate the difference in recovery rate, 100 responses at high stimulus frequencies were analyzed. Before compression, almost uniform amplitudes and latencies were recorded at stimulus frequencies as high as 500Hz, except for a slight increase in latency at 500Hz (Fig. 3). One hour after decompression, a progressive decrease in amplitude was demonstrated at 500Hz. At 100Hz, the amplitude increased and then decreased only slightly. At 200Hz, the amplitude increased transiently and decreased to that of the first response (Fig. 5, upper section).

A progressive increase in latency was observed at 500Hz, whereas at 100Hz the latencies were decreased and then increased only slightly. At 200Hz the latencies decreased transiently and increased to the level of the first response. The response curves of latency were inversely proportional to those of amplitude; the latency decreased with the increment of amplitude, and increased with the decrement of amplitude (Fig. 5, lower section). The area of dorsal column potentials increased or decreased in proportion to the amplitude (Fig. 6, lower section). These changes of amplitude, area and latency were not observed in dorsal root potentials.

These results were also obtained with antidromic potentials.

Discussion

Conduction failures at high stimulus frequencies have been reported in pathological peripheral nerves, i.e., axonal cooling, acute and chronic nerve compression and experimental demyelination. In the CNS, only experimental demyelination was reported to show such a conduction failure.

In this experiment, there was no significant difference in amplitude and latency of dorsal root potential at all frequencies after decompression. From this observation it can therefore be assumed that the changes of amplitudes and latencies of dorsal column potentials at high stimulus frequencies after decompression occurred at the site of compression. At 500Hz, amplitude decreased progressively with each stimulation. Apparently, some dorsal column fibers that could transmit impulses at low stimulus frequencies could not transmit high frequency impulses (500Hz) in the later stage of compression and after decompression. This indicates that the incomplete impairment of spike generation may exist in the CNS fibers injured by mechanical compression which can transmit impulses at low stimulus frequencies.

One of the interesting observations was that the amplitude and area of dorsal column potentials increased at 100Hz after decompression. This suggests that the number of excited fibers increased with repeated stimulation. The impulse conductivity of the CNS fibers injured by compression which were less excitatory at low stimulus frequencies may have been facilitated by a train of moderately high frequency stimuli (100Hz).

Latency increased at 500Hz in the later stage of compression and after decompression. In some pathological fibers, the physiological mechanism of the transmission failure of high frequency impulses has been explained as being due to the prolongation of the refractory period. The fast fibers may be blocked, or the conduction velocity of the fast fibers may be reduced, or both these conditions may be indicated because of the prolongation of initiating spikes. On the other hand, latency was decreased at 100Hz after decompression. It may be that the impulse conductivity of fast fibers was facilitated. The conduction in single fibers must be studied.

References

1. Edwards, M.S.B.; Bolger, C.A. and Jewett, D.L.: Diagnosis of acute nerve compression in the cat with high frequency nerve trains evoked responses. Neurosurg. 12: 1-5, 1983.
2. Davis, F.A.: Impairment of repetitive impulse conduction in experimentally demyelinated and pressure-injured nerves. J. Neurol. Neurosurg. Psychiat. 35: 537-544, 1972.
3. Lehmann, H.J. and Tackmann, W.: Neurographic analysis of trains of frequent electric stimuli in the diagnosis of peripheral nerve disease - Investigations in the carpal tunnel syndrome. Europ. Neurol. 12: 293-308, 1974.
4. McDonald, W.I. and Sears, T.A.: The effects of experimental demyelination on conduction in the central nervous system. Brain 93: 583-598, 1970.
5. Paintal, A.S.: Effects of temperature on conduction in single vagal saphenous myelinated nerve fibers of the cat. J. Physiol. 180: 20-49, 1965.
6. Paintal, A.S.: The influence of diameter of medullated nerve fibers of cats on the rising and falling phases of the spike and its recovery. J. Physiol. 184: 791-811, 1966.
7. Tackmann, W. and Lehmann, H.J.: Relative refractory period of median nerve sensory fibers in carpal tunnel syndrome. Europ. Neurol. 12: 309-326, 1974.

The Effect of Chymopapain on Nerve Impulse Transmission in the Rat

P. Wehling;[*] S.J. Pak; K.P. Cleveland

Abstract

Acute and long-term effects of chymopapain on nervous transmission were studied in rats. The drug was injected into the lumbar spinal canal. Action potentials evoked by electrical stimulation of the sciatic nerve were recorded at the dorsal root entry zone of the lumbar cord. Chymopapain prolongs impulse latency compared to that of control rats treated with isotonic saline. The effect is dose dependent and can last for several weeks.

Introduction

In rare cases, the clinical use of chymopapain for chemonucleolysis can lead to paraplegia (3, 5). In animal experiments, application of chymopapain to the nerve tissue results in demyelination, infiltration, and vascular damage (6, 7). The present study was carried out to quantify the neurophysiological abnormalities caused by injection of chymopapain into the lumbar spinal canal. The aim of the study is to give an assessment of spinal involvement by incorrect injection and possible therapeutic measures.

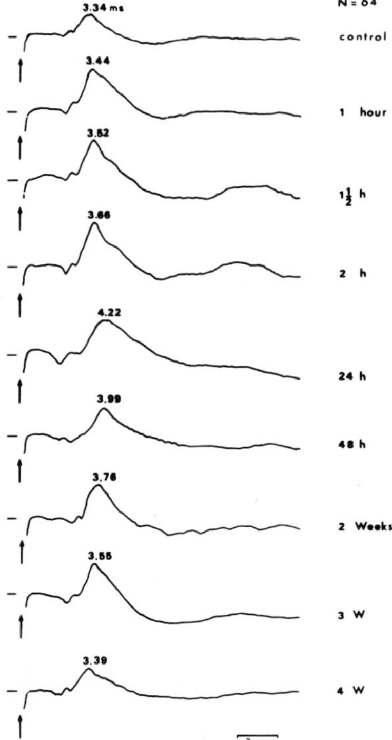

Fig. 1. Chymopapain (500 i.u.) increases latency of action potentials recorded at spinal cord. Latency from stimulus (arrows) to peak is indicated in milliseconds (ms) above the traces. Time elapsed since drug injection is indicated in the right column. Each trace is the average of 64 responses to sciatic nerve stimulation.

* Orthopädische Klinik and Physiologisches Institut der Universität Düsseldorf, Moorenstr. 5, D-4000 Düsseldorf, Federal Republic of Germany

Material and methods

Inbred male Wistar rats (n = 19) were anesthetized using pentobarbital-sodium (40mg/kg i.p.), which does not affect the conduction properties of central and peripheral nerve tissue (2). Anesthetic was reinjected in small doses as required. Body surface temperature was maintained at 30°C by an infrared lamp.

A Picker Myoscript EMG 2002/1 was used for stimulation and recording. The stimulus electrodes were concentric EEG needles placed at the right external malleolus. Stimuli (voltage pulses 0.1ms long) were applied at a frequency of 3Hz. Series of 64 responses were recorded and averaged every 5 minutes. Threshold strength was about 40V. This ensures that the stimulus electrodes were not close enough to the nerve to cause damage. The lumbar signals were picked up by a needle inserted cranial to the spinal process L1 in the midline (cf. ref. 1). The reference electrode was inserted into the skin 2cm to the left of the recording electrode; a needle ground electrode was inserted into the m. glut. max. on the right side. During the long-term study identical recordings were assured by using bone and skin marks. Latencies were determined as the time from stimulus artifact to the first negative peak of the electrically evoked nerve action potentials. Recording began 1 hour before injection and was continued for at least 2 hours after injection.

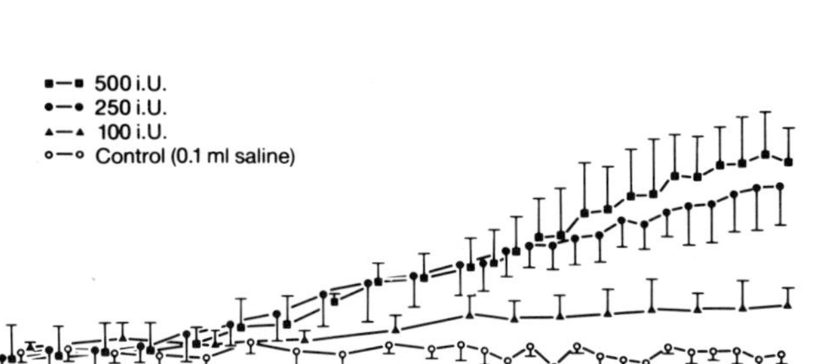

Fig. 2. Latency increase due to chymopapain is dose-dependent. Points at various times after injection of chymopapain or isotonic saline are averages (with standard deviations) from 5 (control), 2 (triangles), and 6 animals (filled circles and squares). Note change in time scale at 50 minutes.

Chymopapain, dissolved in 0.1ml isotonic saline, was injected into the spinal canal between the spinal process L4 and L5 from dorsal. The position of the needle tip was monitored by x-ray in two perpendicular directions. The injection needle was inserted 20 minutes before injection; the injection lasted 1 minute. Doses were 100 i.u. (2 rats), 250 i.u. (6 rats) and 500 i.u. (6 rats). Five control rats received 0.1ml isotonic saline only.

Results

Rats were examined daily for loss of motor function. Three of the high-dose rats developed a temporary paraplegia of the hindlimbs which disappeared 2 to 3 days after injection. Fig. 1 shows representative action potentials recorded before and after chymopapain was applied. The amplitude of the averaged evoked nerve potentials was not significantly changed by the drug. We therefore focused our attention on the latency of the response. In Fig. 1, the latencies to the peak of the first negative component are indicated in ms; arrows mark the stimulus artifacts. In one experiment we demonstrated that succinylcholine does not abolish the L1 responses.

In order to compare results from different animals, we determined the change in latency relative to the mean latency before administration of chymopapain for each animal. Fig. 2 shows the relative change in latency in four groups of animals at various times after application of chymopapain or isotonic saline (the 5 control rats were subjected to exactly the same stimulus and recording procedures). Saline had no effect on the latency (open circles). Chymopapain clearly prolongs the latency at the doses of 250 i.u. and 500 i.u. (squares and filled circles in Fig. 2), the increase being about 10 to 15% relative to the mean latency before application of the drug. In the 2 animals that received a low dose (100 i.u.) of chymopapain, latency increased only slightly (3%; triangles in Fig. 2), but this effect was significant at 2 hours after drug application (p 0.001).

For the long-term study we chose two rats from the high-dose group (500 i.u.) and studied them for 4 weeks after a single dose of chymopapain had been applied. The time course of the relative change in latency is shown in Fig. 3. Points are means of ten measurements on each rat, vertical bars are standard deviation. The greatest increase in latency (20%) occurred 24 hours after application of chymopapain; the latency gradually declined to control levels after that. At 4 weeks after application the latency was not significantly different from that before application of chymopapain.

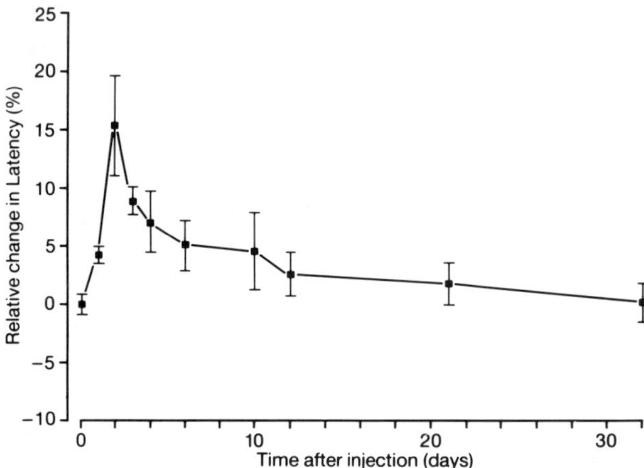

Fig. 3. Chymopapain has a long-term effect on response latency. 500 i.u. drug injected at time zero. Points are means from 2 animals with standard deviations.

Discussion

Chymopapain injected into the spinal cord has a dose-dependent effect on nerve conduction latency. 500 i.u. chymopapain increases response latency for a period of 3 to 4 weeks, and this change of latency is initially accompanied by paraplegia of the hindlimbs. Heininger et al. (4) have described a similar association of increased latency and paraplegia in rats with induced peripheral allergic neuritis.

Thus, our results suggest that the paraplegia observed in patients after incorrectly applied chymopapain may have a similar basis in the prolonged latency of afferent impulses to the spinal cord.

The morphological basis of the increased latency would seem to be essentially the demyelination of afferent nerve fibers. Thus, the rats with induced peripheral allergic neuritis (4) showed demyelination, and degenerative neuropathy in the spinal nerve roots and associated lumbar nerve of chymopapain-treated monkeys was observed by Zook et al. (7). Rydevik et al. (6) found that chymopapain does not acutely block impulse conduction or axonal transport of protein in rabbits, but 0.4% (1ml, 2000 i.u.) chymopapain does induce intraneural edema by affecting the microvessels of the nerve. These authors report degeneration of nerve fibers and increased stimulus threshold for eliciting action potentials.

References

1. Claus, D.; Neundorfer, B.: (1985) Zur Untersuchung der peripheren and zentralen Nervenleitung bei der Ratte, Z. EEG-EMG 14, 160-163.
2. Claus, D.; Weitbrecht, W.; Neundorfer, B.: (1985) Pentobarbital: The influence on somatosensory conduction in the rat. In: J Schramm; S.J. Jones (eds), Spinal Cord Monitoring. Springer-Verlag, Berlin, Heidelberg, pp. 90-94.
3. Dyck, P.: (1985) Paraplegia following chemonucleolysis. A case report and discussion of neurotoxicity. Spine, 10: 359-362.
4. Heininger, K.; Stoll, G.; Linington, C.; Toyka, K.V.; Wekerle, H.: (1986) Conduction failure and nerve conduction slowing in experimental allergic neuritis induced by P2-specific T-cell lines. Ann. Neurol., 19: 44-49.
5. Khalifa, P.; Boissonnas, A.; Giraudet, J.S.; Sereni, D.; Cremer, G.A.; Laroche, C.; Renier, D.; Auberge, T.: (1983) Syndrome de la queue de cheval au decours d'une nucleolyse. Rev. Rhum. Mal. Osteoartic., 50: 311.
6. Rydevik, B.; Branemark, P.I.; Nordborg, C.; McLean, W.G.; Sjostrand, J.; Fogelberg, M.: (1976) Effects of chymopapain on nerve tissue. Spine, 1: 137-148.
7. Zook, B.C.; Kobrine, A.I.: (1986) Effects of collagenase and chymopapain on spinal nerves and intervertebral discs of Cynomolgus monkeys. J. Neurosurg., 64: 474-483.

Diagnosis of Cervical Myelopathy Using Segmental Evoked Spinal Cord Potentials Obtained by Stimulating Finger Surface

K. Shinomiya;* K. Furuya; R. Sato; H. Sato; M. Yokoyama; H. Komori; A. Okamoto

Introduction

Spinal cord monitoring has been widely used for diagnosing the main lesions and pathophysiology of cervical spondylotic myelopathy.

As presented in our paper at the Second International Symposium on Spinal Cord Monitoring, we have been using conductive evoked spinal cord potentials (conductive ESCP) and segmental evoked spinal cord potentials (segmental ESCP) through a recording electrode at the posterior cervical epidural space (2). The conductive ESCP is obtained by stimulating the thoracic spinal cord, and shows the function of the lower extremities. The segmental ESCP is obtained by stimulating the median nerve, and shows the function of the upper extremities.

Recently, in addition to both the above ESCPs, we have been recording a segmental dermatome ESCP resulting from stimulation of the surface of each of the fingers (segmental D-ESCP) to diagnose the spread of intraspinal lesions in reference to the dermatome and to estimate sensory disturbance objectively.

Materials and Methods

We examined a group of 30 patients using the segmental D-ESCP. Eight had ossification of the posterior longitudinal ligament (OPLL), 13 had cervical spondylotic myelopathy, 5 had cervical spondylotic radiculopathy and the other 4 cases had miscellaneous problems.

Our technical method of spinal cord monitoring (Fig. 1) was the same as in the previous paper (3). A recording electrode

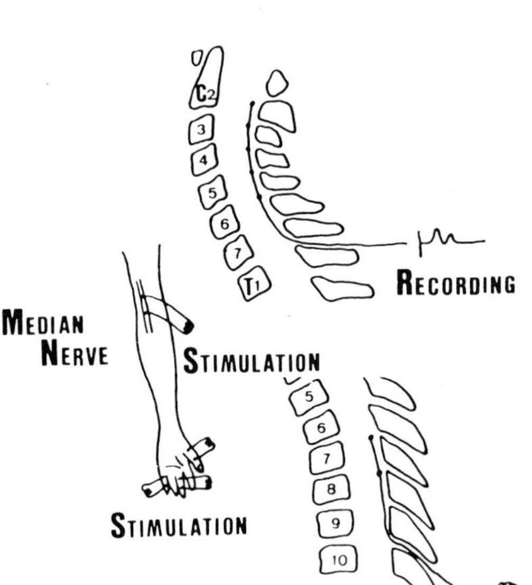

Fig. 1. System of spinal cord monitoring.

* Department of Orthopedic Surgery, Tokyo Medical and Dental University, 1-5-45 Yushima Bunkyo-Ku, Tokyo 113, Japan

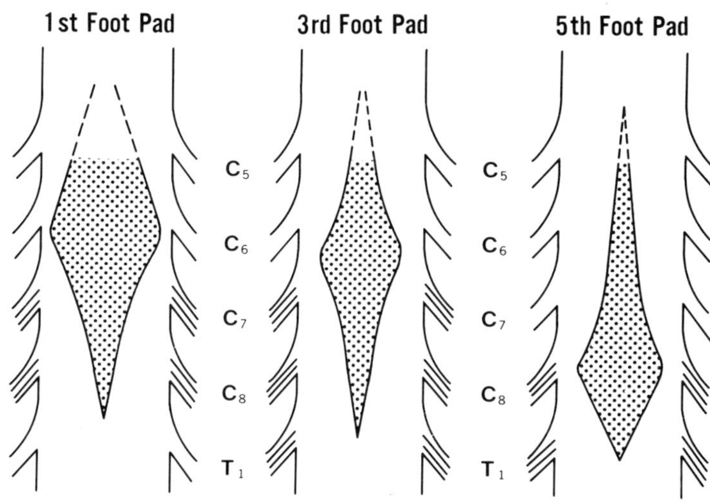

Fig. 2. The areas where the segmental D-ESCPs corresponding to each foot pad in the forelimbs of cats.

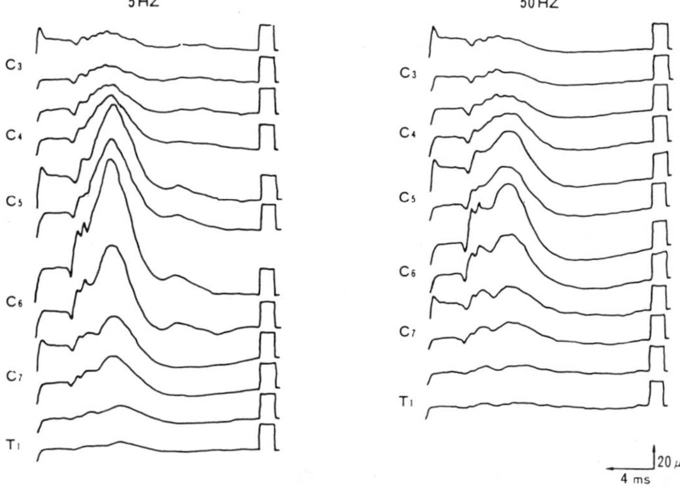

Fig. 3. Change of the segmental D-ESCPs from stimulating the pads of cats by applying a high frequency stimulation.

was inserted into the cervical epidural space through a Tuohy catheter. The recording electrode consisted of a polyethylene tube with 5 platinum bands spaced 15mm apart. This electrode should be placed so that the 5 poles span the main lesion site. It is also important that the recording electrode should be located at the center of the cervical epidural space, in order to record the segmental ESCP equally from both sides.

The stimulating electrode for the conductive ESCP consists of a polyethylene tube with 2 platinum bands, and is inserted into the thoracic epidural space at T8 or T9 through a Tuohy catheter.

Pulses of 0.5msec duration at 13mA were used for the segmental D-ESCP by stimulating the thumb, middle finger and little finger. The band width of the recording system was from 20Hz to 2kHz and 100 to 1000 stimuli were applied and averaged by the DISA 1500 EMG system.

We used monopolar recording and an indifferent electrode was placed in the subcutaneous tissue of the neck. In order to record a clear segmental D-ESCP by stimulating the finger surface, it is important that patients should be placed in a supine position at ease to eliminate EMG activity in the neck muscles. Diazepam and pentazocine were also administered to patients who could not tolerate the electrical stimulation to the finger surfaces.

Results

Prior to this clinical trial, we had generated segmental D-ESCP tracings by stimulating foot pads in the forelimbs of cats. These segmental D-ESCPs from stimulation of the pads of cats consisted of an early negative spike and a negative wave immediately

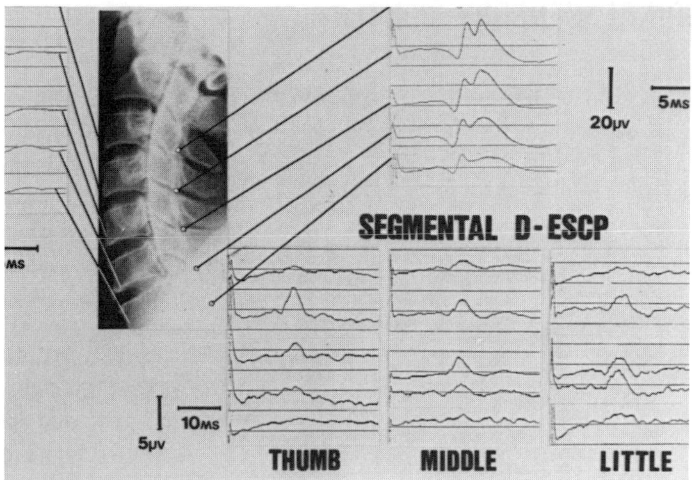

Fig. 4. 60-year-old male, cerebral circulatory disease.

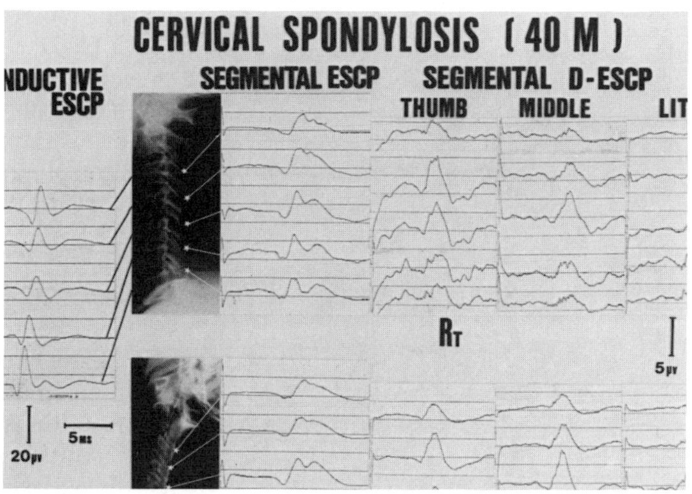

Fig. 5. 40-year-old male, cervical spondylosis.

after the early spike. Sometimes this early negative spike was involved in the early part of the negative wave. Usually the negative wave accounted for most of the segmental D-ESCP.

The areas where the segmental D-ESCPs corresponding to each pad can be recorded were slightly different from each other and the sites of peak amplitude were also different. In particular, a difference of areas of the segmental D-ESCP corresponding to the 1st pad and the 5th pad was clearly shown. These areas were considered to be the intraspinal center of sensation (Fig. 2).

In this basic study the feline segmental D-ESCPs showed a marked decrease of the negative wave when a high frequency stimulation (up to 50/sec) was applied, but the early spike did not show any decrease. Based on these findings, we surmise that the early negative spike originates from the primary afferent nerve and the negative wave originates from the postsynaptic potentials in the posterior horn (Fig. 3). In our clinical examination we were able to record negative spikes and waves similar to those seen in cats.

Case 1. 60-year-old male, cerebral circulatory disease (Fig. 4). There were no myelographic findings and the evoked spinal cord potentials in the cervical region were normal. It was shown that the sites of peak amplitude of the segmental D-ESCP corresponding to each finger were different.

Case 2. 40-year-old male, cervical spondylosis with discogenic pain (Fig. 5). This patient had no sensory disturbance. The segmental D-ESCP showed that the site of peak amplitude corresponding to the thumb was at C4/5, that of the middle finger at C5/6, and that of the little finger at C6/7.

Ten upper extremities in which there were no sensory disturbances were investigated as a control study. Peak amplitude of the segmental D-ESCP by stimulating the thumb converged on C4/5. Sites of peak amplitude converged on C5 to C6 in the case of the middle finger and C6 to C6/7 in the case of the little finger (Fig. 6).

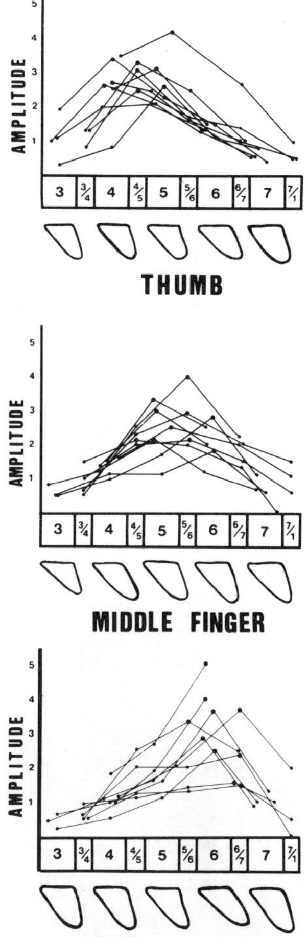

Fig. 6. The peak amplitude of the segmental D-ESCP in ten human upper extremities in which there were no sensory disturbances.

These 10 cases without sensory disturbance were averaged and represented graphically. Areas of the segmental D-ESCP corresponding to each finger partly duplicated each other, but the area corresponding to the thumb was apparently different from that corresponding to the little finger. This showed that there was a sharp distinction between the two intraspinal sensory centers. We considered these results as a normal pattern and have been using them for diagnosing intraspinal spread of myelopathy (Fig. 7).

Case 3. 54-year-old male, cervical spondylotic myelopathy (Fig. 8). Myelography showed an indentation at C3/4. The conductive ESCP and the segmental ESCP also showed a positive change at C3/4. The segmental D-ESCP obtained by stimulating fingers showed a positive wave at C3/4, but a normal pattern at C4 to 7. His sensory disturbance was about 5/10 of normal in all fingers. These findings showed that there was a main lesion in the sensory tract above the intraspinal sensory center.

Case 4. 44-year-old male, myelopathy due to cervical disc herniation (Fig. 9). Myelography, the conductive ESCP and the segmental ESCP showed a marked change at C4/5 and the right segmental ESCP changed more than the left. The segmental D-

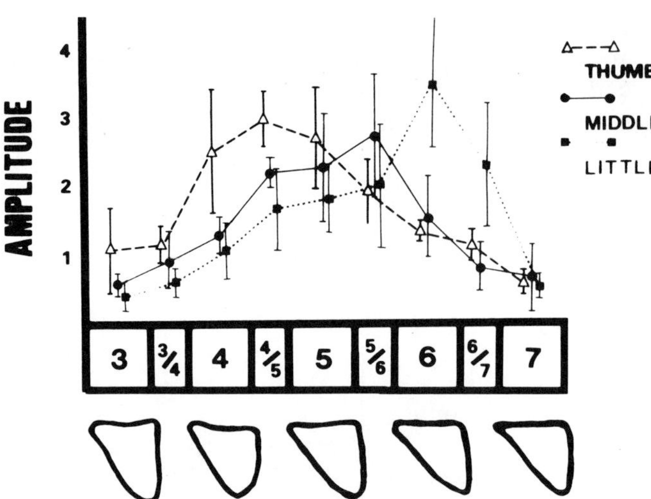

Fig. 7. Areas of the segmental D-ESCP corresponding to each finger.

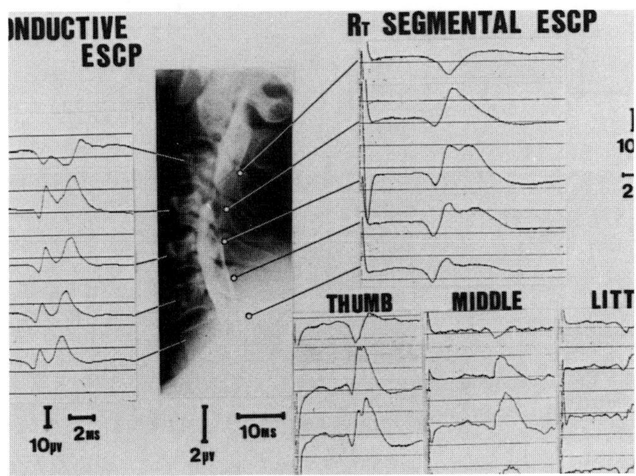

Fig. 8. 54-year-old male, cervical spondylotic myelopathy.

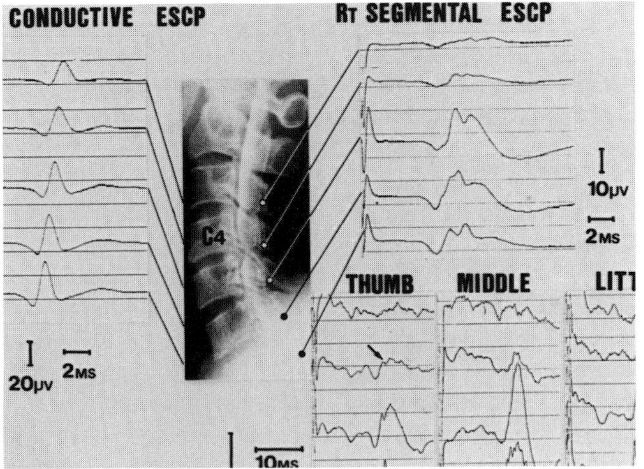

Fig. 9. 44-year-old male, myelopathy due to cervical disc herniation.

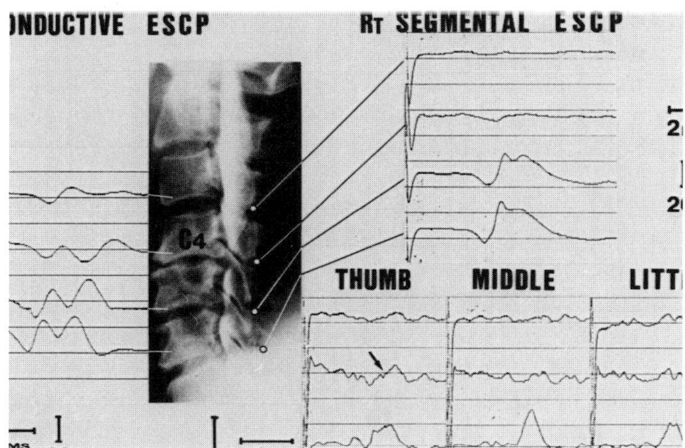

Fig. 10. 58-year-old male, cervical spondylotic myelopathy.

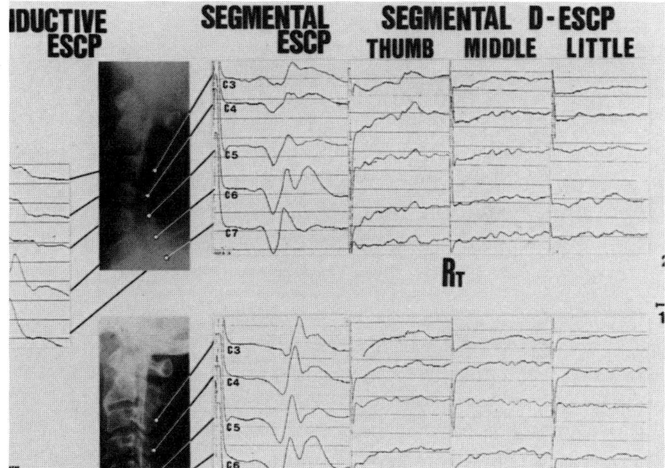

Fig. 11. 68-year-old male, OPLL myelopathy.

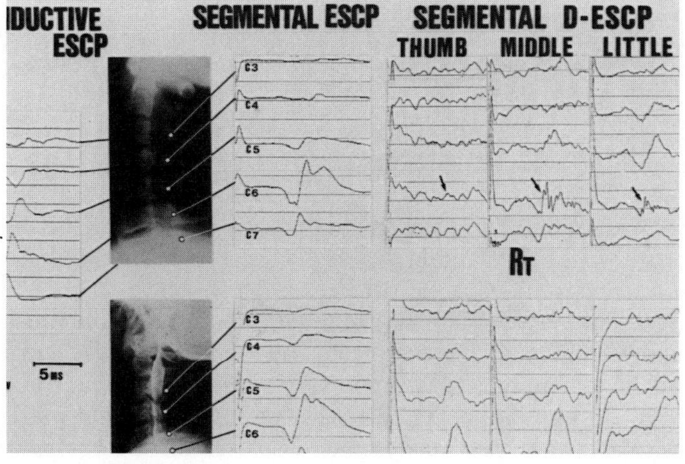

Fig. 12. 61-year-old male, OPLL myelopathy.

ESCPs obtained by stimulating both the middle and the little finger showed a normal pattern, but that from stimulating the thumb decreased at C4/5.

Sensory disturbance was 3/10 of normal in all fingers. According to these findings, sensory disturbance of the middle and little fingers was considered to originate from the sensory tract and that of the thumb from both the sensory tract and the intraspinal sensory center.

Case 5. 58-year-old male, cervical spondylotic myelopathy (Fig. 10). Myelography revealed indentations at multiple levels but both the conductive and the segmental ESCPs showed that the main lesion was at C4/5. Sensory disturbance was 5/10 on the middle finger and 8/10 on the little finger, but there was no sensory disturbance on the thumb. The segmental D-ESCP from the thumb showed a slight decrease at C4/5 in spite of there being no sensory disturbance. It is considered that sensory disturbance might not be found in the case of partial disorder in the intraspinal sensory center.

Case 6. 68-year-old male, OPLL myelopathy (Fig. 11). Myelography revealed a marked compression at C3/4, the discontinuous site of OPLL, but the conductive ESCP and the segmental ESCP showed abnormal waves at C5. The segmental D-ESCP from the thumb could be recorded at C3 and C4, but there was no wave from the middle and little fingers in the whole cervical region. There was a possibility that a disorder in the gray matter had spread caudally beyond C5/6.

Case 7. 61-year-old male, OPLL myelopathy (Fig. 12). Myelography revealed anterior compression due to OPLL at C3 to C5. The conductive ESCP and the segmental ESCP showed a marked change at C4 and C5, but the segmental D-ESCP from the right thumb disappeared and those from the right middle and little fingers showed polyphasic waves at C6, whereas it was considered to be intact by myelography and by conductive ESCP and segmental ESCP on stimulation of the median nerve. This case showed that lesion in the gray matter was more extensive than that diagnosed by myelography or by the conductive ESCP and the segmental ESCP obtained from stimulation of the median nerve.

Discussion

In general, patients who have cervical spondylotic myelopathy or OPLL myelopathy complain of symptoms in both the upper and the lower extremities.

In order to diagnose correctly the spinal cord function for such cases, we have been using a conductive ESCP (by stimulating the thoracic spinal cord), which originates from the tract of the lower extremity, and a segmental ESCP (by stimulating the median nerve), which originates from both the primary afferent and the postsynaptic potentials (1, 4).

We have been continuing with our fundamental study using cats to diagnose sensory disturbance objectively. It was clarified by our fundamental study that the segmental D-ESCP corresponding to each foot pad in the forelimb could be obtained at the epidural area, considered to be the intraspinal sensory center. The negative wave, which accounted for most of the segmental D-ESCP, was considered to consist of postsynaptic potentials.

The characteristics of the segmental D-ESCP obtained by stimulating the finger surface were similar to those of the fundamental study. Therefore, we applied the segmental D-ESCP to the diagnosis of sensory disturbance. However, a record of the correct segmental D-ESCP is not easy in clinical examinations. The segmental D-ESCP consists of a small negative spike and a relatively large negative wave, up to $5\mu V$. Therefore, eliminating EMG activity is most important to record correctly such a low amplitude wave. Electrical stimulation applied is so painful for some patients that EMG activity originating from the neck muscles increases. In such cases pentazocine and diazepam are administered to relieve the tension of the patient and to reduce the

EMG activity. One hundred to 1000 averages are also needed to record clearly such a small potential. The time required for achieving the examination is relatively long and this is one defect of this technique.

However, since much of the information was obtained from the segmental D-ESCP, we have been examining 30 cases for the past year. These 30 cases and 70 cases which had been examined by conductive ESCP and segmental ESCP before 1985, did not show any complications. After numerous clinical examinations, we have determined that changes of the segmental D-ESCP do not show the severity of clinical sensory disturbance, but the spread of lesion in the intraspinal sensory center. Because of these findings, we believe that we can diagnose whether clinical sensory disturbance is due to the gray matter, namely the intraspinal sensory center, or the white matter, the sensory tract above the intraspinal sensory center.

We also found that there was a discrepancy at times between the conductive ESCP or the segmental ESCP from the median nerve and the segmental D-ESCP.

Usually in cases of cervical spondylotic myelopathy that were shown by myelography to be localized lesions, there was no discrepancy. However in patients whose spinal cord had been severely and widely compressed, especially with continuous OPLL, there was a discrepancy at times between the conductive ESCP or the segmental ESCP and the segmental D-ESCP. Namely, the spread of the lesion in the gray matter diagnosed by the segmental D-ESCP was sometimes more extensive than that of the lesion diagnosed by myelography as well as the conductive ESCP and the segmental ESCP. We conclude, therefore, that in cases with severely compressed spinal cord or multilevel lesions, it is most important for surgeons to know the function of the spinal cord and to form an opinion as to the pathophysiology of the jeopardized spinal cord. After such thorough and careful diagnosis, it should be possible to determine correctly the site of the operation and to achieve successful results.

Summary

The segmental D-ESCP (dermatome evoked spinal cord potential) was recorded after stimulating each finger surface for diagnosis of cervical spondylotic and OPLL myelopathy.

According to our fundamental study using cats, the segmental D-ESCP obtained by stimulating the foot pads consists of an early small negative spike originating from the primary afferent nerve and a negative wave originating from the postsynaptic potentials, which accounts for most of the segmental D-ESCP.

In ten human upper extremities in which there were no sensory disturbances, the peak amplitude of the segmental D-ESCP converged on C4/5 from the thumb, on C5 to C6 from the middle finger and on C6 to C6/7 from the little finger. The areas where the segmental D-ESCP could be recorded were determined to be the intraspinal sensory center.

It could be diagnosed whether clinical sensory disturbance was due to the gray matter, namely the intraspinal sensory center, or the white matter, the sensory tract above the intraspinal sensory center.

In patients whose spinal cord had been severely and widely compressed, especially continuous OPLL, the spread of lesion in the gray matter diagnosed by the segmental D-ESCP was sometimes more extensive than that of the lesion diagnosed by myelography, the conductive ESCP and the segmental ESCP. The segmental D-ESCP is considered to be a useful method for diagnosing the spread of intraspinal lesion and forming an opinion as to the pathophysiology of the jeopardized cervical spinal cord.

References

1. Kaneda, A.: An Analysis Of Spinal Cord Potentials Evoked By Median Nerve Stimulation. In: Spinal Cord Monitoring, Springer-Verlag Berlin Heidelberg, pp. 35-42, 1985.
2. Shinomiya, K.: Clinical Study Of Cervical Spondylotic Myelopathy Using Evoked Spinal Cord Potentials. In: Spinal Cord Monitoring, Springer-Verlag Berlin Heidelberg, pp. 290-301, 1985.
3. Shinomiya, K.: Spinal Cord Monitoring Of Spinal Cord Function Using Evoked Spinal Cord Potentials. Fundamental And Clinical Application Of Spinal Cord Monitoring, Saikon Publishing, Tokyo, pp. 161-173, 1984.
4. Willis, W.D.: Evoked Spinal Cord Potentials In The Cat And Monkey: use In Analysis Of Spinal Cord Function. Fundamental And Clinical Application Of Spinal Cord Monitoring, Saikon Publishing, Tokyo, pp. 3-19, 1984.
5. Zimmermann, M.: Fundamentals Of Sensory And Motor Functions Of The Spinal Cord. Spinal Cord Monitoring, Springer-Verlag Berlin Heidelberg, pp. 3-15, 1985.

Descending Recordings

Spinal Cord Potentials (SCPs) Produced by Descending Volleys in the Rat

K. Shimoji;[*] H. Fujioka; Y. Maruyama; H. Shimizu; T. Hokari; T. Takada

Summary

The spinal cord serves an important function in transporting, transmitting and integrating neuronal messages between the brain and the peripheral nervous system. Thus, spinal cord function is influenced not only by peripheral nervous activities but also by supraspinal conditions. Recently, a large amount of information has been accumulated on brainstem control of spinal cord functions. However, it is not yet clear what kinds of peripheral nerve stimulation influence the supraspinal structures which in turn send their descending impulses to the spinal cord. Furthermore, there have been only a few studies on the spinal cord potential (SCP) produced by a descending volley through a feedback loop via the supraspinal structures.

The present study was undertaken to test the effects of descending volleys activated directly or by a feedback on spinal function in terms of SCPs in the rat.

Methods

Twenty-eight Wistar rats, weighing 350-400g, were anesthetized by intraperitoneal injection of pentobarbital sodium, 50mg/kg, and atropine, 0.1mg/kg. Tracheal intubation was carried out via a tracheotomized orifice; and controlled ventilation under muscle relaxation with pancuronium bromide, 0.1mg/kg/hr, was undertaken using the Harvard pump throughout the experiments. Ventilation volume was set at normoventilation by checking arterial carbon dioxide tension ($PaCo_2$) (35-40mmHg). The left femoral

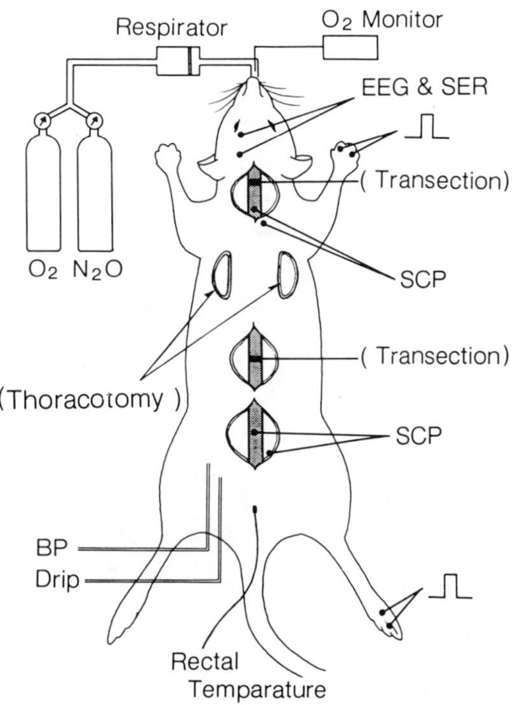

Fig. 1. Schematic presentation of experimental arrangements.

---

* Department of Anesthesiology, Niigata University School of Medicine, 1-757 Asahi-Machi, Niigata 951, Japan

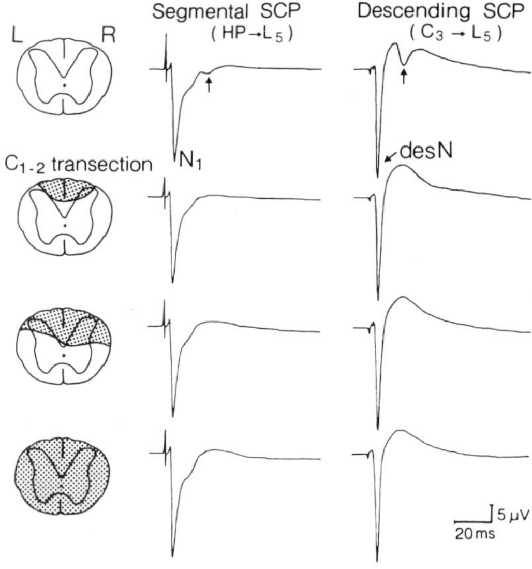

Fig. 2. The SCP directly recorded from the dorsal surface of the spinal cord at the L5 level in response to right hindpaw (HP) stimulation (segmental SCP) and DC stimulation at the C3 level (descending SCP). See the text for explanation.

artery was cannulated for measurements of blood pressure and arterial gases, and the left femoral vein was also cannulated for administration of drugs or electrolyte solution.

The rats were then fixed in a stereotaxic apparatus utilized for brain and spinal cord surgery. A bilateral thoracotomy, 10mm in diameter, was performed to minimize any respiratory movement of the brain and spinal cord (Fig. 1). The spinal cord was exposed at the C1 through C3, T7 through T8 and L1 through L5 vertebral levels. After opening the dura mater, the spinal cord was covered with a pool of mineral oil to protect it from drying. The body temperature of the animals was kept constant at 37-38°C using a small water heating blanket.

A pair of nonpolarizable needle electrodes, 250µ in diameter, were inserted subcutaneously 5mm apart in the right fore- (FP) and hindpaws (HP). In some animals, another stimulating electrode, a platinum wire 250µ in diameter, insulated with a teflon coating except at the bevel of the tips, was also introduced into the periaqueductal gray (PAG) region at zero on the cephalo-

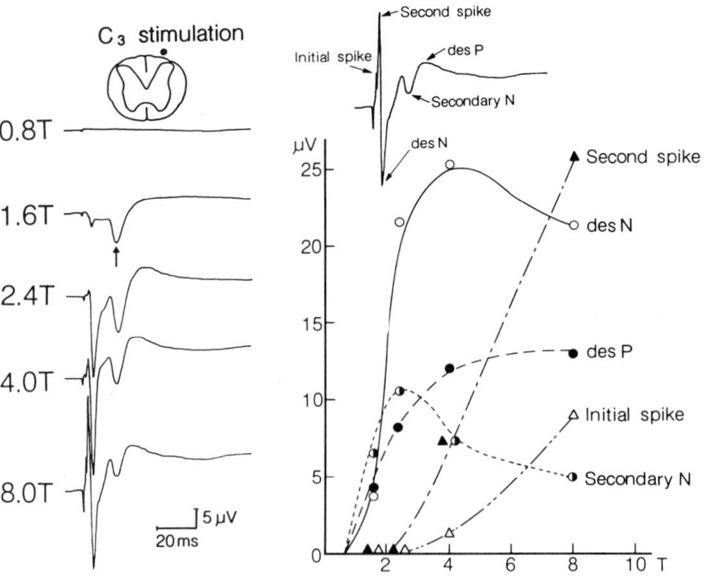

Fig. 3. The effect of graded stimulation of DC on the waveform characteristics of the descending SCP. An example of a record with the stimulation site is shown on the left. Stimulus intensity is represented as multiples of the threshold strength at the start of each trace. The stimulation pulse was delivered 10ms before the start of each sweep, and is visible as a small downward artifact. Amplitude changes (ordinate) of each component in response to the increase in stimulus intensity (abscissa) are shown on the right.

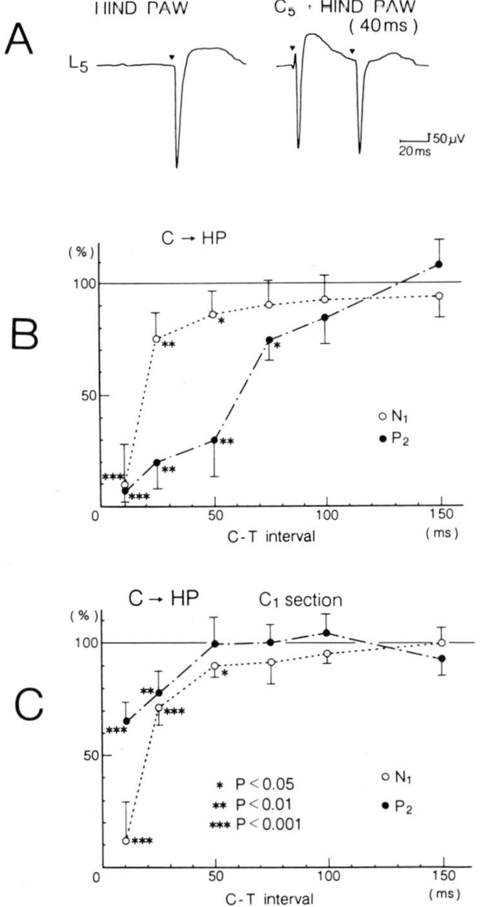

Fig. 4. The effect of dorsal column (DC) stimulation on the segmentally evoked SCP. (A). An example of the segmental SCP from the dorsal surface of L5 in response to hindpaw stimulation (left trace) and the effect of DC stimulation (conditioning) at the C5 level 40ms before hindpaw stimulation (testing) on the segmental SCP (right trace). Note the inhibition of both the N1 and P2 waves of the segmental SCP by the conditioning stimulation. (B). Relationship between the conditioning (C) - testing (T) interval (abscissa) and the amplitudes of the N1 and P2 components, expressed as per cent of the control values, affected by the conditioning stimuli (ordinate) . (C). The same as (B) but after the spinal transection at the C1 level.

caudal axis and 2.0mm on the medio-lateral axis according to the stereotaxic brain map of Pellegrino et al. (1).

Recording electrodes, Ag-AgCl wires, 100μ in diameter, were placed through mineral oil onto the dorsal surfaces of the cervical and lumbar enlargements in midline. In seven rats a glass microelectrode filled with 3 M KCl was inserted into the lamina V of Lexed at the lumbar dorsal horn where the largest negative wave was recorded in response to hindpaw electrical stimulation.

Electrocardiogram, electroencephalogram, arterial blood pressure and rectal temperature were continuously monitored throughout the procedures.

Spinal cord transection was performed at the C1 through C2 and/or T7 through T8 vertebral levels by electrocoagulation. At the termination of the experiments, direct current was applied to the PAG or lamina V through the electrodes, and then the brain and spinal cord were fixed by perfusion with 4% formalin. Histological examinations were carried out for determination of the locations of electrode tips in the PAG and lamina V, and also for checking the extension of spinal sections.

Results and Discussion

1. SCP produced by direct stimulation of the dorsal column.

Fig. 2 demonstrates the SCP (descending SCP), recorded from the dorsal surface of the lumbar enlargement (L5), evoked by stimulation of the ipsilateral dorsal column (DC) of C3 at 2.5 times the threshold (T) strength and the segmental SCP recorded by the same electrode in response to right HP stimulation. Bilateral transection of the dorsal half of the dorsal column at the C1 through C2 level eliminated both the nega-

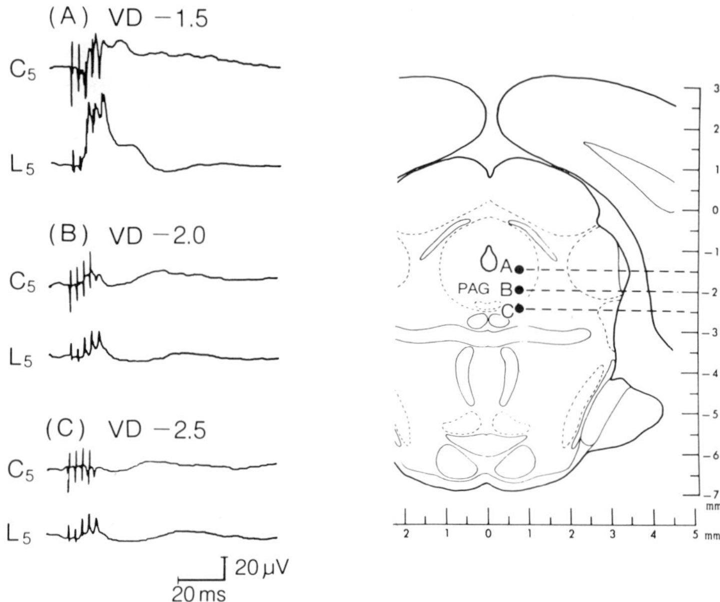

Fig. 5. The SCP produced by PAG stimulation. See the text for explanation. Sites of the stimulating electrode in the PAG are demonstrated on the stereotaxic map (left) of Pellegrino et al. (1).

tive dips imposed on the slow positive waves of the descending and segmental SCPs. Thus, the results indicate that both the negative dips and subsequent secondary positive waves of the descending and segmental SCPs are produced by a feedback loop via the supraspinal structures, sharing afferent volleys in the dorsal column. Even when transection of the cord extended to the dorsal half of the cord, the descending SCP did not change its waveform characteristics. However, when the spinal cord was transected completely, the descending P decreased in amplitude without any change in the descending N. This result suggests that the tonic PAD is constantly produced by the descending volleys coming from the supraspinal structures (2).

Thus, the results have shown that the descending SCP evoked by DC stimulation involves two origins, i.e., one produced by the direct antidromic volley through DC3 and the other evoked by a feedback loop (4-7) conducted through the afferent DC tract. The descending N and following negative dip are thought to be activities of dorsal horn neurons elicited by a direct antidromic volley and by a feedback loop via the supraspinal structures, respectively. The descending P are therefore considered to involve both the directly and indirectly (through a feedback loop) activated PAD. Alternatively, the descending P could reflect an inhibitory post-synaptic potential (IPSP). In fact, such a prolonged IPSP has recently been demonstrated in single neurons of spinal laminae I and II by stimulation of the nucleus magnus and PAG of cats (8). Therefore, such a prolonged IPSP indirectly activated by ascending tracts through the dorsal column could be manifested as descending slow P in the present study.

The differences in the thresholds for evoking the descending SCPs by direct and indirect volleys were tested during the light stage of pentobarbital anesthesia when the EEG showed a spindle wave (Fig. 3). Both the descending N and negative dip were evoked by almost the same stimulus threshold (T) strength. With increase in stimulus strength the descending N, the negative dip and P waves increased up to a stimulus intensity of around 3xT. However, when the stimulus intensity exceeded 3xT, the negative dip decreased gradually with the appearance of another spike potential. This result

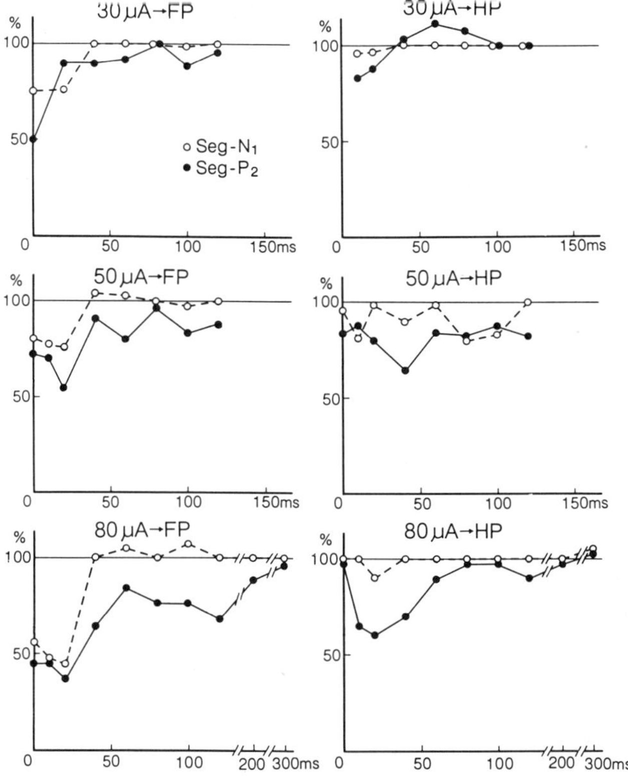

Fig. 6. The effect of PAG stimulation on the segmental SCP. Left column: The effect on the segmental SCP evoked by forepaw (FP) stimulation. Right column: The effect on the segmental SCP evoked by hindpaw (HP) stimulation. Abscissa and ordinate in each graph represent the interval between the two stimulations (PAG and paw) and the change in amplitude of the N1 or P2 wave expressed as a percentage of the control value (without PAG stimulation), respectively.

suggests that a feedback volley descending from the supraspinal structures collides with an antidromic volley activated by an increase in stimulus intensity.

2. The effect of dorsal column stimulation on the segmental SCP.

When electrical stimulation was applied to the cervical DC, it produced not only evoked potentials in the lumbar enlargement but also induced an interaction with the segmental SCP elicited by hindpaw stimulation. The changes in segmental SCP were tested by delivering two volleys to the cervical DC and hindpaw, at various intervals as conditioning and testing stimulations, respectively. Cervical DC stimulation caused an inhibition of the segmental SCP up to 150ms (Fig. 4). When the interval between the conditioning and testing stimuli were made shorter, the degree of inhibition became greater. The inhibition of the P2 wave was significantly greater than that of the N1 wave up to 50ms. In anesthetized human subjects, volleys from the cervical posterior epidural space have been shown to produce an inhibition of the N1 wave for more than 100ms and a transient inhibition for 10-40ms followed by a prolonged facilitation of the P2 wave for more than 100ms, in the segmental SCPs evoked by tibial nerve stimulation (7, 9). Such a biphasic effect on the segmental P2 wave could not be demonstrated in the rat, and therefore, the existence of a species difference or differences in the current distribution of the stimulating pulses within the spinal cord between the rat and man seems likely. When the spinal cord was cut at the C1 through C2 level, inhibition of the P2 wave was attenuated while that of the N1 wave remained unchanged. Thus, the results indicate that the inhibitory action of DC stimulation on the P2 wave in the segmental SCP is enhanced by a feedback mechanism via the supraspinal structures.

3. SCP produced by PAG stimulation.

Burst stimulation (100-300Hz, 4 pulses, 30-100µA) of the PAG produced initial and secondary slow positive waves in both cervical and lumbar regions of the cord dorsum as shown in Fig. 5. When the electrode tips were advanced ventrally, the initial slow wave decreased without any appreciable change in the secondary slow positive wave. The initial slow positive wave began at 12-20ms and peaked at 30-40ms, following the first stimulus pulse with a duration of 45-120ms. The secondary positive wave with a duration of 75-150ms peaked at 95-105ms and 125-135ms in the cervical and lumbar enlargements, respectively, following the PAG stimulation.

The results suggest that PAG stimulation causes the PAD or prolonged IPSP (8) in the cervical as well as lumbar spinal cord, and that there are two mechanisms, an initial one and a delayed one, produced by PAG stimulation. The long latency differences in the peaks of the secondary positive waves between the cervical and lumbar cord, produced by PAG stimulation, may indicate that the descending volleys which produce the secondary positive wave are multisynaptic and/or conducted through small diameter fibers (8, 10, 11).

4. The effect of PAG stimulation on segmental SCP.

PAG stimulation produced inhibition of all components of the segmental SCP and HSP, except the P2 wave which was sometimes affected biphasically; inhibition followed by facilitation, as shown in Fig. 6.

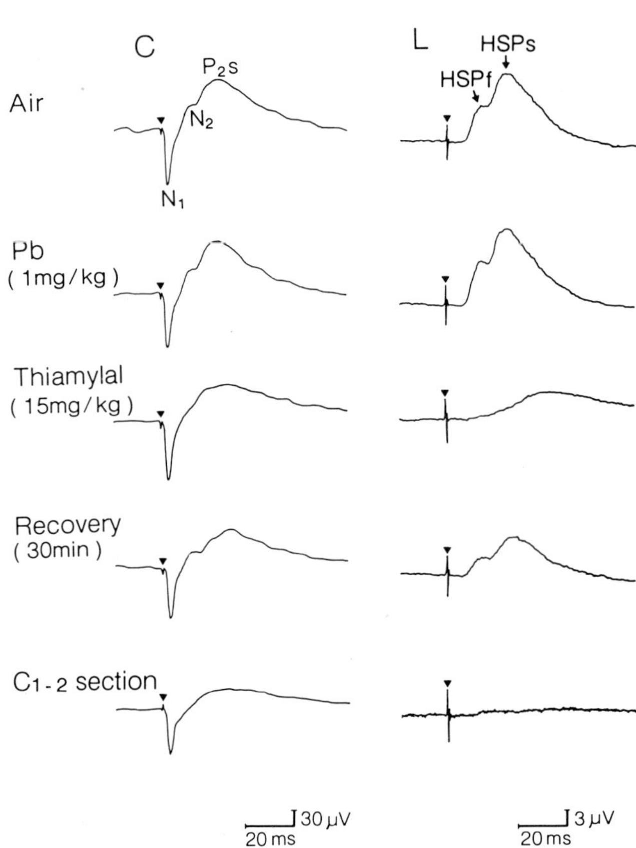

Fig. 7. Heterosegmentally activated slow positive potential (HSP). Left traces: The segmental SCPs recorded from the cervical cord (C5) in response to forepaw stimulation (25 x threshold strength). Right traces: The HSPs produced in the lumbar cord (L5) by stimulation of the forepaw. Top traces were obtained during ventilation with air (Air) under the light stage of pentobarbital anesthesia without any muscle relaxant drug. The second traces were taken 5 minutes after the i.v. injection of pancuronium bromide (pb), 1mg/kg. Note that there is no substantial change in the slow potential, which neglects the electromyogram or mechanogram artifact.

This biphasic effect on the P2 wave has also been observed as a result of epidural stimulation in anesthetized humans (7, 9). Similar inhibition was seen in the SCPs produced not only by noxious but also non-noxious electrical stimuli. The inhibitory effects of PAG stimulation were more prominent on the P2 than on the N1 of the segmental SCP. The degree of inhibition and its duration were prolonged as the stimulus strength was increased.

5. Heterosegmentally activated slow positive wave (HSP) via a feedback loop.

Moderate to intense (more than 2xT) stimulation of the fore- and hindpaws produced a long-latency slow positive wave or heterosegmentally activated slow positive wave (HSP) at the lumbar and cervical regions of the cord dorsum, respectively (Fig. 7). The HSP was very sensitive to thiamylal anesthesia, as was the second component of the P2 wave in the segmental SCP. It consisted of an initial slow positive wave (HSPf) superimposed by another positive one (HSPs), or was sometimes interrupted by a negative wave. With spinal transection at the C1 through C2 level, the HSP disappeared almost completely, leaving a trace of the initial component, with simultaneous abolishment of the second component of the P2 wave.

Thus, most of the components of the HSP are considered to be produced by a feedback loop via the supraspinal structures. A small part of the initial component of the HSP might be evoked by a reflex arch between the upper and lower extremities, since

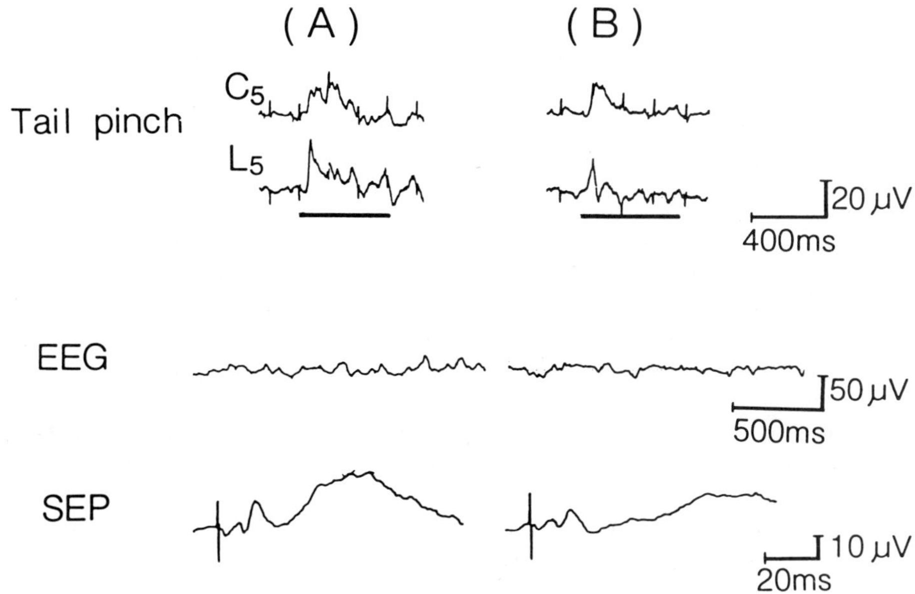

Fig. 8. Slow cord dorsum potential produced by natural stimuli from a feedback loop via the supraspinal strucutres. A single pinch stimulation (tail pinch) with forceps produced a prolonged slow positive wave in cervical as well as lumbar cords during the light stage of pentobarbital anesthesia (A). Inhalation of 70% nitrous oxide mixed with 30% oxygen suppressed pinch-evoked slow positive waves (B). The electroencephalogram and somatosensory evoked potential from the contralateral somatosensory area of the scalp in response to forepaw stimulation were also recorded.

spinal transection at the T1 level or additional injection of anesthetic eliminated the potential.

The cord dorsum positive potential (P wave) in a spinal animal evoked by peripheral nerve stimulation is believed to reflect the primary afferent depolarization (PAD), a causative agent of presynaptic inhibition, which can also be recorded from the dorsal root as a slow negative wave, DRPv by Lloyd's terminology. As mentioned before, a prolonged IPSP has been demonstrated in the dorsal horn (laminae I and II) of the cat by PAG and nucleus raphe magnus stimulation (8). A similar IPSP caused by a feedback loop can also be assumed to have occurred in the present study. Therefore, it is suggested that the HSP as well as the second component of the segmental P2 wave reflect the PAD or prolonged IPSP elicited by a feedback loop via the supraspinal structures.

Bilateral lesion of the dorsal column at the C1 through C2 or T1 level slightly but significantly enhanced the HSP, while the second component of the positive wave activated by dorsal column stimulation was abolished. This result therefore suggests that the HSP does not pass through the dorsal column (DC) but is tonically inhibited by that system, whereas the negative dip and the secondary component of the P2 wave are produced by the tracts that travel through the DC.

6. Slow positive cord dorsum potential evoked by natural stimuli.

Single-touch stimulation of hairs or skin did not evoke any significant potential deflection in the segmental or heterosegmental area of the spinal cord. However, a pinch or prick applied to the skin provoked a long lasting positive potential in both the segmental and heterosegmental areas of the dorsal cord (Fig. 8). These long lasting positive potentials evoked by natural stimuli were very vulnerable to the action of anesthetic drugs, and were eliminated by spinal transection at the C1 through C2 level. Thus, the slow positive potential evoked by pinch is also thought to be elicited by a feedback loop via the supraspinal structures in the same way the HSP or the negative dip with the subsequent secondary component of the P2 wave of the segmental SCP provoked by an intense electrical stimulation of the paws.

A majority of wide dynamic range (WDR) neurons (8 of 11) simultaneously recorded from the lamina V of the dorsal horn also responded to segmental as well as heterosegmental stimulation. Most WDR (7 of 8) neurons which responded to both forms of stimulation were augmented by segmental skin pinch or electrical stimulation, while they were inhibited by heterosegmental skin pinch or intense electrical stimulation. The inhibition of firing of the WDR neurons by electrical stimulation lasted from approximately 20 to 120ms and showed a similar time course of that of the HSP. The WDR neuronal activities are therefore suggested to be inhibited by the PAD or IPSP activated through a feedback loop via the supraspinal structures. Thus, the WDR neuron might be regulated not only by wide sensory modalities converging from the same segmental region, but also by the sensory inputs from the heterosegmental area which travel through a feedback loop via the supraspinal structures.

Summary

The spinal cord potentials (SCPs) produced in the cervical and lumbar regions of the cord dorsum in response to several descending volleys were studied in rats lightly anesthetized with pentobarbital.

Dorsal column (DC) stimulation produced a directly (antidromically) activated slow negative (desN) and positive wave (desP) superimposed by the negative (dip) and secondary positive wave. Later, two components (the negative dip and the secondary positive wave) were demonstrated to be activated by a feedback loop via the supraspinal structures. The ascending tract that activated the desN and desP waves may therefore be conducted through the DC.

DC stimulation inhibited the segmental (seg) N1 and P2 waves by more than 120ms. The inhibition was more pronounced in the P2 wave than in the N1 wave. The inhibition by DC stimulation of the P2 was decreased by cord transection at the C1 through C2 level.

PAG stimulation produced two slow positive waves in the cervical as well as in lumbar regions of the cord dorsum, and prolonged inhibition of the segmental SCPs. The

P2 wave, however, was sometimes affected biphasically, i.e., inhibition followed by facilitation.

Fore and hindpaw stimulation evoked the slow positive waves and heterosegmental slow positive waves (HSPs), in the lumbar and cervical regions of the cord, respectively. The HSP was almost abolished by transection of the cord at the C1 through C2 level, indicating the the HSP is mostly activated by a feedback loop via the supraspinal structures.

The most wide dynamic range (WDR) neurons were inhibited by heterosegmental stimulation. The time course of this inhibition coincided with that of the HSP, suggesting that the inhibition is presynaptically activated by descending volleys through a feedback loop via the supraspinal structures.

References

1. Pellegrino, L.J.; Pellegrino, A.S.; Cushman, A.J.: A Stereotaxic Atlas of the Rat Brain, 2nd Ed. New York: Plenum Press, 1979.
2. Wall, P.D.: The substantia gelatinosa. Trends in Neuro Sciences, 3: 221-224, 1980.
3. Foreman, R.D.; Beall, J.E.; Applebaum, A.E.; Coulter, J.D.; Willis, W.D.: Inhibition of primate spinothalamic tract neurons by electrical stimulation of dorsal column or peripheral nerve. In: J.J. Bonica; D. Albe-Fessard (eds), Advances in Pain Research and Therapy, Vol. 1, Raven Press, New York, 1976, pp. 405-410.
4. Tang, A.H.: Dorsal root potentials in the chloralose-anesthetized cat. Exp. Neurol., 25: 393-400, 1969.
5. Shimoji, K.; Ito, Y.; Ohama, K.; Sawa, T.; Ikezono, E.: Presynaptic inhibition in man during anesthesia and sleep. Anesthesiology, 43: 388-391, 1975.
6. Shimoji, K.; Matsuki, M.; Ito, Y.; Masuko, K.; Maruyama, M.; Iwane, T.; Aida, S.: Interactions of human cord dorsum potential. J. Appl. Physiol., 40: 79-84, 1976.
7. Shimizu, H.; Shimoji, K.; Maruyama, Y.; Matsuki, M.; Kuribayashi, H.; Fujioka, H.: Human spinal cord potentials produced in lumbosacral enlargement by descending volleys. J. Neurophysiol., 48: 1108-1120, 1982.
8. Light, A.R.; Casale, E.J.; Menetrey, D.M.: The effects of local stimulation in nucleus raphe magnus and periaqueductal gray on intracellularly recorded neurons in spinal laminae I and II. J. Neurophysiol., 56: 555-571, 1986.
9. Shimoji, K.; Shimizu, H.; Maruyama, Y.; Matsuki, M.: Dorsal column stimulation in man: Facilitation of primary afferent depolarization. Anesth. Analg., 61: 410-413, 1982.
10. Wessendorf, M.W.; Proudfit, H.K.; Anderson, E.G.: The identification of serotonergic neurons in the nucleus raphe magnus by conduction velocity. Brain Res., 214: 168-173, 1981.
11. Zahs, K.; Lakos, S.; Basbaum, A.I.: Immunoreactive serotonergic axons in the dorsolateral funiculus (DLF) of the cat and rat. Soc. Neurosci. Abstr., 11: 125, 1985.

Intradural Spinal Recordings (Particular Reference to Invasive Methods)

C. Ertekin[*]

Definitions

Electrical responses recorded from the vicinity of the spinal cord can vary depending upon the sites of stimulation and recording and other variables in the techniques used. The variations in the recording and stimulation techniques not only produce different configurations and changes in potential parameters, but also the source of electrical generation of the spinal cord potentials could differ in detail. Such technical variations have prevented any standardization of terminology of spinal cord evoked potential (SCEP). This paper does not intend to offer any solution, but it will present more simplified definition of SCEPs, according to the different methods used.

SCEPs can be divided into two main groups; ascending SCEPs and descending SCEPs. The term "ascending SCEP" may cover all the electrical responses recorded near the spinal cord either invasively or non-invasively in an ascending direction of the impulse volley following stimulation of a peripheral nerve or spinal cord or root below the site of recording electrode. Segmental and conducted SCEPs are the examples for ascending SCEPs. On the other hand, when we say descending SCEP, we mean that the recording electrode at the cord level is caudal to the stimulation site such as the electrical stimulation of the scalp over the motor cortex and epidural recording from the cervical and thoracic cord levels (14). In this case, SCEP is clearly related to the activities of the descending spinal tracts. However, if both of the recording and stimulation electrodes are placed close to the surface of spinal cord either epidurally (64, 109, 110, 127) or intrathecally (41, 120, 121), stimulation of either electrode produces SCEPs which can not be easily defined as they are neither purely ascending nor descending due to possible orthodromic and antidromic activation of the afferent and efferent spinal pathways (Fig. 1). In order to make a simpler definition; ascending and descending SCEPs obtained by both spinally oriented electrodes will be included and discussed in the group of ascending SCEPs (Table 1).

Segmental SCEP

If the recording electrode is close to the vertebral levels corresponding to lumbosacral or cervical enlargements, stimulation of the nerves in the lower or upper extremities will produce "segmental SCEPs" respectively at the lumbosacral or cervical cord levels. Segmental SCEP is the most easily recorded potential regardless of the methods used. This is probably because of the distance between stimulation and recording sites, which permits less temporal dispersion of the peripheral afferent volley.

---

* 1438 sok. No: 5\1, Alsancak., Izmir-Turkey

In addition to that, a very dense electrical activity with less spatial dispersions is recorded in the pre- and postsynaptic segmental spinal structures. Although the origin of this activity is poorly understood, and the results are somewhat variable depending on the recording methods applied, the configuration of the segmental SCEP is quite similar to the other various methods such as the intrathecal recording (39, 40, 79, 89, 129, 130), the epidural recording (18, 57, 111, 112, 113), the skin-surface recording (23, 25, 30, 33, 38, 59, 61, 83, and others). A common configuration of segmental SCEP, whichever method is used, is a prominent negative sharp wave and followed by a positive wave and sometimes a second negative wave which can not be constantly recorded. Prominent negative sharp waves sometimes start with a small positive dip and some spiky wavelets appearing on the rising phase of the negative sharp wave (Fig. 2).

Segmental SCEPs in man are similar to the "cord dorsum" or "intermediary cord" potentials obtained from the animal studies (3, 5, 6, 9, 10, 11, 12, 19, 28, 36, 48, 49, 51, 52, 53, 71, 101, 128, 131, 132). The precise nature and origin of the different components of the human segmental SCEP is not known exactly, and interpretation of the origin of the segmental SCEP has been controversial. However, one can say that the segmental SCEP is composed of the presynaptic electrical activity in the dorsal root and dorsal column fibers (namely the early spiky wavelets and the rising phase of the first negative sharp wave) and the segmental postsynaptic activity of the dorsal horn (i.e., the following part of the negative sharp wave, positivity and second negativity).

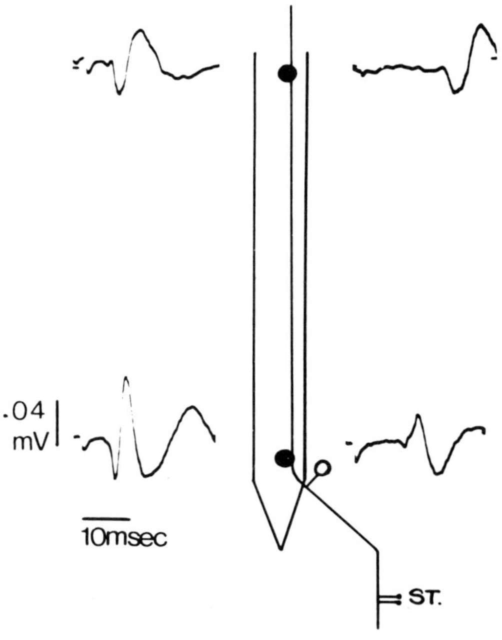

Fig. 1. Human segmental and conducted SCEPs recorded intrathecally. On the right column: Lumbosacral segmental SCEP recorded at the T11-12 intervertebral (i.v.) level (lower trace); conducted SCEP recorded at the C6-7 i.v. level (upper trace) after stimulation of the posterior tibial nerve at the popliteal fossa. On the left column: Spinally evoked ascending SCEP (upper trace) and descending SCEP (lower trace) obtained by stimulating and recording from both intrathecal electrodes situated in T11-12 and C6-7 i.v. levels. The records are single oscilloscopic sweeps from a normal adult subject. A downward deflection of the trace is positive in this and in all the subsequent figures (calibration: 10msec and 0.04mV). Conduction velocities along the spinal cord are calculated 39ms on the right and 46 and 49ms on the left.

Conducted SCEP

When the recording electrode is placed far above the level of the lumbosacral cord, generally above the T9-T10 vertebral levels; adequate stimulation of a mixed nerve in the leg will produce SCEPs with increasing latencies toward the cranial direction. Upper recording levels cause not only longer latencies but also smaller amplitudes of response with polyphasic shapes in some recordings. Such features fit with the recording of the impulse volley travelling from the afferent long spinal tracts. This kind of potential is often called "conducted SCEP."

Unlike the segmental SCEP, the conducted SCEP is not easily obtained and is severely attenuated by skin surface recording in spite of good resolution techniques and different frequency filters. This fact is especially obvious in the high thoracic and cervical vertebral levels after stimulation of a leg nerve (24, 25, 26, 27, 61, 62, 73, 74, 96, 103, 106, 107, 117, 125, and others). The attenuation and/or absence of conducted SCEP at high thoracic and cervical levels recorded from the skin surface in adult normal subjects can be attributed to the long distance between the electrical generation sources and the recording sites, and the great temporal and spatial dispersion of the afferent volley travelling along the multiple afferent spinal pathways. Some indirect methods have been developed to calculate the afferent conduction along

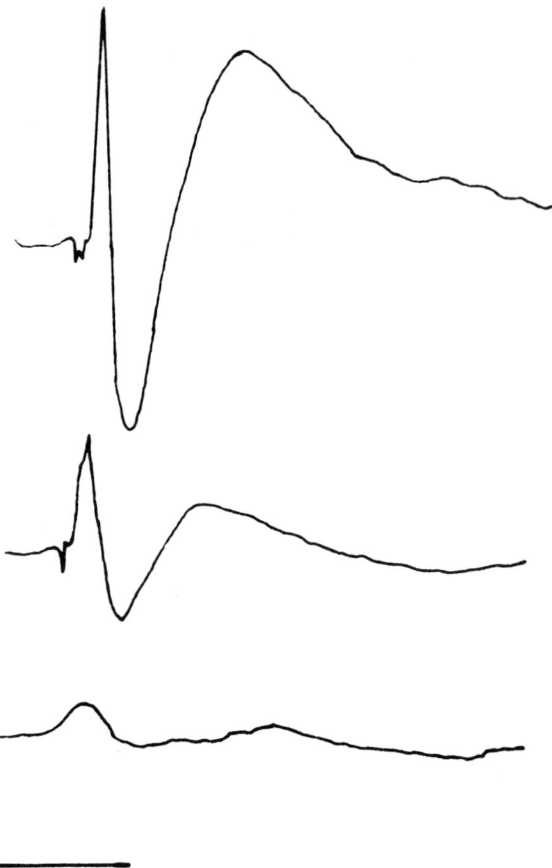

Fig. 2. Subdural (upper trace), epidural (middle trace) and skin surface recordings of lumbosacral segmental SCEPs from the same normal subject. All traces are averaged responses. Posterior tibial nerve is stimulated at the level of the popliteal fossa (calibrations: 25msec and 10mV). Note the similarity and predominance of the first negative sharp wave from all recordings; absence of following positivity in skin surface recording and less-defined appearance of the following second negative slow wave in epidural recording.

the spinal cord in conjunction with the somatosensory cortical evoked potentials and other variables (34, 35, 37, 58).

Conducted SCEPs have been clearly obtained by invasive techniques either epidurally or intrathecally (41, 42, 60, 77, 78, 89, 112, 130). While only diphasic or triphasic conducted potentials with a conduction velocity of about 60-70ms could be recorded on the skin surface (23, 26, 27, 61), it has been demonstrated that at least three components of conducted potentials appear with different activation thresholds and different conduction velocities (30-80ms) within the spinal cord when epidural technique is used (60, 77). Similarly, conducted SCEPs have been elicited from the cervical cord intrathecally after the stimulation of the posterior tibial nerve, and the recorded response was found to be higher in amplitude and generally triphasic in shape with a conduction velocity of about 30-50ms (41, 42, 89) (Fig. 3). The results obtained from the different recording techniques mentioned above, are not in accord with each other. It is therefore concluded that the conducted SCEPs have somewhat different origins and are not easily applicable to each other in various recording techniques.

Spinally evoked conducted SCEP

When two electrodes are placed close to the spinal cord epidurally or intrathecally, it is possible to evoke complex cord potentials in either the sites of stimulation or of recording. Such a recording method is often used for spinal cord monitoring. The stimulating electrode is introduced into the posterior epidural space of the upper thoracic or cervical spinal cord depending on the operating site. Descending SCEPs are recorded near the conus medullaris (109, 110, 120, 121, 122, 127). Although the descending SCEP in this method is preferred during intraoperative monitoring, the reverse direction of stimulation and recording condition could also be used (4, 64, 65, 108,

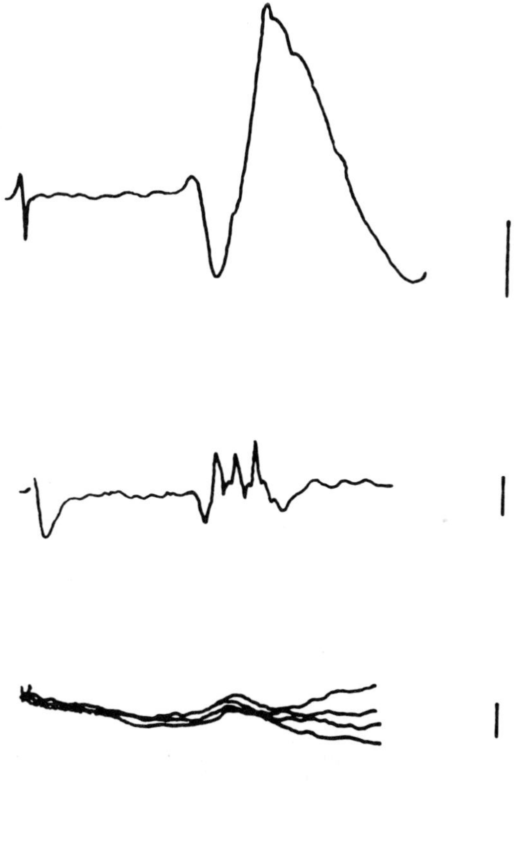

Fig. 3. Subdural (upper trace at the C6-7 i.v.), epidural (middle trace at the T1-2 i.v.) and bipolar skin surface (lower trace at the C6-7 i.v.) recorded conducted SCEPs in different normal adult subjects. All of them are averaged responses (calibrations: 10msec and (30, 1 and 0.2)mV respectively from top to the bottom). Note the different configuration of the responses and the difficulties to obtain at the skin surface recording.

115, 116). In the latter condition, the lower spinal cord and/or conus medullaris/cauda equina is stimulated and the ascending SCEPs are recorded from the upper thoracic or cervical levels. The spinally evoked ascending SCEP consists of two waves; the initial response with prominent negativity, and a second late response with longer duration. The Second response appears with stronger intensities of stimulation. The spinally evoked descending SCEP has also a similar configuration with two waves, a faster response with higher amplitude, longer duration and higher threshold (Fig. 4). However when the stimulus strengths are intense enough, other earlier and later components appear in the records and make the shape of the SCEP more complex, especially in intrathecal recordings (41) (Fig. 5). The initial di- or triphasic waves are conducted about with 70-80ms and the second wave with the ranges of 40-50ms (64, 109, 127). The spinally evoked descending SCEP was found to be slightly faster in the first component than that of the ascending SCEP (8, 41, 64). It was concluded that the two components of ascending SCEPs are likely to be conducted through different parts of

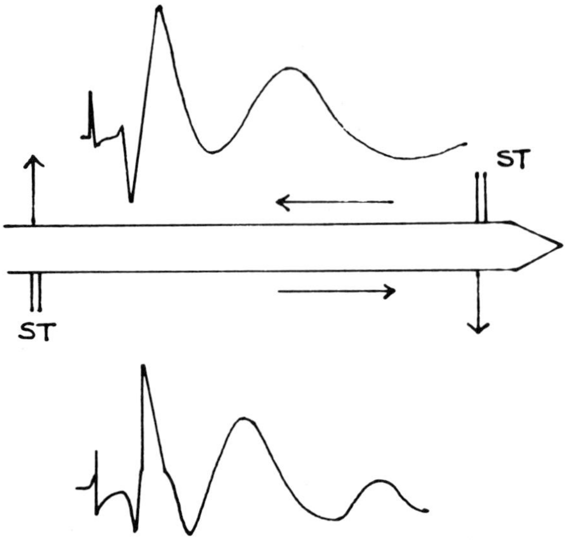

Fig. 4. Spinally evoked ascending (upper trace) and descending (lower trace) SCEPs obtained by stimulating and recording from both intrathecal electrodes situated in C6-7 and T12-L1 i.v. levels (calibrations: 25msec and 0.1mV). Note the first predominantly negative wave with higher amplitude and the following negative component with longer duration and smaller in amplitude in both potentials.

the cord; the initial wave in the lateral tracts and the later wave in the posterior columns, whereas the descending SCEPs could also be involved with the volleys from the descending pathways situated in the lateral and anterolateral columns (8, 56, 102, 124).

Descending motor SCEP

It may be important to evaluate the motor as well as the sensory tracts in the spinal cord. However, all the current electrophysiological techniques depend upon recording activity in the sensory pathways at the peripheral, spinal and cortical levels. On the other hand, corticospinal tracts are often and predominantly involved in many neurological disorders such as multiple sclerosis and spinal tumors. There are numbers of clinical reports showing the failure of monitoring using the sensory cord and cortical potentials to reveal motor deficits produced during operations (95). It has also been shown that experimentally induced cord ischemia can result in damage to descending systems which may not be detected by monitoring of the SCEP (7). Therefore a method which could evaluate the electrophysiological status of the descending tracts, especially for the pyramidal tract, was needed. Such a method has been recently developed (13, 14, 68, 69, 70). Descending motor pathways are electrically stimulated through by a pair of electrodes on the intact scalp (86, 87) and the descending motor volley is then recorded from the cervical and thoracic spinal cord with bipolar epidural electrodes. This has advantages over the other methods in that only single scalp stimuli are needed to produce clear motor volleys rather than averaging many separate responses. The potential is di- or triphasic in shape with very high amplitude and with conduction velocities of 50-74ms (13, 14). Another way to activate the corticospinal tract is to place an electrode on the scalp overlying the motor cortex and a second one on the hard palate (70). Activation of the corticospinal tract is also possible by percutaneous stimulation of the cord at different vertebral levels and by EMG recordings from the extremity or pelvic floor muscles (22, 80, 118).

It has recently been shown that descending lumbosacral cord potentials with very short onset latencies were recorded intrathecally after stimulation of the median nerve, and it was concluded that this descending response must originate in the descending and very fast-conducting propriospinal pathways. It appears to be the first direct evidence in man of the interlimb reflex action between the arm and leg which is important in the coordination of movement and posture (98, 99) (Fig. 6).

Recording of SCEPs from different levels in depth

SCEPs can be recorded with different levels of depth in relation to the spinal cord. As mentioned before, three fundamental levels of recording can be possible for the segmental and conducted SCEPs (Fig. 7):

1. Skin-surface recording.

2. Epidural recording and other recordings close to the epidural space.

3. Subdural (or intrathecal) recording.

1. Skin-surface recording - in this method, the leads are placed on the skin at separated intervertebral levels along the midline. After stimulation of the peripheral nerve, the segmental and conducted SCEPs

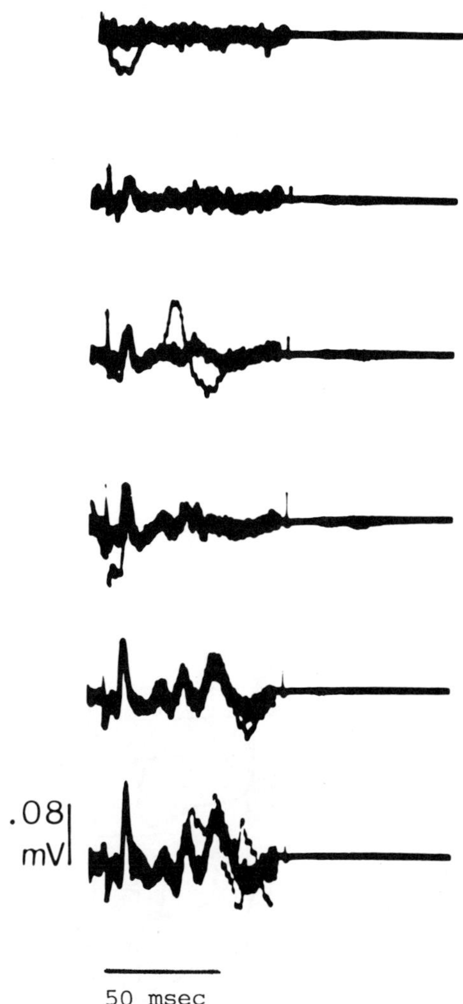

Fig. 5. Spinally evoked descending SCEPs obtained intrathecally (from C6-7 to T11-12 i.v. levels). The stimulation strength is increased from 0 to 10 volts at the C6-7 i.v. level. Twenty oscilloscopic traces are superimposed (calibrations: 50msec and 0.08mV). Note the lower threshold for the first negative wave and the appearance of later waves in complexity by increasing of stimulus intensity.

are averaged and recorded either in a bipolar mode between two successive skin leads (23, 25, 26) or with a common reference electrode (61, 83, and others). The amplitude of SCEPs are somewhat enhanced by using a reference electrode distant from the active vertebral lead such as spina iliaca, knee, hand, scapulae, sternum, midfrontal scalp, etc., though this technique also increases background noise and nonconducted segmental responses contaminate all the recordings along the axis of vertebral leads. Both high and low frequency responses are partly attenuated and the components of the SCEPs are not well differentiated owing to the long distance between the spinal cord and skin-surface electrode. The amplitude of the first negative sharp wave of the segmental SCEP becomes 10 times smaller in the skin surface in comparison to those obtained intrathecally (42). Nevertheless, a correlation can be seen in the latencies of the first

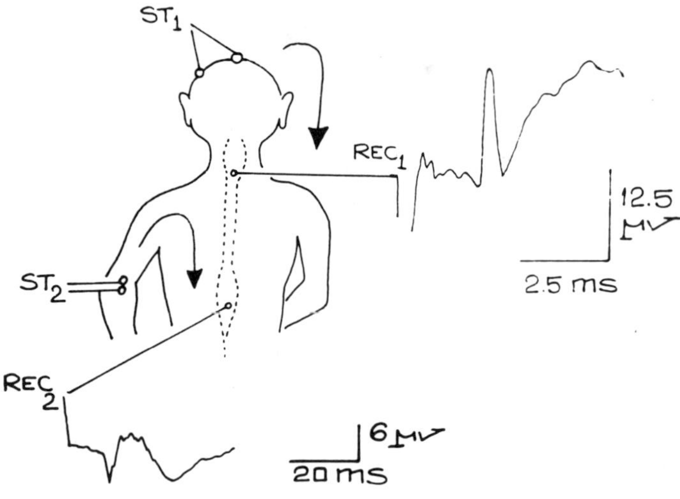

Fig. 6. Human descending SCEPs. On the upper right: Descending motor SCEP to scalp stimulation, recorded from posterior epidural space of the cervical segment (ST1-REC1) (reproduced from the Fig. 4, upper trace of Boyd et al., 1986). On the lower left: Descending SCEP recorded intrathecally at the lumbosacral level with the stimulation of the median nerve at the elbow (ST2-REC2) (reproduced from Fig. 2, bottom trace of Sarica and Ertekin, 1985).

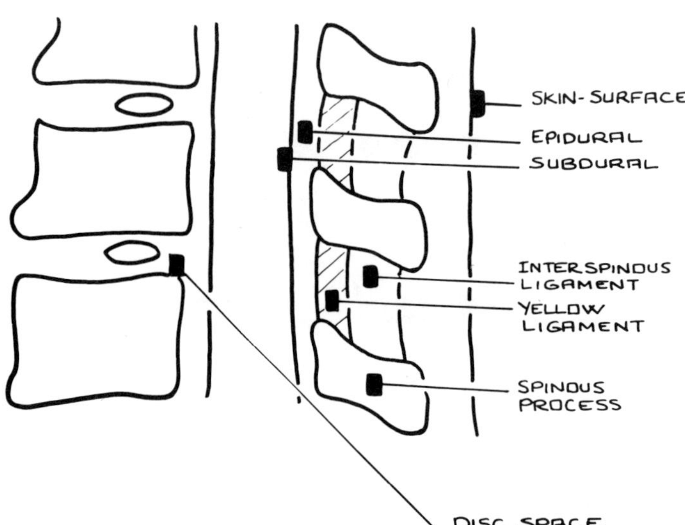

Fig. 7. Schematic diagram of various recording sites and techniques for SCEPs.

negative peak of the segmental SCEP in both recordings. The difficulty of obtaining conducted SCEPs, especially at high thoracic and cervical levels, after the stimulation of the leg nerve has often been reported. In spite of the disadvantages, skin surface recordings are of course noninvasive, and their ease of application to assess spinal cord function makes them useful for investigation of medical neurological problems. However, it is not a reliable method for spinal cord monitoring.

2. Epidural recording - SCEPs can be produced from the epidural level by introducing a 16-gauge Tuohy needle and then inserting a catheter electrode through the Tuohy needle (111, 114). Other techniques with an approximately equivalent recording level were also described as a needle electrode placed in the interspinous ligament (54, 75, 76); Kirshner's wire electrode or a single Stainman pin inserted into the spinous processes of human; and feline vertebrae (17, 85, 90, 105); and recordings made by needle electrodes which were inserted anteriorly into the disc space and posteriorly to the yellow ligament (108).

The initial negative component of the segmental SCEP recorded epidurally was found to have an amplitude of about 1/3 of those obtained intrathecally (40). Although all the components of segmental responses coincide with their peak latencies in both recording techniques, the second slower negative wave can not be clearly seen in the epidural recording.

On the other hand, conducted SCEPs are satisfactorily and successfully recorded from the leads located epidurally. At the upper thoracic level, the afferent volley is split up into at least three components with conduction velocities ranging from 40-80ms. Slower potentials require higher intensities of the stimulation of the peripheral leg nerve, but all components are stable in amplitude and latency at the stimulation frequencies up to 20/sec (60, 77, 78, 130). Some authors have preferred to stimulate the cauda equina by inserting a needle electrode intrathecally in order to provide higher responses at the high thoracic and cervical levels (76, 81, 108, 114).

3. Subdural recording - subdural recording is seldom performed due to the possible risk and discomfort to the patient, although no permanent or serious complications resulting from this technique have been seen or reported yet (39, 40, 41, 42, 43,

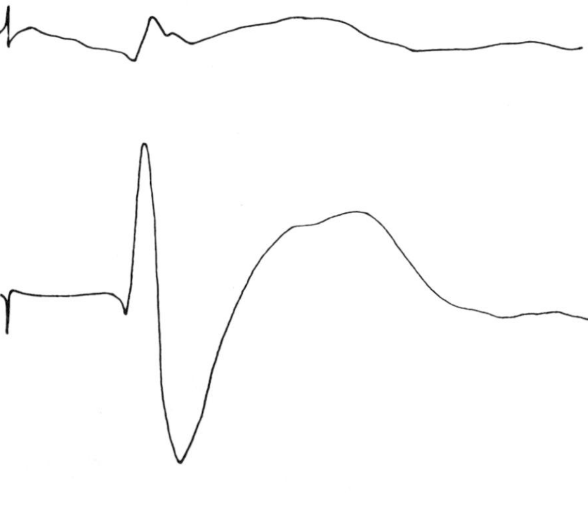

Fig. 8. Comparison of the lumbosacral segmental SCEPs after stimulation of the sural nerve (upper trace) and of the posterior tibial nerve (lower trace) at the level of the ankle of the same leg. One-hundred-twenty-eight responses are averaged (calibrations: 50msec and 18mV) (Ertekin et al., 1984).

44, 46, 79). However, great effort and rigid precautions should be undertaken as described previously (40); and the method should be restricted to special clinical problems where it is necessary. The subdural technique has recently gained special interest during spinal cord surgery (20, 89, 130).

In nonsurgery patients, teflon-coated stainless steel needles are normally introduced into the subarachnoidal space with a technique similar to the spinal tap. During the insertion of the sterilized needle electrode, rectangular electrical pulses (0.1ms, 1 or 2mA) are delivered at a rate of 1/sec. As soon as the subarachnoidal space is entered, motor responses in the form of contraction of muscles in a segmental manner are observed in the limbs according to the level of the intrathecal electrode (39, 40).

After the exposure of the spinal cord, recording electrodes are positioned directly on the dorsal surface of the cord during open spinal surgery (89, 130), or an 18-gauge spinal needle is inserted intrathecally using a lateral approach between C1-C2 vertebrae during percutaneous cervical cordotomy (20).

The recording of SCEPs by such intraoperative techniques allows an accurate localization of the site of pathology which is necessary for optimal neurosurgical intervention. It also provides an observation of the continuity and capacity of the spinal cord components after neurosurgical lesions or repairs (89, 130). The subdural method would facilitate the study of the so-called "killed end" or "spinal cord evoked injury potential" in patients with segmental spinal lesions demonstrated experimentally

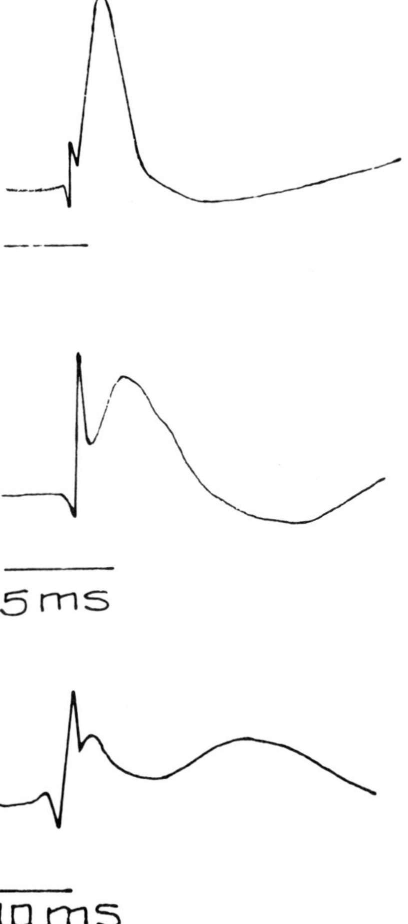

Fig. 9. Cord dorsum potentials at the L7-S1 from a decapitated cat after stimulation of the sural nerve (upper trace); the stimulation of the low threshold fibers in muscle nerve (middle trace) and the stimulation of the high threshold fibers in muscle nerve (lower trace) (reproduced and redrawn from the different figures of Bernhard, 1951, 1952, 1953). Note the size of afferent spike is considerably higher in muscle nerve stimulation in comparison to N1 and N2 slower waves, while the reverse is the case in stimulation of the cutaneous nerve.

(Schramm et al., 1983). One of the advantages of the subdural technique is to register the spontaneous ongoing cord potentials (45, 46, 100). Since the spontaneous cord potentials originate from the interneurons facilitated by peripheral afferent inputs and inhibited by suprasegmental descending pathways (46, 100), they may contribute to spinal cord monitoring by observing the status of gray matter of spinal cord in addition to the monitoring of afferent and efferent tract activities from the white matter of the spinal cord.

Both segmental and conducted SCEPs recorded subdurally vary in shape and size according to the spatial relationship between the position of the electrode tip and the part of spinal cord and roots within the vertebral canal. When the tip of the recording electrode is behind the dorsal funiculus, the segmental input from the periphery produces typical segmental SCEP (Fig. 1 and Fig. 2). There is a general tendency to compare the components of the segmental SCEP recorded in man with those obtained from animal studies, namely the first spiky wavelets sometimes seen at the beginning or on the rising phase of the negative sharp wave are considered to be equal to intramedullary afferent spikes; first and second negative waves are thought to be the postsynaptical N1 and N2 intermediary cord dorsum potentials of animal records. The application of data obtained from animal studies may not mean that the origin of the human segmental SCEP is exactly the same. There is no general agreement about the source of components; for example the so-called N1 wave of the initial negative sharp wave recorded by various techniques from the human subjects could be interpreted as either presynaptic activity with the afferent volleys from the dorsal roots and dorsal column afferent nerve fibers (1, 2, 26, 31, 32, 40, 43, 47, 63) or postsynaptic at the dorsal horn level (30, 38, 50, 61, 79, 82, 83, 89, 112, 125, 126).

When the subdural electrode is located far above the segmental level of the cord, conducted SCEPs can also be recorded with a variable shape and conduction characteristics due to the position of the tip of the electrode in relation to the posterior or posterolateral location. In short segments of the spinal cord the conducted SCEP with di- or triphasic shapc and very high amplitude has been recorded with maximal speeds of 30-50ms along with a rapid decline of the amplitude of the conducted SCEP activated by either posterior tibial nerve or dorsal root stimulation (79, 89). The maximal conduction velocity of the afferent volley passing throughout the dorsal funiculus between the lumbosacral and cervical enlargements was also found to be in the range of 30-50ms (41). Recordings from the surface of the human spinal cord over the lateral column revealed conduction velocities 50ms or faster (89). The configuration of conducted SCEP and the values of conduction velocity are somewhat different according to the method used. Epidurally recorded conducted SCEPs seem to be composed of afferent volleys from the faster component of dorsolateral tracts and the slower component from the dorsal funiculus. This is probably because the epidurally oriented leads have a capacity to record a larger field from the angular

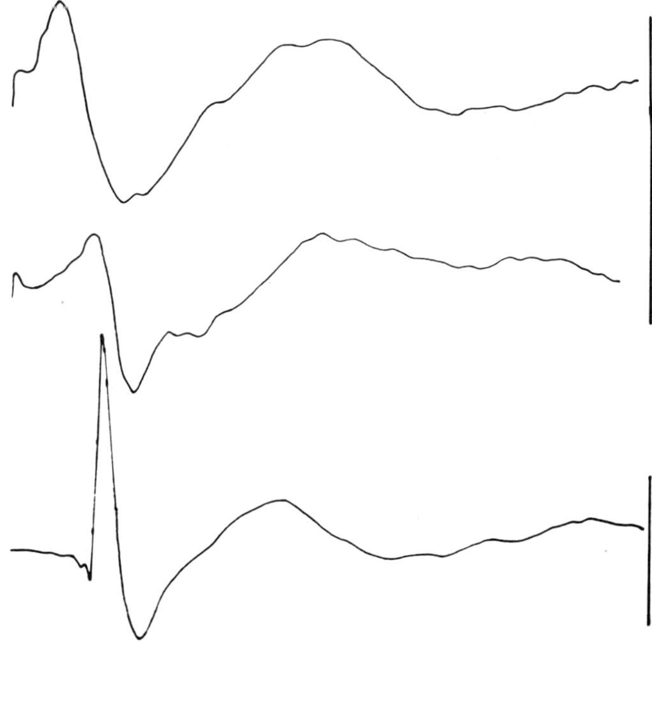

Fig. 10. Lumbosacral segmental SCEPs evoked by patellar tendon tapping (upper trace), Achilles tendon tapping (middle trace) and evoked by electrical shocks to the posterior tibial nerve at the popliteal fossa (lower trace) of the same leg. These responses were intrathecally recorded and 128 responses were averaged (calibrations: 50msec and 20mV). Note the time sequences of responses fit very well between the Achilles tendon and posterior tibial SCEPs; the very small amplitude of tendon responses is probably due to great temporal dispersion of the proprioceptive afferents to tapping (Ertekin et al., 1984).

view of the electrodes; intrathecally oriented leads close to the dorsal column might not be able to pick up the afferent volley passing through the dorsolateral column. Thus, conduction velocities of 30-50ms are reasonable to compare to those obtained from the dorsal funiculus of the experimental physiology (15, 84, 93, 123) and are in accord with the morphometry of myelinated fibers of the fasiculus gracilis in man (91). The faster conduction velocities of the initial components of epidurally recorded polyphasic conducted SCEP in man may correlate with the faster conduction velocities along the dorsolateral afferent pathways determined experimentally (28, 29, 48, 55, 72, 92, 97, 101, 119).

Since the conduction velocity of the proximal segments of the sciatic nerve is about 60ms (16, 41) the afferent nerve conduction velocity becomes slower within the central afferent fibers of the posterior funiculus in man. This fact has been repeatedly determined in experimental physiology (72, 93).

Effect of peripheral stimulation variables on SCEPs

SCEPs are usually obtained by electrical stimulation of the mixed nerves, for instance the posterior tibial and the lateral popliteal nerves in the lower limbs, and the median and ulnar nerves in the upper limbs. Conventional bipolar or monopolar surface or needle stimulating electrodes are used. Therefore the SCEP is a mixture of the peripheral volleys coming from the muscle and other deep afferents and from the skin afferent nerves. The segmental SCEP is very small with poorly defined late components after stimulation of the pure sensory nerve such as the sural nerve compared to SCEP recorded after stimulation of the mixed nerve (26, 39, 47, 94, 113) (Fig. 8). It can be proposed that the contribution of the exteroceptive afferents to the segmental SCEP is considerably small and the segmental SCEP is mainly produced by stimulation of the proprioceptive muscles and other deep afferent nerve fibers. It was experimentally

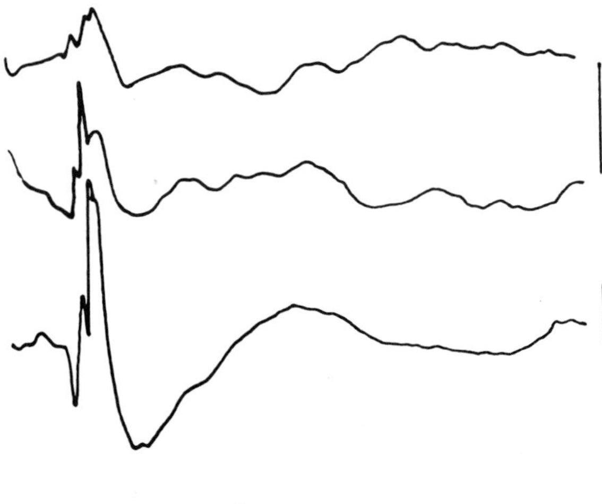

Fig. 11. Segmental SCEPs obtained intrathecally from the T11-12 i.v. level by stimulation of the posterior tibial nerve at the popliteal fossa. Upper trace: At the weak electrical shocks sufficient to evoke only soleus H-Reflex. Middle trace: At the intensity to produce maximal H-Reflex. Lower trace: At the supramaximal electrical stimulation with maximal M-response. Sixty-four (lower) or 128 (others) responses are averaged (calibrations: 50msec and 5mV).

found that the afferent spike of cord dorsum potential was indeed very small compared to the following slow N1 wave; whereas stimulation of mixed or muscle nerves produced afferent spikes with a higher amplitude in comparison to the following negative waves (6, 9, 10, 11, 21, 52, 53, 72, 101) (Fig. 9). In spite of the experimental data, supramaximal stimulation of the sural nerve could not produce clear and well-defined slow potentials similar to N1 and N2 of animal studies (47). This observation may also point out that animal data are not easily and exactly applicable to human studies. In a practical point of view, a stimulus intensity well above the motor threshold (2-3 times even more) is generally ideal in order to obtain the segmental SCEP recorded superficially (66, 73, 94, 126, 133).

Amplitudes of SCEPs are increased with stimulation at more proximal sites (39, 40, 41, 94). Bilateral stimulation of the homologue nerves at the same site results in SCEPs greater in amplitude and better defined at all the recording levels than they are after the stimulation of either single nerve (26, 126). These findings are obviously due to the more dense afferent volley excited from the periphery.

A stimulus rate less than 10Hz provides efficiently obtained and reproducible recordings. The higher frequency stimulations involve a risk of attenuating the components related to postsynaptic activities of the segmental SCEP (38, 50). It also causes discomfort to the patient.

Depending on the individual peripheral nerve stimulated, the magnitude of a SCEP can also vary. For instance, stimulation of the posterior tibial nerve at the popliteal fossa produces greater response than the peroneal nerve at the same stimulating level (26, 39). If clinical problems are required to investigate the effect of segmental afferent innervation, it is necessary to stimulate different peripheral nerves for their supply of afferent roots. For instance the ulnar nerve for C7-8 spinal roots, the musculocutaneous nerve for C5-6 roots.

Monitoring of segmental and conducted SCEPs by using the H-Reflex (30, 33, 38, 41, 47, 79, 94) did not seem to be practical for clinical purposes. Neither of the SCEPs obtained by stimulating the tendon reflexes (47, 88) could be useful practically due to very small size and variability (Fig. 10). The selective and precise contribution of the group IA afferents to segmental and conducted SCEPs is a matter of debate which is probably related to the insensitivity of the methods presently applied in man (Fig. 11).

Clinical applications

At the end of this review is a summary of the current and possible future clinical applications of spinal cord potentials with various recording techniques. A summary related with the topic is shown on Table 2.

Table 1: A simple classification of the SCEP according to the site of stimulation and recording.

A. Ascending spinal cord evoked potentials (SCEP)	1. Segmental SCEP 2. Conducted SCEP 3. Spinally evoked ascending (and descending) SCEP
B. Descending spinal cord evoked potentials (SCEP)	1. Descending motor SCEP 2. Peripherally evoked descending SCEP

Table 2: Current and possible future applications of spinal cord potentials with different recording techniques.

Techniques:	Current and possible future applications:
I. Skin-surface Cerebral Evoked Potentials	A. Medical neurological disorders for gross recording (with topographical diagnosis and prognosis. B. Maturational studies in childhood.
II. Epidural and similar invasive recordings: A. By peripheral and/or spinal cord stimulation	A. Spinal Cord monitoring for afferent tracts
B. By descending motor SCEP	B. Spinal cord monitoring for efferent tracts (i.e., pyramidal tract).
III. Subdural recording during open spinal cord surgery	A. For accurate localization of the site of pathology and for the effect of lesion on repair. B. Spinal cord monitoring during open spinal cord surgery.

References

1. Abbruzzesse, M.; Favale, E.; Leandri, M.; Ratto, S.: Spinal components of the cerebral somatosensory evoked response in normal man: The "S wave". Acta. Neurol. Scandinav. 58, 213-220, 1978.
2. Anziska, B.J.; Cracco, R.Q.: Short latency SEPs to median nerve stimulation: Comparison of recording methods and origin of components. Electroenceph. Clin. Neurophysiol. 52, 531-539, 1981.
3. Austin, G.M.; McCouch, G.P.: Presynaptic component of intermediary cord potential. J. Neurophysiol. 18, 441-451, 1955.
4. Baba, H.; Shima, I.; Tomita, K.; Omeda, S.I.; Sawada, Y.; Yugami, H.; Ugaji, U.; Shinoda, K.; Yashimizu, N.; Masuyama, S.; Yonezawa, K.; Tsuji, S.; Nomura, S.: Clinical usefulness of spinal cord evoked potentials. In: J. Schramm; S.J. Jones (eds.). Spinal Cord Monitoring. Springer-Verlag Berlin Heidelberg, New York, Tokyo, 1985, pp. 245-249.
5. Barron, D.H.; Matthews, B.H.C.: The interpretation of potential changes in the spinal cord. J. Physiol. (Lond.) 92, 276-321, 1938.
6. Beall, J.E.; Applebaum, A.E.; Foreman, R.D.; Willis, W.P.: Spinal cord potentials evoked cutaneous afferents in monkeys. J. Neurophysiol. 40, 199-211, 1977.
7. Bennett, M.H.: Effect of compression and ischemia on spinal cord evoked potentials. Exp. Neurol. 80, 508-519, 1983.
8. Bennett, M.H.; Akin, O.N.: Extraspinal stimulation and recording. A method for operative monitoring. In: J. Schramm; S.J. Jones (eds.). Spinal Cord Monitoring. Springer-Verlag Berlin, Heidelberg, New York, Tokyo, 1985, pp. 51-58.
9. Bernhard, C.G.: On the analysis of the spinal cord potentials. Univ. Leeds. Med. Mag. 21, 1-10, 1951.
10. Bernhard, C.G.: The cord dorsum potentials in relation to peripheral source of afferents stimulation. Cord. Spring Harbor Symposia on Quantitative Biology 17, 221-232, 1952.
11. Bernhard, C.G.: The spinal cord potentials in leads from the cord dorsum in relation to peripheral source of afferent stimulation. Acta Physiol. Scand. 29, Suppl. 106, 1-29, 1953.
12. Bernhard, C.G.; Widen, L.: On the origin of the negative and positive spinal cord potentials evoked by stimulation of low threshold cutaneous fibres. Acta Physiol. Scand. 29, Suppl. 106, 42-54, 1953.
13. Boyd, S.G.; Cowan, J.M.A.; Rothwell, J.C.; Webb, P.J.; Marsden, C.D.: Monitoring spinal motor tract function using cortical stimulation: A preliminary report. In: J. Schramm; S.J. Jones (eds.): Spinal Cord Monitoring. Springer-Verlag Berlin, Heidelberg, New York, Tokyo, 1985, pp. 227-230.
14. Boyd, S.G.; Rothwell, J.C.; Cowan, J.M.A.; Webb, P.J.; Morley, T.; Asselman, P.; Marsden, C.D.: A method of monitoring function in corticospinal pathways during scoliosis surgery with a note on motor conduction velocities. J. Neurol. Neurosurg. Psychiat. 49, 251-257, 1986.
15. Brown, A.G.: Cutaneous afferent fibre collaterals in the dorsal columns of the cat. Exp. Brain Res. 5, 293-305, 1968.
16. Buchthal, F.; Rosenfalck, A.: Evoked action potentials and conduction velocity in human sensory nerves. Brain Res. (Special issue) 3, 1-122, 1966.
17. Bunch, W.H.; Scarff, T.B.; Trimble, J.: Spinal cord monitoring. J. Bone Joint Surgery. 65-A, 707-710, 1983.
18. Caccia, M.R.; Ubiali, E.; Andreussi, L.: Spinal evoked responses recorded from the epidural space in normal and diseased humans. J. Neurol. Neurosurg. Psychiat. 39, 962-972, 1976.
19. Campbell, B.: The distribution of potential fields within the spinal cord. Anat. Rec. 91, 77-88, 1945.
20. Campbell, J.A.; Lipton, S.: Intraspinal evoked potentials in man during cervical cordotomy. In: S. Homma; T. Tamaki (eds.): Fundamentals And Clinical Applications Of Spinal Cord Monitoring. Saikon Publ. Tokyo, 1984, pp. 245-252.
21. Coombs, J.S.; Curtis, D.R.; Landgren, S.: Spinal cord potentials generated by impulses in muscle and cutaneous afferent fibres. J. Neurophysiol. 19, 452-467, 1956.
22. Cowan, J.M.A.; Rothwell, J.C.; Dick, J.P.R.; Thompson, P.D.; Day, B.L.; Marsden, C.D.: Abnormalities in central motor pathway conduction in multiple sclerosis. Lancet. 2, 304-307, 1984.
23. Cracco, R.Q.: Spinal evoked response: Peripheral nerve stimulation in man. Electroenceph. Clin. Neurophysiol. 35, 379-386, 1973.
24. Cracco, J.B.; Cracco, R.Q.: Spinal somatosensory evoked potentials: Maturational and clinical studies in evoked potentials. Ann. N.Y. Acad. Sci. 388, 526-537, 1982.
25. Cracco, J.B.; Cracco, R.Q.; Graziani, L.J.: The spinal evoked response in infants and children. Neurology (Minneap.) 25, 31-36, 1975.
26. Cracco, R.Q.; Cracco, J.B.; Sarnowski, R.; Vogel, H.B.: Spinal evoked potentials. In: J.E. Desmedt (ed.): Clinical use of cerebral, brainstem and spinal somatosensory evoked potentials. Prog. Clin. Neurophysiol. Vol. 7, pp. 87-104, Karger, Basel, 1980.
27. Cracco, J.B.; Cracco, R.Q.; Stalove, R.: The spinal evoked potentials in man: A maturational study. Electroenceph. Clin. Neurophysiol. 46, 58-64, 1979.

28. Cracco, R.Q.; Evans, B.: Spinal evoked potential in the cat: Effect of asphyxia, strychnine, cord section and compression. Electroenceph. Clin. Neurophysiol. 44, 187-201, 1978.
29. Cusick, J.F.; Myklebust, J.; Larson, S.J.; Sances, A.: Spinal evoked potentials in the primate: Neural substrate. J. Neurosurg. 49, 551-557, 1978.
30. Delbeke, J.; McComas, A.J.; Kopec, S.J.: Analysis of evoked lumbosacral potentials in man. J. Neurol. Neurosurg. Psychiat. 41, 293-302, 1978.
31. Desmedt, J.E.; Cheron, G.: Central conduction in man: Neural generators and interpeak latencies of the far field components recorded from neck and right or left scalp and earlobes. Electroenceph. Clin. Neurophysiol. 50, 382-403, 1980.
32. Desmedt, J.E.; Cheron, G.: Prevertebral (esophageal) recording of subcortical somatosensory evoked potentials in man: The spinal P13 component and the dual nature of the spinal generators. Electroenceph. Clin. Neurophysiol. 52, 257-275, 1981.
33. Dimitrijevic, M.R.; Larsson, L.E.; Lehmkul, D.; Sherwood, A.: Evoked spinal cord and nerve root potentials in humans using a noninvasive recording technique. Electroenceph. Clin. Neurophysiol. 45, 331-340, 1978.
34. Dorfman, L.J.: Indirect estimation of spinal cord conduction velocity in man. Electroenceph. Clin. Neurophysiol. 42, 26-34, 1977.
35. Dorfman, L.J.; Bosley, T.M.: Age-related changes in peripheral and central nerve conduction in man. Neurology (Minnap.) 29, 38-44, 1979.
36. Eccles, J.C.: Presynaptic inhibition in the spinal cord. Progress in Brain Research, Vol. 12. In: J.C. Eccles; J.P. Schade (eds.): Physiology Of Spinal Neurons, Elsevier, Amsterdam/London/New York, 1964, pp. 65-91.
37. Eisen, A.; Burton, K.; Larsen, A.; Hoirch, M.; Calne, D.: A New Indirect Method For Measuring Spinal Conduction Velocity In Man.
38. El-Negamy, E.; Sedgwick, E.M.: properties of a spinal somatosensory evoked potential recorded in man. J. Neurol. Neurosurg. Psychiat. 41, 762-768, 1978.
39. Ertekin, C.: Human evoked electrospinogram. In: J.E. Desmedt (ed.): Developments In Electromyography And Clinical Neurophysiology, Vol. 2, Karger, Basel, 1973, pp. 344-351.
40. Ertekin, C.: Studies on the human evoked electrospinogram, Part 1: The origin of the segmental evoked potentials. Acta Neurol. Scandinav. 53, 3-20, 1976a.
41. Ertekin, C.: Studies on the human evoked electrospinogram, Part 2: The conduction velocity along the dorsal funiculus. Acta Neurol. Scandinav. 53, 21-38, 1976b.
42. Ertekin, C.: Comparison of the human evoked electrospinogram recorded from the intrathecal, epidural and cutaneous levels. Electroenceph. Clin. Neurophysiol. 44, 683-690, 1978a.
43. Ertekin, C.: Evoked electrospinogram in spinal cord and peripheral nerve disorders. Acta Neurol. Scandinav. 57, 329-344, 1978b.
44. Ertekin, C.; Mutlu, R.; Sarica, Y.; Uckardesler, L.: Electrophysiological evaluation of the afferent spinal roots and nerves inpatients with conus medullaris and cauda equina lesions. J. Neurol. Sci. 48, 419-433, 1980.
45. Ertekin, C.; Sarica, Y.; Uckardesler, L.: Studies on the human spontaneous electromyelogram (EMyeloG): I. Normal subjects. Electroenceph. Clin. Neurophysiol. 55, 13-23, 1983.
46. Ertekin, C.; Sarica, Y.; Uckardesler, L.: Studies on the human spontaneous electromyelogram (EMyeloG): II. Patients with peripheral nerve, root and spinal cord disorders. Electroenceph. Clin. Neurophysiol. 55, 24-33, 1983.
47. Ertekin, C.; Uckardesler, L.; Sarica, Y.: Contribution of proprioceptive afferent inputs to the segmental evoked electrospinogram (EspG) in man. In: S. Homma; T. Tamaki (eds.): Fundamentals And Clinical Application Of Spinal Cord Monitoring. Saikon Publ. Tokyo, 1984.
48. Feldman, M.H.; Cracco, R.Q.; Farmer, P.; Mount, F.: Spinal evoked potential in the monkey. Ann. Neurol. 7, 238-244, 1980.
49. Fernandez de Molina, A.; Gray, J.A.B.: Activity in the dorsal spinal gray matter after stimulation of cutaneous nerves. J. Physiol. (Lond.) 137, 126-140, 1957.
50. Ganes, T.: Synaptic and nonsynaptic component of the human cervical evoked response. J. Neurol. Sci. 55, 313-326, 1982.
51. Gasser, H.S.; Graham, H.T.: Potentials produced in the spinal cord by stimulation of dorsal roots. Am. J. Physiol. 103, 303-320, 1933.
52. Gelfan, S.; Tarlov, I.M.: Differential vulnerability of spinal cord structures to anoxia. J. Neurophysiol. 18, 170-188, 1955.
53. Gelfan, S.; Tarlov, I.M.: Physiology of spinal cord, nerve root and peripheral nerve compression. Am. J. Physiol. 185, 217-226, 1956.
54. Hahn, J.F.; Lesser, R.; Klem, G.; Lueders, H.: Simple technique for monitoring intraoperative spinal cord function. Neurosurgery. 9, 692-695, 1981.
55. Happel, L.T.; LeBlanc, H.J.; Kline, D.G.: Spinal cord potentials evoked by peripheral nerve stimulation. Electroenceph. Clin. Neurophysiol. 38, 349-354, 1975.

56. Harada, Y.; Takemitsu, Y.; Atsuta, Y.; Imai, M.: Determination of the pathways of ascending and descending conductive spinal cord evoked potentials (SCEP). In: S. Homma; T. Tamaki (eds.): Fundamentals And Clinical Application Of Spinal Cord Monitoring. Saikon Publ. Tokyo, 1984, pp. 33-43.

57. Hattori, S.; Saiki, K.; Kawai, S.: Diagnosis of the level and severity of cord lesion in cervical spondylotic myelopathy-spinal evoked potentials. Spine 4, 478-485, 1979.

58. Hume, A.L.; Cant, B.R.: Conduction time in central somatosensory pathways in man. Electroenceph. Clin. Neurophysiol. 45, 361-375, 1978.

59. Jones, S.J.: Short latency potentials recorded from the neck and scalp following median nerve stimulation in man. Electroenceph. Clin. Neurophysiol. 43, 853-863, 1977.

60. Jones, S.J.; Edgar, M.A.; Ransford, A.O.: Sensory nerve conduction in the human spinal cord: Epidural recordings made during scoliosis surgery. J. Neurol. Neurosurg. Psychiat. 45, 446-451, 1982.

61. Jones, S.J.; Small, D.G.: Spinal cord and subcortical evoked potentials following stimulation of the posterior tibial nerve in man. Electroenceph. Clin. Neurophysiol. 44, 299-306, 1978.

62. Kakigi, R.; Shibasaki, H.; Hashizume, A.; Kuroiwa, Y.: Short latency somatosensory evoked spinal and scalp recorded potentials following posterior tibial nerve stimulation in man. Electroenceph. Clin. Neurophsiol. 53, 602-611, 1982.

63. Kaneda, A.; Yamaura, I.; Kamikozuru, M.; Shinomiya, K.; Sato, H.; Yokoyama, M.: An analysis of spinal cord potentials evoked by median nerve stimulation. In: J. Schramm; S.J. Jones (eds.): Spinal Cord Monitoring. Springer-Verlag Berlin Heidelberg, New York, Tokyo, 1985, pp. 35-42.

64. Kotani, H.; Hattori, S.; Kawai, S.; Saiki, K.; Yamasaki, H.; Omete, K.: Evaluation of spinal cord function using segmental and conductive spinal evoked potentials. In: S. Homma; T. Tamaki (eds.): Fundamentals And Clinical Application Of Spinal Cord Monitoring. Saikon Publ. Tokyo, 1984, pp. 253-268.

65. Kotani, H.; Hattori, S.; Senzoku, F.; Kawai, S.; Saiki, K.; Yamasaki, H.; Omete, K.: Evaluation of cord function in cervical spondylosis by a combined method using segmental and conductive spinal evoked potentials (SEP). In: S. Schramm; S.J. Jones (eds.): Spinal Cord Monitoring. Springer-Verlag Berlin Heidelberg, New York, Tokyo, 1985, pp. 274-283.

66. Lastimosa, A.C.B.; Bass, N.H.; Stanback, K.; Norvell, E.E.: Lumbar spinal cord and early cortical evoked potentials after tibial nerve stimulation: Effects of stature on normative data. Electroenceph. Clin. Neurophysiol. 54, 499-507, 1982.

67. Lesser, R.P.; Lueders, H.; Hahn, J.; Klem, G.: Early somatosensory potentials evoked by median nerve stimulation. Intraoperative monitoring. Neurology (Minneap.) 31, 1519-1523, 1981.

68. Levy, W.J.: Spinal evoked potentials from the motor tracts. J. Neurosurg. 58, 38-44, 1983a.

69. Levy, W.J.; York, D.H.: Evoked potentials from the motor tracts in humans. Neurosurg. 12, 422-429, 1983b.

70. Levy, W.J.; York, D.H.; McCaffrey, M.; Tanzer, F.: Motor evoked potentials from transcranial stimulation of the motor cortex in humans. Neurosurg. 15, 287-302, 1984.

71. Lloyd, D.P.C.; McIntyre, A.K.: On the origins of dorsal root potentials. J. Gen. Physiol. 32, 409-443, 1949.

72. Lloyd, D.P.C.; McIntyre, A.K.: Dorsal column conduction of group I muscle afferent impulses and their relay through Clarke's column. J. Neurophysiol. 13, 39-54, 1950.

73. Lueders, H.; Andrish, J.; Gurd, A.; Weiker, G.; Klem, G.: Origin of far field subcortical potentials evoked by stimulation of the posterior tibial nerve. Electroenceph. Clin. Neurophysiol. 52, 336-344, 1981.

74. Lueders, H.; Dinner, D.S.; Lesser, R.P.; Klem, G.: Origin of far field subcortical evoked potentials to posterior tibial and median nerve stimulation: A comparative study. Arch. Neurol. (Chic.) 40, 93-97, 1983.

75. Lueders, H.; Gurd, A.; Hahn, J.; Andrish, J.; Weiker, G.; Klem, G.: A new technique for intraoperative monitoring of spinal cord function. Multichannel recording of spinal cord and subcortical evoked potentials. Spine 7, 110-115, 1980.

76. Lueders, H.; Hahn, J.; Gurd, A.; Tsuji, S.; Dinner, D.; Lesser, R.; Klem, G.: Surgical monitoring of spinal cord function: Cauda equina stimulation technique. Neurosurgery 11, 482-485, 1982.

77. Macon, J.B.; Poletti, C.E.: Conducted somatosensory evoked potentials during spinal surgery. Part 1: Control conduction velocity measurements. J. Neurosurg. 57, 349-353, 1982(a).

78. Macon, J.B.; Poletti, C.E.; Sweet, W.H.; Ojeman, R.G.; Zervas, N.T.: Conducted somatosensory evoked potentials during spinal surgery. Part 2: Clinical applications. J. Neurosurg. 57, 354-359, 1982(b).

79. Magladery, J.W.; Porter, W.E.; Park, A.M.; Teasdall, R.D.: Electrophysiological studies of nerve and reflex activity in normal man. IV. The two-neurone reflex and identification of certain action potentials from spinal roots and cord. Bull. Johns Hopkins Hosp. 88, 499-519, 1951.

80. Marsden, C.D.; Merton, P.A.; Morton, H.B.: Percutaneous stimulation of spinal cord and brain: Pyramidal tract conduction velocities in man. J. Physiol. 328, 61P, 1982.

81. Maruyama, Y.; Shimizu, H.; Fujioka, H.; Shimoji, K.; Takahashi, H.; Homma, T.; Yasukawa, K.; Nakamura, T.: Spinal cord function monitoring by spinal cord potentials during spine and spinal surgery. In: S. Homma; T. Tamaki (eds.): Fundamentals And Clinical Application Of Spinal Cord Monitoring. Saikon Publ. Tokyo, 1984, pp. 191-201.

82. Matsukado, Y.; Yoshida, M.; Goya, T.; Shimoji, K.: Classification of cervical spondylosis or disc protrusion by preoperative evoked spinal electrogram. J. Neurosurg. 44, 435-441, 1976.

83. Matthews, W.B.; Beauchamps, M.; Small, D.G.: Cervical somatosensory evoked responses in man. Nature (Lond.) 252, 230-232, 1974.

84. McDonald, W.I.; Sears, T.A.: The effects of experimental demyelination on conduction in the central nervous system. Brain 93, 583-598, 1970.

85. McNeal, D.R.; Swank, S.; Satomi, K.; Passoff, T.: Comparison of spinal evoked potentials recorded from bone and the epidural space. In: S. Homma; T. Tamaki (eds.): Fundamentals And Clinical Application Of Spinal Cord Monitoring. Saikon Publ. Tokyo, 1984, pp. 77-86.

86. Merton, P.A.; Morton, H.B.: Stimulation of the cerebral cortex in the intact human subject. Nature 285, 227, 1980.

87. Merton, P.A.; Morton, H.B.; Hill, D.K.; Marsden, C.D.: Scope of a technique for electrical stimulation of human brain, spinal cord and muscle. Lancet 2, 597-600, 1982.

88. Monster, A.W.: The human electrospinogram and its use in the study of reflex response. In: E. Desmedt (ed.): Clinical uses of cerebral, brainstem and spinal somatosensory evoked potentials. Prog. Clin. Neurophysiol. Vol. 7, Karger-Basel, 1980, pp. 118-125.

89. Nashold, B.S.; Ovelman-Levitt, J.; Sharpe, R.; Higgins, A.C.: Intraoperative evoked potentials recorded in man directly from dorsal roots and spinal cord. J. Neurosurg. 62, 680-693, 1985.

90. Nordwall, A.; Axelgaard, J.; Harada, Y.; Valencia, P.; McNeal, D.R.; Brown, J.D.: Spinal cord monitoring using evoked potentials recorded from feline vertebral bone. Spine 4, 486-494, 1979.

91. Ohniski, A.; O'Brien, P.C.; Okazaki, H.; Dyck, P.J.: Morphometry of myelinated fibers of fasiculus gracilis of man. J. Neurol. Sci. 27, 163-172, 1976.

92. Oscarsson, O.: Differential course and organization of uncrossed and crossed long ascending spinal tracts. Progress in Brain Research. In: J.C. Eccles; J.P. Schade (eds.): Vol. 12. Physiology Of Spinal Neurons. Elsevier. Amsterdam, London, New York, pp. 164-176.

93. Petit, D.; Burgess, P.R.: Dorsal column projection of receptors in cat hairy skin supplied by myelinated fibers. J. Neurophysiol. 31, 849-855, 1968.

94. Phillips, L.H.; Daube, J.R.: Lumbosacral spinal evoked potentials in humans. Neurology (Minneap.) 30, 1175-1183, 1980.

95. Raudzens, P.: Intraoperative monitoring of evoked potentials. Ann. NY Acad. Sci. 388, 308-326, 1982.

96. Riffel, B.; Stohr, M.; Korner, S.: Spinal and cortical evoked potentials following stimulation of the posterior tibial nerve in the diagnosis and localization of spinal cord diseases. Electroenceph. Clin. Neurophysiol. 58, 400-407, 1984.

97. Rossini, P.M.; Greco, F.; De Palma, L.; Pisano, L.: Electrospinogram of the rabbit: Monitoring of the spinal conduction in acute cord lesions versus clinical observation. Eur. Neurol. 19, 409-413, 1980.

98. Sarica, Y.; Ertekin, C.: Descending lumbosacral cord potentials (DLCP) evoked by stimulation of the median nerve. Brain Research. 325, 299-301, 1985a.

99. Sarica, Y.; Ertekin, C.: Descending lumbosacral cord potentials (DLCP) evoked by stimulation of the median nerve. In: J. Schramm; S.J. Jones (eds.): Spinal Cord Monitoring. Springer-Verlag Berlin Heidelberg, New York, Tokyo, 1985b.

100. Sarica, Y.; Eti, M.; Ertekin, C.: Studies on the human spontaneous electromyelogram (EMyeloG). IV. Spontaneous spinal cord activity of the cat relevant to the human spontaneous EmyeloG. Electroenceph. Clin. Neurophysiol. 55, 168-179, 1983.

101. Sarnowski, R.J.; Cracco, R.Q.; Vogel, H.B.; Mount, F.: Spinal evoked response in the cat. J. Neurosurg. 43, 329-336, 1975.

102. Satomi, K.; Nishimoto, G.I.: Effects of selective spinal cord transections on evoked spinal potentials in cats. In: S. Homma; T. Tamaki (eds.): Fundamentals And Clinical Application Of Spinal Cord Monitoring. Saikon Publ. Tokyo, 1984, pp. 87-98.

103. Schiff, J.A.; Cracco, R.Q.; Rossini, P.M.; Cracco, J.B.: Spine and scalp somatosensory evoked potentials in normal subjects and patients with spinal cord disease: Evaluation of afferent transmission. Electroenceph. Clin. Neurophysiol. 59, 374-387, 1984.

104. Schramm, J.; Krause, R.; Shigeno, T.; Brock, M.: Experimental investigation on the spinal cord evoked injury potentials. J. Neurosurg. 59, 485-492, 1983.

105. Schroeder, F.W.; Hyman, A.A.; Powell, S.H.; Boetner, R.B.: Spinal evoked potentials recorded from canine vertebral bone: Response alterations to anesthesia, hypotension, temperature. In: S. Homma; T. Tamaki (eds.): Fundamentals And Clinical Application Of Spinal Cord Monitoring. Saikon Publ. Tokyo, 1984, pp. 135-144.

106. Seyal, M.; Emerson, R.G.; Pedley, T.A.: Spinal and early scalp recorded components of the somatosensory evoked potential following stimulation of the posterior tibial nerve. Electroenceph. Clin. Neurophysiol. 55, 320-330, 1983.
107. Seyal, M.; Gabor, A.J.: The human posterior tibial somatosensory evoked potentials: Synapse dependent and synapse independent spinal components. Electroenceph. Clin. Neurophysiol. 62, 323-331, 1985.
108. Shimada, N.; Matsuda, H.; Hirose, T.; Kondo, M.; Hashimoto, T.; Nakajima, K.; Abe, J.; Shimazu, A.: Intraoperative evoked spinal cord potential for diagnosing the level of cervical myelopathy: Recording through the disc space and the yellow ligament. In: S. Homma; T. Tamaki (eds.): Fundamentals And Clinical Application Of Spinal Cord Monitoring. Saikon Publ. Tokyo, 1984, pp. 287-297.
109. Shimizu, H.; Shimoji, K.; Maruyama, Y.; Sato, Y.; Harayama, H.; Tsubaki, T.: Slow cord dorsum potentials elicited by descending volleys in man. J. Neurol. Neurosurg. Psychiat. 42, 242-246, 1979.
110. Shimizu, H.; Shimoji, K.; Maruyama, Y.; Matsuki, M.; Kuribayashi, H.; Fujioka, H.: Human spinal cord potentials produced in lumbosacral enlargement by descending volleys. J. Neurophysiol. 48, 1108-1120, 1982.
111. Shimoji, K.; Higashi, H.; Kano, T.: Epidural recording of spinal electrogram in man. Electroenceph. Clin. Neurophysiol. 30, 236-239, 1971.
112. Shimoji, K.; Kano, T.; Higashi, H.; Morioka, T.; Henschel, E.O.: Evoked spinal electrograms recorded from epidural space in man. J. App. Physiol. 33, 468-471, 1972.
113. Shimoji, K.; Matsuki, M.; Shimizu, H.: Wave-form characteristics and spatial distribution of evoked spinal electrogram in man. J. Neurosurg. 46, 304-313, 1977.
114. Shimoji, K.; Maruyama, Y.; Shimizu, H.; Fujioka, H.; Taga, K.: Spinal cord monitoring: A review of current techniques and knowledge. In: J. Schramm; S.J. Jones (eds.): Spinal Cord Monitoring. Springer-Verlag Berlin, Heidelberg, New York, Tokyo, 1985, pp. 16-28.
115. Shinomiya, K.; Furuya, K.; Yamaura, I.; Kamikozuru, M.; Sato, H.; Nakai, O.; Sato, R.; Kaneda, A.: Spinal cord monitoring of spinal cord function using evoked spinal cord potentials. In: S. Homma; T. Tamaki (eds.): Fundamentals And Clinical Application Of Spinal Cord Monitoring. Saikon Publ. Tokyo, 1984, pp. 161-173.
116. Shinomiya, K.I.; Furuya, K.; Yamaura, I.; Sato, H.; Kaneda, A.; Yokoyama, M.: Clinical study of cervical spondylotic myelopathy using evoked spinal cord potentials. In: J. Schramm; S.J. Jones (eds.): Spinal Cord Monitoring. Springer-Verlag Berlin Heidelberg, New York, Tokyo, 1985, pp. 290-301.
117. Small, M.; Matthews, W.B.: A method of calculating spinal cord transit time from potentials evoked by tibial nerve stimulation in normal subjects and in patients with spinal cord disease. Electroenceph. Clin. Neurophysiol. 59, 156-164, 1984.
118. Snooks, S.J.; Swash, M.: Motor conduction velocity in the human spinal cord: Slowed conduction in multiple sclerosis and radiation myelopathy. J. Neurol. Neurosurg. Psychiat. 48, 1135-1139, 1986.
119. Snyder, B.G.E.; Holliday, T.A.: Pathways of ascending evoked spinal cord potentials of dogs. Electroenceph. Clin. Neurophysiol. 58, 140-154, 1984.
120. Tamaki, T.; Tsuji, H.; Inoue, S.; Kobayashi, H.: The prevention of iatrogenic spinal cord injury utilizing the evoked spinal cord potential. Int. Orthop. 4, 313-317, 1981.
121. Tamaki, T.; Noguchi, T.; Takano, H.; Tsuji, H.; Nakagawa, T.; Imai, K.; Inoue, S.: Spinal cord monitoring as a clinical utilization of the spinal evoked potential. Clin. Orthop. 184, 58-64, 1984.
122. Tamaki, T.; Takano, H.; Takakuwa, K.; Tsuji, H.; Nakagawa, T.; Imai, K.; Inoue, S.: An assessment of the use of spinal cord evoked potentials in prognosis estimation of injured spinal cord. In: J. Schramm; S.J. Jones (eds.): Spinal Cord Monitoring. Springer-Verlag Berlin Heidelberg, New York, Tokyo, 1985, pp. 221-226.
123. Taub, T.; Bishop, P.O.: The Spinocervical tract: Dorsal column linkage, conduction velocity, primary afferent spectrum. Exp. Neurol. 13, 1-21, 1965.
124. Toyoda, A.; Kanda, K.: Origins of spinal cord potentials evoked by stimulation of the cat spinal cord. In: S. Homma; T. Tamaki (eds.): Fundamentals And Clinical Application Of Spinal Cord Monitoring. Saikon Publ. Tokyo, 1984, pp. 99-111.
125. Tsuji, S.; Luders, H.; Lesser, R.P.; Dinner, D.S.; Klem, G.: Subcortical and cortical somatosensory potentials evoked by posterior tibial nerve stimulation: Normative values. Electroenceph. Clin. Neurophysiol. 59, 214-228, 1984(a).
126. Tsuji, S.; Luders, H.; Dinner, D.S.; Lesser, R.P.; Klem, G.: Effect of stimulus intensity of subcortical and cortical somatosensory evoked potentials by posterior tibial nerve stimulation. Electroenceph. Clin. Neurophysiol. 59, 229-237, 1984(b).
127. Tsuyama, N.; Tsuzuki, N.; Kurokawa, T.; Imai, T.: Clinical application of spinal cord action potential measurement. Int. Orthop. 2, 39-46, 1978.
128. Wall, P.D.: Presynaptic control of impulses at the first central synapse in the cutaneous pathway: Progress in Brain Research, Vol. 12. In: J.C. Eccles; J.P. Schade (eds.): Physiology Of Spinal Neurons. Elsevier, Amsterdam, London, New York, 1964, pp. 92-115.

129. Whittle, I.R.; Johnston, I.H., Besser, M.: Spinal cord monitoring during surgery by direct record-
     ing of somatosensory evoked potentials. Technical note. J. Neurosurg. 60, 440-443, 1984.
130. Whittle, I.R.; Johnston, I.H.; Besser, M.: Recording of spinal somatosensory evoked potentials for
     intraoperative spinal cord monitoring. J. Neurosurg. 64, 601-612, 1986.
131. Willis, W.D.: Evoked spinal cord potentials in the cat and monkey: Use in the analysis of spinal
     cord function. In: S. Homma; T. Tamaki (eds.): Fundamentals And Clinical Application Of Spinal
     Cord Monitoring. Saikon Publ. Tokyo, 1984, pp. 3-19.
132. Willis, W.D.; Weir, M.A.; Skinner, R.D.; Bryan, R.N.: Differential distribution of spinal cord field
     potentials. Exp. Brain Res. 17, 169-176, 1973.
133. Yamada, T.; Machida, M.; Kimura, J.: Far field somatosensory evoked potentials after stimulation
     of the tibial nerve. Neurology (Minneap.) 32, 1151-1158, 1982.

Corticospinal Direct Response to Stimulation of the Exposed Motor Cortex in Humans

Y. Katayama;[*] T. Tsubokawa; S. Maejima; T. Hirayama; T. Yamamoto

Introduction

In experimental animals, it has been demonstrated that a corticospinal direct (D) response to stimulation of the motor cortex can be recorded from the lateral column of the spinal cord or the spinal epidural space (7, 9, 15). Recently, it has been reported that identical responses can be recorded in humans with transcranial brain stimulation (1, 10, 11). Such a technique would be of great clinical value for monitoring pyramidal tract (PT) functions during intracranial and intraspinal surgery. However, it may not be certain whether spinal cord responses to transcranial brain stimulation really represent impulses mediated by PT neurons since, to the best of our knowledge, no detailed data regarding corticospinal D response to stimulation of the exposed motor cortex of humans is currently available for comparison (10). We have previously compared corticospinal D response to stimulation of exposed motor cortex and spinal cord responses to transcranial brain stimulation in the cat, and we have found that spinal cord responses to transcranial brain stimulation are easily confused with responses other than corticospinal D response (7). However, because of the tremendous difference in brain volume between the cat and humans, any conclusion must await studies in humans. In our neurosurgery department, we have five years' experience recording corticospinal D response to stimulation of the motor cortex while exposed for intracranial surgery. We made these recordings for the purpose of identifying the motor cortex and protecting it from surgical damage. The data obtained appears to provide useful information for comparison of corticospinal D response and spinal cord responses to transcranial brain stimulation in humans. We, therefore, summarize herein our experience of recordings of human corticospinal D response during intracranial surgery.

Methods

Data analyzed in the present study were those from 20 patients who showed no motor deficits before surgery, indicating that PT functions were intact. A pair of flexible, platinum wire electrodes insulated except their tip (Medtronic Co. M-8483) were inserted into the epidural space of the cervical and/or upper thoracic levels before surgery. The subjects were placed in the lateral position and 18-gauge Touhy needles were inserted into the epidural space at the upper thoracic level. The electrode was inserted with a stylette through Touhy needles and advanced to the appropriate position under x-ray control. The Touhy needles were then removed and the electrodes were fixed by adhesive tape on the skin (10, 22, 23). The response to stimulation of the

* Department of Neurological Surgery, Nihon University School of Medicine, 30 Oyaguchikami-Machi, Itabashi-Ku, Tokyo 173, Japan

motor cortex exposed during surgery was recorded monopolarly from the electrodes, with a reference electrode placed at the paravertebral muscles. The scalp of the forehead was grounded. During intracranial surgery, the motor cortex and other cortical areas were directly stimulated by multicontact plate electrodes (Medtronic Co. M-3586). The response evoked by stimulation of various cortical areas was investigated by delivering stimuli to various pairs of electrodes. The 4 electrodes were 5mm in diameter and spaced 5mm apart. Thus, interpolar distance for stimulation varied from 10 to 30mm. Stimuli were applied by monophasic square wave pulses of 0.2-0.5msec duration delivered at 4Hz. The delivery of stimuli were triggered by electrocardiograms. Signals from electrodes were fed to an amplifier with bandpass range of 5Hz to 5KHz and averaged for 32-128 sweeps with a signal processor.

Results

The response recorded from the spinal epidural space to stimulation of the exposed motor cortex (Fig. 1A) invariably had a large negative wave (N1) preceded by a small positive wave (P1). These two waves covaried on the basis of the amplitude, the latency and the refractory period. The P1-N1 response followed double pulse stimulation with interstimulus intervals more than 500Hz (Fig. 1A). Slight facilitatory effects occurred by antecedent conditioning stimulation with interstimulus intervals longer than the refractory period (Fig. 1A). As the recording site was moved caudally, the latency of the P1-N1 response increased, the amplitude decreased and the duration increased (Fig. 1B). The difference in peak latency of the N1 wave over the known distance between two recording sites was used to calculate the conduction velocity (Fig. 1B). The calculated conduction velocity was within a range between 50 and 75m/s. When each recording site was plotted with roentgenographically estimated stimulation-recording distance on the abscissa and the latency on ordinate, the regression line which reflects the conduction velocity, crossed the ordinate near zero, i.e., the stimulation site (Fig. 1B).

The lowest intensity for evoking the P1-N1 response was as low as 2mA, when stimulation electrodes were appropriately placed on the motor cortex. As the stimulation intensity was raised, the amplitude increased and then saturated at certain levels of intensity. It was also found that, as the stimulation intensity was raised, the latency decreased, without changes in conduction velocity between the two recording sites, and then became constant with stimulation intensity over certain levels.

At the same level of stimulation intensity, the longer interpolar distance, within a range between 10 and 30mm, produced much larger P1-N1 response. This was observed even if both poles of the stimulation electrodes were appropriately placed on the motor cortex. In such cases, the amplitude remained approximately the same after reversing the polarity of stimulation. Thus, when both poles were placed appropriately on the motor cortex, stimulation artifacts could be removed with a signal averaging technique by applying anodal currents alternately to one and then the other pole of stimulation electrodes. Such a technique, however, was rarely necessary when patients were appropriately grounded. When only one pole of stimulation electrode was placed on the motor cortex, the amplitude was slightly larger with the anode than with a cathode applied to the motor cortex.

Even with the interpolar distance of 30mm, the P1-N1 response was recorded only when at least one pole was placed precisely on or very close to the motor strip. With the interpolar distance of 10mm, the P1-N1 response was obtained only from stimulation of restricted areas of the motor cortex. The focus giving the response with an interpolar distance of 10mm was found in the motor cortex within areas 0-8cm lateral to the midline, i.e., hand, trunk and thigh areas. Stimulation of these areas with two poles

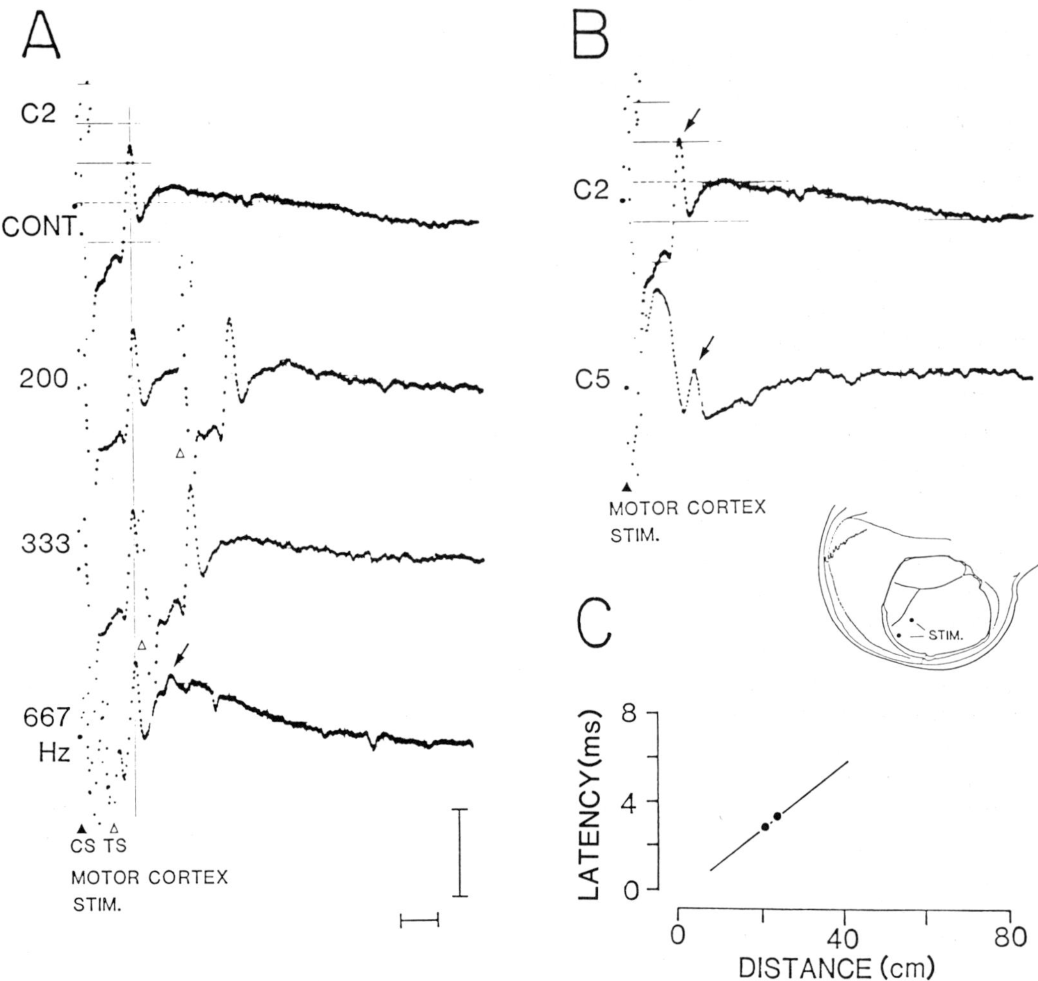

Fig. 1. A: Representative examples of responses recorded from the spinal epidural space (C2) to stimulation of the exposed motor cortex. Upward deflection corresponds to negativity. The effects of double pulse stimulation with various interstimulus intervals (200-667Hz) are also shown. Conditioning stimulus (CS); test stimulation (TS). B: Representative examples of responses recorded from different levels of the spinal epidural space (C2 and C5). Arrows indicate N1 wave. Note that the amplitude of the P1-N1 response decreases and the duration increases as the recording site is move caudally. Two recording sites are also plotted with the stimulation-recording distance on abscissa and the latency on ordinate. Calibrations, 2ms; 10mV.

with the interpolar distance of 20 or 30mm places in parallel to the gyrus always evoked large P1-N1 response. Stimulation of the medial motor cortex was not examined.

The P1-N1 response was recorded independently of the depth of anesthesia, and was relatively resistant to the physical condition of the motor cortex. While three negative waves (N2, N3, N4) were sometimes seen following the initial P1-N1 response, these later responses were vulnerable to changes in physical condition of the cortex. The threshold of these responses was higher than the P1-N1 response. In cases in which stimulation electrodes were placed epidurally over the motor cortex and recordings were made without anesthesia (see next chapter), the initial P1-N1 response, and the later N2, N3 and N4 responses were seen. In such recordings, large muscle responses recorded from the reference electrode placed in the paravertebral muscles were also observed. The progressive administration of thiopental (50-250mg, i.v.) clearly attenuated the N2, N3 and N4 responses, and the muscle responses. In contrast, the initial P1-N1 response showed only a slight depression in amplitude.

Discussion

Since the initial P1 and N1 waves recorded in the present study covaried on the basis of the amplitude, the latency and the refractory period, these two waves most likely represent a single synchronous impulse approaching toward and arriving at the recording site, respectively. Five lines of evidence indicate that the P1-N1 response is mediated by a fiber tract without intervening synapses. First, the regression line between the stimulation-recording distance and the latency, which reflects the conduction velocity, has been shown to pass near the stimulation site, suggesting no synaptic delay in latency. Second, it has been demonstrated that the P1-N1 response is capable of following double pulse stimulation with interstimulus intervals of more than 500Hz. Third, only slight facilitatory effects of double pulse stimulation have been seen. Finally the P1-N1 response, in contrast to the later N2, N3 and N4 responses, has been shown to be clearly resistant to surgical doses of anesthetics. These observations are all consistent with the interpretation that the P1-N1 response is not a synaptically mediated potential. The large and synchronous shape of the response is also a characteristic expected from a volley mediated by a simple fiber tract without intervening synapses. The observation that the P1-N1 response is limited to stimulation of the motor cortex indicates that a fiber tract mediating this response originates only in the motor cortex. No fiber tracts other than those of PT neurons are known to have such a direct projection from the motor cortex to the spinal cord. Thus, the present study appears to have demonstrated that a corticospinal D response identical to the response described for experimental animals (7, 9, 15) can be recorded in humans as well.

There are many similarities between the P1-N1 response recorded in the present study and corticospinal D response reported for experimental animals (7, 9, 15). The conduction velocity of the P1-N1 response, i.e., 50-75m/s, is similar to that obtained for corticospinal D response or PT neurons in the cat and monkey (2, 7, 8, 12, 16). This range of conduction velocity is consistent with the conduction velocity for fast PT neurons (4, 21). The later N2, N3 and N4 responses recorded in the present study resemble indirect (I) responses reported for experimental animals (7, 9, 15). The I responses have been shown to be due to repetitive activation of PT neurons in the monkey (9).

The observation that the corticospinal D response in humans is limited to the motor cortex provides us the means to identify the motor cortex during intracranial surgery. The focus giving corticospinal D response with an interpolar distance of 10mm for stimulation is restricted to the hand, trunk and thigh areas of the motor cortex. Unlike motor evoked potentials recorded from muscles, the corticospinal D response is resistant to surgical doses of anesthetics and is unaffected by muscle relaxants. The use of this technique during intracranial surgery would, therefore, help to reduce neurological complications resulting from unnecessary damage to the motor cortex (24).

The amplitude of corticospinal D response decreases as the recording sites are moved caudally. This may be in part because the volley becomes less synchronous as it travels down the cord (8). This interpretation is supported by the fact that the duration of the corticospinal D response increases as the recording sites are moved caudally. The decrease of the amplitude, however, is also in part due to a reduction in density of PT neurons. For example, when the hand area of the motor cortex is stimulated, the majority of PT neurons activated are absent in the thoracic level of the spinal cord. The present study has demonstrated that, if both poles for stimulation are placed appropriately on the motor cortex, a longer interpolar distance produces a larger corticospinal D response even with the same stimulation intensity. This may be due to recruitment of larger number of axons with longer interpolar distances. Thus, stimulation with longer interpolar distances, up to 30mm, may activate PT neurons projecting to different levels of the spinal cord simultaneously. The use of longer interpolar distance is, therefore, of value for intraoperative monitoring of PT in cases where it is not known which portion of PT will be damaged during surgery.

Physiological characteristics of the corticospinal D response demonstrated in the present study are largely in agreement with those recently reported for spinal cord responses to transcranial brain stimulation in humans (1, 10, 11). While one group of investigators has reported a conduction velocity of 70-100m/s for the PT (10) using transcranial brain stimulation as well as stimulation of the lateral spinal tracts, other groups of investigators have reported a conduction velocity of 50-74m/s (1). The latter values are similar to those obtained in our study. Our own observation regarding conduction velocity of spinal cord responses to transcranial brain stimulation are consistent with the conduction velocity of corticospinal D response investigated in the present study. Activities in the PT mediating corticospinal D response may also be responsible for the muscle responses recorded from reference electrodes placed in the paravertebral muscles. Such responses have also been shown to be evoked by single transcranial brain stimulation (5, 13, 14, 19). The intraspinal conduction velocity estimated from muscle responses from transcranial brain stimulation and transcutaneous cord stimulation is also in agreement with the value obtained in the present study (13, 19). Thus, we believe that the present study has provided evidence supporting the proposition that spinal cord responses to transcranial brain stimulation may be identical to corticospinal D response. It should, however, be mentioned that transcranial brain stimulation needs very high stimulation intensity which may unavoidably activate brain areas other than the motor cortex.

References

1. Boyd, S.G.; Rothwell, J.C.; Cowan, J.M.A.; Webb, P.J.; Morley, T.; Asselman, P.; Marsden, C.D.: A method of monitoring function in corticospinal pathways during scoliosis surgery with a note on motor conduction velocities. J. Neurol. Neurosurg. Psychiat. 49: 251-257, 1986.
2. Brookhart, J.M.; Morris, R.E.: Antidromic potential recordings from the bulbar pyramid of the cat. J. Neurosurg. 11: 387-398, 1948.
3. Chang, H.T.: Dendritic potential of cortical neurons produced by direct electrical stimulation of the cerebral cortex. J. Neurophysiol. 14: 1-21, 1951.
4. Deschenes, L; Labelle, A.; Landry, P.: Morphological characteristics of slow and fast pyramidal tract cells in the cat. Brain Res. 178: 251-274, 1979.
5. Hassan, N.F.; Rossini, P.M.; Cracco, R.Q.; Cracco, J.B.: Unexposed motor cortex activation by low voltage stimuli. In: C. Morocutti; P.A. Rizzo (eds.): Evoked Potentials: Neurophysiological And Clinical Aspects. Amsterdam, Elsevier, pp. 107-113, 1985.
6. Hern, J.E.C.; Landgren, S.; Phillips, C.G.; Porter, R.: Selective excitation of corticofugal neurones by surface stimulation of the baboon's motor cortex. J. Physiol. (Lond.), 161: 73-99, 1962.
7. Katayama, Y.; Tsubokawa, T.; Sugitani, M.; Maejima, S.; Hirayama, T.; Yamamoto, T.: Assessment of spinal cord injury with multimodally evoked spinal cord potentials. Part 1. Localization of lesions in experimental spinal cord injury. Neuro-Orthopedic 1: 130-141, 1986.
8. Lance, J.W.: Pyramidal tract in spinal cord of cat. J. Neurosphysiol. 17: 253-270, 1954.
9. Landau, W.M.; Bishop, G.H.; Clare, M.H.: Site of excitation in stimulation of the motor cortex. J. Neurophysiol. 29: 1206-1222, 1966.
10. Levy, W.J., Jr.; York, D.H.: Motor tract evoked potentials. In: R.H. Nodar; C. Barber (eds.): Evoked Potentials II. Butterworth, Boston, pp. 363-372, 1984.
11. Levy, W.J.; York, D.H.; McCaffrey, M.; Tannzer, F.: Motor evoked potentials from transcranial stimulation of the motor cortex in humans. Neurosurg. 15: 287-301, 1984.
12. Lloyd, D.P.C.: The spinal mechanism of the pyramidal system in cats. J. Neurophysiol. 4: 525-546, 1941.
13. Marsden, C.D.; Merton, P.A.; Morton, H.B.: Percutaneous stimulation of spinal cord and brain: Pyramidal tract conduction velocities in man. J. Physiol. 328: 61, 1982.
14. Merton, P.A.; Morton H.B.: Stimulation of the cerebral cortex in the intact human subject. Nature 285: 227, 1980.

15. Patton, H.D., Amassian, V.E.: Single and multiple unit analysis of cortical stage of pyramidal tract activation. J. Neurophysiol. 17: 345-363, 1954.
16. Philips, C.G.; Porter, R.: Corticospinal Neurons. London: Academic press, p. 268, 1977.
17. Purpura, D.P.; McMurty, J.G.: Intracellular activities and evoked potential changes during polarization of motor cortex. J. Neurophysiol. 28: 166-185, 1965.
18. Rank, J.B., Jr.: Which elements are excited in electrical stimulation of mammalian central nervous system. A review. Brain Res. 98: 417-440, 1975.
19. Rossini, P.M.; Marciani, M.G.; Caramia, M.; Hassan, N.F.; Cracco, R.Q.: Transcutaneous stimulation of motor cerebral cortex and spine: Noninvasive evaluation of central efferent transmission in normal subjects and patients with multiple sclerosis. In: Alan R. Liss, Inc. (ed.): Evoked Potentials, pp. 76-84, 1986.
20. Shimoji, K.; Higashi, H.; Knao, T.: Epidural recording of spinal electrogram in man. Electroenceph. Clin. Neurophysiol. 30: 236-239, 1971.
21. Takahashi, K.: Slow and fast groups of pyramidal tract cells and their respective membrane potentials. J. Physiol. 28: 908-924, 1965.
22. Tsubokawa, T.: Clinical value of multimodality spinal cord evoked potentials for prognosis of spinal cord injury. In press.
23. Tsubokawa, T.; Katayama, Y.; Maejima, S.; Hirayama, T.; Yamamoto, T.: Assessment of spinal cord injury with multimodality evoked spinal cord potentials. Part 2. Correlation with neurological outcome in clinical spinal cord injury. Neuro-Orthoped. In press.
24. Tsubokawa, T.; Yamamoto, T.; Hirayama, T.; Maejima, S.; Katayama, Y.: Clinical application of corticospinal evoked potentials as a monitor of pyramidal function. Nihon Univ. J. Med. 28: 27-37, 1986.

Usefulness of Motor Evoked Potentials (Pyramidal D-Response) for Assessment of Spinal Cord Injury

T. Tsubokawa;[*] Y. Katayama; S. Maejima; T. Hirayama; T. Yamamoto

Introduction

In spinal cord injury, recognition of the severity and anatomical extent of damage provides important information for clinical management and for early prediction of outcome. In order to obtain such information, the direct response of the pyramidal tract to activation of the motor cortex, which can be recorded from the epidural space of the spinal cord in humans as well as experimental animals, is recorded in combination with recordings of conventional spino-spinal evoked potentials and periphero-spinal evoked potentials in acute periods of cervical cord injury.

In the present study, the clinical value of recording pyramidal direct (D) responses during acute periods of spinal cord injury is studied, with special reference to a correlation between pyramidal D response, conventional spino-spinal evoked potential, periphero-spinal evoked potentials and the outcome of motor functions.

Methods

Nineteen patients with acute cervical spinal cord injury were studied. We recorded pyramidal motor potentials, spino-spinal cord and periphero-spinal evoked potentials. Ten cases had complete paraplegia and nine cases had partial paraplegia upon arrival at our clinic. During x-ray examination, just after admission, a pair of flexible platinum wire electrodes insulated except their tip (Medtronic Co. M-8483) were inserted into the epidural space at locations both cranial

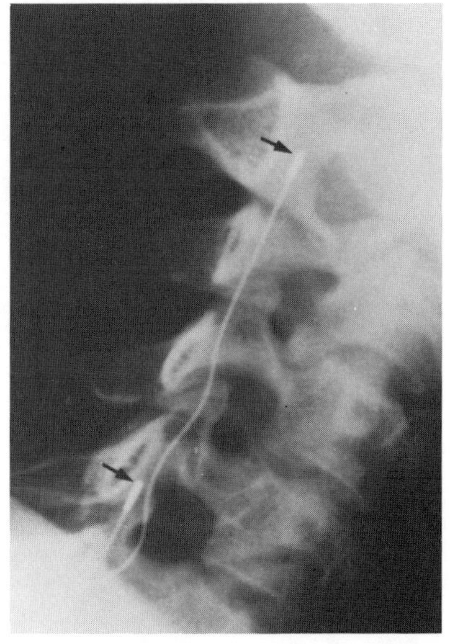

Fig. 1. Epidural flexible platinum electrode inserted into the epidural space cranially and caudally to the level of injured cord.

and (Medtronic Co. M-8483) were inserted into the epidural space at locations both cranial

* Department of Neurological Surgery, Nihon University School of Medicine, 30-1 Oyaguchikami-Machi, Itabashi-Ku, Tokyo 173, Japan

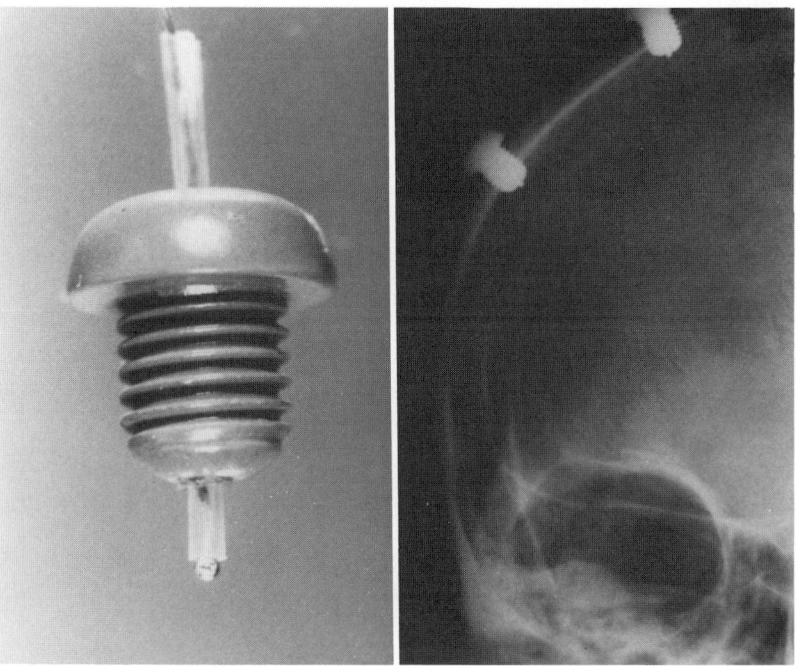

Fig. 2. Stimulating electrode and a pair of electrodes applied on the skull as shown in craniogram.

and caudal to the level of injured cord by using a Touhy needle (Fig. 1). Spino-spinal evoked potentials were recorded from one of those two electrodes in response to the stimulation of the other, with reference electrodes placed at the paravertebral muscles. Stimuli were applied by monophasic square wave pulses of 0.2-0.5msec duration. Spinal evoked potential was also recorded in response to median or ulnar nerve stimulation as periphero-spinal evoked potentials. The pyramidal D response was recorded from epidural electrodes which were placed rostral and caudal to the level of injury. For stimulating electrodes for the pyramidal motor potentials, a pair of small skull holes were drilled at the skull area overlying the motor cortex and screw electrodes were inserted so that their tips were in contact with the dura overlying the motor cortex (Fig. 1). In some cases, transcranial brain stimulation was employed using needle electrodes inserted into the pericranium at the area overlying the motor cortex. Stimuli were applied by monophasic square pulses of 0.2-0.5msec duration triggered by electrocardiograms.

Signals from the electrodes were fed to an amplifier with a bandpass range of 5Hz to 5KHz and averaged for 32-128 sweeps with a signal processor.

Recording of each evoked potential was repeated at predetermined intervals until the 4th day post-injury. During this period, monitorings of evoked potentials were also conducted when surgery for the cord injury was performed. The outcome of each patient was assessed at 6 months after the injury by using the neurological status evaluation method (Sunnybrook).

Table 1

Changes of the Multimodality Evoked Potentials in Patients with Complete Spinal Cord Injury.

Name	Age	Location of injury (complication)	Pretreatment Segmental SEP P_1	N_1	P_2	Pretreatment Conduction SEP P_1N_1	P_2N_2	Pretreatment Cortically evoked SEP (D-wave) CDSEP	Pretreatment Trans-cranially evoked SEP (D-wave) TDSEP	Treatment in acute stage	Just after Segmental SEP P_1	N_1	P_2	Just after Conduction SEP P_1N_1	P_2N_2	Just after Cortically evoked SEP (D-wave) CDSEP	Just after Trans-cranially evoked SEP (D-wave) TDSEP	Neurological improvement or deterioration — Neurological score (pre → post)	Motor power (post)	Sensory (post)
K.M.	11	C5.6	−	−	−	−	−		−	laminectomy & cooling	−	−	−	−	−		−	1→1	0	1
M.T.	31	C5.6.7	−	−	−	−	−		−	laminectomy, cooling, vertectomy & fusion	−	−	−	−	−		−	1→2	1	1
K.O.	46	C3.4(Head injury)	−	−	−	−	−		−	laminectomy, cooling verterectomy & fusion	+	−	−	−		−	−	1→1	0	0
A.O.	22	C5(Head injury)	−	−	−	−	−	−	−	laminectomy, cooling, vertectomy & fusion	−	−	−	−	−	−	−	1→1	0	1
S.K.	66	C4.5.6	−	−	−	−	−	−	−	Halo-device	+	±	−	−	±	−		1→2	0	1
M.K.	22	C4.5.6	±	−	−	−	−		−	Halo-device	−	−	−	−	−	−	−	1→1	0	1
Y.S.	45	C4.5.6	−	−	−	±	±		±	laminectomy, cooling, verterectomy & fusion	+	±	±	±	+		±	1→3	3	1
M.N.	26	C4	+	±	±	+	±		±	laminectomy, cooling, verterectomy & fusion	+	+	±	−	±		+	1→6	6	2
K.H.	29	C5(Head injury)	±	−	−	±	−	±		laminectomy, cooling, verterectomy & fusion	+	+	+	+	±	±		1→5	5	1
S.Y.	45	C5.6	+	±	−	+	±	±		laminectomy, cooling, verterectomy & fusion	+	+	±	−	±	+		1→7	5	1

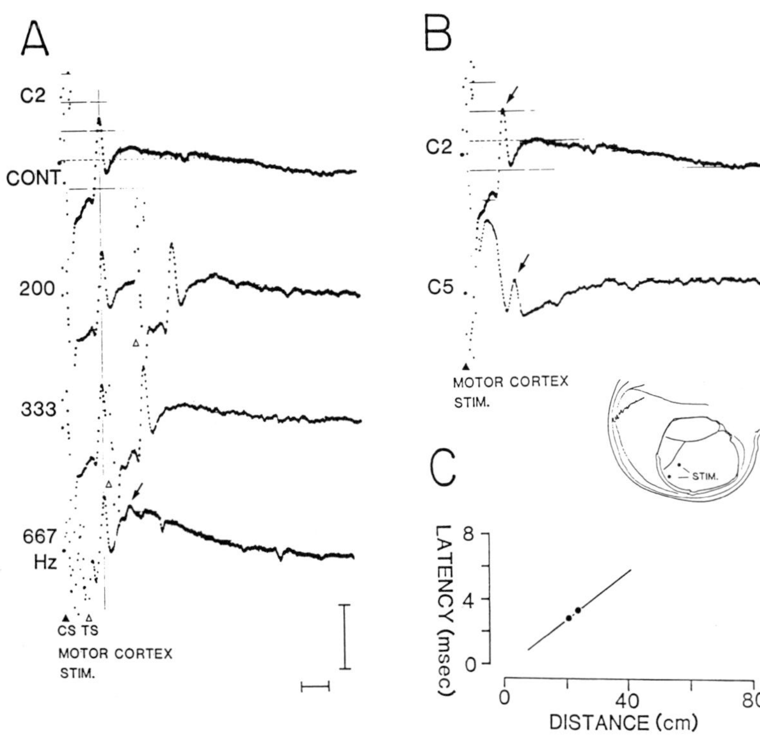

Fig. 3. Pyramidal motor potentials recorded at the epidural space at the level of C2 and C3. In this case, the motor cortex was stimulated under direct vision during cranial surgery.
A. The effects of the potential by high frequency stimulation.
B. The calculation of conduction velocity of latency difference of the pyramidal potentials recorded at different two points.
C. The relationship between distance between two recording sites and latency.

Results

1. Normal wave of spinal cord evoked potentials: Normal ascending spino-spinal potential had two negative waves (N1 and N2). Conduction velocity for N1 (via the dorsolateral column) 70-90m/sec and N2 (via the dorsal column) conduction velocity showed wide variations. Normal descending spino-spinal potential showed characteristics essentially identical to the ascending spino-spinal potentials, except N2 conduction velocity.

Pyramidal motor potentials had an initial negative wave followed by polyphasic waves. While N1 was resistant to intravenous administration of thiopental, following polyphasic waves were clearly diminished in amplitude by thiopental. N1 followed repetitive stimulation of more than 500Hz. Conduction velocity of N1, calculated from differences in latency between different recording sites was 50-69m/sec (Fig. 3).

2. Changes induced by cervical spinal cord injury.

A. Complete paraplegia on admission:

Ten were neurologically diagnosed as having a complete cord injury with complete paraplegia, although this was not certain in three patients because they were unconscious from associated head injuries. In this group, laminectomy and local cooling was immediately performed in eight patients and vertebrectomy and fusion were undertaken in seven cases within two weeks after local cooling. Two other patients were conservatively treated with application of a Halo device (Table 1).

Among the 10 neurologically complete cord injury cases, no evoked potentials could be recognized in five (Table 1, Fig. 4, cases 1-5); some responses in spino-spinal evoked

Table 2

Changes of the Multimodality Evoked Potentials in Patients with Incomplete Spinal Cord Injury.

Pretreatment and treatment

Name	Age	Location of injury (complication)	Segmental SEP P_1	N_1	P_2	Conduction SEP P_1N_1	P_2N_2	Cortically evoked SEP (D-wave) CDSEP	Transcranially evoked SEP (D-wave) TDSEP	Treatment in acute stage
T.O.	28	C4.5	+	+	+	−	+	−	−	verterectomy.fusion&cooling
K.N.	19	C4.5	+	+	+	−	+	−		Halo-device
S.T.	48	C5.6(spondylosis)	+	+	−	±	±		−	Discotomy&fusion
O.Y.	18	C5.6	+	+	−	±	±		±	Halo-device
K.Y.	20	C5.6	+	±	−	+	±		±	Halo-device
M.M.	34	C7	+	±	−	−	±		+	verterectomy.fusion, cooling&Halo-device
S.N.	30	C3	+	+	−	±	−	rt-stim.(+) lt-stim.(−)		Halo-device
M.S.	54	C5.6.7	+	−	−	±	−	rt-stim.(−) lt-stim.(+)		laminectomy.cooling &anterior fusion
T.T.	55	C4.5(spondylosis) −	+	−	−	+	−		+	removal of fracture fragment

Just after treatment in the acute stage, and neurological improvement or deterioration

Name	Segmental SEP P_1	N_1	P_2	Conduction SEP P_1N_1	P_2N_2	Cortically evoked SEP (D-wave) CDSEP	Transcranially evoked SEP (D-wave) TDSEP	Neurological score pre post	Motor power (post)	Sensory (post)
T.O.	+	+	+	−	+	−	−	2−5	3	2−1
K.N.	+	+	+	−	+	−	−	4−5	3	3
S.T.	+	+	+	+	+		−	4−5	3	3
O.Y.	+	+	±	±	+		+	4−7	6 hand skill	2
K.Y.	+	±	−	±	±		±	4−7	4	1
M.M.	+	+	−	±	+		+	4−7	5	1
S.N.	+	+	−	±	±	rt-stim.(+) lt-stim.(+)		rt:10−10 lt:4−8	rt-hand 7 lt-hand 4	1
M.S.	+	+	+	±	+	rt-stim.(−) lt-stim.(+)		rt:4−4 lt:6−10	rt-hand 5 lt-hand 6	1−2
T.T.	+	−	−	+	−		+	6−9	7	1

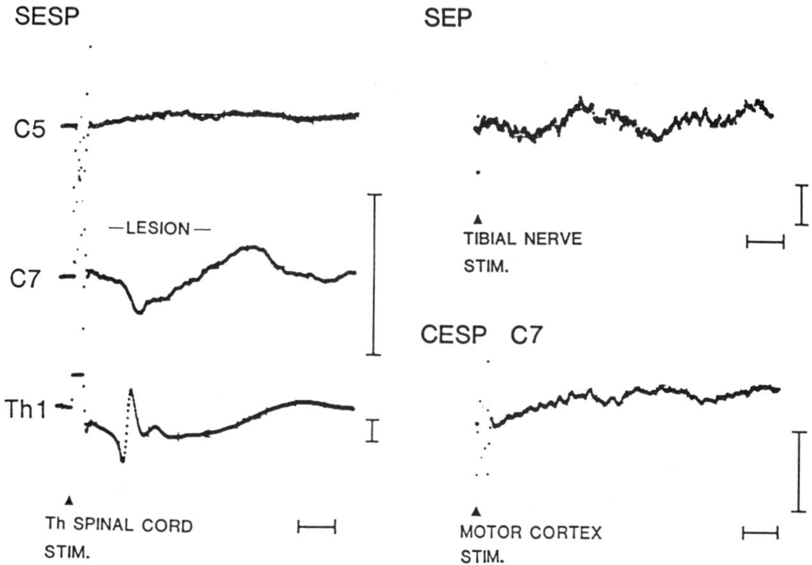

Fig. 4. Recording of multimodality evoked potentials (SESP: spino-spinal evoked responses; SEP: somatosensory evoked potentials; CESP: the pyramidal motor potentials) on complete cervical cord injury.

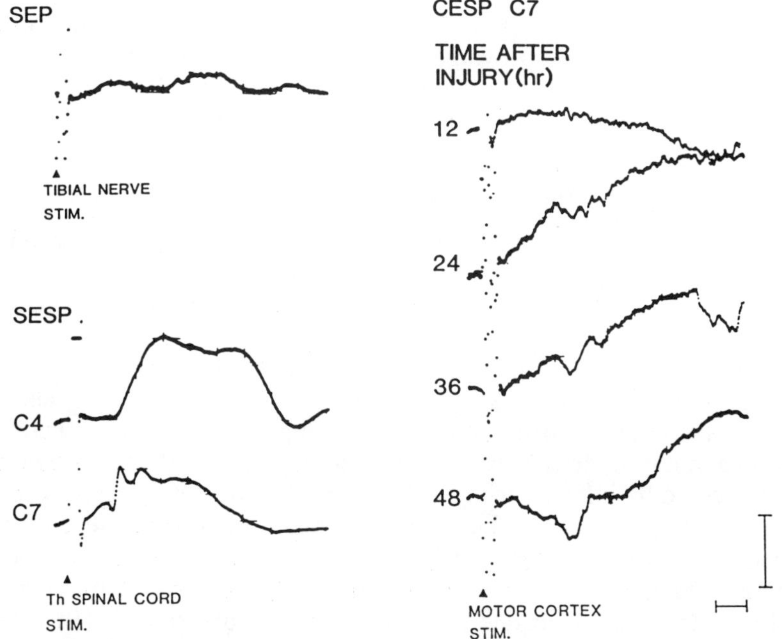

Fig. 5. The recording of multimodality evoked potentials in incomplete cervical cord injury. By using neurological examination, the extent of the lesion at C7 was difficult to evaluate. As shown on this figure, this lesion was central cord injury type, according to the analysis of the multimodality spinal cord evoked potentials. Pyramidal motor potentials were only traceably recorded at 12 hours after injury, but the potentials were recorded with 50% amplitude and a slight increase of latency comparing to normal wave form. Its amplitude was gradually increased at 48 hours after injury. Motor function was also improved correlating with the recovery of the potentials.

potentials having simple configuration or with low amplitude and longer latency were recorded in the other five (cases 6-10) without recording of any pyramidal motor potentials. Case 7, 8 and 9 had traceable response with low amplitude and increased latency in both the pyramidal motor potentials and spino-spinal evoked potentials in acute periods. The neurological improvement of motor functions in these three cases was remarkable and in two of the three, active movement of both hands against moderate resistance was seen.

B. Incomplete paraplegia at admission:

In the incomplete cervical cord lesion group, four patients were conservatively treated with application of a Halo device (Table 2), three with anterior surgery, one with laminectomy and local cooling and one with posterior surgery for removal of laminar fracture fragments. Four cases (Table 2, cases 1-4) were neurologically diagnosed as having a partial cord injury, but the location or extent of the lesion in the cord could not be identified by neurological examination. Cases 5 and 6 were diagnosed as having central cord injury by neurological examination, and cases 7 and 8 were Brown-Sequard type injury and case 9 was posterior cord injury caused by fracture of the lamina (Table 2).

In 4 cases (Table 2, cases 1-4) whose lesions were not neurologically diagnosed, traceable responses were seen in spino-spinal evoked potentials. The pyramidal motor responses could not be recorded even after treatments in these cases and no neurological improvement of the motor functions, except case 4, whose pyramidal motor potentials showed recovery in amplitude following intensive treatment during the acute period. In the central cord syndrome group (cases 5, 6), a pyramidal motor potential was recorded in case 6, and a very small but traceable response was recorded in case 5. Their motor functions were improved (Fig. 5). As for the patient with a posterior cord injury (case 9), pyramidal motor potentials were recorded with a low amplitude and simple configuration. Neurological improvement was remarkable with almost normal motor function.

Discussion

Based on evoked potential data, a group of patients who were completely paraplegic on admission appear to include two categories of lesions. Patients associated with the disappearance of all evoked potentials recorded and no recovery of these potentials (cases 1-5) may have had complete cord lesions. The outcome of this group of patients was paraplegia without exception. On the other hand, patients showing detectable activity in initial and/or later recordings of at least some evoked potentials (cases 6-10) may have had incomplete cord lesions even though these patients were completely paraplegic on admission. In this experimental study, we have demonstrated that the extent of lesions could be inferred from various combinations of changes in multimodality evoked spinal cord potentials. Theoretically, 4 types of incomplete cord lesions, i.e., anterior, central, posterior and Brown Sequard type cord syndromes, may be identified.

According to experimental observations previously reported, while pyramidal tract fibers are located in a relatively deep portion of the lateral funiculus (LF). Spino-spinal evoked potentials appears to be mediated by fibers located superficially within the LF and DF. The presence of traceable responses in spino-spinal evoked potentials in association with no responses in pyramidal motor potentials such as seen in case 6 may represent central cord syndrome. This pattern of changes has been demonstrated to occur also in experimentally produced central cord lesions in cats. Similar inference can be made also for a group of patients showing incomplete paraplegia on admission. The presence of responses in spino-spinal evoked potentials in association with the absence of responses in pyramidal motor potentials, such as seen in 4 cases whose lesions were not neurologically diagnosed (Table 2, cases 1-4), may indicate that such lesions actually belonged to a category of central cord syndrome.

Differences in the degree of changes and recovery in evoked potentials from case to case may be related to the severity of the syndrome. The outcome of motor function in cases in which no detectable activity was seen in pyramidal motor potentials is paraplegia. On the other hand, cases in which detectable activity was eventually seen in pyramidal motor potentials later became capable of walking, indicating that these cases had a moderate form of the central cord syndrome. This correlation between the data of pyramidal motor potentials and the outcome suggests that the degree of changes in evoked potentials is in fact related to the severity of damages in the pyramidal tract.

Because of their origin in the pyramidal tract fibers, it is not surprising that the recovery of activity in pyramidal motor response is intimately correlated to later recovery of motor functions. In contrast, recovery of activity in spino-spinal evoked response can occur without significant recovery of motor functions. In this respect, they provide more direct information. It must be emphasized that the recovery of activity in pyramidal motor potentials preceded clinical recovery of motor functions in all the cases investigated. This indicates that recordings of pyramidal motor potentials are useful for early prediction of outcome.

Summary

The present study has demonstrated that the recordings of spino-spinal evoked potentials are useful for determining the extent and the nature of spinal cord injury, and pyramidal motor potentials are necessary to determine the severity of the injury and thereby the outcome. What should be investigated in future studies is the period of sequential recordings necessary for providing data highly correlated with outcome. We carried out in the present study sequential recordings for 4 days after injury, which appears sufficient at present to provide data correlated with the outcome.

References

1. Yamamoto, T.; Tsubokawa, T.; Hirayama, T.; Maejima, S.; Katayama, Y.: Noninvasive monitoring of corticospinal evoked responses from the spinal epidural space in the cat. Nihon Univ. J. Med. 27: 17-21, 1985.
2. Tsubokawa, T.; Yamamoto, T.; Hirayama, T.; Maejima, S.; Katayama, Y.: Clinical application of corticospinal evoked potentials as a monitor of pyramidal function. Nihon Univ. J. Med. 28: 27-37, 1986.
3. Marsden, C.D.; Merton, P.A.; Morton, H.B.: Percutaneous stimulation of spinal cord and brain: Pyramidal tract conduction velocities a man. J. Physiol. 328: 61, 1982.
4. Merton, P.A.; Morton, H.B.: Stimulation of the cerebral cortex in the intact human subject. Nature 285: 227, 1980.
5. Rowed, D.W.: Value of somatosensory evoked potentials for prognosis in partial cord injury. In: Tator CH. (ed.) Early Management Of Acute Spinal Cord Injury. Raven Press, New York, 1982.
6. Katayama, Y.; Tsubokawa, T.; Sugitani, S.; Maejima, S.; Hirayama, T.; Yamamoto, T.: Assessment of spinal cord injury with multimodality evoked spinal cord potentials. Part 1. Localization of lesions in experimental spinal cord injury. Neuro-Orthopedics 1: 130-141, 1986.

Spinal Cord Potentials (SCPs) Produced by Descending Volleys in Man

K. Shimoji;[*] H. Fujioka; Y. Maruyama; H. Shimizu; T. Hokari; T. Takada

Introduction

Experiments in animals have demonstrated that dorsal root potentials (DRPs) or cord dorsum positive waves can be produced by stimulation of the cerebral cortex or brain stem (1-5). Direct stimulation of the dorsal surface of the cord has also been shown to evoke cord dorsum negative-positive complexes over the lumbosacral enlargement in experimental animals (6).

Previous studies in this laboratory have shown that epidural stimulation at the rostral segment of the cord produces a series of potential changes in the human lumbosacral enlargement, which is very similar in waveform to the segmental SCP (Fig. 1) (7-9). The potential (the descending SCP) consists of initial positive spikes followed by a sharp negative and slow positive wave. Based on the characteristics of the waveform and electrophysiological findings in each component of the descending SCP, we suggested that the spikes, a sharp negative and slow positive wave reflect the action potentials of the descending volleys through the cord, synchronized activity of the interneurons in the dorsal horn and primary afferent depolarization (PAD), respectively (9).

This paper surveys our results obtained so far on the descending SCPs in man.

Waveform characteristics of SCPs produced by descending volleys

Specimen records of Figure 1 (B) represent the SCPs recorded with the same electrodes as those of Fig. 1 (A) situated in the posterior epidural space (PES) at the lumbar enlargement in response to epidural stimulation of the cervical cord at graded stimulus strength. The similarity between the negative-positive complex evoked by descending volleys and the N1 through P2 complex elicited by the segmental nerve stimulation suggests that the origins of these slow descending N and P waves of SCPs are similar to those of the N1 and P2 waves of segmentally evoked SCPs. The segmental N1 and P2 waves in man show approximately the same characteristics as the slow negative and positive waves of the cord dorsum potential in animals, which are believed to be produced by the excitation of interneurons and PAD, respectively (10-14).

The descending N and P waves, however, reach their peaks 0.8-1.0ms faster than the N1 and P2 waves after the arrival of the volleys (measured from the positive peak of the initial spike). This may indicate that the descending volley is more synchronous than that in the peripheral nerve. In addition, the descending N wave always had one peak, while the segmentally evoked N1 wave was sometimes subdivided into two peaks, as reported in the cat (15), the monkey (16), and also in man (17). This suggests that the

* Department of Anesthesiology, Niigata University School of Medicine, 1-757 Asahi-Machi, Niigata 951, Japan

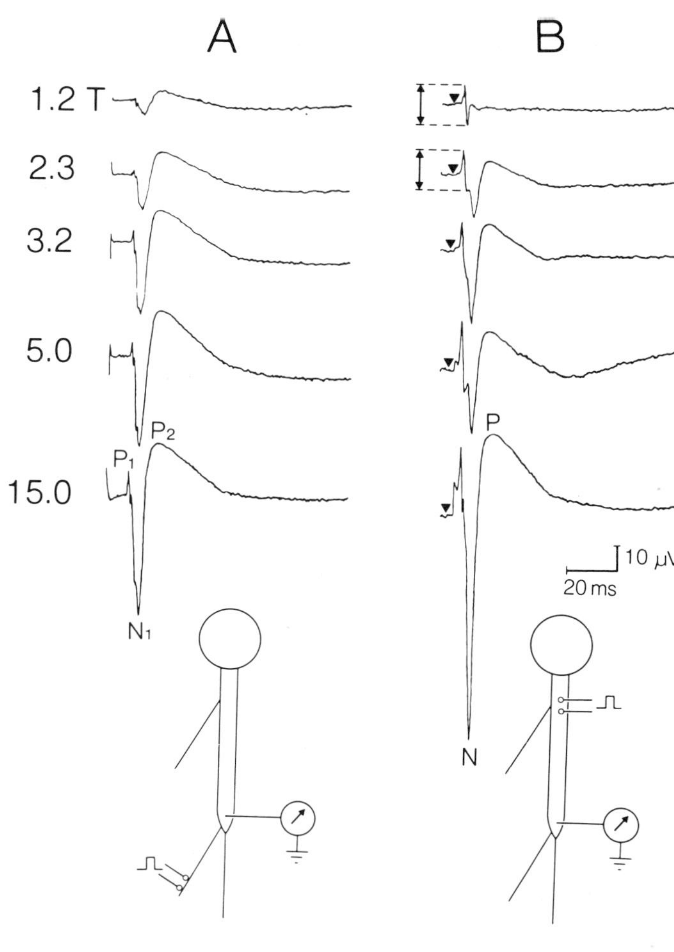

Fig. 1. Comparative demonstration of the SCPs, recorded from the posterior epidural space of the lumbar enlargement, evoked by segmental volley (A) and by descending volley (B) (9). The segmental volley was elicited by tibial nerve stimulation at the popliteal space while the descending volley was elicited by epidural stimulation of cervical spinal cord. Recording of the SCPs was carried out for spinal cord monitoring during spine surgery in a patient with scoliosis under general anesthesia (neuroleptanesthesia with complete muscle relaxation). Note that the initial spike potential produced by a descending volley (marked by scale) has the same amplitude as between 1.2 and 2.3 times the threshold strength (T). This may indicate that the descending impulse, which is slower than the maximum conducting volley along the cord, is responsible for production of the subsequent slow negative-positive complex.

dorsal column, which has no A fibers, conveys more homogenous volleys than the peripheral nerve.

Previous studies have shown that a single electrical pulse applied from the PES produces a tingling or a vibratory sensation, but no unpleasantness, in the wakeful subject (18-20). When the stimulus strength is increased, a muscle twitch is provoked in the segmental area but not beyond, even by an intense stimulation (15T). Therefore, it is more likely that the effective stimulating pulses from the PES spread to superficial tracts of the dorsolateral funiculus but not to the corticospinal pathway, under the present experimental conditions. Nevertheless, some effects on motor neuron pools in the lumbosacral enlargement cannot be entirely ruled out. Thus, the dorsal column and/or the dorsolateral funiculus might be the main tracts contributing to the provocation of the descending N and P waves in the PES of lumbosacral enlargement.

Effects of double shocks to rostral cord from the PES

When double shocks are applied to the cervical enlargement from the PES, the SCP produced in the lumbosacral enlargement by the second volley should be affected by the preceding volley. To test this, we calculated the amplitude of each component of the SCP produced by the second stimuli [Fig. 2 (A)] as compared with that produced by a simple summation [Fig. 2 (B)]. When the two stimulating volleys were separated by more than 200ms, both the descending N and P waves produced by the testing volley

Fig. 2. Effects of double shocks to the cervical PES on the descending N and P waves (9). (A). Specimen records of the descending SCPs evoked in the lumbosacral enlargement by the double shocks to the tibial nerve. (B). Simple summation of the two descending SCPs evoked separately at intervals shown on the left of A, which were recorded for references to A. (C). Time course of the conditioned changes in three components of the descending SCP.

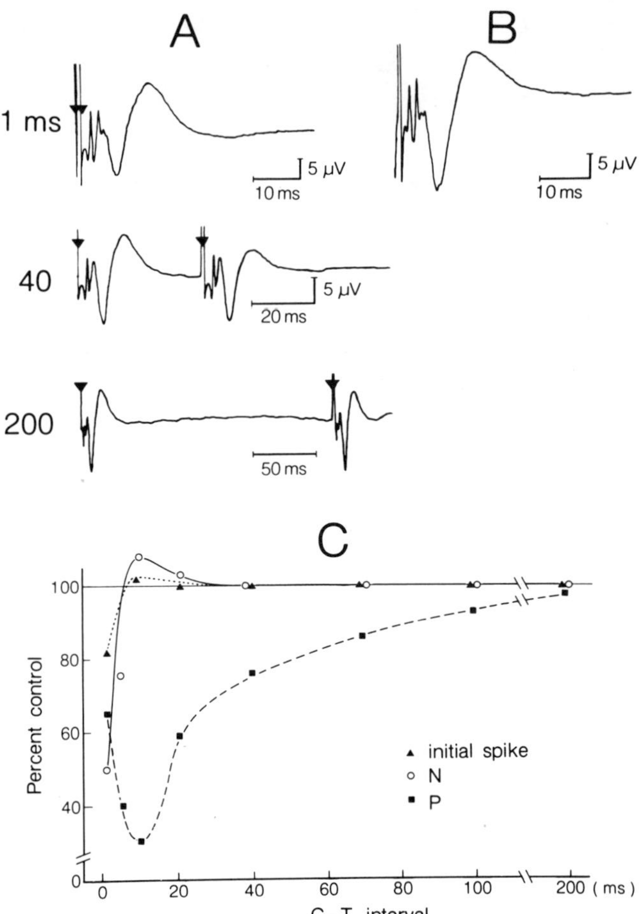

were minimally affected by the preceding (conditioning) volley. However, when the interval between two volleys was reduced, the amplitude of the slow P wave produced by the second volley was inhibited. The maximum inhibition of the P wave elicited by the testing volley was obtained at interstimulus intervals of 8-12ms [Fig. 2 (C)] in the four subjects tested. The N wave produced by the second volley decreased when the conditioning-testing interval was shortened to less than 10-70ms. The initial spike potential was barely affected when the two stimuli were separated by more than 5ms. The recovery curves of these potentials constructed by this conditioning-testing experiment [Fig. 2 (C)] also indicate that the initial spike potential of the SCP produced by the descending volley is a compound action potential of a certain spinal tract without interposing synapses along the tract. The recovery curves of the subsequent N and P waves indicates that both these two potentials are transsynaptic in origin, but possess different characteristics.

Interaction between the descending and segmental SCPs

When the segmental and descending volleys are delivered at an interval to produce two SCPs in the same time period, there is a partial occlusion of the amplitudes of the slow negative and positive waves (Fig. 3). The amount of occlusion becomes greater when the two stimuli are delivered at a higher intensity (9). Moreover, there are greater occlusions between the descending P and P2 waves than between the descend-

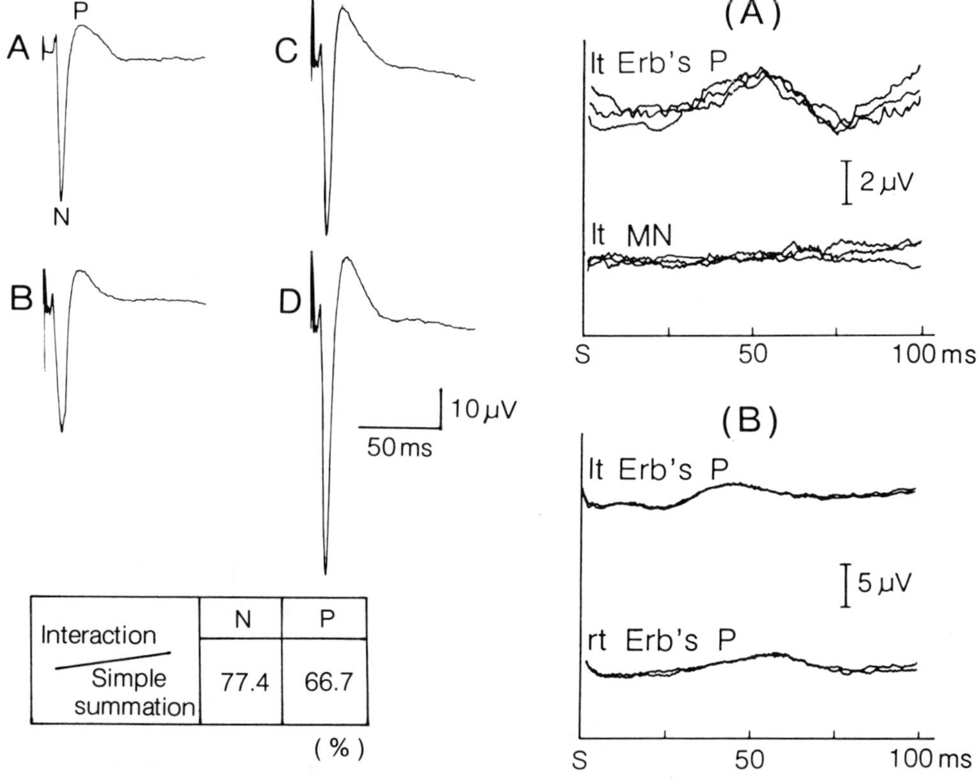

Fig. 3. Interaction between the descending and segmental SCPs. (A). The SCP produced in the PES at T12 vertebral level in response to epidural stimulation of the cervical cord at C6 vertebral level. (B). The SCP recorded with the same electrode in A in response to epidural stimulation of the cauda equina at L4 vertebral level (segmental SCP). (C). The interacted SCP evoked by simultaneous epidural stimulations of the cervical cord and the cauda equina. The timing of both volleys were adjusted so that the initial spike potentials of the two SCPs were produced in the same time period. (D). The SCP constructed by simple summation of the descending (A) and segmental (B) SCPs. Note that both negative and positive potentials are occluded by simultaneous stimulations of the cervical cord and cauda equina (C). The ratio of the interacted N and P potentials to those of simple summation is also demonstrated in the bottom table.

Fig. 4. The SCP produced in the lumbar enlargement in response to stimulation at the Erb's point. The Erb's point was stimulated with two subcutaneous Ag-AgCl needle electrodes (200 in diameter and 10mm in length), inserted 10mm apart. Three averaged responses (N = 50) were superimposed in (A). Note that a slow positive wave with long latency is clearly demonstrated by stimulation of the Erb's point but not by stimulation of the median nerve at the wrist. (B). The SCPs produced in the lumbar enlargement by stimulations of both sides in a patient with herpes zoster in the right T1 segment. Right Erb's point stimulation also produced a slow positive wave but its peak latency was prolonged in comparison to that by left stimulation, seemingly due to the nerve degeneration.

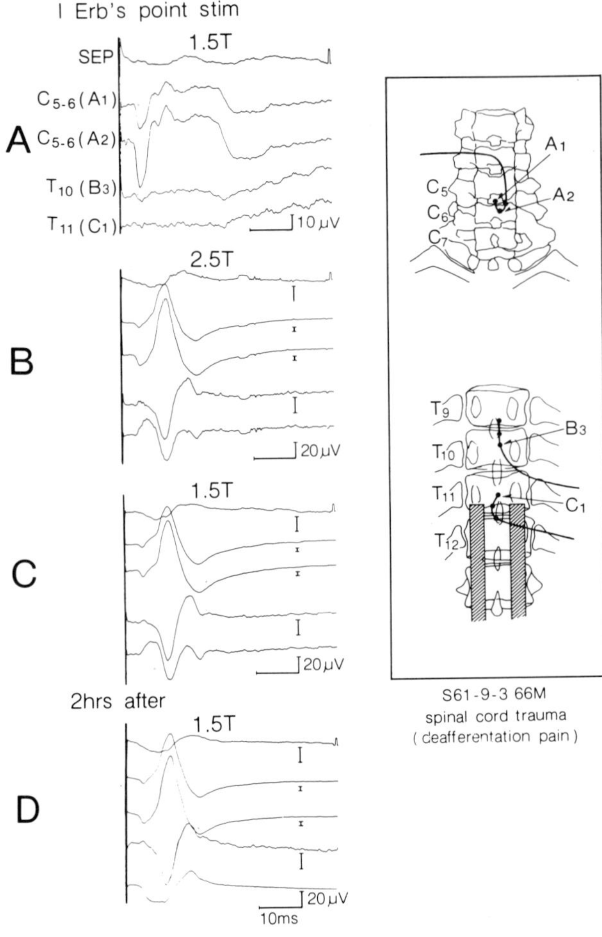

Fig. 5. The SCPs recorded from the lumbar enlargement (HSPs) and segmental SCPs recorded from C5 through C6 vertebral level in response to Erb's point stimulation in a patient with spinal cord lesion at the S4 spinal segment (incomplete transection). Note the voltage calibration (sensitivity is reduced in B through D due to enhancement of the amplitude). See the text for details.

ing N and N1 waves. For instance, occlusions of the negative and positive waves amounted to about 77 and 66%, respectively, of those produced by a simple summation (Fig. 3). There was a linear addition of the initial spike potentials without occlusion. Thus, a greater occlusion in the slow positive waves than in the negative waves may indicate that there are more common elements in the paths producing PADs or, perhaps, that a presynaptic terminal may only be depolarized to a limit set by the equilibrium potential of the transmitter generating PAD with both the segmental and descending volleys than in those producing the synchronous interneuronal activities (9).

The effect of dorsal column stimulation on interneurons in the dorsal horn has been reported by several investigators (6, 21, 22). Interneurons with a wide response range in the cat (23) or the hair-activated and low-threshold spinothalamic tract cells in the monkey (6) are excited and then inhibited by single dorsal column volleys. The inhibition has been shown to be very powerful, lasting for about 150ms, and sometimes the initial excitation is absent (24). The stimulus effects were previously thought to be due to a descending rather than an ascending volley, since there was no change in the effect when the cord was functionally interrupted rostrally by a cold block (21). The same argument might be applied in the present study as long as the strength of the epidural stimulation was within the range of the conditioning-test experiment (Fig. 3). However, a more intense stimulation can activate a feedback loop via supraspinal structures, as suggested by a graded stimulation experiment and the animal experiments in this laboratory (9). Even after the end of epidural stimulation, there is a prolonged inhibition of pain without any effect on the SCP (18), which suggests the activation of a certain inhibitory system mediated through supraspinal structures. Such an inhibition might not be reflected as a change in the segmental SCP evoked even by an intense stimulation of the peripheral nerves, probably because the segmental SCP reflects the activity primarily in low-threshold sensory afferents (9).

The SCP produced in lumbosacral enlargement by stimulation of brachial plexus

The bottom traces of Fig. 4 (A) show the specimen records from the PES of the lumbar enlargement in response to median nerve stimulation at the wrist in a wakeful subject. Any potentials were hardly noticeable in the lumbar enlargement. Thus, peripheral nerve stimulation at the distal site in the upper extremity can hardly evoke any potential change in the caudal segments of the spinal cord in a normal man. This, however, does not mean that peripheral nerve stimulation at a distal site in the upper extremity has no influence on spinal function in the lumbar enlargement. It is rather more likely that the electrical activity does exist but is barely demonstrated due to the temporal and spatial dispersion of the potential when the distal site of a peripheral nerve is stimulated. Therefore, it is predicted that when a more rostral site of the peripheral nerve is stimulated, a potential deflection could be more clearly demonstrated in the lumbosacral enlargement in man, similar to that observed in the rat (25).

When the stimulation electrodes are placed at a more rostral site, i.e., Erb's point, a slow potential change is clearly demonstrated in the wakeful subject, as shown in the top traces in Fig. 4 (A). The waveform of this slow potential is very similar to that of the heterosegmental slow potential (HSP) produced in the lumbar cord of the rat by stimulation of the forepaw (25). Therefore, the HSP is thought to be not specific to the rat. The present results might indicate that a large amount of synchronized afferent volleys are needed to evoke the HSP. In the rat, paw stimulation alone might be enough to provoke sufficient synchronized volleys in the cord, since the distance between the forepaw and the cord is very much smaller than that in man.

Fig. 5 demonstrates the simultaneous recordings of the segmental SCPs from the PES at C5 through C6 vertebral level and heterosegmentally activated slow potentials (HSPs) at T11 vertebral level in a patient with a spinal cord injury at the S4 spinal segment. Stimulation at Erb's point in a moderate strength produced a typical segmental SCP in the PES of C5 through C6 vertebral level. It must be also noted that the two positive summits preceded by two small negative dips were also superimposed on the slow positive (P2) wave. The small negative wave recorded over the lumbar enlargement is thought to be the far field potential of the N1 wave, since both the latency and duration of the potential were the same as those of the N1 wave. When the stimulus strength was increased from 1.5T to 2.5T, a large positive potential, preceded by the N1 and followed by another negative wave, was recorded in the cervical enlargement. On the other hand, slow positive-negative-positive deflections were produced in the lumbar enlargement. Surprisingly, even when stimulus strength was decreased to 1.5T, the waveforms of the SCPs recorded from the two levels did not return to the former patterns, but showed the same patterns as those previously elicited by a stronger stimulation (2.5 x T).

During the stimulation at 2.5T, the subject complained of a slight pain at the stimulus site, but remained quiet. The enhanced SCPs did not return to the control patterns during the whole testing period (about 2 hours). So far, we have not seen such a fundamental change in the SCP pattern during our short-term observation of other subjects. As far as we know, there have been no similar reports on an evoked potential in the literature. Our results are reminiscent of the "long-term potentiation" produced in the hippocampus by tetanic stimulation in experimental animals (26). It is conceivable that some plastic changes which developed locally or supraspinally became manifest by repetition of a stronger stimulation. Such a prolonged increase in the evoked potential in the present study is also reminiscent of a long-term synaptic potentiation, even in the superior cervical ganglion removed from the rat, by the tetanic preganglionic stimulations (27).

Further, a weak electrical stimulus (6mA, 0.1ms) applied to the cervical cord dorsum from the PES in this subject produced a descending SCP with a characteristic pattern. Several negative-positive-going potentials, following the initial spike potential, were then replaced by a large positive slow potential. We have not so far observed such patterns in the descending SCPs in normal wakeful subjects. A series of negative-positive potential could be divided into at least three negative components, suggesting that the latter two components are produced by descending volleys through feedback loops via supraspinal structures. Similar potential changes, having several negative-positive components, have also been observed in the segmental SCP evoked by the burst stimuli in normal volunteer in our previous study (28). Thus, a pathologically hyperactive circuit involving a supraspinal structure developed slightly rostral to the transected site on the cord, might probably induce the repetitive impulses responding to the descending feedback volleys to produce such several negative-positive-going potentials preceding the large positive slow wave, which is also the result of the summation of several slow positive waves.

Summary

Spinal cord potentials (descending SCPs) produced in the posterior epidural space (PES) of the lumbosacral enlargement in response to descending volleys were studied in man and surveyed with reference to the data from our laboratory.

Epidural stimulation of the cervical cord evokes the initially positive spikes followed by sharp negative (desN) and slow positive (desP) waves in the PES of lumbosacral enlargement.

The waveform characteristics of the descending spike, desN and desP, are very similar to those of segmentally activated spike potential (P1), sharp negative (N1) and slow positive (P2) wave, respectively.

Several physiological features of the descending spike, desN and desP, are akin to those of the P1, N1 and P2, respectively, suggesting that the descending spike, desN and desP, reflect action potentials of descending volleys, synchronized activities of dorsal horn interneurons and primary afferent depolarization (PAD), respectively.

There are partial occlusions between the desN and N1, and also between the desP and P2. This indicates that there are common elements in the paths producing the corresponding negative and slow positive waves.

Percutaneous stimulation at the Erb's point produces slow potential changes in the lumbosacral enlargement, which are similar in waveform to the heterosegmental slow potentials (HSPs) observed in the rat. These slow potential changes are suggested to be elicited by descending volleys through a feedback loop via supraspinal structures.

Percutaneous stimulation at the Erb's point in a patient with a spinal cord injury at the S4 segment evoked a series of slow negative-positive-going potentials in the cervical and lumbar enlargements. An increase in stimulus strength led to a profound enhancement of these potentials, which did not return to the control for at least 2 hours, even when the stimulus strength was decreased. The results suggest that some plastic changes develop in a feedback loop terminating at the dorsal horn in patients with deafferentation pain caused by cord trauma.

References

1. Andersen, P.; Eccles, J.C.; Sears, T.A.: Presynaptic inhibitory action of cerebral cortex on the spinal cord. Nature (Lond.), 194: 740-741, 1962.
2. Carpenter, D.; Engberg, I.; Lundberg, A.: Presynaptic inhibition in the lumbar cord evoked from the brain stem. Experientia, 18: 450-451, 1962.

3. Besson, J.M.; Rivot, J.P.: Spinal interneurones involved in presynaptic controls of supraspinal origin. J. Physiol. (Lond.), 230: 235-254, 1973.

4. Engberg, I.; Lundberg, A.; Ryall, R.W.: Reticulospinal inhibition of transmission in reflex pathways. J. Physiol. (Lond.), 194: 201-223, 1968.

5. Martin, R.F.; Haber, L.H.; Willis, W.D.: Primary afferent depolarization of identified cutaneous fibers following stimulation in medial brain stem. J. Neurophysiol., 42: 779-790, 1979.

6. Foreman, R.D.; Beall, J.E.; Applebaum, A.E.; Coulter, J.D.; Willis, W.D.: Effects of dorsal column stimulation on primate spinothalamic tract neurons. J. Neurophysiol., 39: 534-546, 1976.

7. Shimizu, H.; Shimoji, K.; Maruyama, Y.; Sato, Y.; Harayama, H.; Tsubaki, T.: Slow cord dorsum potentials elicited by descending volleys in man. J. Neurol. Neurosurg. Psychiatry, 42: 242-246, 1979.

8. Shimizu, H.; Shimoji, K.; Maruyama, Y.; Sato, Y.; Kuribayashi, H.: Interaction between human evoked electrospinograms elicited by segmental and descending volleys. Experientia, 35: 1199-1200, 1979.

9. Shimizu, H.; Shimoji, K.; Maruyama, Y.; Matsuki, M.; Kuribayashi, H.; Fujioka, H.: Human spinal cord potentials produced in lumbosacral enlargement by descending volleys. J. Neurophysiol., 48: 1108-1120, 1982.

10. Barron, D.H.; Matthews, B.H.C.: The interpretation of potential changes in the spinal cord. J. Physiol. (Lond.), 92: 276-321, 1938.

11. Bernhard, C.G.; Widen, L.: On the origin of the negative and positive cord potentials evoked by stimulation of low threshold cutaneous fibres. Acta. Physiol. Scand. 29, Suppl. 106: 42-54, 1953.

12. Eccles, J.C.; Kostyuk, P.G.; Schmidt, R.F.: Central pathways responsible for depolarization of primary afferent fibres. J. Physiol. (Lond.), 161: 237-257, 1962.

13. Koketsu, K.: Intracellular potential changes of primary afferent nerve fibers in spinal cord of cats. J. Neurophysiol., 19: 375-392, 1956.

14. Carpenter, D.O.; Rudomin, P.: The organization of primary afferent depolarization in the isolated spinal cord of the frog. J. Physiol. (Lond.), 229: 471-493, 1973.

15. Austin, G.M.; McCouch, G.P.: Presynaptic component of intermediary cord potential. J. Neurophysiol., 18: 441-451, 1955.

16. Beall, J.E.; Applebaum, A.E.; Foreman, R.D.; Willis, W.D.: Spinal cord potentials evoked by cutaneous afferents in the monkey. J. Neurophysiol., 40: 199-211, 1977.

17. Maruyama, Y.; Shimoji, K.; Shimizu, H.; Kuribayashi, H.; Fujioka, H.: Human spinal cord potentials evoked by different sources of stimulation and conduction velocities along the cord. J. Neurophysiol., 48: 1098-1107, 1982.

18. Shimoji, K.; Kitamura, H.; Ikezono, E.; Shimizu, H.; Okamoto, K.; Iwakura, Y.: Spinal hypalgesia and analgesia by low-frequency electrical stimulation in the epidural space. Anesthesiology, 41: 91-94, 1974.

19. Shimoji, K.; Matsuki, M.; Shimizu, H.; Iwane, T.; Takahashi, R.; Maruyama, M.; Masuko, K.: Low-frequency, weak extradural stimulation in the management of intractable pain. Brit. J. Anesth., 49: 1081-1086, 1977.

20. Shimoji, K.; Shimizu, H.; Maruyama, Y.; Matsuki, M.: Dorsal column stimulation in man: Facilitation of primary afferent depolarization. Anesth. Analg., 61: 410-413, 1982.

21. Handwerker, H.O.; Iggo, A.; Zimmermann, M.: Segmental and supraspinal actions on dorsal horn neurons responding to noxious and non-noxious skin stimuli. Pain, 1: 147-165, 1975.

22. Hillman, P.; Wall, P.D.: Inhibitory and excitatory factors influencing the receptive fields of lamina 5 spinal cord cells. Exp. Brain Res., 9: 284-306, 1969.

23. Martin, G.F.; Humberston, A.O.; Laxson, C.; Panneton, W.M.: Evidence for direct bulbospinal projections to laminae IX, X and the intermediolateral cell column. Studies using axonal transport techniques in the North American opossum. Brain Res., 170: 165-171, 1979.

24. Mendell, L.: Properties and distribution of peripherally evoked presynaptic hyperpolarization in cat lumbar spinal cord. J. Physiol. (Lond.), 226: 769-792, 1972.

25. Shimoji, K.; Maruyama, Y.; Shimizu, H.; Fujioka, H.; Taga, K.: Spinal cord monitoring -- A review of current techniques and knowledge. In: J. Schramm; S.J. Jones (eds), Spinal Cord Monitoring, Springer-Verlag, Berlin, 1985.

26. Sastry, B.R.; Goh, J.W.; Auyeung, A.: Associative induction of post tetanic and long-term potentiation in CA1 neurons of rat hippocampus. Science, 232: 988-990, 1986.

27. Brown, T.H.; McAfee, D.A.: Long-term synaptic potentiation in the superior cervical ganglion. Science, 215: 1411-1413, 1982.

Anesthesia Influence on Recordings

Effects of Anesthetic Drugs on Spinal Cord Monitoring: An Update

W. T. Frazier[*]

Introduction and plan

The plan of this brief paper is to do the following:

A. Summarize the material presented at the 1984 meeting in Erlangen, including some impressions of the consensus (and/or lack thereof) in the discussion periods.

B. Review some of the major papers (and abstracts) appearing since the time of preparation for the 1984 meeting.

C. Synthesize the above in a general status of the current "state-of-the-art," including the major limitations.

D. Present preliminary data on a possible change in our recommended anesthetic plan (i.e., one of several possible alternates when signals seem depressed by the usual drugs).

E. Speculate on directions of future developments in anesthetic plans for spinal cord monitoring (SCM), including possible relationship to current emerging theories of general anesthesia.

Summary of the status of anesthetic effects on spinal cord monitoring as of the 1984 meeting

Material was presented on both animal work and human work at the 1984 meeting. Even though animal work is quite relevant to our basic understanding of these phenomena, I will be emphasizing mainly the human work due to its immediate clinical applicability.

In one of the introductory papers, Shimoji et al., (1) presented their results from epidural recordings in humans (i.e., spinal cord potentials - SCP). In addition to reviewing the neurophysiology of such recordings, they summarized their work as well as that of several other groups in regard to the pharmacological effect on these spinal cord potentials (2, 3, 4, 5, 6, 7). It is of special interest to note that there are drug effects even at these early processing levels. In particular, in their Table I, they summarized the large number of pharmacological effects on the major early components of the SCP (i.e., P1, N1, and P2). The most prominent component of the segmental SCP (presumably coming from the synaptic neural elements at this level) is the N1 wave. In regard to this first sharp negative component (N1, thought to represent activity of inter-neurons in the dorsal horn), the predominant effects noted were: a) increases in amplitude caused by drugs with "strong hypnotic effects" -- thiopental, ketamine, and

---

* Department of Anesthesiology, Emory University School of Medicine, 475 Woodruff Memorial Building, Atlanta, GA 30322

the halogenated agents (halothane, enflurane, and isoflurane); b) minimal amplitude changes with diazepam and nitrous oxide (N_2O); and c) depression of N1 amplitude by morphine, fentanyl, and presumably larger doses of N_2O.

Since N1 synaptic activity in the dorsal horn might be related to stimulation of cells or origin of the spino-thalamic tract, it is tempting to call attention to the apparent correlation between analgesic drug administration and depression of such activity. The noted increase in N1 by somewhat sedative drugs is also tempting to correlate with the clinical observation of apparent hyperalgesia in low doses of such drugs. Thus, by this brief summary it is appropriate to emphasize that even at the earliest cord synaptic levels it may be possible to distinguish between drugs (presumably in their moderate dose range) which are known to have different effects on the whole organism. That is, there are different effects caused by sedatives and analgesics, keeping in mind that any of these drugs -- be they analgesics or sedatives (of either intravenous or gaseous type) may, in higher doses, make the patient unconscious and unresponsive enough to allow such an agent to function as a general anesthetic.

The next effects of depressant/anesthetic drugs were presented by Thurner et al. (8). In these patients, a standard induction was carried out with flurnitrazepam, etomidate, and pavulon. After this, patients were placed on 60% N_2O and then 20 minutes later were given two supplemental anesthetic plans, the effects of which were as follows: a) when enflurane was added in increasing dose of .5, 1, and 1.5%, there was in general an increase in central conduction time, and a decrease in the amplitude of the major EP components; b) when fentanyl was added in the range of 1.8 to 3.6 to 7.2 μg/kg there were essentially no changes in central conduction time, and in general a decrease in the amplitude of the major components. The major conclusions were that "enflurane induces does-dependent latency delay and amplitude attenuation in both subcortical and cortical somatosensory evoked potentials. fentanyl-benzodiazapine anesthesia causes no apparent alteration of subcortical responses. Cortical potentials are attenuated in amplitude, but little affected in latency."

Koht and his collaborators (9) presented material from a large group of patients in which they did a thiopental and fentanyl induction and then used a maintenance plan of fentanyl (20-60 μg/kg), muscle relaxation, and thiopental drip (1.5mg/kg/hr) or low concentrations of enflurane, isoflurane, or halothane (0.25-0.5 MAC). Of their series of 381 successful intraoperative monitoring cases, they mention that there were five instances where the SSEP was very significantly depressed due to the inhalation agents used, and there were an additional three cases where the SSEP was lost due to bolus injection of one of the intravenous agents.

The report presented by Frazier et al., (10) summarized the results obtained in 156 consecutive cases using a basic technique of a thiopental and fentanyl induction (fentanyl 5 μg/kg) followed by a maintenance program of 66% N_2O and continuous fentanyl infusion (usually at 2μg/kg/hr). Variations on this basic theme were: a) occasionally having to increase the fentanyl induction and continuous infusion rates to accommodate for analgesic tolerance; b) occasional test use of low dose halothane, enflurane, or isoflurane; and c) in the latter part of the series, the N_2O percentage was reduced from 66 to 50% with a substitution of a continuous infusion of thiopental at 2mg/kg/hr. From the data presented at that time, it was concluded that in a very preliminary way it looked as though the reduction of N_2O from 66 to 50% (with the attendant increase in the thiopental level), resulted in somewhat generally better recording conditions. Also it was noted that use of very low dose halogenated agent concentrations (e.g., 1/8 to 1/4% isoflurane) on top of the remainder of the standard plan had definite but only moderate depression of the amplitudes of the evoked potentials. In a general way, it was noted that it seemed as though each of the drugs would be likely to have a dose response curve -- that is, minimal effects at low doses and increasing effects at higher doses; and therefore, no one drug was absolutely bad or absolutely good.

In the debate which followed the above presentations, it is fair to say that there were certain general conclusions: 1) fentanyl was probably one of the better agents to use although there was not a clear consensus as to whether one should use moderate or high dose (with the attendant disadvantages of having to use other agents when a moderate dose was used and having to deal with profound respiratory depression or possibly inadequate levels of anesthesia when only high dose fentanyl was used); 2) there seemed to be general agreement that low dose thiopental was a reasonable supplemental plan, but perhaps no really hard numbers as to what constituted low to moderate to high doses; 3) it seemed likely that the higher doses of N_2O were more depressing than the drugs which could be used to substitute either completely or partially for N_2O (i.e., isoflurane, high dose fentanyl, or thiopental); 4) low dose halogenated agents (particularly enflurane or isoflurane rather than halothane) may be allowable, but certainly the dose needs to be kept moderately low; 5) signals recorded with stimulation of the arms were easier to obtain than signals from the legs (probably due to a number of factors); and 6) obtaining intraoperative evoked potentials is much more difficult in the presence of pre-existing neurological lesions interfering with sensory signal transmission from the lower legs.

The debate which followed the above presentations also left open several important areas for further investigations and clinical experience: 1) there was no consensus as to whether or not the anesthetic plan should provide for a "wake-up" test; some groups felt that the yield of satisfactory intraoperative monitoring was such that one could give up the option of doing the wake-up test (i.e., allowing higher doses of such drugs as fentanyl and/or drugs with slow kinetics such as moderate dose isoflurane); 2) there was no clear answer as to the relative importance of pharmacokinetics versus pharmacodynamics; that is, for example, the use of N_2O (despite its known moderate EP amplitude depression) can be advantageous since it can be loaded and off-loaded quite rapidly (2-5 minutes) and therefore is very helpful in terms of rapid anesthetization, rapid performance of the wake-up test, and rapid re-awakening at the end of the case. In contrast, the possible pharmacodynamic advantages of drugs such as isoflurane or even high dose fentanyl (in combination with halogenated agents or thiopental) would make it more difficult to do a wake-up test and might result in a longer awakening period (therefore making it more difficult to do an early neurological exam); 3) there generally seemed to be some who were satisfied with the anesthetic plans they were using and others who felt we definitely needed new drugs; 4) there was a general consensus that there was a need for a test of the function of the descending motor tracts. That is, there had been at that time several anecdotal and a few published reports (11) of anterior spinal artery syndromes such that the patients awakened paralyzed at the end of the case without any obvious changes in the classical somatosensory evoked potentials presumably traveling over the more posterior parts of the spinal cord.

"New" papers

Since the time of preparation for the meeting in Erlangen in 1984, there have appeared several abstracts and papers which have bearing on the problem of the effects of anesthetics on spinal cord monitoring.

With a basal anesthetic of nembutal, McPherson, Johnson, and Traystman (12) added either N_2O or fentanyl, and then subsequently added fentanyl or N_2O. They observed that after the nembutal, adding 50% N_2O caused no changes in upper and lower extremity EP amplitudes, and that then adding fentanyl did not depress the EP amplitude (but did latencies). In turn, with the dogs anesthetized with nembutal and fentanyl, adding 50% N_2O resulted in moderate decreases in upper extremity EPs but no changes in the lower extremities. They called attention to the differential upper and lower extremity effects of N_2O (+ fentanyl, on top of nembutal), noted the failure of Narcan (a narcotic antagonist) to reverse the N_2O effects, and urged that ". . . the effects of alteration of N_2O concentration must be considered during the monitoring of SEP using upper extremity stimulation if narcotic is present."

The study of Peterson et al., (13) centered on the median nerve somatosensory evoked potential in patients who were premedicated with morphine and phenergan and underwent a barbiturate induction. These patients were placed on 60% N_2O and then had added amounts of either halothane (0.5-1.0-1.5 MAC in N_2O) or isoflurane (0.5-1.0-1.5 MAC in N_2O). At the 1.5 MAC level with either plan, N_2O was then removed. Their results indicated that both halothane and isoflurane (in N_2O) depressed the EP amplitude, with isoflurane depression exceeding that of halothane at equal MAC levels. They also noted that when the N_2O was removed at the 1.5 MAC level, there was an increase in the evoked potentials by 30-40%. It was our observation of their work that it is reasonable to call attention to the fact that when the halothane or isoflurane were increased from 1.0 to 1.5 MAC (in the presence of 60% N_2O), there was a decrease in the SSEP by approximately the same amount as the subsequent increase when 60% N_2O was removed (at the 1.5 MAC level), indicating in a semi-quantitative way that the depressive effects of adding halothane and isoflurane (from 1.0 to 1.5 MAC) are of the same order of magnitude as increases in EPs caused by removing N_2O (in the presence of 1.5 MAC of halothane and isoflurane).

Drummond et al., (14) intended to see if one could use protective levels of thiopental all the way to an isoelectric EEG and still have some usable neurophysiological index of brain activity with the use of median nerve somatosensory evoked potentials. Their results showed moderate depression of the median nerve SSEP at moderate doses of thiopental, but that there were still useful levels of the median nerve SSEP at the higher doses. There was some question if these results also held true for the lower limbs. Obviously there were some caution in regard to the use of such high dose barbiturate levels in patients with significant cardiovascular disease. In regard to the applicability of this work in the operating room, they would try to use a primarily barbiturate anesthetic plan (even at doses lower than an isoelectric EEG).

Patients involved in anesthetic induction with thiopental and moderately high dose fentanyl (25µg/kg, 1 bolus) were divided into two groups (15); those who received N_2O for a period of time after which the nitrous was exchanged for enflurane or isoflurane (.25 - 1%, adjusted on clinical grounds "to maintain hemodynamic stability"), and those patients in which the basal anesthetic was enflurane or isoflurane (.25 - 1%) for a period of time and then exchanged with 50% N_2O. "Additional thiopental (25mg) was given in response to movement or hemodynamic responses to stimulation" (but not within 10 minutes of SEPs). They recorded EPs from both upper and lower limbs. They concluded that the depressant effects of N_2O exceeded those of enflurane or isoflurane (on top of a single injection/high dose fentanyl technique). Nevertheless, it is interesting to note that in their Group 3 (with preoperative neurological deficits), there were no significant differences with N_2O (versus enflurane) for the lower extremities, the most frequent site for stimulation during spinal cord monitoring. Other questions which I would have regarding the advocacy of this anesthetic plan relate to whether or not one would usually redose fentanyl for such long cases (or continuously infuse), or if one did not, what the blood levels would be in the latter part of the case (e.g., likely to be falling quite rapidly in the middle of the case and reaching possibly much lower levels near the end).

Sloan and Koht's (16) patients were done with moderate to high dose fentanyl (10-20 µg/kg) with the addition of a halogenated agent. The fentanyl loading dose was supplemented with occasional extra doses (to a total of 20-50 µg/kg). Either enflurane or isofluranc were added at the .25 MAC level. Their potentials were from both the upper and lower extremities. Near the end of the case, they did a washout of either the enflurane or isoflurane (taking about 30 minutes) and then added 50% N_2O to a maximal effect at approximately 7-10 minutes. They reported that their results were a consistent major decrease in cortical SSEP caused by 50% N_2O after washout of the halogenated agent (in the presence of probably significant fentanyl levels, no supplementation having occurred within 30 minutes). However, no data were presented in regard to SSEP changes during the washout of the halogenated agent. Also since the fentanyl administration was in essence a single (early) bolus technique, it would be interesting to know the blood level late in the case (e.g., at the time of enflurane/isoflurane washout and N_2O loading). Also, it would be interesting to know the dynamics of the washout of the halogenated agents after a prolonged administration of such fat soluble agents since there were no reported airway or blood levels of these anesthetics. It is also important to point out that this is once again a study of an agent (N_2O) by adding it on top of changing and significant (but unknown) levels of another very potent agent (i.e., "high" dose fentanyl).

Koht et al., (17) noted that a large loading dose of sufentanil (5 μg/kg) resulted in depression to about 60% of baseline and that the EPs were stable over the next 30 minutes.

Bird et al., (18) reported SEPs from the arms and were recorded after thiopental induction and in the presence of 70% N_2O. Within 1-7 minutes, they gave sufentanil in a 1 μg/kg bolus followed by infusion. After 75 minutes they added a second bolus (.5 μg/kg) and increased the infusion rate. They noted a steady decrease (e.g., 20-50%) in EP amplitudes during the initial sufentanil infusion and noted that the second bolus of sufentanil depressed the SSEP an insignificant amount. They concluded that sufentanil depresses the SSEP but that these changes are slow in the presence of continuous infusion (and minimal at higher additional levels), and therefore could be dissociated from the above changes due to other intraoperative problems (presumably causing abrupt changes).

Kalkman et al., (19) recorded median nerve somatosensory evoked potentials in patients initially on 100% oxygen and then given Etomidate in a 0.3mg/kg bolus followed by an infusion for 10 minutes. An alternate plan was to substitute Midazolam at 0.2mg/kg for Etomidate followed by possibly a 0.1mg/kg repeat dose if needed. Their very interesting results showed a sizable increase of the SSEPs with Etomidate (2.5 - 3 times larger for the N1 through P1 and the P1 through N2 components). They also noted a 1/3 to 1/2 decrease in these components with Midazolam. It is also of interest to note that 9 out of 10 of the patients had significant myoclonic movements with Etomidate.

Kochs et al., (20) reported on median nerve somatosensory evoked potentials in patients. They also addressed the effects of etomidate as well as diprivan in doses of 0.3mg/kg or 2mg/kg respectively. The results noted at approximately 2 minutes of administration of the drugs were as follows: For Etomidate, there was a 2 to 11 times increase in the N20 and P25 signals, with a 73% incidence of myoclonus; with diprivan, there was a 10 to 80% increase in the N20 through P25 components, without myoclonus. With both drugs, there was a return to control at approximately 6 minutes.

Synthesis of the older and newer material

Looking back at the various anesthetic plans which have been presented in the literature to date, it is clear that fentanyl (and now perhaps sufentanil) continue to be thought of as very useful agents and minimally depressing (even at higher dose levels) of the somatosensory evoked potentials from either the upper or lower extremities. Secondly, it is clear that N_2O does depress SSEPs; however, it is not clear in my opinion at what dose the changes are clinically significant; nor is it clear in any given situation how N2O compares to one of the other inhaled gases such as the volatile halogenated agents (halothane, enflurane, and isoflurane). In those basic and clinical investigations proposing to compare N2O to one of the halogenated agents, it is almost universally the case that these studies are performed in conjunction with the utilization of other agents such as fentanyl and/or thiopental, and are not done in such a way as to make it clear that equipotent doses are on board at the times the conclusions are being drawn.

It seems likely that low dose thiopental continuous infusion (after usual induction/loading doses) is useful and not prominently depressing of the SSEPs. It is important to note that useful SSEPs (e.g., for monitoring in the ICU) are still recordable at very high barbiturate levels.

It is important to point out that many of the studies trying to evaluate the relative utility of a particular set of anesthetic agents involve adding new drugs on top of previously existing drugs, with or without a thorough attempt to "washout" the previous drug. Therefore, there are always residual questions regarding the circumstances under which one is giving the "new" drug as regards to how completely the previous drug has been washed out. Furthermore, when statements are made in the literature regarding the comparative effects of drugs, it is possible that the order in which the drugs are given is important and that there are significant residual levels or other drugs remaining. In addition, when one has an anesthetic plan made up of multiple agents (for example: A, B, and C), and is then trying to substitute agent D for agent C to test its relative effects, it is entirely possible that one is dealing with a considerably higher level of total anesthetic than was really needed (a super MAC dose) or else that when one is trying to remove agent C before giving D, there is a transient period of inadequate anesthesia. In the former case (the super MAC case), testing the two drugs, C and D, really would be at a significantly higher total level of depression of the nervous system, and may yield different effects than at a level near 1 MAC.

There are still several unresolved questions in regard to the anesthetic management for cases involving somatosensory evoked potentials (particularly of the lower extremities). It is still not clear if many of the anesthetic plans which are being advocated would allow for a "wake-up" during the case or even if many institutions still desire that this test be feasible during the case. Secondly, it is not clear how many of the anesthetic plans proposed would affect the condition of the patient at the end of the case as it relates to the rapidity of awakening or the ease of performing a reliable neurological exam soon thereafter. It is also apparent that many anesthesia teams manage the intraoperative hemodynamic changes by additional bolus doses of sedatives or analgesics, and that others manage some of these changes by drugs more directly acting on the vasculature and its control systems (for example, beta blockers and/or vasodilators). Certainly the issue has seldom been addressed as to whether or not the anesthetic requirements from an analgesic/sedative point of view are varying greatly during SCM on a moment to moment basis in response to varying levels of surgical stimulation (21). It is also clear that there is no consensus as to whether or not the blood levels of the various anesthetic agents (in particular, intravenous agents such as fentanyl and sufentanil) should be kept at a relatively steady level by means of a loading dose and continuous infusion or rather whether it is permissible to give a very large bolus dose loading dose at the beginning of the case, thus allowing the agents to gradually decrease in its blood/brain levels near the end of the case.

In a very general way, it seems likely that there could be a relatively common agreement on at least three points: 1) that with the exception of Etomidate and possibly diprivan, all other commonly used anesthetic agents (whether partial or full anesthetic) would be likely to have a depressive dose-response curve such that with the increased dose there would be an expected continuously (although not necessarily linearly) depressed SSEP amplitude; furthermore, there might be some low doses which have minimal effects (perhaps even a "threshold") and some higher doses which could exhibit a "ceiling" effect (22); 2) at low doses (e.g., 1/8 - 1/4 MAC) the effects of the halogenated agents may be surprisingly small on the depression of EPs, keeping in mind that these effects may very well depend on a combination with other drugs (e.g., various levels of N_2O and/or intravenous agents such as Fentanyl); 3) by virtue of the plans presented by many anesthesia teams, it seems likely that most will be relying of a "balanced" technique; that is, a combination of various drugs leading to a synergistic or additive desirable anesthetic depression of wakefulness and reflex activity, while maintaining a minimal to moderate degree of depression of SSEPs.

The need for a balanced technique comes from the commonly recognized disadvantages of employing many of the single agents. Certainly it is recognized that one could not do the case completely with N_2O since 1 MAC N_2O exceeds 100% and even in lower doses exceeds the lower doses which will usually provide safe levels of oxygenation (e.g., 60-75%). It is also clear that due to the significant depression of SSEPs with the higher doses of N_2O, that most groups are limiting their use of it to 50% or less. It is also likely that high dose fentanyl technique is not sufficient anesthesia for major surgery unless supplemented by other agents (23). Certainly if fentanyl is almost the sole agent, there is no possibility of a wake-up test during the case nor will emergence at the end of the case be very rapid. With such a high dose technique, there would be the attendant problems of some muscle rigidity and the profound respiratory depression at the end of the case (22), to say nothing about the possibility of intraoperative awareness should blood levels be allowed to drop too low (e.g., with a single bolus at the beginning of a long case).

It is also unlikely that any of the halogenated agents (for example: halothane, enflurane, or isoflurane) would be used widely as the sole anesthetic agent since their presence in 1 MAC concentrations in many patients would result in unsuitable hemodynamic depression. Furthermore, even though perhaps useful SSEPs could be obtained in some patients at the higher concentrations of these agents, it is likely that there would be too much depression in many others. It is also probable, for long cases involving major spinal surgery, the use of these fairly lipid soluble agents would result in a prolonged waking time, and would result in intraoperative conditions not very conducive to doing a wake-up test.

In regard to the use of barbiturates (mainly thiopental), it is commonly recognized that the higher dose ranges result in dangerous depression of the cardiovascular system, making it very difficult to do a wake-up test and leading to a very slow emergence. It seems likely that the use of benzodiazepines (for example, valium or medazolam) will be limited to supplemental use either in the preoperative period or in very low doses during the case.

The use of the relatively new drug, Etomidate, is a most interesting event. The two papers referred to in the above review (and the third, contained in this volume) are most interesting in that this work clearly shows relative facilitation of the SSEPs under certain circumstances. This is a most useful and interesting set of findings, but should be tempered by the well recognized disadvantages of Etomidate in regards to adrenal suppression when used at significant levels for even relatively short cases (24), and due to the high incidence of myoclonic episodes.

I think that it is clear that most investigators would feel that although we have reached a reasonable plateau in the development of anesthetic plans for use during spinal cord monitoring (albeit without a solid consensus), there is very likely a need for new anesthetic agents. It is always possible that such agents will come along by empirical discovery. However, it is very likely in the long run that the development of suitable agents will at least be partially tied to continued progress in our understanding of the specific neuropharmacological mechanisms involved in the kind of general anesthesia state necessary to allow for simultaneous successful recordings of somatosensory evoked potentials.

Preliminary data on ketamine

As stated in our report at the 1984 meeting, there is a significant incidence of failure to achieve signals large enough to make SCM reliable. Although there is not agreement as to the actual incidence of such monitoring failures, there does seem to be agreement that it does occur apart from intraoperative surgical causes and is more common with pre-existing neural deficits. The concern to improve the reliability of SCM leads to the trial of drugs already in the traditional anesthetic armamentarium (or combinations thereof) as noted by the papers referred to above. (and material presented by McPherson and Ducker in this volume) utilizing Etomidate.

We have therefore tried on several occasions to improve our signals by substituting intravenous ketamine for 50% N_2O. We have kept the remainder of the anesthetic plan the same: a) fentanyl -- 5 μg/kg bolus, plus continuous infusion of 2 μg/kg/hr (increased for tolerance as indicated); b) thiopental (after recovery from induction dose) -- 1mg/kg bolus, plus infusions of 2mg/kg/hr. This dose of fentanyl (alone) is sufficient to produce adequate post-surgical analgesia in most patients (with a pCO_2 of less than 48 Torr). This dose of thiopental (alone) is sufficient to provide "light" sedation (e.g., during surgical procedures performed with regional anesthesia). The method has been to turn off the N_2O (and flush out rapidly with a high flow technique), giving the ketamine "bolus" in divided doses over 5 - 10 minutes (1.5mg/kg) and then starting the infusion (5 - 6mg/kg/hr). This plan is an attempt to keep total anesthetic at the same level, matching the "washout" of N_2O to the increasing levels of ketamine. In some instances, we have (in the same patient) also substituted 0.5% isoflurane for the 50% N_2O, again trying to match the "wash-in" and "washout" phases and reversing to the original anesthetic circumstances (e.g., N_2O -- ketamine -- N_2O -- isoflurane -- N_2O). Gaseous anesthetic concentrations have been monitored by mass-spectrometry to document wash-in/washout.

Although very preliminary, our initial impressions are that substituting ketamine for 50% N_2O has resulted in "retrieval" of usable signals and that isoflurane has not been as useful (although perhaps slightly better than an equivalent amount of N_2O). Typical results of these clinical trials are shown in Fig. 1 and Fig. 2. Ketamine is generally regarded as capable of being a total anesthetic for short cases and has a spectrum of actions somewhat similar to N_2O (e.g., analgesia at sub-anesthetic doses, "sleep" at higher doses, possible requirement of other drugs for control of blood pressure, etc.). When used as the sole anesthetic (and without proper premedication -- e.g., with tranquilizers), there has been concern about more disorientation in a wake-up period than with more conventional agents. There is reason to think that ketamine may be the most specific general anesthetic in that at usual doses airway reflexes remain relatively intact and some cardiovascular responses (e.g., to strong surgical stimuli) are relatively intact. Fig. 1 and Fig. 2 present typical results with ketamine substitution/supplemental ("retrieval" of a poor SSEP and augmentation of a fair SSEP respectively).

Theories of anesthesia in relation to spinal cord monitoring

As mentioned in the above synthesis, some individuals consider it likely that future successes in neurological monitoring during the general anesthetic state will depend upon the development of new anesthetic drugs. Anesthesia in the last several decades has faced a similar developing pharmacological technology in the area of muscle relaxants. Several years ago when the use of a nondepolarizing muscle relaxant was indicated, the only practical agent available to the anesthesia team was Curare. Curare,

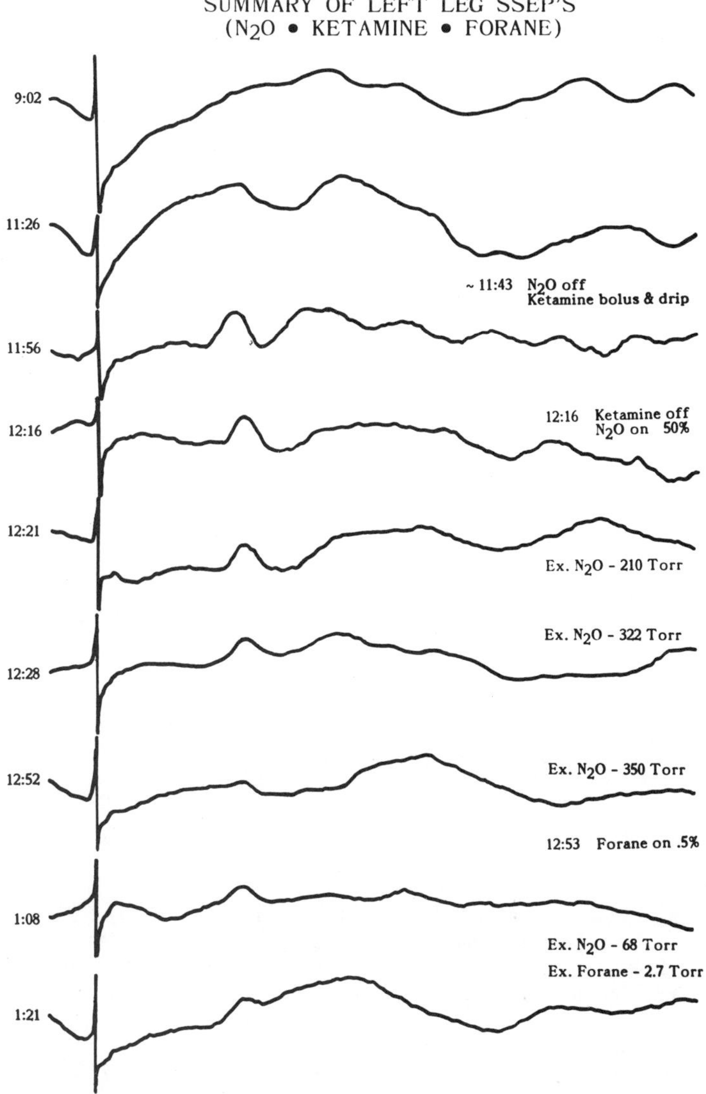

Fig. 1. Retrieval of poor single leg SSEPs by substitution of Ketamine or Isoflurane (see text for method) for 50% N2O (in conjunction with basal plan of continuous infusion of Fentanyl and Thiopental). Control signals (over ~2.5 hours) showed minimal signal (e.g., signals at 9:02 and 11:26). N2O was turned off at 11:43 and i.v. Ketamine begun (bolus plus infusion). Within 5-10 minutes, a usable P38 signal appeared and remained for ~20 minutes at which time the Ketamine was turned off and the N2O turned back on (at 12:16). The P38 signal progressively diminished to near zero and at 12:53, exchange of Isoflurane 0.5% for the 50% N2O was begun. Within 15 minutes, a usable P38 had returned at an amplitude approximately one-half its size with the Ketamine and remained stable as the Forane rose to an exhaled concentration of ~0.4% Epoch length: 175ms. Signal at extreme left is shock artifact. N2O and Forane concentration are exhaled concentrations (in Torr; dry atmospheric pressure of ~700 Torr) as measured by airway mass spectrometry. Successive records are from 9:02 a.m. to 1:21 p.m.

though very useful, has well known undesirable effects of histamine release (with consequent potential decreases in blood pressure and bronchoconstriction) and partial ganglionic blockade (causing hypotension also) due to its effects on the sub-type of the cholinergic/nicotinic receptor present in the sympathetic ganglia. That is, other than its histamine releasing effects, the chief disadvantage of Curare was the fact that it was not a pure N1 (i.e., neuromuscular junction) receptor blocker, but also an N2 (ganglionic) receptor blocker. In recent years, newer nondepolarizing muscle relaxants have become available which are highly specified for the N1 receptor and, therefore, do not result in significant degrees of ganglionic blockade (e.g., Vecuronium and Atracurium).

Similarly, if there is a particular set of neurons and synapses in the nervous system which have to be modified to produce the minimal necessary state of general anesthesia, it is possible that drugs could be developed to have these more specific actions

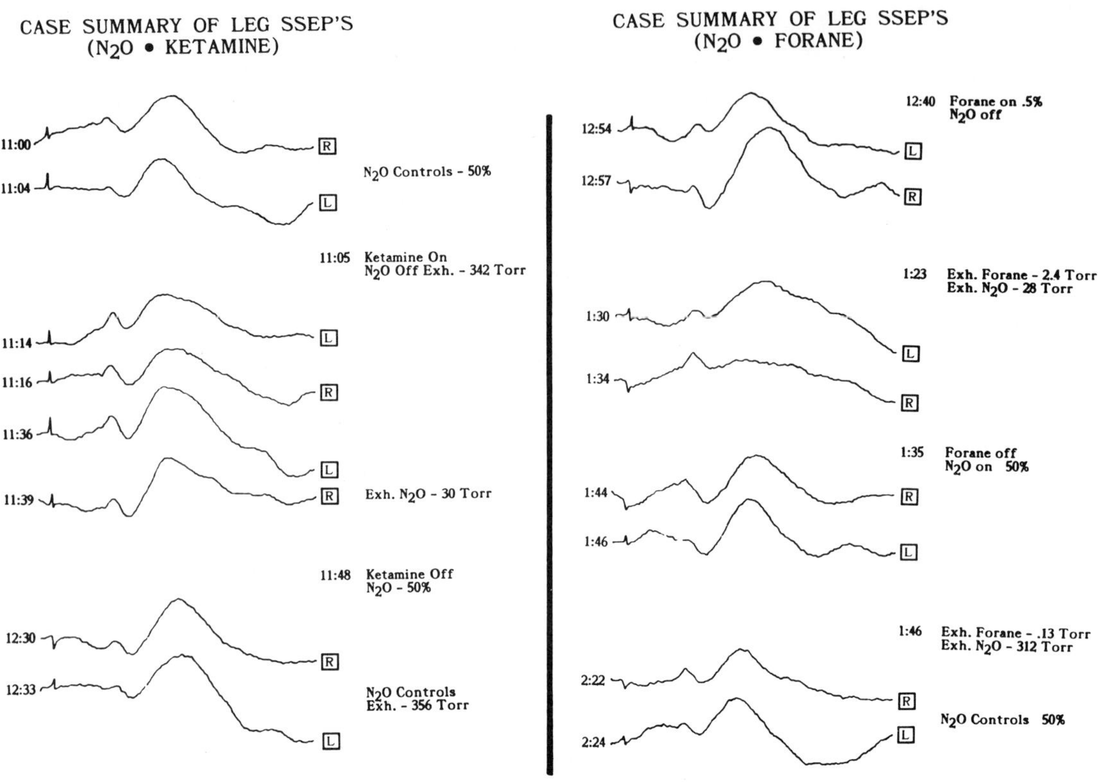

Fig. 2. Improvement of individual right (R) and left (L) leg SSEPs by substitution of i.v. Ketamine for 50% N2O (see text for method).

In the left panel: After more than 1 hour of stable control SSEPs (e.g., those at 11:00 and 11:04), N2O was washed out while a bolus of Ketamine was given and infusion begun. By 11:14, the P38 component of the R and L leg signals approximately doubled in amplitude; at 11:48, the Ketamine was turned off, the N2O was turned back on, and the P38 returned to the same amplitude as approximately 1.5 hours earlier.

In the right panel: After a stable control period (P38 still depressed), exchange of Isoflurane (Forane) 0.5% for N2O - 50% was begun. Over the next ~ 30 minutes, there was no significant change in P38, but the later components markedly diminished. AT 1:35, Forane was turned off, N2O turned on, and over the next 10 minutes the late components returned to their former size with no significant change in P38. Signals at 2:22 - 2:24 (50% N2O) were approximately the same as at 11:00 - 11:04 (~ 2 hours 20 minutes span). Epoch length, etc., same as in Fig. 1.

without interfering with the transmission of somatosensory information to a convenient location in the brain (i.e., sensory cortex) with minimal degradation. In retrospect, it has been rather fortunate for animal neurophysiological and neuropharmacological research that it does seem to be possible to depress the central nervous system to the point of nonwakefullness while still being able to trace signals from the periphery to the sensory cortex. However, I think those involved in clinical anesthesia would still desire to have drugs more specific than those currently available for the management of most cases, and in particular the management of those cases in which it is desirable to do neurophysiological monitoring.

In a very general and highly over simplified way, I would propose that we may divide the drugs that we commonly have available into relatively non-specific and relatively specific categories. I would place the gaseous agents (both N_2O and the commonly used halogenated agents) in the category of non-specific. By virtue of their physical structure and very wide distribution as gases soluble in both the aqueous and nonaqueous phases of the nervous system, their actions are likely to be very widespread. Although these drugs may have some relatively specific effects (e.g., analgesia at low doses of N_2O with the patient still awake, and perhaps some EEG depression unique to isoflurane as opposed to enflurane and halothane), it is still likely that these drugs depress most parts of the nervous system. Although these newer halogenated agents are far better than most of the gaseous/volatile anesthetic agents available in previous years, they do have widespread effects on neural tissue and muscular tissue leading to some degrees of muscle relaxation, vasodilation, and ultimately depression of the myocardium.

In contrast to the gaseous/volatile agents, it seems to be relatively clear that we will have to look to intravenous agents for fairly specific pharmacological/anesthetic action. In most instances, pharmacologists consider that the specificity of intravenous drugs follows from their interaction with "receptors" at various locations on the neurons. Such locations need not be postsynaptic as classically is the case with the non-depolarizing muscle relaxants, but may be presynaptic or even intracellular and may be involved in any one of the myriad of neuronal processes involved in spontaneous activity, signal conduction, transmitter action (synthesis, release, action, re-uptake, etc.), and can involve not only modification of excitatory processes, but also cellular processes resulting in tonic spontaneous activity and in the various kinds of neuronal inhibition (synaptic, presynaptic, etc).

In regard to specificity of drugs useful to the anesthesia team, I would submit that the following intravenous agents have useful degrees of specificity: fentanyl, benzodiazepines, etomidate, diprivan, and ketamine. Thiopental may be the drug which resides mainly in the middle of what may be a specific/non-specific continuum - due to perhaps its lipid solubility as well as its possible receptor actions (e.g., at the GABA receptor complex) (25).

At this point in time it is relevant to raise as an issue whether or not there is a commonly accepted theory of general anesthesia. In the history of this subject, the only prominent and lasting theory has been that of correlating the potency of anesthetic agents with their lipid solubility. However, in the last few years, there are new proposals being made regarding issues such as specific receptors and interactions of anesthetics with proteins (26). Also we are finally approaching an era where there may be a "neuropharmacological" theory of anesthesia (27) as opposed to the previously existing physical theories of anesthesia (i.e., the lipid solubility correlations). Calling attention to the incompleteness of lipid solubility theories is work showing a similar correlation between anesthetic potency and another physical chemical event (i.e., the depression of the light production of the enzyme luciferase (isolated from bacteria and fireflies) (25). Thus it can be seen that the correlation between anesthetic potency and lipid solubility or protein enzyme inhibition will increasingly be seen as logically and phenomenologically incomplete since these theories do not address the issue of the neurological substrate (i.e., neural circuitry) of the waking/anesthetized states.

In an evolution from these physical chemical correlation theories of anesthesia, it has become more apparent in recent years that one has to begin to look at basic neuropharmacological mechanisms for the explanation of the general anesthetic state. In particular, it has been of interest to correlate the potency of anesthetics with their measured action on certain neurons (for example, their ability to hyperpolarize cells). On the cellular neuronal level, it is recognized that hyperpolarization of a cell

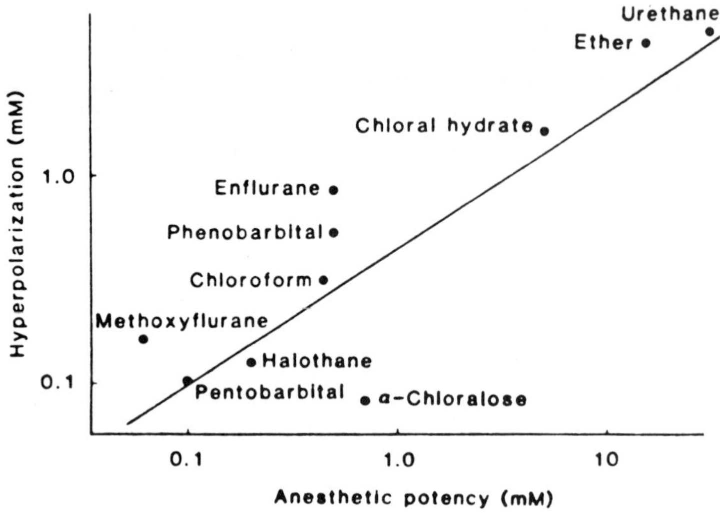

Fig. 3. Relative hyperpolarizing effects of various anesthetics (on motor neurons in the isolated frog spinal cord, by sucrose gap recording) plotted against the anesthetic concentration of the drug (from several literature sources); divergence of α-chloralose (causing regression line to shift downward at lower doses) is speculated to be possibly related to some superimposed excitatory effects which this drug may show; from Figure 1-B of Nicoll and Madison (Ref. 27).

decreases the probability of its firing (i.e., it moves it farther away from its firing threshold). Also at the synaptic level, hyperpolarization of the presynaptic and postsynaptic elements and the coincident changes in membrane impedance may result in some instances in decreased synaptic efficacy. An example of this can be seen in Fig. 3 (from Fig. 1-B) (28) where there is a distinct correlation between the hyperpolarization of certain neurons and the anesthetic potency of such drugs as methoxyflurane, halothane, pentobarbital, etc. In terms of directness (i.e., the effects on neurons) and in terms of simplicity, this theory relating anesthetic potency of agents to their ability to hyperpolarize cells is particularly attractive and has been elaborated on by several workers (26). In terms of comparative anesthetic effects, it has been shown in invertebrate cells that drugs such as thiopental tend to hyperpolarize individual neurons (with coincident decreases in membrane resistance and PSP amplitudes, probably due to increased potassium conductance) while the effects of ketamine are quite different (e.g., variable polarization, diminution of some EPSPs, and minimal resistance changes) (29). It is commonly accepted that in most cells it is the resting level of potassium conductance (gK) which controls the membrane potential and therefore controls the probability of a cell and its axon firing. That is, one simple theory of general anesthesia can be that many agents result in a generalized increase in the resting potassium conductance and, therefore, hyperpolarize most, if not all, of the cells in the nervous system. Subsequent to that, it may be that at low doses, cellular activity supporting a tonic resting waking state is depressed (more so than activity in other circuits (e.g., sensory connections to the cortex) thus resulting in a facilitation of the naturally occurring transition from the waking to the non-waking or sleeping state while leaving signal transmission to the cortex relatively intact. Although it is unlikely that such a simple explanation could account for the bulk of the neurophysiological correlates of general anesthesia, this theory has the advantage of directness (i.e., coming from data collected on neurons) and simplicity.

However simple the theory of anesthetic drugs acting through changes in resting potassium conductance (and therefore changes in resting membrane potential), it is likely that such effects are very complicated since there are many kinds of neurons (silent, spontaneously active, excitatory, inhibitory, etc.) and there are an enormous number of connections and possible circuits involved in the processing of sensory information, the maintenance of the waking or non-waking states, and in the production of reflexive and higher order movements.

In contrast to this theory of the anesthetic state (i.e., changes in potassium conductance resulting in hyperpolarization of most neurons in the brain), it is attractive to place the main future attention on the notions of specific receptors involved in the maintenance of the waking state and the naturally occurring sleep states. Of the drugs listed above as being relatively specific, there is now a considerable amount of evidence that the central actions of many of these drugs may be related to fairly specific receptors. In particular there is a great deal of evidence that drugs such as the benzodiazepines (e.g., valium, medaolam, etc.), thiopental, and etomidate can have fairly specific and direct actions on the benzodiazepine/chloride receptor complex in the central nervous system (24) (although they may have other relatively non-specific effects also). Furthermore, there is evidence to indicate that ketamine has a fairly specific blocking action at a particular sub-type of a widely occurring CNS receptor (the NMDA sub-type of the excitatory amino acid receptors) (30, 31).

Given the preliminary results with etomidate, ketamine, and diprivan (as described above), it is interesting to speculate that we are collectively on the right track by using drugs (e.g., ketamine) which may block neural transmission in certain circuits (e.g., those most necessary in supporting the waking state) or by using drugs which may disinhibit certain cortical neurons involved with SSEPs (e.g., by use of Etomidate). Meanwhile, we are increasingly avoiding those drugs which may depress almost all neuronal activity (e.g., high dose N_2O and high dose halogenated agents). We are beginning to take steps to be much more specific neuropharmacologically.

Summary

It has been my intent in the above review, presentation of data, and synthesis to call attention to the fact that there has probably been progress in the development of anesthetic techniques for spinal cord monitoring in the last five years. However, as of this date, there is no clear consensus as to what the one best anesthetic plan is (although there is probably an increasing consensus as to what are the agents to avoid). Furthermore, I have tried to support the view that it is likely that most anesthesia personnel would welcome the development of new anesthetic agents which would be more specific in their neurological action and which would thus allow a more successful differentiation between the production of the anesthetized state and the depression of SSEPs. As in most areas of pharmacology, a firm basis of drug therapy consists of a knowledge of mechanisms, specificity of drug action, a balance of complimentary (or synergistic) effects, and a minimization of toxicity. We are just beginning to learn something about mechanisms (at the neuronal circuit level), and we still await the development of very specific anesthetics with negligible side effects.

References

1. Shimoji, K.; Maruyama, Y.; Shimiju, H.; Fujioka, H.; Taga, K.: Spinal cord monitoring - a review of current techniques and knowledge. In: J. Schramm; S.J. Jones (eds.): Spinal Cord Monitoring. Springer-Verlag, Berlin, Heidelberg, New York, Toronto, p. 16-28, 1985.
2. Kailda, R. et al.: Effects of Diazepam on evoked electrospinogram and evoked electromyogram in man. Anesthesia and Analgesia, 60: 197-200, 1981.

3. Kano, T.; Shimoji, K.:   The effects of ketamine and neuroleptanalgesia on the evoked electrospinogram and electromyogram in man. Anesthesiology, 40: 241-246, 1974.

4. Maruyama, Y. et al.: Effects of Morphine on human spinal cord and peripheral nervous activities. Pain, 8: 63-73, 1980.

5. Shimoji, K. et al.: Evoked spinal electrograms recorded from epidural space in man. J. Applied Physiol., 33: 468-471, 1972.

6. Shimoji, K. et al.: The effects of Thiamylal Sodium on electrical activities of the central and peripheral nervous systems in man. Anesthesiology, 40: 234-240, 1974.

7. Shimoji, K. et al.: Presynaptic inhibition in man during anesthesia and sleep. Anesthesiology, 43: 388-391, 1975.

8. Thurner, F.; Schramm, J.; Romstck, J.: Effects of fentanyl and enflurane on cortical and subcortical SEP during general anesthesia in man. In: J. Schramm; S.J. Jones (eds.): Spinal Cord Monitoring. Springer-Verlag, Berlin, Heidelberg, New York, Toronto, pp. 82-89, 1985.

9. Koht, A.; Sloan, T.; Ronai, A.; Toleikis, J.: Intraoperative deterioration of evoked potentials during spinal surgery. In: J. Schramm; S.J. Jones (eds.): Spinal Cord Monitoring. Springer-Verlag, Berlin, Heidelberg, New York, Toronto, pp. 161-166, 1985.

10. Frazier, W.T. et al.: Anesthetic technique for spinal cord monitoring. In: J. Schramm; S.J. Jones (eds.): Spinal Cord Monitoring. Springer-Verlag, Berlin, Heidelberg, New York, Toronto, pp. 69-81, 1985.

11. Carlson, C.A.: Somatosensory evoked potential monitoring in aortic coarction surgery: Failure to predict adverse neurologic outcome. Tenth Annual Gulf/Atlantic Resident's Conference, Atlanta, Georgia, 1984.

12. McPherson, R.W.; Johnson, R.M.; Traystman, R.J.: Synergistic effects of N_2O and fentanyl or the somatosensory evoked potential in dogs. Anesthesiology, Vol. 59, No. 3, A317, September, 1983.

13. Peterson, D.O.; Drummond, J.C.; Todd, M.M.: Effects of halothane and isoflurane on somatosensory evoked potentials in dogs. Anesthesiology, Vol. 61, No. 3A, A344, September, 1984.

14. Drummond, J.C.; Todd, M.M.; Sang, U.H.: The effect of high dose sodium thiopental on brain stem auditory and median nerve somatosensory evoked responses in humans. Anesthesiology, 62: 249-254, 1985.

15. McPherson, R.W.; Mahla, M.; Johnson, R.; Traystman, R.J.: Effects of enflurane, isoflurane, and nitrous oxide on somatosensory evoked potentials during fentanyl anesthesia. Anesthesiology, 62: 626-633, 1985.

16. Sloan, T.B.; Koht, A.: Depression of cortical somatosensory evoked potentials by nitrous oxide. British J. Anaesth., 57: 849-852, 1985.

17. Koht, A.; Kumovec, M.A.; Sloan, T.B.; Carlvin, A.O.: The effects of sufentanil on median nerve somatosensory evoked potentials. Anesth. Analg., 65: S1-S170, S81, 1986.

18. Bird, J.; Donegan, J.; Rupp, S.; White, P.; Sassler, D.: Effects of sufentanil bolus and two steady state infusions on median nerve evoked potentials. Anesthesiology, Vol. 65, No. 3A, A341, September, 1986.

19. Kalkman, C.J.; van Rheineck, A.T.; Hesselink, E.M.; Bovil, J.G.:   Effects of Etomidate or Midazolam on median nerve somatosensory evoked potentials. Anesthesiology, Vol. 65, No. 3A, A356, September, 1986.

20. Kochs, E. et al.: Alterations of somatosensory evoked potentials by etomidate and diprivan. Anesthesiology, Vol. 65, No. 3A, A353, September, 1986.

21. Ausems, M.E.; Hug, C.C.; Stanski, D.R.; Burm, A.G.L.: Plasma concentrations of alfentanil required to supplement nitrous oxide anesthesia for general surgery. Anesthesiology, 65: 362-373, 1986.

22. Murphy, M.R.; Hug, C.C.: The anesthetic potency of fentanyl in terms of its reduction of enflurane MAC. Anesthesiology, 57: 485-488, 1982.

23. Hug, C.C.: What are the roles of narcotic analgesia in anesthesia? Refresher Courses in Anesthesiology (American Society of Anesthesiologists, Inc.), 9: 71-83.

24. Wanscher, M.; Tonnesen, E.; Huttel, M.; Larsen, K.: Etomidate infusion and adrenocortical function - a study in elective surgery. ACTA Anesthesiology Scandinavia, 29: 483-485, 1985.

25. Haefely, W.; Polc, P.: Physiology of GABA Enhancement by benzodiazepines and barbiturates. Benzodiazepine/GABA Receptors and Chloride Channels: Structural and Functional Properties. Alan R. Liss, Inc.

26. Miller, K.W.: The nature of the site of general anesthesia. In: J.R. Smythies; R.J. Bradley (eds.): International Review of Neurobiology, Academic Press, 27: 1-61, 1985.

27. Krnjevic, K.: Cellular and synaptic effects of general anesthetics. In: S.H. Roth; K.W. Miller (eds.): Molecular and Cellular Mechanisms of Anesthetics, Plenum Publishing Corporation, pp. 1-16, 1986.

28. Nicoll, R.A.; Madison, D.V.: General anesthetics hyperpolarize neurons in vertebrate central nervous system. Science, 217: 1055-1057, 1982.

29. Frazier, W.T.: Effects of thiopental and ketamine on neurons in the isolated abdominal ganglion of apylsia. Fifth International Congress on Pharmacology, A-428, p. 72, July, 1972.
30. Lodge, D.; Arnis, N.A.: Effects of Phencyclidine on excitatory amino acid activation of spinal interneurons in the cat. Europ. J. Pharm., 77: 203-204, 1982.
31. Arnis, N.A.; Berry, S.C.; Burton, N.R. et al.: The dissociative anesthetics, ketamine and phencyclidine, selectively reduce excitation of central mammalian neurons by N-methyl-aspartate. British J. Pharm., 79: 565-575, 1983.

Sites, Rates and Filters that Best Eliminate Background Noise and Variability during Cortical Evoked Potentials Spinal Cord Monitoring

M. R. Nuwer;[*] E. G. Dawson

Introduction

Understanding the physiology of the posterior column and associated pathways is essential in planning for spinal cord monitoring. The physiology of these important pathways can be investigated in terms of the system's response to simple stimuli. These stimuli can vary in several important ways including the location of the stimulation site, duration of each stimulus pulse, stimulus intensity, stimulus repetition rate, sites of the principal recording electrode and its corresponding reference electrode, and the effects of the high and low analog filters. We have studied the physiology of the posterior column and related pathways in a systematic way (Nuwer and Dawson, 1984a, 1984b, 1984c). The details of our studies have been published previously. These studies provide basic knowledge about physiology. They also were designed so that

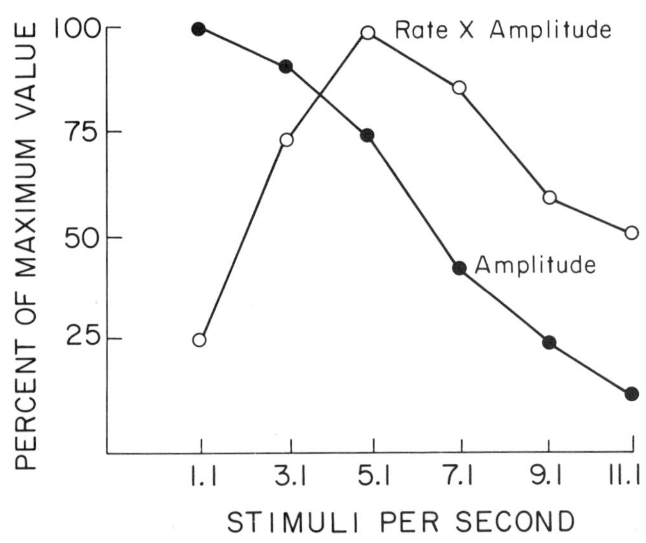

Fig. 1. Effects of increasing the rate of stimulus presentation in one patient. As rate is increased, EP amplitude decreases. The product Rate X Amplitude helps to compare the advantageous increase in speed of testing and the disadvantageous loss of amplitude. In this patient the rate 5.1 per second appeared to be the best compromise between speed and attenuation. A 30Hz low filter was used (Nuwer and Dawson, 1984a).

[*] UCLA Department of Neurology, Reed Neurological Research Center, 710 Westwood Plaza, Los Angeles, CA 90024

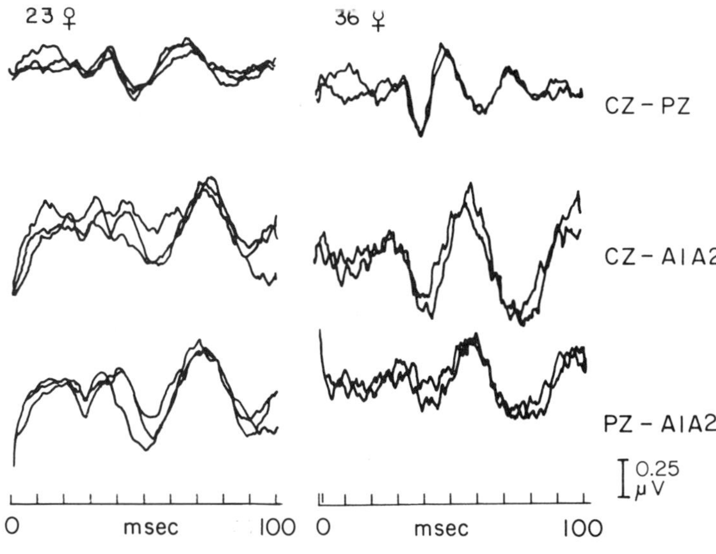

Fig. 2. Bipolar recordings are usually less noisy and more reproducible than referential recordings made in the OR. These two patients show the typical differences in the two techniques. Stimulation here was to one posterior tibial nerve in each patient and the 30Hz low filter was used. Linked ears was used as the reference for the referential recordings (Nuwer and Dawson, 1984a).

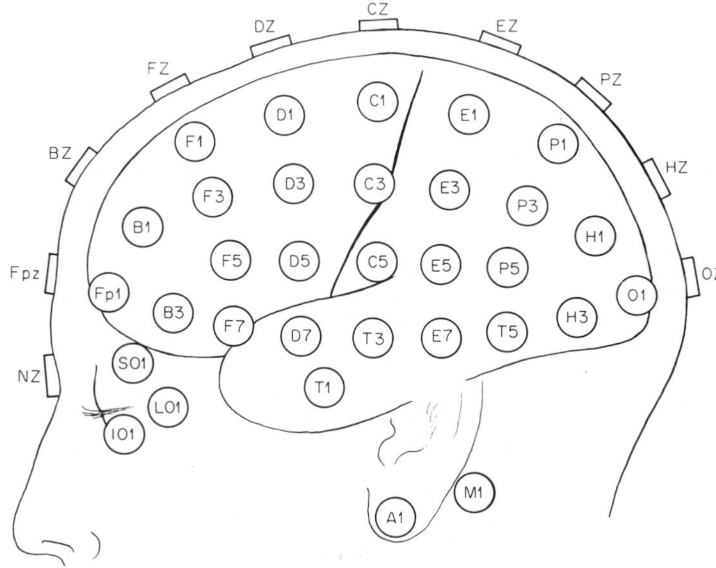

Fig. 3. Alphabetic nomenclature for electrode sites over the left hemisphere. This location system is based on the 10% system described by Chatrian et al. (1985). The alphabetic nomenclature used here is an alternative to the prime and double prime nomenclature suggested by Chatrian et al. The electrode sites on the right scalp are given by even numbered designations corresponding to the odd numbered designations seen here for the left hemisphere. The site Ez is located just behind the vertex (Nuwer, 1986).

useful, clinical information can be extracted. From this clinical information we have been able to make recommendations regarding optimizing the methodology used in clinical spinal cord monitoring with cortical somatosensory EPs.

During the initial years of cortical somatosensory spinal cord EP monitoring, recommendations were made by a number of investigators regarding stimulus and recording parameters. When we first began to use these methods in the operating room, we were disappointed to see the great amount of background noise, artifact and unpredictable signal variability that occurred. This noise and variability substantially interfered with the task of monitoring. Since one of us (MRN) had had long clinical experience in outpatient EP testing, we considered that the noise and variability in our recordings were not due to simple inexperience. Our choice of stimulus and recording techniques was quite in line with choices commonly discussed at meetings and described in the literature up to that point. Other clinicians had also noted great difficulty in performing

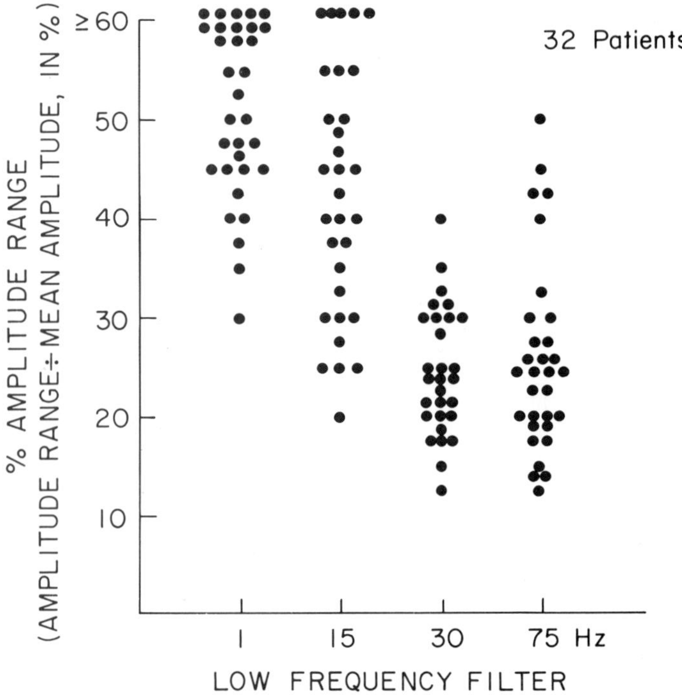

Fig. 4. Increasing the low frequency filter up to 30Hz enhances the reproducibility of the intraoperative EPs, i.e., the percent amplitude range is reproduced. However, the use of a 75Hz filter may so attenuate the EP that the amplitude range actually increases. Compare to Fig. 5 (Nuwer and Dawson, 1984a).

monitoring at the scalp, encouraging them to develop epidural or intervertebral ligament recording techniques. In doing so, Lueders et al. (1982) declared that such scalp monitoring simply could not be done adequately. Other investigators (Jones, 1982a, 1982b, Jones et al., 1983) have made similar remarks. Typical recommendation then had included stimulation rates of 0.8-2.0 per second, low settings of the high and low analogue filters (e.g., 1 and 1,000Hz), and use of referential rather than bipolar recording technique. Variability of cortical recordings was noted to be especially great when filters were set low (e.g., Schramm et al., 1979). On the basis of our studies of the physiology along the important involved pathways, we were able to describe clinical recommendations which substantially differed from those previous techniques.

The experience with these cortical spinal cord monitoring techniques in our institution now extends over more than 300 cases, including 160 cases of application in scoliosis surgery. Overall, noise, artifacts, and meaningless variability have been constrained to quite low amounts in most cases. Clinical experience is further detailed elsewhere in this volume (Chapter by More, Dawson and Nuwer). Details of the basic physiology of these pathways are reviewed below, and are based on experiences with the initial 115 patients (Nuwer and Dawson, 1984a, 1984b, 1984c).

Stimulation

Stimulus site and intensity.

In 25 patients we directly compared EPs from unilateral peroneal nerve stimulation at the knee to EPs from ipsilateral posterior tibial nerve stimulation at the ankle, intraoperatively. The latency for posterior tibial nerve stimulation was significantly longer than for peroneal stimulation, reflecting the shorter peripheral pathway traversed by the latter. No significant EP amplitude difference was seen between the two sites. The degree of noise and random amplitude or latency changes appeared comparable for the two sites. For practical purposes we chose the posterior tibial nerve site for routine monitoring. The ankle is more accessible under surgical drapes.

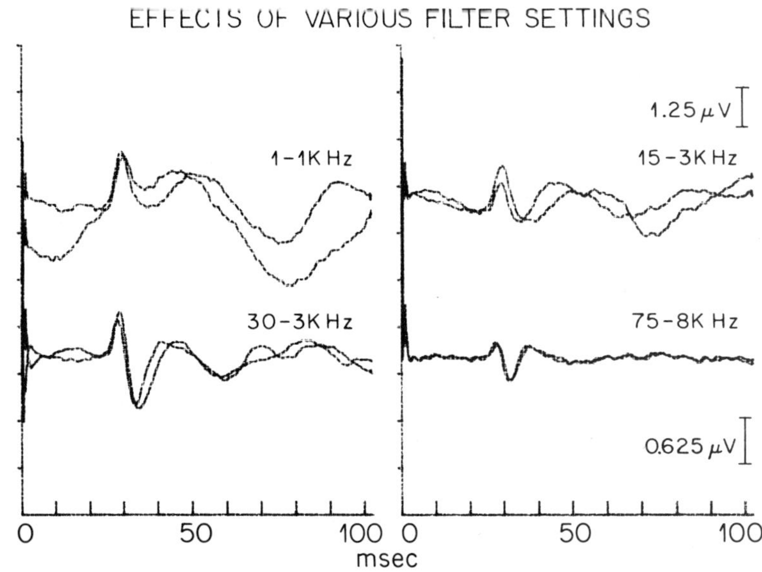

Fig. 5. Effects of 4 different filter settings during intraoperative recordings, taken from a single scalp channel (Cz-Pz). Variability is greatest in the 1Hz channel and is reduced by higher filter settings. Each pair of EPs is a typical set of 2 consecutive recordings in this one patient. The amplitude scale is doubled for the lower two EPs. Compare to Fig. 4 (Nuwer and Dawson, 1984c).

Stimulus intensity was gradually increased in a systematic way on 6 patients. The EP amplitude recorded at the scalp gradually increased and then plateaued. This plateau was usually reached at about 20mA intensity. This corresponds to approximately twice motor threshold in a typical patient.

Stimulus rates.

Stimulation rates of 1.1, 3.1, 5.1, 7.1, 9.1 and 11.1 per second were studied systematically in 10 patients. As the EP stimulation rate increased, the EP amplitude decreased in all patients. This gradual decrease of amplitude is likely to have contributed to previous recommendations to use a very slow stimulus repetition rate during monitoring; yet, monitoring situations require the quick collection of data. An optimal tradeoff between quickness and amplitude attenuation can be measured by the product of Rate X Amplitude. The optimal rate would be one at which Rate X Amplitude was at its maximum. The rate 5.1 per second was optimal in 7/10 patients; 7.1 per second was optimal in 2/10 patients; and 3.1 per second was optimal in the remaining 1/10 patient (Fig. 1). The rate of stimulation around 5 per second appears to yield EPs without excessive amplitude loss and do so much more quickly than the previously recommended rates of 2 per second or less.

Recording

Sites and references.

EPs were recorded using both a bipolar and a referential technique in 24 patients. The bipolar techniques used closely spaced electrodes from several scalp sites near Cz and Pz. The referential recordings used ear, forehead or shoulder references. The referential recordings yielded EPs which were higher in amplitude in 22/24 patients. The EP latency was shorter in the referential channel in 20/24 patients. These referential channels sometimes also recorded early far field peaks which were not seen at all in the bipolar channels (because of cancellation effects). However, the amount of random background variability was greater using the referential technique for 10/24 patients (Fig. 2), and it was comparable in amplitude for the two techniques in the remaining 13/24 patients. In no case was the random background noise and variability lower using

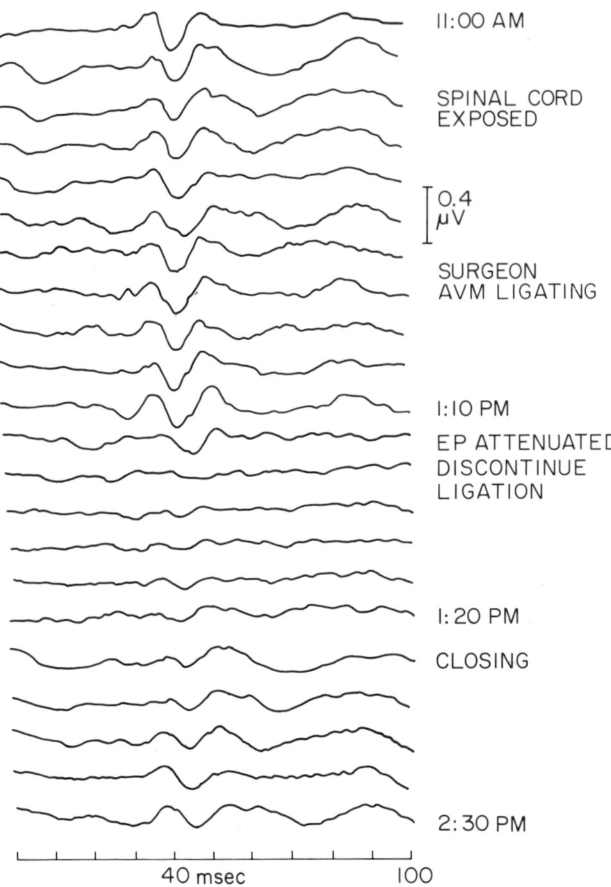

Fig. 6. The disappearance of EPs in this patient warned the surgeon against further ligation of a thoracic cord arteriovenous malformation (AVM). When the EP disappeared most of the proposed ligation had been accomplished; further intervention was stopped. EPs returned to normal amplitudes in about 1 hour, although latencies remained more delayed. The baseline EP in this patient had an abnormal delay in latencies, probably due to his AVM. These EPs showed more variability (30%) than was seen in most patients. Stimulation was to one peroneal nerve. A 30Hz filter was used. The patient awoke without any new neurological impairment (Nuwer and Dawson, 1984a).

the referential technique. Some of the increase in the noise variability in the referential channels appeared to be from movement artifact, residual muscle artifact and occasionally interference from other electrical devices in the OR. Movement artifact occurred because it was considered impractical to repeatedly stop the operation to perform EPs. The amplitude of the EP was often less in the bipolar channel than in the referential technique. Overall, however, the bipolar technique appeared to yield reproducibility superior to references placed at the nose, at the ears or on the shoulders. Subsequent experience with variations on these techniques has suggested that the best reference is at Fz. This site seems to have an optimal trade off between the elimination of noise seen with bipolar technique, and the better EP amplitudes seen with referential techniques. In individual monitoring situations, a neck reference is often helpful in finding and following the subcortical peaks. When these peaks are present and the channel has low noise, such a referential technique can be quite advantageous. However, noise and artifact and other random variability sometimes do not become problems until an operation is already half done, and so a wise monitoring team will often choose to follow an Fz reference, other bipolar channels and a neck reference channel. In this way they can make use of the best channels for an individual patient at each point in the operation.

In 12 patients, isopotential maps of the EP fields (topographic EP maps) were created, using 12-15 channels per patient at locations near and at Cz and Pz. This was done for the major EP peaks around 30-40msec after stimulation. The 12 patients so studied showed no cases with a maximum Ep peak amplitude at Cz. Instead, the maxima were generally posterior to Cz, sometimes slightly ipsilateral to the side of stimulation. Chatrian et al., (1985) described an extension to the 10-20 system which has been called the 10% system. An alternate, alphabetic nomenclature for this 10% system has also been described and is shown in Fig. 3. Among the 12 patients studied with topographic maps, the early major positive scalp peak had its maximum at Pz in 5, Ez in 4, E1 in 2, P1 in one and E2 in one. Since these were EPs from left-sided stimulation, the E1 and P1 maxima represent a paradoxical ipsilateral (left-sided) EP peak localization, believed due to the interhemispheric fissure location of the generator site (Cruse et al., 1982).

For bipolar recordings near the vertex, the amplitude difference between the recording sites is more relevant than the absolute amplitudes at those sites. In 28 patients we compared the amplitude of E1-E2 to the amplitude of Cz-Pz. The bipolar EP amplitude was larger in the transverse derivation (E1-E2) in 14/28 cases. It was larger in the anterior-posterior derivation (Cz-Pz) in 12/28 cases. In the remaining 2 cases the amplitude was equal in the two derivations.

Filter effects.

The goal of filtering data is the removal of unwanted noise while keeping most of the desired signal. Fig. 4 and 5 demonstrate the effect of various settings of the low filter. The amount of random background variability was studied and measured as a "percent amplitude range". This demonstrated that the reproducibility of the EPs was usually optimal with low filtering at 30Hz. Much of the slower, later, taller components of the EP were eliminated, and the reproducibility of the remaining faster, earlier components was improved. At filter settings below 30Hz the EP continued to show considerable meaningless variability. At filter settings of 75Hz (or higher) the EP became so attenuated that it was difficult to find at all in some cases, leading to much random variability due only to background noise.

The effects of changing the high frequency filter were also examined, but no significant effect was seen on the percent amplitude range, or on mean amplitude or mean latency. the settings tested included 1,000, 3,000 and 8,000Hz.

Conclusions

Studies of the physiology of the somatosensory pathways produce data and conclusions about the nature of the generators, variability of response and effects of changing the stimulus. With knowledge of this physiology in hand, the clinician can help decide how to approach clinical spinal cord monitoring with cortical EPs. Improved reproducibility of the scalp EP can be otained by using closely place bipolar scalp recording channels rather than reference leads away from the scalp. Stimulus rates around 5.1 per second seem to provide an optimal trade off between quickness and EP amplitude loss. Moderate amounts of low frequency filtering with a 30Hz setting appear to also provide an optimal trade off between noisy open filters and low amplitude results at even more restricted filter settings. Overall recommendations are reviewed in Table 1. Most of these recommendations have become popular since we first began to encourage their use in 1982.

With these techniques, reproducible scalp EPs should be expected. Loss of EPs (Fig. 6) should raise very great concern about clinical problems. Among our initial 106 patients, EPs were lost only in 4 cases. In 2 cases the disappearance was transient (less than 5 minutes) and then the EPs returned to 100% of the previous level with only a mild (1-2msec) latency delay. In the other two cases the disappearance lasted 30-45 minutes. Amplitudes returned to 100% of the baseline only after 1-2 hours and the latencies remained mildly delayed (by 2-3msec). All four of these patients awoke without any neurological deficit. These cases probably represented true, transient spinal cord impairment rather than false positives. Technical causes of such change were checked and ruled out in each case. In the two cases of prolonged EP change, the surgeons did alter their operation on the basis of the EP information. In this way the EPs may have prevented postoperative paraparesis or paraplegia in these patients. Other clinical results are discussed elsewhere in this volume (Chapter by More, Dawson and Nuwer).

Table 1: Simplified set-up for scalp recordings

Stimulation

Nerves: Posterior tibial at ankle
Side: Alternate left and right
Intensity: Above motor threshold
Electrodes: Subdermal needles
Pulse width: 200 microseconds
Repetition rate: 5.1 per second at each leg

Recordings

Channels: E1 - E2
Ez - Fz
Lumbar - sacral
Available alternative channels:
Cervical - FpZ
Filters: 30 - 3000Hz
60Hz filter: Off
Timebase: 60msec
Sample size: 300 - 500
Measure: Latencies and amplitudes of most prominent peaks between 30 and 45msec.

References

1. Chatrian, G.E.; Lettich, E.; Nelson, P.L. (1985): Ten percent electrode system for topographic studies of spontaneous and evoked EEG activities. Am. J. EEG Technol. 25: 83-92.
2. Cruse, R.; Klem, G.; Lesser, R.P.; Lueders, H. (1982): Paradoxial lateralization of cortical potentials evoked by stimulation of posterior tibial nerve. Arch. Neurol. 39: 222-225.
3. Jones, S.J. (1982a): Clinical applications of short latency somatosensory evoked potentials. Ann. NY Acad. Sci. 388: 369-387.
4. Jones, S.J. (1982b): Somatosensory evoked potentials: The abnormal waveform. In: A.M. Haliday (ed.): Evoked Potentials In Clinical Testing. Edinburgh, Churchill Livingstone, pp. 429-470.
5. Jones, S.J.; Edgar, M.A.; Ransford, A.O.; Thomas, N.P. (1983): A system for the electrophysiological monitoring of the spinal cord during operations for scoliosis. J. Bone Joint Surg. 65B: 134-139.
6. Lueders, H.; Gurd, A.; Hahn, J.; Andrish, J.; Weiker, G.; Klem, G. (1982): A new technique for intraoperative monitoring of spinal cord function: Multichannel recording of spinal cord and subcortical evoked potentials. Spine 7: 110-115.
7. More R.C.; Nuwer, M.R.; Dawson, E.C. (19xx): True and false positive amplitude attenuations during evoked potential spinal cord monitoring. In: (the present volume).
8. Nuwer, M.R. (1986): Evoked Potential Monitoring In The Operating Room. Raven, New York.
9. Nuwer, M.R.; Dawson, E. (1984a): Intraoperative evoked potential monitoring of the spinal cord: Enhanced stability of cortical recordings. Electroencephalogr. Clin. Neurophysiol. 59: 318-327.
10. Nuwer, M.R.; Dawson, E. (1984b): Somatosensory evoked potential monitoring: Measurement of variability. In: R.H. Nodar; C. Barber (eds.): Evoked Potentials II. Butterworth, Boston, pp. 510-513.
11. Nuwer, M.R.; Dawson, E. (1984c): Intraoperative evoked potential monitoring of the spinal cord: A restricted filer, scalp method during Harrington instrumentation for scoliosis. Clin. Orthop. 183: 42-50.
12. Schramm, J.; Hashizume, K.; Fukushima, T.; Takahashi, H. (1979): Experimental spinal cord injury produced by slow, graded compression: Alterations of cortical and spinal evoked potentials. J. Neurosurg. 50: 48-57.

Effect of Isoflurane on Human Median Nerve Evoked Potentials

Satwant K. Samra[*]

Abstract

This paper reviews the recent literature dealing with effect of isoflurane on different components of human median nerve evoked potentials. These effects are compared to effects of other intravenous and inhalation anesthetics. Changes in peak latencies and amplitude of EP (N9), N13, N20 and conduction times EP-N13 and EP-N20 after administration of 0.5%, 1%, 1.5%, and 2% end-tidal concentration of isoflurane have been studied. Peak latencies of all components increased after all concentrations of isoflurane. No significant change in EP-N13 conduction time was noted, while N13-N20 conduction time increased significantly and this increase was dose related. Amplitude of EP and N13 did not show significant change with 1% and 1.5% isoflurane when compared to control values. Amplitude of N20 decreased significantly following isoflurane anesthesia when compared to control values and this amplitude reduction was dose related. With 2% isoflurane, amplitude reduction is so great that in a significant number of patients N20 may not be discernible. When compared to other commonly used inhalation anesthetics, effect of isoflurane on amplitude of cortical components of MN-SSEP's is similar to that produced by halothane, less marked than that produced by enflurane anesthesia. When compared to commonly used intravenous anesthetics, effect of 1% isoflurane on latency and amplitude of N20 is comparable to that produced by 75µg/ml of thiopental, but is more marked when compared to 55-63µg/kg fentanyl given intravenously.

Introduction

Although the technique of signal averaging was defined by Dawson (5) more than three decades ago, it is only in the last ten years that developments in computer technology have resulted in the availability of portable equipment which has made recording of evoked potentials a practical reality in clinical medicine. In addition to being utilized as diagnostic and prognostic tests in patients with neurological diseases, evoked potentials are being monitored, with increasing frequency, to monitor the integrity of neural pathways at risk in patients undergoing surgery under general anesthesia. While qualitative neurophysiological effects of general anesthetics have been well recognized (Clark et al., 1971; Clark and Rosner, 1973) for a long time, information regarding quantitive effects of commonly used anesthetics is relatively new. It has been shown that many commonly used intravenous anesthetics, analgesics, and sedatives alter short latency median nerve somatosensory evoked potentials (MN-SSEP's). Until recently,

[*] Department of Anesthesiology E-91, University of Texas Medical Branch, Galveston, TX 77550-2778

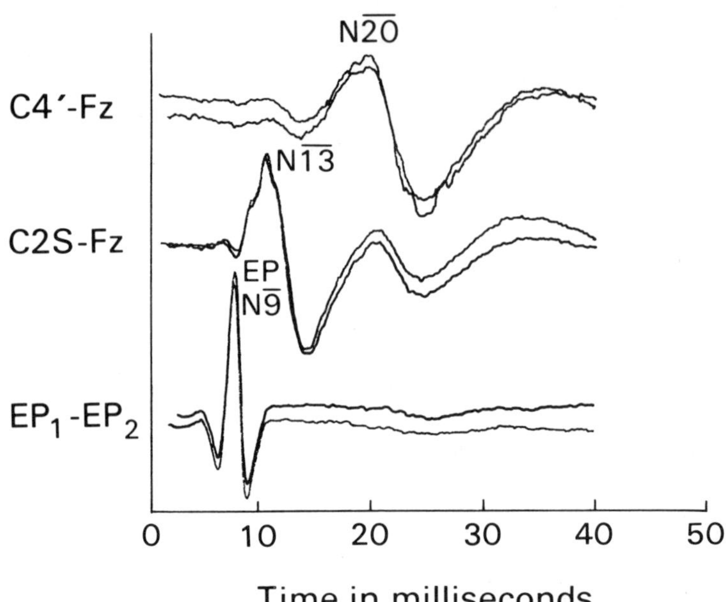

Figure 1: Typical traces of MN-SSEP's. The points at which measurements of peak latency of different evoked potentials were made in each derivation have been identified. Negative wave forms in this and other figures are represented by an upward deflection. Reproduced from Samra, Vanderzant, Domer and Sackellares (17) with permission from authors and Anesthesiology, published by J.B. Lippincott Co., Philadelphia, PA.

volatile anesthetics have been claimed to cause greater attenuation of cortical evoked potentials leading to common recommendation that these drugs should be entirely avoided in patients undergoing intraoperative monitoring of evoked potentials, despite the fact that to date no comparative study between all inhalation anesthetic agents and intravenous agents has been published. Only in recent years have some investigators studied the effects of inhalation anesthetics on somatosensory evoked potentials (McPherson et al., 1985; Peterson et al., 1986; Pathak et al., 1987). Due to common use of polypharmacology in today's practice of anesthesiology, most of these studies have investigated the effect of one or another inhalation anesthetic in addition to either an inhalation (nitrous oxide) or intravenous (fentanyl) anesthetic. We have recently studied the effect of isoflurane alone (without any other intravenous or inhalation anesthetic) on human median nerve evoked potentials (Samra et al., 1987). In this paper, this data will be reviewed and results compared to other published reports dealing with quantitative changes in latency and amplitude of cortical component (N20) of median nerve evoked potentials.

Methods

The protocol was approved by the committee to review investigations involving human beings. Fifteen consenting adult patients (six men and nine women) scheduled for elective surgery under general anesthesia were studied. Surgical procedures included exploratory laparotomy (4), mastectomy (1), vulvectomy (1), and lumbar laminectomy (9). The patients' mean age was 35 years (range 20-52) with a mean weight of 60kg (range 50.9-92.0) and a mean height of 170cm (range 156.5-187.5). All patients were evaluated preoperatively by an anesthesiologist and a neurologist and found free of systemic disease (ASA P.S. I) and neurologic disorders other than lumbar radiculopathy in patients undergoing laminectomy.

Control NM-SSEP's were recorded on the morning of surgery in unpremedicated patients. Anesthesia was induced with intravenous thiamylal sodium (5-7mg/kg) followed by either succinylcholine (1mg/kg) or pancuronium bromide (0.1mg/kg) to facilitate endotracheal intubation. Heart rate, ECG, systemic blood pressure, esophageal temperature, and end-tidal concentrations of carbon dioxide and isoflurane were continuously monitored during surgery. All patients were mechanically ventilated

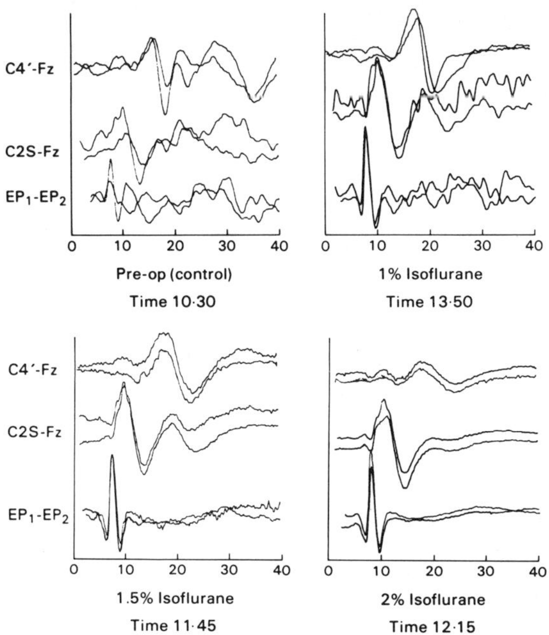

Figure 2: Left MN-SSEP's monitoring with isoflurane anesthesia. The effect of different concentrations of isoflurane in one representative patient is shown. EP (EP$_1$-EP$_2$ derivation) and N13 (C2S - Fz derivation) show lack of change in latency and morphology while change in amplitude of N20 (C4' - Fz derivation) with increase in end-tidal isoflurane from 1% to 2% is obvious. Reproduced from Samra, Vanderzant, Domer and Sackellares (17) with permission from authors and Anesthesiology, published by J.B. Lippincott Co., Philadelphia, PA.

to maintain an end-tidal carbon dioxide tension of 35-40mmHg. Anesthesia was maintained with 100% oxygen and isoflurane. MN-SSEP's were recorded at 0.5, 1.0, 1.5, and 2% end-tidal isoflurane concentrations. Intraoperative recording of MN-SSEP's began at least 20 minutes after injection of thiamylal sodium. End-tidal isoflurane concentration was held constant for at least 10 minutes before recording MN-SSEP's. The dose of isoflurane was administered according to needs of the patient (determined by changes in systemic blood pressure and heart rate) in keeping with the level of surgical stimulation. Therefore, not all patients could be studied at all concentrations, and the

TABLE 1: Amplitude (mean ± SD) of Different Peaks of MN-SSEPs at Different Concentrations of Isoflurane

End-tidal Isoflurane (volume %)	EP		N13		N20	
	N	Value (µvolt)	N	Value (µvolt)	N	Value (µvolt)
Control	13	2.49 ± 1.50*	13	2.34 ± 0.75*	13	3.07 ± 2.00*
0.5	7	2.67 ± 1.05	7	2.38 ± 0.72	7	1.50 ± 0.84
1.0	13	2.28 ± 1.64*	13	2.19 ± 0.91*	13	1.54 ± 1.06*
1.5	13	2.19 ± 1.67*	13	2.16 ± 0.84*	13	1.07 ± 0.78*
2.0	10	2.44 ± 1.78	10	2.00 ± 0.93	7	0.96 ± 0.63
P value		.59		0.65		0.0009
Pairwise Comparisons						
Control vs. 1		NS		NS		Significant
Control vs. 1.5		NS		NS		Significant
1 vs 1.5		NS		NS		Significant

NS = Not significant
* = Values included in statistical analysis

TABLE 2: Comparative Effects of Anesthetics on MN-SSEPs

Source	Anesthetic	Dose	Increase in Latency of N20	Change in Amplitude of N20	Loss of Cortical EP
Drummond et al. (1985)	Sodium Thiopental	35 µg/ml	8.0%	↓ 33.0%	None
		51 µg/ml	10.5%	↓ 43.0%	
		66 µg/ml	12.0%	↓ 52.0%	
		75 µg/ml	15.0%	↓ 57.0%	
Schubert et al. (1986)	Fentanyl	55–63 µg/kg	5.0%	↓ 35.0%	None
McPherson et al. (1986)	Etomidate	0.4 mg/kg	–	↑ 170.0%	None
Kochs et al. (1986)	Etomidate	0.3 mg/kg	6.0%	↑ 624.0%	None
	Diprivan	2.0 mg/kg	9.2%	↑ 52.0%	None
Kalkman et al. (1986)	Etomidate	0.3 mg/kg	7.1%	↑ 164.0%	None
	Midazolam	0.2 mg/kg	0.5%	0%	None
Sebel et al. (1984)	Nitrous Oxide	10%	1.7%	↓ 25.0%	
		30%	2.0%	↓ 30.5%	
		50%	0.4%	↓ 43.5%	2/5
Peterson et al. (1986)	Halothane with 60% N₂O	0.5 MAC	5.0%	↓ 61.0%	
		1.0 MAC	8.4%	↓ 70.0%	None
		1.5 MAC	13.2%	↓ 80.0%	
	Isoflurane with 60% N₂O	0.5 MAC	7.0%	↓ 52.0%	
		1.0 MAC	12.0%	↓ 85.0%	
		1.5 MAC	19.5%	Not discernible	5/7
	Enflurane with N₂O	0.5 MAC	6.2%	↓ 52.0%	
		1.0 MAC	17.0%	↓ 85.0%	4/7
		1.5 MAC	Not recordable	Not recordable	7/7
Samra et al. (1987)	Isoflurane	0.5%	11.3%	↓ 49.0%	
		1.0%	14.0%	↓ 50.0%	
		1.5%	18.0%	↓ 65.0%	1/14
		2.0%	17.5%	↓ 69.0%	3/10

order of exposure to different concentrations of isoflurane varied. Recordings with 0.5% isoflurane were made either prior to surgical incision or at the end of surgical procedure. We could record MN-SSEP's at stable end-tidal isoflurane concentrations of 0.5% in seven, 1.0% in all 15, 1.5% in 14, and 2% in ten patients. We obtained recordings in only three patients at all four end-tidal concentrations of isoflurane, whereas 13 of 15 patients were successfully studied at both 1% and 1.5% concentrations. All patients were interviewed 24 hours after surgery. None had recall of intraoperative events, and all patients expressed willingness to participate in a similar study again if necessary.

Stimulation and recording parameters for MN-SSEP's followed the American Electroencephalographic Society Guidelines (1). Either the left or right median nerve was stimulated at the wrist, and recordings were made from electrodes located at Erb's point over the brachial plexus, the spinous process of the second cervical vertebra (C-2S) and the contralateral sensory cortex (C3' or C4' of the international 10-20 System).

Intensity of stimulation current varied among patients. Each patient's sensory threshold (minimal current the patient could feel) and motor threshold (when movement of thumb was visible) were determined in preoperative studies. We used a stimulus intensity equal to motor threshold for recording of control traces. Stimulus intensity was increased to twice the motor threshold after induction of anesthesia. The effect of altering stimulus intensity alone was evaluated in three patients in whom MN-SSEP's were recorded with variable stimulus intensity, while end-tidal concentration of isoflurane and other variables were held constant. At least two averages were obtained for each recording to assure reproducibility, and a mean of two readings were used for measurement of peak latencies and amplitude of Erb's point potential (EP), cervical potential (N13), and scalp potential (N20). From these values conduction times EP-N13, N13-N20, and EP-N20 were calculated.

Statistical Analysis

Mean and standard deviation of values for various peak latencies, conduction times, and amplitude were calculated. Numerical data were subjected to repeated-measures analysis of variance using Hotelling's T-square test to determine statistical significance. A P value less than 0.05 was considered significant. This statistical approach is most appropriate for studies involving repeated measurements in the same individual at different times but its disadvantage is that it is applicable only to a data set in which all measurements have been made at all time points. Because it was not possible to administer all concentrations of isoflurane to all patients, values obtained in 13 patients that could be studied at both 1.0% and 1.5% end-tidal isoflurane concentrations were included for repeated-measures analysis of variance.

Results

Satisfactory traces were obtained in all patients studied. All patients were hemodynamically stable, and mean arterial pressure was within 15% of control readings. Maximum esophageal temperature change during the study period was 0.5°C. Figure 1 shows a typical trace, along with identification of various components of MN-SSEP's for measurement of peak latencies and calculation of conduction times. Amplitude was measured from each identified negative peak to the next positive peak. The effect of isoflurane on morphology of various MN-SSEP's of a representative patient is shown in Figure 2. EP and N13 peaks remained well defined and without significant increase in their peak latencies. Contralateral scalp responses (N20) were reduced in amplitude but were evident in all patient after end-tidal concentrations of 0.5 and 1.0%; in 13 out of 14 patients after 1.5%; and in seven out of ten patients after 2% isoflurane. N20 latency and conduction time EP-N20 progressively increased with increasing concentration of isoflurane. Mean increase in latency of N20 in seven patients after 2% isoflurane was 3.7msec, the range being 2.1-5.0msec. Mean EP-N20 conduction time increased by 3.2msec, with a range of 1.7-4.4msec. MN-SSEP's components later than N20 were less durable and could not be identified after administration of 0.5% isoflurane.

Mean latency of EP increased from 10.2msec (control) to 10.9 and 10.8msec after 1% and 1.5% isoflurane respectively. Difference from control values was statistically significant while that between 1% and 1.5% isoflurane was not significant. Mean latency of N13 increased from 13.93msec (control) to 14.65 and 14.57msec after 1% and 1.5% isoflurane respectively. Once again, difference from control values was significant while that between 1% and 1.5% isoflurane was not significant. Maximum increase in latency was noted in N20 component where it increased from 19.45msec (control) to 22.21 and 22.94msec with 1% and 1.5% isoflurane respectively. Increase in N20 latency with both concentrations of isoflurane was statistically significant when com-

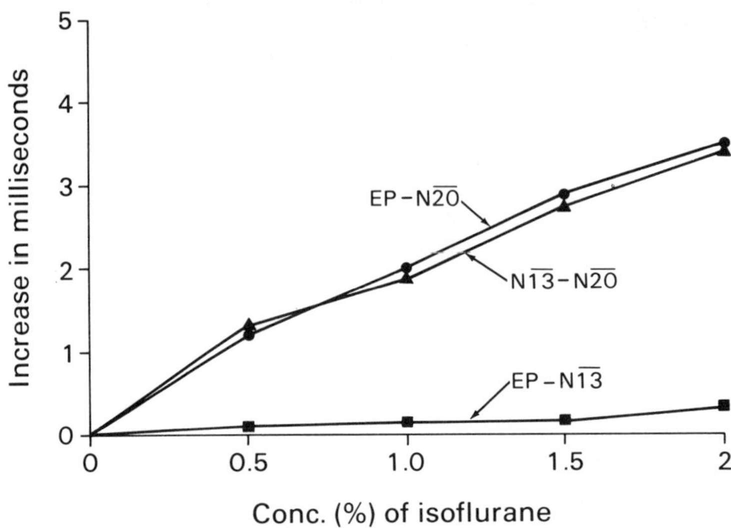

Figure 3: Increase in conduction time EP-N13 was not signficant. Conduction times N13-N20 and EP-N20 showed significant increase with all concentrations of isoflurane when compared to control values. Difference between 1% and 1.5% isoflurane was also significant.

pared to control values and the difference between 1% and 1.5% isoflurane too was statistically significant.

Changes in conduction times are diagramatically shown in Figure 3. There was not significant change in conduction time EP-N13 with 1% or 1.5% isoflurane, while conduction time N13-N20 increased significantly compared with control values. N13-N20 conduction times recorded at 1% and 1.5% concentrations of isoflurane also differed significantly. Similarly, EP-N20 conduction time increased significantly after administration of isoflurane when compared with control values, and the difference between 1% and 1.5% isoflurane was also significant.

Table 1 shows the numerical values for amplitude of different components of MN-SSEP's. There was no significant change in amplitude of EP and N13 with 1% or 1.5% isoflurane compared with control values, nor was the difference between two concentrations significant. By contrast there was a remarkable reduction of amplitude of N20 with isoflurane anesthesia compared with control values, and the difference between 1% and 1.5% isoflurane was statistically significant. There was a further decrease in amplitude of N20 after 2% isoflurane. We wish to emphasize that the effect of isoflurane on amplitude of N20 was highly variable and as already mentioned, in three out of ten patients studied with 2% isoflurane, N20 could not be identified. We noted that the predominant effect of change in stimulus intensity (in three patients studied) was a change in amplitude with minimal change in latency (Fig. 4). A slight change in latency can be attributed to a better description of an individual component, which allows more precise measurement of latency.

Discussion

The objectives of this study were to 1) determine if MN-SSEP's can be successfully recorded in patients anesthetized with isoflurane; 2: quantitate the effect of Isoflurane on change in latency and amplitude of MN-SSEP's; 3) compare the effect of isoflurane on MN-SSEP's with different neural generators, thereby studying the differential effects of isoflurane on different parts of the nervous system; and 4) to compare our results with published information dealing with quantitative effects of other intravenous and inhalation anesthetics.

Our patients were free of neurologic disease, with the exception of those with lumbar radiculopathy undergoing lumbar laminectomy. All were undergoing surgical procedures that do not have any effect on neural pathways assessed by MN-SSEP's. Anesthesia was maintained with oxygen and isoflurane following induction with thiamylal. Shimoji et al., (1974) have shown that thiamylal (5mg/kg) has a significant effect on both scalp and spinal evoked response. However, in their study SSEP's had returned to control values after 8-10 minutes. In our study the minimum interval between administration of thiamylal and recording of MN-SSEP's was 20 minutes. Other significant factors to be considered in the design of this study are time-dependent changes and those related to change of technical parameters such as stimulus intensity. The order of different concentrations of isoflurane in our patients was varied to nullify any effect that duration of anesthesia alone might have had. We observed similar effects of anesthetic concentration (0.5% and 1.5%) in five individual patients, regardless of whether MN-SSEP's were recorded at the beginning or toward the end of 3-4 hours of anesthesia.

Despite the increased use of SSEP's for intraoperative monitoring, there are no definite criteria regarding choice of stimulus intensity that should be used for the recording of SSEP's during surgery (American Electroencephalographic Society Guidelines, 1984). Most laboratories use stimulus intensities of either motor threshold or just above motor threshold. The study of McPherson et al., (1985) of anesthetics on SSEP's used stimulus intensity of motor threshold in unanesthetized patients and arbitrarily increased it to three times the motor threshold in anesthetized patients. Drummond, Todd and Sang (1985) used stimulus intensity of motor threshold +20%. However, the effect of stimulus intensity alone on somatosensory evoked potentials has been previously investigated in detail in unanesthetized patients. Lesser and co-workers (10) reported that in awake volunteers, motor threshold stimulation gave consistently submaximal responses, while a sum of motor plus sensory threshold gave potentials that were consistently close to maximal amplitude; and when stimulus intensity was increased, no remarkable change was seen in absolute latency of N9 and N18, while latency of N13 decreased slightly. Nuwer and Dawson (13) while studying the effects of different stimulation and recording parameters on SSEP's in anesthetized patients, reported that as stimulus intensity was gradually increased, the amplitude of SSEP's increased to reach a plateau at about 20ma. They did not comment on any change in latency. Tsuji et al., (21) have studied the effect of stimulus intensity on subcortical and cortical evoked potentials after stimulation of posterior tibial nerve. They have recommended the use of stimulus intensity three times the sensory threshold.

We felt that use of such a strong stimulation current (while awake) was not justified in our volunteer patients in whom this monitoring was not medically indicated. Therefore, like previous investigators, we recorded MN-SSEP's with stronger current after induction of anesthesia. We noted that the predominant effect of change in stimulation intensity was a change in amplitude with a minimal change in latency (Figure 4). Therefore, some of the effect of isoflurane on amplitude of MN-SSEP's in our study as well as previously published studies might have been counter balanced by change in stimulus intensity. Our results showed that latencies of all measured components (EP, N13, and N20) increased significantly after isoflurane (1% and 1.5%) when compared with control values. A comparison between recordings made after 1% isoflurane and 1.5% isoflurane showed no significant difference between peak latencies of EP and N13, while latency of N20 shows a significant difference. Two explanations for these observations are: 1) this difference in increase in latency of various peaks with different generator sources represents differential effects of isoflurane at different parts of nervous systems; or 2) an increase in peak latencies of EP and N13 does not represent an effect of isoflurane because it is not dose related, but is due to other physiologic and/or technical variables. Physiologic variables that are known to affect sensory evoked potentials (hemodynamic stability, $PaCO_2$, and temperature) were carefully controlled in this study. While the maximum decrease in esophageal temperature noted was 0.5°C, changes of temperature of the upper limb itself might have contributed to slight increases in latency of EP. Positioning of the upper limb, which was outstretched during surgery, could also have accounted for minor increases of peak latencies. Conduction times are independent of both limb position and limb temperature. Conduction time EP-N13 did not change significantly with 1% and 1.5% isoflurane when compared with control values. Conduction time N13-N20 increased significantly with isoflurane when compared with control values, and the difference in values with 1% and 1.5% isoflurane was also statistically significant. The total conduction time EP-N20 showed similar changes because N13-N20 conduction is a major contributor to this measurement. These data suggest that isoflurane has differential effects on the human nervous system.

Generator sources of median nerve evoked potentials have been postulated based on animal, clinical, and clinicopathologic studies. The state of present knowledge on this subject has been critically examined by Emerson and Pedley (7), who concluded that EP is the afferent volley in the brachial plexus at the Erb's point. N13 is a postsynaptic potential recorded maximally near the cervicomedullary junction, with near and far field components. Its generator is probably dorsal gray matter of the rostral cervical spinal cord or nucleus cuneatus. N20 is probably the first cortical response to sensory input that may have more than one generator. No significant change in EP-N13 conduction time and a significant change in N13-N20 conduction time in our study suggests a greater impairment of synaptic transmission compared with afferent fiber conduction by isoflurane. This mechanism explains the difference in

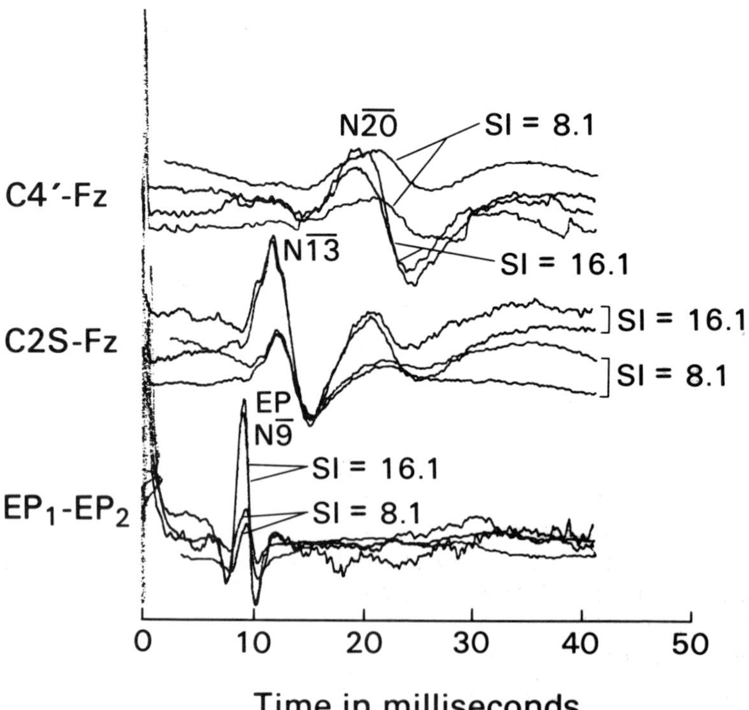

Figure 4: Effect of stimulus intensity (SI) on MN-SSEP's in one patient anesthetized with isoflurane (1.5% end-tidal). All recordings were made in duplicate to document reproducibility. Increase in stimulus intensity resulted in increase in amplitude without significant change in latency of various MN-SSEP's components. Reproduced from Samra, Vanderzant, Domer and Sackellares (17) with permission from authors and Anesthesiology, published by J.B. Lippincott Co., Philadelphia, PA.

isoflurane's effect on amplitude of various components of MN-SSEP's. EP and N13, with generator sources in peripheral nerve and spinal cord, do not involve multiple synapses, and their amplitude is not affected by increasing concentrations of isoflurane. By contrast N20 with its postulated generator source in thalamocortical radiation or sensory cortex involves multiple synapses and shows progressive decrease in amplitude with increasing concentrations of isoflurane. Similar differential effects with thiamylal in humans have been previously reported by Shimoji et al., (20).

Within the last five years several well designed clinical investigations of effects of anesthetics on sensory evoked potentials have been published (Sebel et al., 1984; Pathak et al., 1984; McPherson et al., 1985; Peterson et al., 1986; McPherson et al., 1986; Bird et al., 1986; Kochs et al., 1986; Kalkman et al., 1986; Pathak et al., 1987; Samra et al., 1987). Effect of all commonly used anesthetics (both inhalation and intravenous agents) has been studied in separate studies. A comparison of data collected in separate studies is not scientifically valid because of differences in patient population, experimental design and technical details of evoked potential recordings and interpretation. Yet it is valuable for clinicians involved in the care of patients undergoing intraoperative evoked potential recordings to understand the differences and similarities among different anesthetics regarding their effect on latency and amplitude of components of evoked potentials being monitored. Table 2 showing the effect of different anesthetics on latency and amplitude of cortical (N20) component of MN-SSEP's has been compiled by normalizing the data in eight published clinical studies (6, 8, 9, 12, 16, 17, 18, 19).

Table 2 has been derived from clinical studies which used similar methodology (similar patient selection and evoked potential recording techniques) and provided the numerical values for measurements of latency and amplitude of N20 both for control readings and after different drug doses. From each investigation mean changes in latency and amplitude resulting from effect of anesthetic used have been converted to a percentage change from control values in that investigation. This method of presentation is expected to provide the clinicians useful information about comparative effects of different anesthetics. Until recently, it was believed that all anesthetics result in an increase in latency and decrease in amplitude of cortical evoked potentials. Within the last year, it has been shown in three separate investigations (8, 9, 12) that etomidate, an intravenous induction agent, produces a marked increase in amplitude of N20 with minimal increase (6-7%) in latency. The only other drug which shares this property is diprivan which in one study (9) has been shown to increase the amplitude of N20 by 52% along with increase in latency (9.2%). With the exception of these two drugs, all other anesthetics studied so far have been shown to decrease the amplitude of N20. Intravenous drugs thiopental and fentanyl, result in 33-57% dose related decrease in amplitude, but total loss of N20 has not been observed even with doses adequate for surgical anesthesia. On the contrary, all inhalation anesthetics have been shown to result in total loss of N20 in some patients at higher concentrations. While administration of etomidate has been shown to increase the amplitude of N20 (which is claimed to be the effect of propylene glycol used as preservative in commercially available etomidate) the utility of this drug for maintenance of anesthesia remains questionable for two reasons: 1) this drug is used as an induction agent and it is not known how rapid or slow the recovery may be if used as a continuous infusion for maintenance of anesthesia; and 2) marked increase in amplitude of N20 may mask injury to neural pathway at risk. Similarly use of fentanyl or pentothal as a sole anesthetic agent presents the problem of delayed awakening (hence a delay in neurological evaluation) and respiratory depression in the postoperative period. Added to that is the disadvantage that fentanyl has no amnesic properties and even when used in high doses, fentanyl anesthesia may be associated with intraoperative recall. Therefore, use of potent inhalation anesthetics in low concentrations (1 MAC or less) combined with low dose of fentanyl offers the advantages of hemodynamic stability, rapid recovery, and minimal but predictable alterations in latency and amplitude of N20.

Since the publication of our study (17), we have used oxygen-isoflurane anesthesia, supplemented by small doses (5-10µg/kg) of fentanyl in fifty neurosurgical patients in whom somatosensory evoked potentials (following median nerve and posterior tibial nerve stimulation) were being recorded. Satisfactory recordings were obtained in all cases.

Bibliography

1. American Electroencephalographic Society: Guidelines for clinical evoked potential studies. J. Clin. Neurophysiol., 1984; 1: 3-53.
2. Bird, J.; Donegan, J.; Rupp, S. et al.: Effect of sufentanil bolus and two steady state infusions on median nerve evoked potentials. Anesthesiology, 1986; 65: A-341.
3. Clark, D.L.; Hosick, E.; Rosner, B.S.: Neurophysiological effects of different anesthetics in unconscious man. J. Appl. Physiol., 1971; 31: 884-891.
4. Clark, D.L.; Rosner, B.S.: Neurophysiologic effects of general anesthetics. I. The electroencephalogram and sensory evoked responses in man. Anesthesiology, 1973; 38: 564-582.
5. Dawson, G.D.: A summation technique for the detection of small evoked potentials. Electroencephalogr. Clin. Neurophysiol., 1954; 6: 65-84.
6. Drummond, J.C.; Todd, M.M.; Sang, U.H.: The effect of high dose sodium thiopental on brain stem auditory and median nerve somatosensory evoked responses in humans. Anesthesiology, 1985; 63: 249-254.
7. Emerson, R.G.; Pedley, T.A.: Generator sources of median somatosensory evoked potentials. J. Clin. Neurophysiol., 1984; 1: 203-218.
8. Kalkman, C.J.; Leyssius, A.T.V.; Hesselink, E.M. et al.: Effect of etomidate or midazolam on median nerve somatosensory evoked potentials. Anesthesiology, 1986; 65: A-356.
9. Kochs, E.; Treede, R.D.; Roewer, N. et al.: Alterations of somatosensory evoked potentials by etomidate and diprivan. Anesthesiology, 1986; 65: A-353.
10. Lesser, R.P.; Koehle, R.; Lueders, H.: Effect of stimulus intensity on short latency somatosensory evoked potentials. Electroencephalogr. Clin. Neurophysiol., 1979; 47: 377-382.
11. McPherson, R.W.; Mahla, M.; Johnson, R.; Traystman, R.J.: Effects of enflurane, isoflurane, and nitrous oxide on somatosensory evoked potentials during fentanyl anesthesia. Anesthesiology, 1985; 62: 626-633.
12. McPherson, R.W.; Sell, B.; Traystman, R.J.: Effect of thiopental, fentanyl and etomidate on upper extremity somatosensory evoked potentials in humans. Anesthesiology, 1986; 584-589.
13. Nuwer, M.R.; Dawson, E.: Intraoperative evoked potential monitoring of the spinal cord: Enhanced stability of cortical recordings. Electroencephalog. Clin. Neurophysiol., 1984; 59: 318-327.
14. Pathak, K.S.; Brown, R.K.; Cascorbi, N.F.; Nash, C.L.: Effects of fentanyl and morphine on intraoperative somatosensory cortical evoked potentials. Anesth. Analg., 1984; 63: 833-837.
15. Pathak, K.S.; Ammadio, M.; Kalamchi, A. et al.: Effect of halothane, enflurane and isoflurane on somatosensory evoked potentials during nitrous oxide anesthesia. Anesthesiology, 1987; 66: 753-757.
16. Peterson, D.O.; Drummond, J.C.; Todd, M.M.: Effects of halothane, enflurane, isoflurane, and nitrous oxide on somatosensory evoked potentials in humans. Anesthesiology, 1986; 65: 35-40.
17. Samra, S.K.; Vanderzant, C.W.; Domer, P.A. et al.: Differential effects of isoflurane on human median nerve somatosensory evoked potentials. Anesthesiology, 1987; 66: 29-35.
18. Schubert, A.; Peterson, D.O.; Drummond, J.C.; Saidman, L.J.: The effect of high dose fentanyl on human median nerve somatosensory evoked responses. Anesth. Analg., 1986; 65: S136.
19. Sebel, P.S.; Flynn, P.J.; Ingram, D.A.: Effect of nitrous oxide on visual, auditory and somatosensory evoked potentials. Br. J. Anesth., 1984; 56: 1403-1407.
20. Shimoji, K.; Tatsuhiko, K.; Nakashima, N.; Shimizu, H.: The effects of thiamylal sodium on electrical activities of the central and peripheral nervous systems in man. Anesthesiology, 1974; 40: 234-240.
21. Tsuji, S.; Luders, H.; Dinner, D.S.; Lesser, R.P.; Klem, G.: Effect of stimulus intensity on subcortical and cortical somatosensory evoked potentials by posterior tibial nerve stimulation. Electroencephalogr. Clin. Neurophysiol., 1984; 59: 229-237.

Evoked Potential Monitoring of Anesthetic and Operative Manipulation

J. Danto;[*] M. Cataletto; M. Wolpin

Introduction

Intraoperative monitoring during surgical procedures, such as orthopedic surgery, relies on a fundamental that the system being measured is at risk to the surgical procedure itself, and that these risks will manifest themselves in changes in the somatosensory evoked potential. A factor critical to the adequate monitoring of these potentials is the understanding of the effects generated by anesthetic control.

There are two general categories of anesthetic related effects. One is the patient's status including temperature, hemodilution, and blood pressure; and the other is those changes induced by the anesthetic agents themselves.

Methods and Procedures

The patients were all stimulated over the posterior tibial nerve at the ankle, behind the medial malleollus using 9mm disc electrodes on a bar with an interdisc distance of 30mm. Stimuli were 100μsec rectangle pulses, over motor threshold, at a repetition rate of 5.11/sec. If the morphology was not clear enough to monitor accurately, the repetition rate was dropped to 2.82/sec.

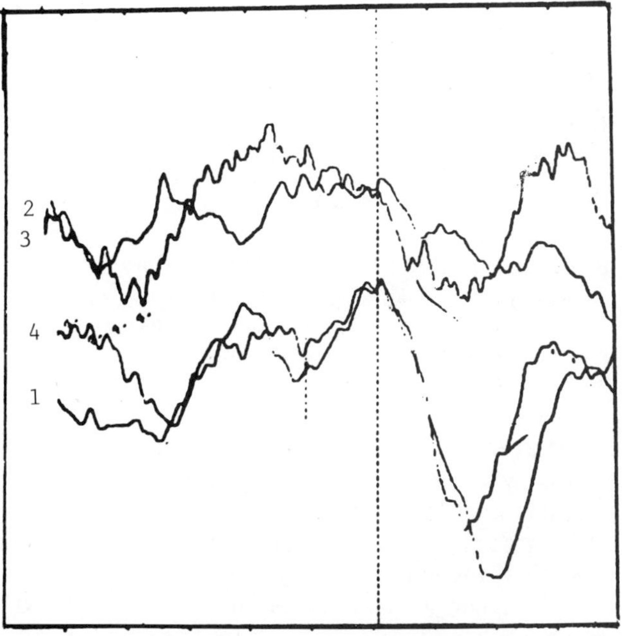

Fig. 1. SSEPs during resection of spinal tumor. Trace 1: dura covered. Traces 2 and 3: dura exposed. Trace 4: Dura covered with a pattie.

The stimulus level could be increased once the patient was anesthetized.

* Department of Surgery, Maimonides Medical Center, 214 Engle Street, Englewood NJ 07631

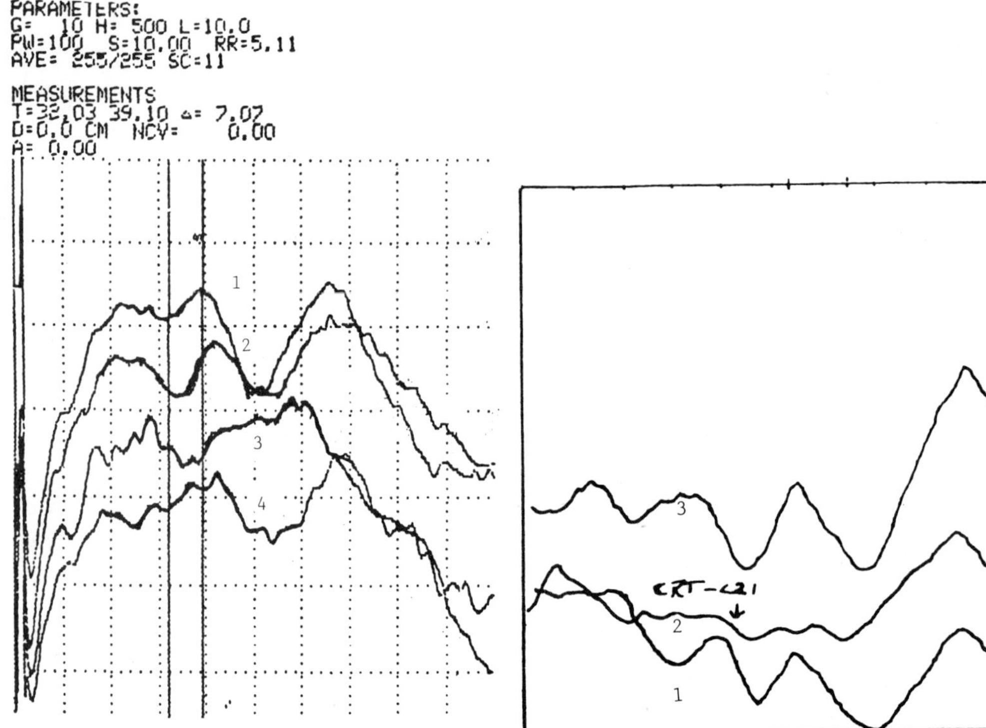

Fig. 2. Effects of body temperature on SSEP. Trace 1: baseline SSEP with rectal temperature of 37°C. Trace 2: Slight shift in SSEP latency with drop in temperature to 33°C. Trace 3: Further deterioration in SSEP latency and intensity with temperature reduced to 26°C. Trace 4: Beginning of restoration of normal latencies and morphology as temperature elevated to baseline levels.

Fig. 3. Effects of hemodilution on SSEP. Trace 1: hematocrit = 32. Trace 2: hematocrit < 21. Trace 3: hematocrit restored to 30.

The recording electrodes were placed as follows:

One pair of electrodes were placed in each popliteal fossa to measure peripheral conduction. The active cortical electrode was placed approximately 2cm posterior to the vertex or on the forehead. A neutral ground site was most often the shoulder or occasionally the upper leg.

The signals were amplified through a filter band from 10Hz to 500Hz with 150 to 200 events collected for each average. The gain was sufficient for a 1μV/div display.

Data was collected from each leg separately and both legs were measured during the procedure. Occasionally one leg would show no change while the other would reflect changes from the procedure, hence the alternating stimulus paradigm. The popliteal fossa potential (usually recorded at 9-10ms) gave good indication of stimulus integrity during the procedure, and additionally provided information that helped differentiate direct spinal cord manipulation from changes of anesthetic/patient status.

The first example depicts the effects of temperature on the SSEP. Figure 1 shows evoked potentials measured during surgical correction of a lateral spondylolisthesis. The first trace shows the baseline potential in response to unilateral posterior tibial nerve stimulation. The second and third traces show the change in potential when the spinal dura itself was exposed to the air with no other changes in anesthetic level. The

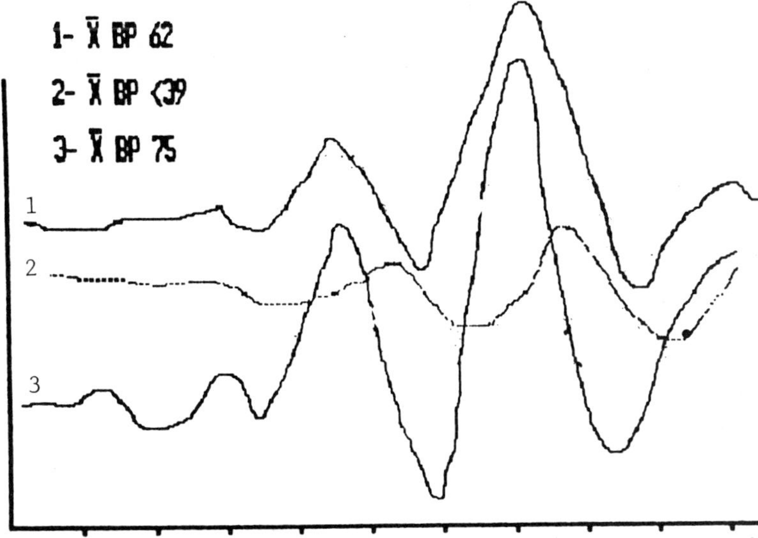

Fig. 4. Effects of blood pressure changes on SSEP. Trace 1: Baseline posterior tibial nerve SSEP with a mean blood pressure of 62. Trace 2: SSEP loss of amplitude and latency with diminished blood pressure to a mean of <39. Trace 3: Restoration of amplitude and latency of SSEP with increase in blood pressure to mean of 75.

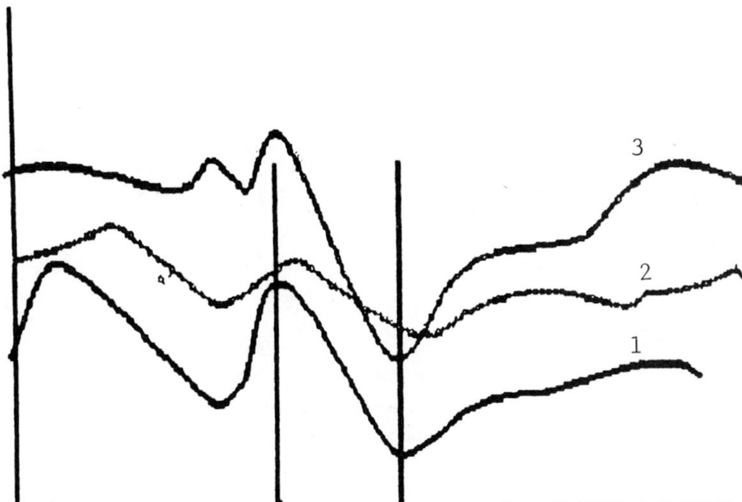

Fig. 5. Effects of Halothane on SSEP. Trace 1: baseline posterior tibial nerve SSEP. Trace 2: SSEP 8 minutes after introduction of 1.0 Halothane. Trace 3: SSEP 10 minutes after Halothane discontinued.

fourth trace shows the same condition moments later when the open dura was covered. This has been duplicated in our laboratory with temperature controlled douches of the spinal cord of experimental animals. As was the case with the clinical experience, reducing the temperature shows a relatively consistent prolonging of latencies of the SSEP.

Figure 2 demonstrates a similar phenomenon in a patient who was undergoing a cardiothoracic procedure and demonstrated the first wave form as a baseline; the third wave form, when the body temperature was reduced to 26°C; and the final trace later in the procedure when the body temperature was raised to a normal level. Published data (Sebel et al.) suggest that an 0.1ms shift in latency may occur for every degree centigrade of body temperature shift.

The second effect described is hemodilution. Without adequate perfusion to the spine the nerve cells apparently reduce their exogenous function and focus on maintenance of cell stability.

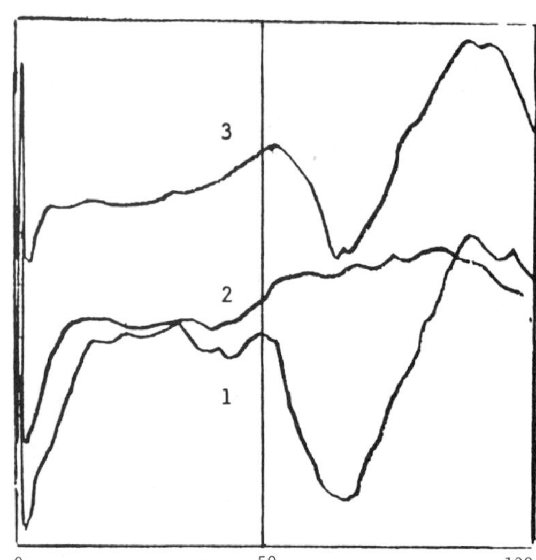

Fig. 6. Effects of Fentanyl on SSEP. Trace 1: baseline SSEP. Trace 2: five minutes after introduction of a bolus of Fentanyl. Trace 3: SSEP five minutes later.

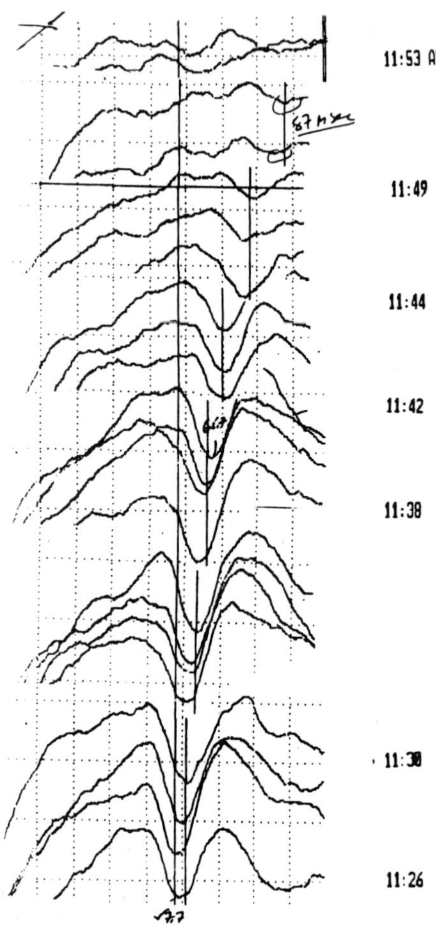

Fig. 7. Effects of Aortic cross-clamp on SSEP. The early traces (11:26) show the baseline SSEP as the Aorta was clamped. The uppermost trace reflects the loss of SSEP 23 minutes after cross-clamp.

Figure 3 reflects the somatosensory evoked potentials for a 13 year old girl undergoing surgical correction of congenital scoliosis. Baseline data are reflected in the first trace followed by a deterioration of the wave form apparently reflected by hemodilution. The patient's hematocrit had dropped significantly to 12 from a normal level of 32. The third trace shows the recovery after transfusions restored the patient to the normal hematocrit level.

Just as the quality of blood supply is critical for the nerve cell/spinal cord to maintain normal conduction abilities, the quantity of blood and blood pressure affect cellular function.

Figure 4 demonstrates the effect of blood pressure changes on the evoked potential. The first trace is the mean somatosensory evoked potential of an 80 year old male admitted for a femoral bypass procedure. The second trace shows the SSEP after a drop in blood pressure to a mean of below 40. The third trace shows the restoration of normal latencies when the blood pressure again reached a mean of 75.

The second component of anesthetic control that will effect evoked potentials are those induced by the anesthetic agents themselves. These act in several different ways

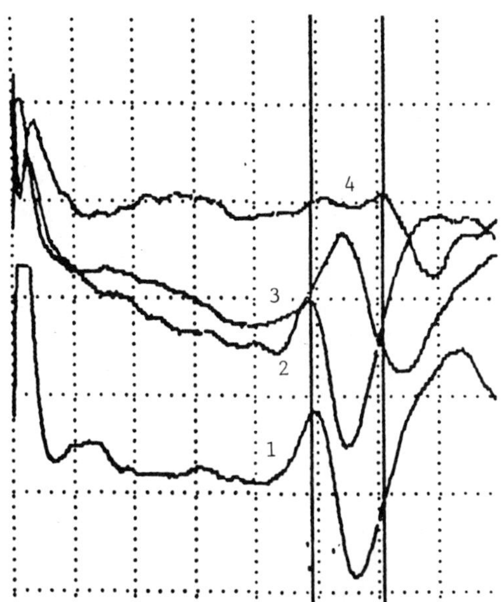

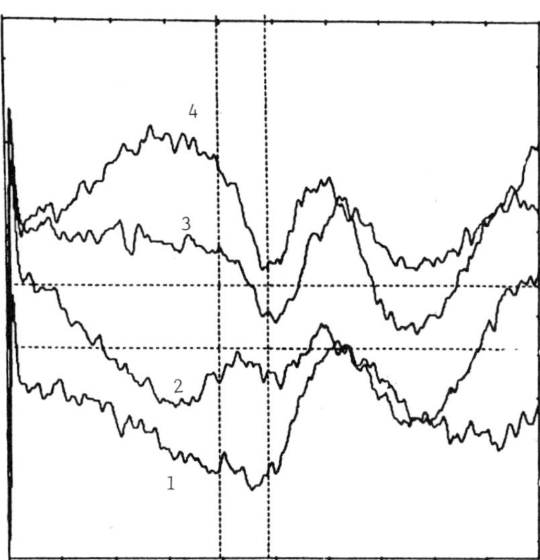

Fig. 8. Effects of Aortic cross-clamp on SSEP (Critical intercostal - no distal perfusion). Trace 1: baseline posterior tibial nerve SSEP prior to cross clamping. Trace 2: cross-clamp plus 2 minutes. Trace 3: cross-clamp plus 4 minutes. Trace 4: cross-clamp plus 8 minutes.

Fig. 9. Effects of retractor placement on SSEP. Trace 1: Baseline potential with hip dislocated. Trace 2: Retractor pressing on Sciatic nerve. Trace 3: Removal and replacement of retractor. Trace 4: New baseline two minutes later.

based on their pharmacokinetics. Halothane, one of the more common inhalation agents, is a halogenated aliphatic hydrocarbon which is often used in many procedures both as an induction and as a maintenance agent. The cerebral metabolic oxygen consumption $(CMRO_2)$ is decreased based on a dose related response. It is used to induce anesthesia in patients who have been monitored preoperatively. The dramatic changes can be seen in Figure 5.

Enflurane, a halogenated methyl-ethyl ether, can also be used both for induction and maintenance of anesthetic levels. Similar to Halothane, Enflurane produces a decrease in the $CMRO_2$ of the brain to a considerably higher degree than the Halothane, while the CBF increase is noted but not to the same extent. Once Enflurane is added as an anesthetic agent, the changes are noted shortly afterwards as a shift in latency. It is interesting that Isoflurane, an isomer of Enflurane, has no CNS seizure-like activity which can be seen as an Enflurane effect.

Nitrous Oxide is an inorganic anesthetic agent most often used as a supplement to other agents and has a potentiating effect on SSEP.

The second group of anesthetic agents is the narcotic group characterized by Morphine and Fentanyl. Morphine, an opiate alkaloid, is used most frequently as a preoperative medication. When used in anesthetic doses, it can decrease the $CMRO_2$ of the brain by as much as 40%.

Fentanyl is a synthetic narcotic which is structurally related to Merpidine and is 50 to 100 times more potent than Morphine with similar affects on the SSEP. Sufentanyl, another agent in the narcotic group, stronger than Fentanyl, also has an effect on SSEP, lowering amplitude and increasing latency. These changes may be seen in Figure 6. The lower trace represents the pre-induction baseline potential with the middle trace reflecting the status five minutes after a bolus of Fentanyl and the third trace the return to baseline five minutes later.

The difficulty this presents to the surgeon and/or anesthesiologist is that these changes are either indistinguishable from changes induced by the surgical procedure, or they degrade the wave form to such a point that the changes induced by the surgical procedure are no longer readable.

The second aspect of this presentation is to demonstrate some of the changes that are actually seen by the surgical procedure. The first of these shows the effects of cross clamping the aorta during a resection of a thoraco-abdominal aneurysm. There are two types of changes that have been reported by researchers such as Cunningham et al. The first is a slow insidious change over approximately 20-25 minutes which represents a deterioration of the wave form following peripheral ischemia. This may be seen in Figure 7. The second, more critical phenomenon, is the wave form morphology within a 5-8 minute period, which is more typical of a compromise of intercostal supply. This may be seen in Figure 8. Following the restoration of normal blood supply, both these potentials recovered to normal.

The last category are those effects precipitated by the surgical manipulation itself. An example is the change in somatosensory evoked potential caused by distraction using an outrigger during the Harrington rod procedure. The last figure shows a similar advantage of intraoperative monitoring to monitor the surgical procedure is the positioning of a retractor during a hip arthroplasty. This figure (9) shows the deterioration of the potential caused purely by the placement of the retractor and when the retractor was released, the potential returned to normal.

Summary

These data are presented to demonstrate the changes in somatosensory evoked potential that can be precipitated by anesthetic control as well as by surgical control. They underscore the need for close cooperation between the anesthetic monitoring and surgical monitoring procedures to ensure the availability of good feedback from evoked potential data.

Effects of Intravenous Anesthetic Induction Agents on Somatosensory Evoked Potentials: Thiopental, Fentanyl, and Etomidate

R.W. McPherson;[*] B. Sell; R.J. Traystman

Introduction

Monitoring of somatosensory evoked potentials (SEP) appears useful in decreasing intraoperative neurological injury in both spinal and cranial procedures. Changes in both SEP wave size (amplitude) and delay following stimulation (latency) may be used to assess injury. Because the period of risk is frequently unclear, environmental changes must be well described so that any wave form changes are not falsely attributed to neurological injury. Changes in blood gases, arterial blood pressure, and intermittent anesthetic drug administration (intravenous drug) or concentration changes (volatile gases) are frequent during operation and must frequently be considered as causes of wave form changes.

Modern anesthesia consists of a combination of volatile anesthetic gases (halothane, enflurane, isoflurane) which are combined with intravenous agents such as narcotics and hypnotic drugs. Varying the proportion of each component allows modulation of speed of awakening, stable blood pressure, etc.

Anesthetic gases, including nitrous oxide, depress the scalp recorded SEP waves (4, 6, 7). Although suppression of scalp recorded SEP waves is unimportant in patients who have waves of normal size, such depression may make SEP monitoring difficult or impossible in patients who have abnormally small wave forms. Intravenous drugs such as narcotics and barbiturates are used to induce anesthesia, maintain anesthesia, or to supplement anesthetic gases such as halothane, enflurane, or isoflurane. The amount of anesthetic gas required to maintain stable anesthesia may be substantially decreased by supplementation with intravenous agents such as barbiturates or narcotics, thus avoiding high concentration of volatile anesthetic gases which severely depress even normal wave forms. Since cardiovascular and electroencephalograph (EEG) effects of anesthetic induction agents appear dose related, drug effects on the SEP can also reasonably be expected to be dose related.

In the present study, we studied the largest doses of these drugs usually administered; i.e., an amount of drug sufficient to induce anesthesia. Smaller doses (about 25% of induction dose) are used to supplement pre-existing anesthesia. We chose to study three drugs of different classes: Narcotics (fentanyl - Sublimaze®), barbiturate (thiopental - Pentothal®), and a member of the imidazole group of com-

---

* Department of Anesthesiology and Critical Care Medicine, Johns Hopkins Hospital, 600 North Wolfe Street, Baltimore, MD 21205

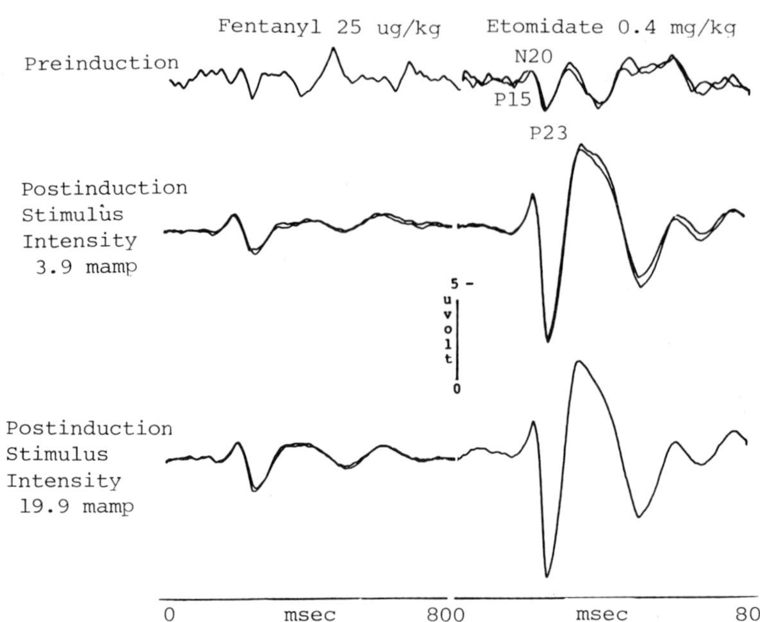

Fig. 1. Scalp recorded waves are shown in a patient who received fentanyl (25µg/kg) for his initial operation and etomidate (0.4mg/kg) for a second operation three weeks later.

pounds (etomidate - Amidate®), which are routinely used to induce anesthesia. The drug doses were chosen for equivalency for rapidly producing unconsciousness.

Methods

Upper extremity somatosensory evoked potentials (SEP) were studied in 27 patients who had neurologically normal upper extremities. Blood pressure, pulse, temperature, and end tidal carbon dioxide tension were monitored and maintained within normal limits. Nine patients received thiopental (Pentothal; 4mg/kg, IV), nine patients received fentanyl (Sublimaze; 25 µg/kg, IV), and nine patients received etomidate (Amidate; 0.4mg/kg, IV). Immediately after loss of consciousness, the patients received pancuronium (0.1mg/kg, IV bolus) for muscle paralysis. The patients were mechanically hyperventilated to maintain end expiratory carbon dioxide tension 35-40mmHg.

SEPs were generated using a Nicolet Med. 80. The median nerve of the non-dominant hand was stimulated using sterile 23 gauge needles placed following localization of the point of maximum twitch. The responses were recorded over the second cervical vertebra and the contralateral somatosensory cortex (C4' or C3') with active electrodes referenced to the forehead (FPz). Stimuli at motor threshold intensity of 200µSec duration were delivered at 5.9/sec and averaged by the computer. In the awake patient, 256-512 stimuli were averaged, whereas in the anesthetized patient, averaging 128 stimuli was adequate to produce stable waves. Replicate waves were obtained in each study period.

We evaluated early scalp recorded waves and subcortical waves. In text and figures, we will indicate for each wave a polarity (N = negative, P = positive), and a nominal latency. Thus N20 is a negative wave occurring 20mSec following stimulation. Several characteristic waves were recorded at the level of the spinal cord (N14) and the scalp (N20, P23). We also evaluated the conduction time from spinal cord to cortex (central conduction time - CCT).

SEP waves were evaluated before and approximately 2 minutes after induction of anesthesia with either fentanyl, thiopental, or etomidate.

Data in text and figures are presented as mean $\pm$ SEM. Analysis of variance (ANOVA) was used to assess the effect of drug administration on the SEP. $P < .05$ was

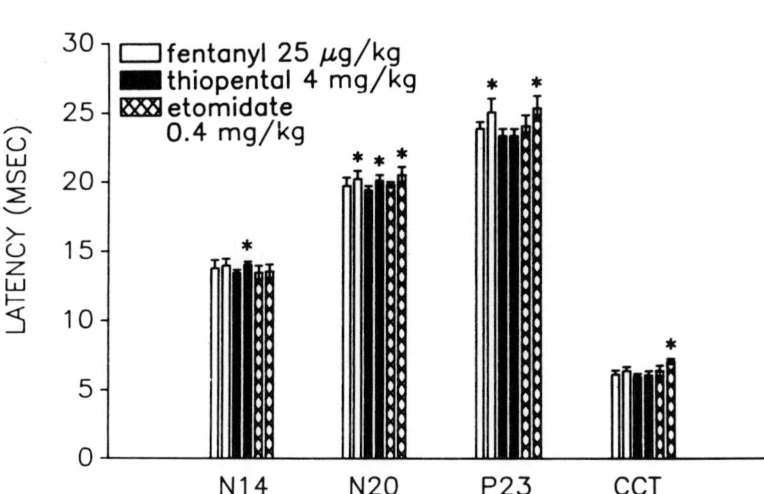

Fig. 2. Latency of waves recorded over the spinal cord (N14) and contralateral somatosensory cortex (N20, P23) following median nerve stimulation in patients receiving fentanyl (N = 9), thiopental (N = 9), or etomidate (N = 9) is shown. Central conduction time (CCT) was computed as the difference between N20 and N14 latencies. Mean $\pm$ SEM. * = P < .05.

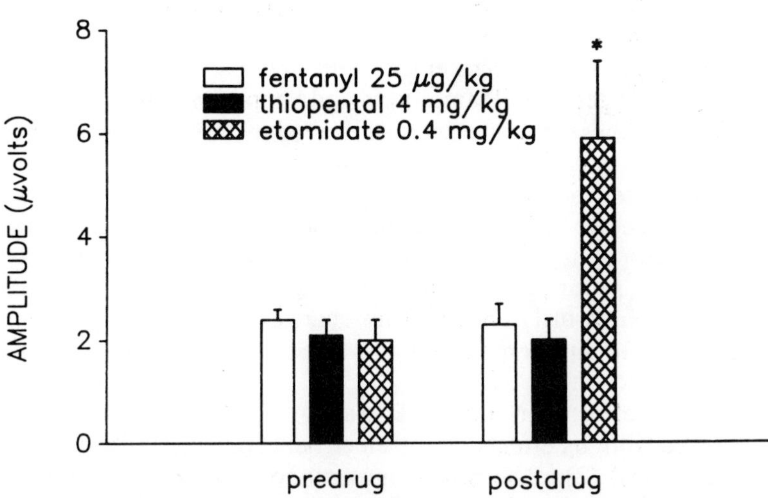

Fig. 3. Amplitude of the major early scalp recorded waves (N20 through P23) in patients receiving fentanyl (N = 9), thiopental (N = 9), and etomidate (N = 9) is shown. Mean $\pm$ SEM. * = P < .05.

considered significant and a Duncan's multiple range test was performed to identify the different values.

Results

Satisfactory data was obtained from all patients prior to and after induction of anesthesia. The three groups were comparable in age (about 45 years) and height (about 68 inches). Anesthesia induction did not change arterial blood pressure in any of the three groups.

Fig. 1 shows scalp recorded SEP waves in a patient, not included in the study, who received fentanyl (25μg/kg, IV bolus) for induction of anesthesia for his initial surgery and 3 weeks later received etomidate (0.4mg/kg, IV bolus) for induction of anesthesia. On both occasions, the patient received pancuronium (0.1mg/kg) immediately following loss of consciousness. It can be seen in the figure that fentanyl had minimal effects on SEP amplitude and latency, whereas etomidate caused a fourfold increase in the amplitude of the scalp recorded waves.

Fig. 2 shows the effects of the three drugs on latency of waves at the level of the spinal cord (N14), and somatosensory cortex (N20, P23). We also determined the central conduction time (CCT) from cervical spinal cord (N14) to cortex (N20). The effect of the drugs on the waves recorded at the spinal cord level was slight, with an increase in the latency found only with thiopental. The N20 wave latency was increased by fentanyl, thiopental, and etomidate. P23 latency was also increased by fentanyl and etomidate. The lack of changes in N14 latency coupled with small changes in N20 caused CCT to be unchanged in patients receiving fentanyl and thiopental and to be slightly increased in patients who received etomidate.

The effect of the three drugs on wave amplitude is shown in Fig. 3. Wave amplitude was comparable prior to drug administration. Neither fentanyl nor thiopental administration altered wave amplitude, whereas etomidate increased wave amplitude to about 300% of the pre-induction value.

Discussion

Our data show that large doses of fentanyl and thiopental cause only minor changes in SEP latency and amplitude. Etomidate, on the other hand, causes a dramatic increase in wave amplitude associated with slight increases in wave latency. These results are important for two reasons. First, anesthesia induction agents with only small SEP effects are identified (fentanyl, thiopental). Second, drug-produced changes in wave amplitude due to etomidate may affect intraoperative SEP monitoring.

Our results concerning thiopental are in agreement with previous investigators (1) who demonstrated only minor changes in SEP amplitude and latency in doses sufficient to induce anesthesia. In view of these minor effects, the smaller doses of barbiturate frequently used to supplement anesthetic gases (approximately 25% of the induction dose) would be expected not to alter the SEP.

Fentanyl has previously been shown to produce an increase in lower extremity latency (5). In patients receiving multiple drugs for premedication, extremely large doses of fentanyl appears to slightly increase latency with a modest decrease in amplitude (8). Since we found only slight increases in latency and no change in amplitude, smaller doses of fentanyl routinely used to supplement volatile anesthetic gases (1 - 5μg/kg) would be expected to have no effect on the SEP.

Etomidate caused dramatic effects on the SEP wave. The increase in wave amplitude can not be predicted from its effects on the electroencephalogram (since it is similar to barbiturate in its effects) (2). The wave augmentation caused by etomidate may enhance or inhibit intraoperative monitoring. If etomidate is administered by intermittent bolus, the drug-induced amplitude increases followed by a decay of drug effect with return to pre-drug level may make diagnosis of injury related wave changes difficult. However, etomidate may be administered by continuous infusion to maintain stable anesthesia in neurological patients (3). The stability of etomidate blood levels with a continuous infusion suggest that stable augmentation of SEP wave amplitude may be possible.

In conclusion, thiopental and fentanyl cause only minor changes in early scalp recorded SEP waves and thus should be satisfactory to supplement anesthetic gases with little changes in wave form. On the other hand, etomidate administration may cause wave form changes and make diagnosis difficult, and thus should be avoided when wave form stability is important.

References

1. Drummond, J.C.; Todd, M.M.; Hoi Sang, U.: The effects of high dose sodium thiopental on brainstem auditory and median nerve somatosensory evoked responses in humans. Anesth., 63: 249-254, 1985.
2. Ghoneim, M.M.; Yamada, T.: Etomidate: A clinical and electroencephalographic comparison with thiopental. Anesth. Analg., 56: 479-485, 1977.
3. Lees, N.W.; Glasser, J.; McGroarty, P.J.; Miller, B.M.: Etomidate and fentanyl for maintenance of anaesthesia. Br. J. Anaesth., 53: 959-961, 1981.
4. McPherson, R.W.; Mahla, M.; Johnson, R.; Traystman, R.J.: Effects of enflurane, isoflurane, and nitrous oxide on somatosensory evoked potentials during fentanyl anesthesia. Anesth., 62: 626-633, 1985.
5. Pathak, K.S.; Brown, R.H.; Cascorbi, H.F.; Nash, C.L.: Effects of fentanyl and morphine on intraoperative somatosensory evoked potentials. Anesth. Analg., 63: 833-837, 1984.
6. Peterson, D.O.; Drummond, J.C.; Todd, M.M.: Effects of halothane, enflurane, isoflurane, and nitrous oxide on somatosensory evoked potentials in humans. Anesth., 65: 35-40, 1986.
7. Samra, S.K.; Vanderzant, C.W.; Domer, P.A.; Sackellares, J.C.: Differential effects of isoflurane on human median nerve somatosensory evoked potentials. Anesth., 66: 29-35, 1987.
8. Schubert, A.; Drummond, J.C.; Peterson, D.O.; Saidman, L.J.: The effect of high dose fentanyl on human median nerve somatosensory evoked response. Can. J. Anaesth., 34: 35-40, 1987.
9. Sebel, P.S.; Bovill, J.G.; Wauduier, A.; Rog, P.: Effect of high dose fentanyl anesthesia on the electroencephalogram. Anesth., 55: 203-211, 1981.

Augmentation of Somatosensory Evoked Potential Waves in Patients with Cervical Spinal Stenosis

R. W. McPherson;[*] T. B. Ducker

Introduction

Intraoperative monitoring of somatosensory evoked potentials (SEP) is thought to decrease the risk of neurological injury during corrective spinal surgery (4, 10) and intracranial vascular procedures (20). Diagnosis of change in neurological function as revealed by changes in SEP waves may be hindered by small waves.

Anesthesia is generally considered detrimental to intraoperative monitoring because halothane (1, 16), enflurane (16), isoflurane (18), and nitrous oxide (12) decrease the size of scalp recorded waves. In doses sufficient to induce anesthesia, thiopental causes only minor changes in SEP latency and amplitude (3, 13). Large doses of fentanyl in unpremedicated patients cause only minor changes in latency and amplitude (13), whereas fentanyl administration in anesthetized patients may increase latency (15). In contrast, etomidate (Amidate), increases SEP amplitude by about 400% in circumstances in which both fentanyl and thiopental do not alter the SEP amplitude (13). The increase in SEP amplitude is surprising since the effects of etomidate on the electroencephalogram are similar to thiopental (7).

In order to determine if this SEP wave augmentation by etomidate can be used to enhance the ability to monitor intraoperative SEP, we used etomidate to induce anesthesia in two patients with absent or greatly decreased preoperative SEP waves. We also determined the time course of SEP wave augmentation in two groups of neurologically normal patients.

Methods

This study was approved by the institutional review committee and each patient gave prior written permission. Hemodynamic monitoring was conducted based on the clinical status of each patient. Blood pressure, pulse, temperature, pulse oximetry, and end tidal carbon dioxide were measured in each patient. Following induction of anesthesia, ventilation was controlled to maintain the end tidal carbon dioxide tension in the normal range (35-40mmHg).

In two patients with symptomatic cervical stenosis and twelve patients without neurological abnormality of the upper extremity, anesthesia was induced with etomidate (0.4mg/kg, iv bolus) followed by pancuronium (0.1mg/kg, iv bolus). No other

[*] Department of Anesthesiology and Critical Care Medicine, Johns Hopkins Hospital, 600 North Wolfe Street, Baltimore, MD 21205

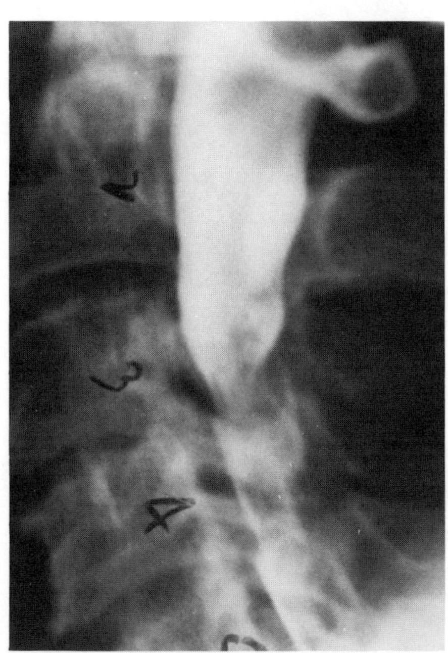

Fig. 1. The preoperative cervical myelogram of patient 1 is shown. Severe stenosis at several levels is shown.

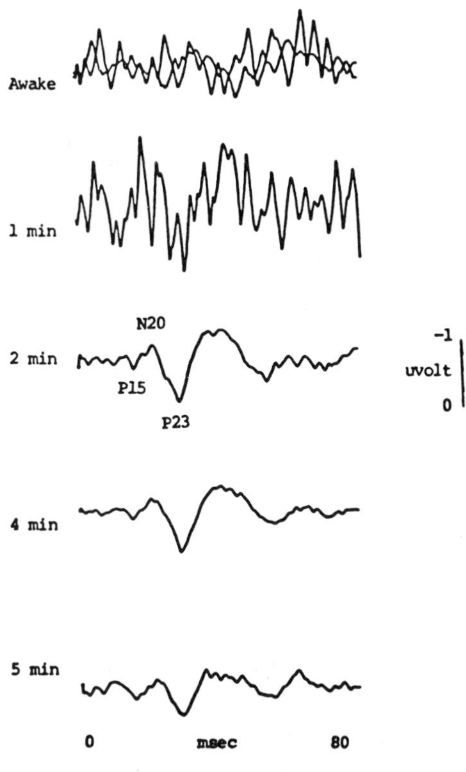

Fig. 2. Scalp recorded somatosensory evoked potential waves for patient 1 following stimulation of a median nerve are shown. Prior to anesthesia, 512 stimuli were delivered at 5.9sec and averaged. Following induction of anesthesia, 128 stimuli were delivered at 5.9sec and averaged. The same stimulus intensity was used before and after etomidate administration. The times refer to intravenous administration of etomidate (0.4mg/kg, iv bolus).

anesthetic agent was administered until the study was completed (approximately 6 minutes following injection of etomidate). In an additional group of patients with neurologically normal upper extremities (N = 12), etomidate (0.1mg/kg) was administered during stable anesthesia consisting of thiopental (0.4mg/kg, iv bolus) plus fentanyl (10 μg/kg, iv bolus) for induction of anesthesia and isoflurane (0.5 - 1.0%) administered without nitrous oxide for approximately 30 minutes. For 10 minutes prior to study and for the period of study (approximately 6 minutes) the isoflurane concentration remained constant and no intravenous drugs were administered.

In the two patients with cervical stenosis, the SEP following stimulation of the more symptomatic upper extremity was studied, whereas in the neurologically normal patients, the nondominant upper extremity was stimulated. All patients awoke promptly at the end of operation and no patient had intraoperative deterioration of neurological function. Table 1 shows SEP monitoring parameters and waves which were evaluated.

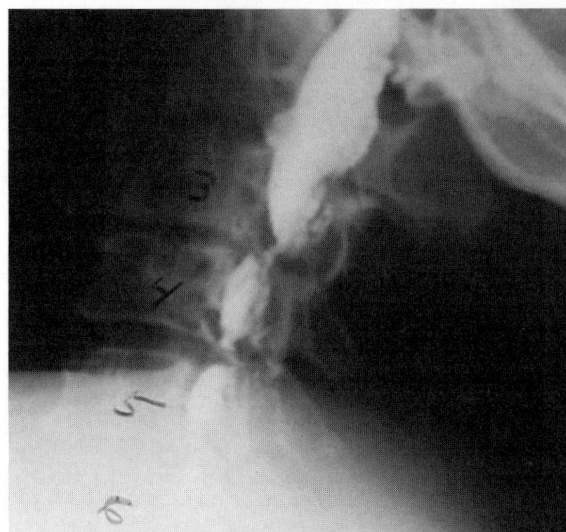

Fig. 3. The preoperative cervical myelography of patient 2 is shown. Severe stenosis at several levels is shown.

Patient 1

This 62 year old black male had profound spastic quadriparesis and urinary incontinence. He first noted numbness and clumsiness of both hands one year prior to admission. He had a one month history of spastic weakness of the lower extremities. Past medical history and review of systems was noncontributory.

Abnormal physical findings were confined to the nervous system. He was without neck pain, limitation of range of motion of the neck, or symptoms of cervical arthritis. Cerebral function and cranial nerve examination were normal. Physical examination showed weakness of triceps, wrist extensors and flexors, and hand grip. Fine motor movement was practically absent in both upper extremities with the right side more involved than the left. The lower extremities showed spastic paraparesis with a gait which required assistance. Sensory deficits were profound, with complete loss of proprioception and position sense in the lower extremities and hypalgesia which extended up to the T3 level. Superimposed was a

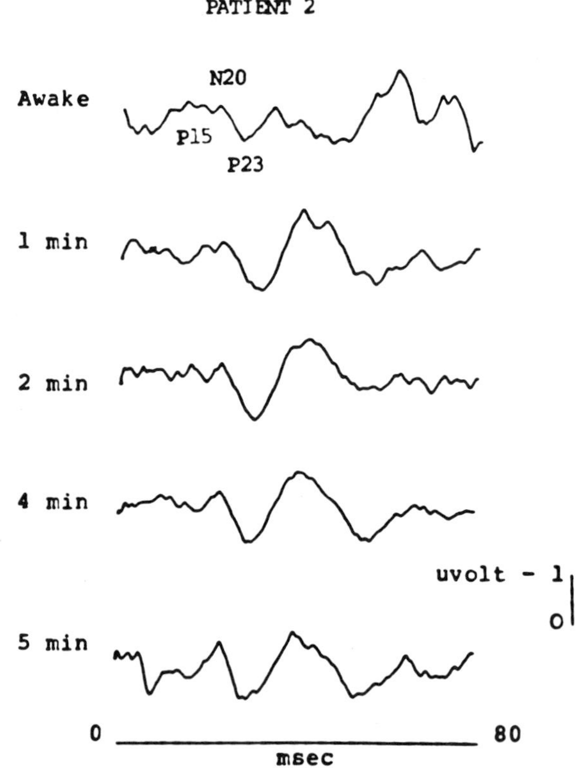

Fig. 4. Scalp recorded somatosensory evoked potential waves in patient 2 are shown. Prior to anesthesia, 512 stimuli were delivered at 5.9sec and averaged, while after anesthesia induction 128 stimuli (5.9sec) were averaged. The same stimulus intensity was used before and after etomidate administration.

hypalgesia of both upper extremities. Preoperative myelography (Fig. 1) showed marked spondylitic ridging and stenosis with significant defects at multiple levels, most severe at C3 through C4. The spinal block at C3 through C4 was nearly complete, for only after placing the chin down for 10-15 minutes did any of the myelographic material pass the C3 through C4 level. Preoperative SEP showed complete loss of all activities in the lower extremities and practically no scalp recorded responses following stimulation of either median nerve. Cervical laminectomies were performed at C3 through C7 levels in the prone position with skeletal traction. Severe constriction with angulation was found at the C3 through C4 level. Fig. 2 shows intraoperative scalp recorded SEP determined by stimulation of the right median nerve (more symptomatic upper extremity). Baseline SEP were performed without premedication, and despite a distinct thumb twitch, no scalp recorded waves were recorded (Fig. 2, baseline). The figure shows

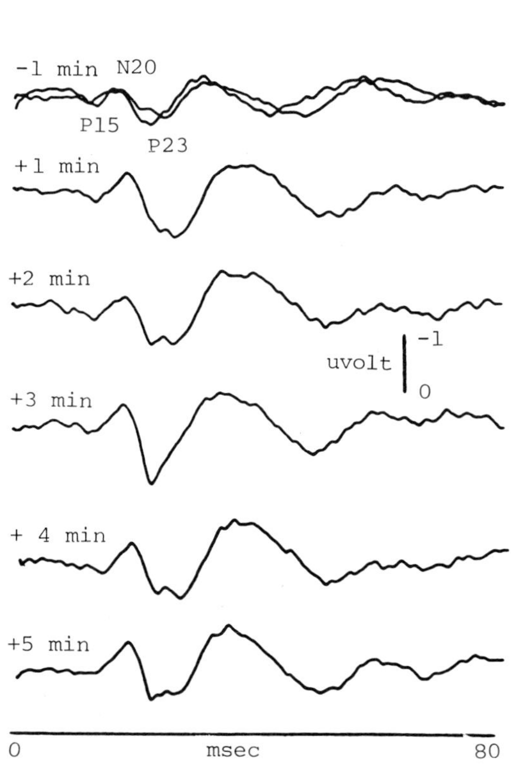

Fig. 5. The effect of 0.1mg/kg etomidate in a patient receiving 1% isoflurane is shown. The wave tripled in size within one minute of bolus etomidate injection.

that a wave of complex morphology but of prolonged latency (N20 latency about 23msec) was measurable 2 minutes after the injection of etomidate into a peripheral vein. Thereafter etomidate was administered by continuous infusion until decompression was complete with a stable waveform recordable until approximately 5 minutes following discontinuation of the etomidate infusion. Following completion of surgery, the patient awakened promptly and neurologic status was unchanged by surgery.

Four months following surgery, the patient had good control of his upper extremities, but had residual clumsiness in both hands. His previous distal hypalgesia was associated with a burning paresthesia. His lower extremities showed only modest improvement, and he could ambulate with the aid of a cane. Proprioception and position sense remained markedly impaired in the lower extremities, but position sense for foot movement was approximately 70% accurate. Evoked potentials from the lower extremities remained absent while potentials from the median nerves were recorded (not shown).

Patient 2

This 68 year old black male had a 9 month history of progressive loss of use of both upper and lower extremities following an automobile accident. He had a cervical myelopathy including numbness and clumsiness in the upper extremities, loss of power of the big muscles of the arms and hands, and stiffness and difficulty with his gait. In the two months prior to admission he had increasing difficulty in bladder control.

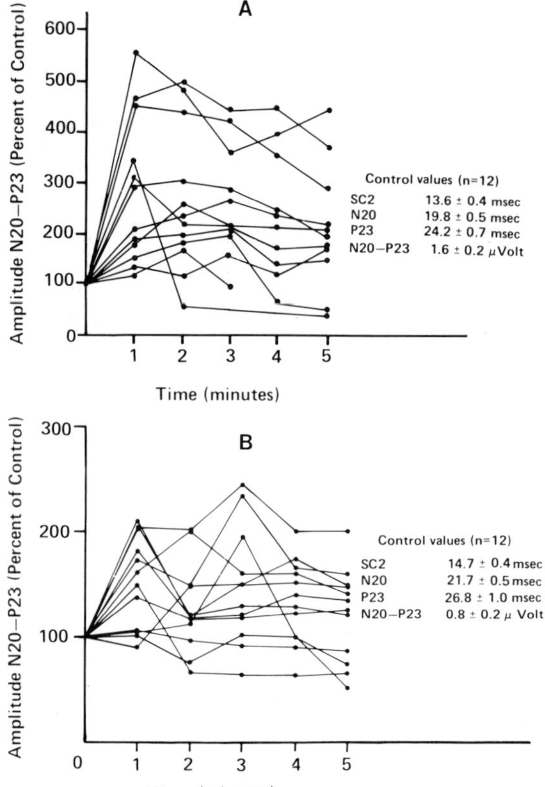

Fig. 6. Changes in N20 through P23 amplitude in 12 patients who received etomidate (0.4mg/kg) are shown in panel A. Changes in N20 through P23 amplitude in 12 patients who received 0.1mg/kg etomidate are shown in panel B.

Physical examination revealed normal cerebral and cranial nerve function. Motor examination revealed quadriparesis, with the left upper extremity being more involved than the right, with weakness in deltoids, biceps, and triceps on that side, and bilateral loss of power, grip and dexterity. Lower extremities had adequate power, but movements were awkward. Deep tendon reflexes were hyperactive with bilateral Hoffman and Babinski signs. Sensory examination demonstrated diffuse distal lower extremity sensory loss which extended to the midthoracic area. Proprioception was diminished in the lower extremities. Cervical myelogram (C1 through C2 puncture) showed severe cervical spondylosis with multiple extradural compressions seen from C3 through C4 to C6 through C7 levels (Fig. 3). Preoperative SEP could not be obtained from the lower extremities and were poor from the upper extremities.

Cervical laminectomy was performed in the prone position with skeletal traction. Decompression extended from C3 through C7 with the area with the most severe constriction found at C3 through C4. As shown in Fig. 4, small amplitude SEP waves were recorded prior to etomidate administration, with larger waves obtained immediately following drug administration. The post-operative course was uneventful, and there was a definite but modest improvement in gait, hand function and strength in the upper extremities so that he was completely independent. There were persistent moderate residual disabilities in hand function.

Results

In both groups of neurologically normal patients, the SEP was determined before and each minute for five minutes after intravenous bolus administration of etomidate. This time period was chosen because observations in patients 1 and 2 suggested that the waves began to return to predrug size within 5 minutes of drug administration. We chose to end the study at 5 minutes because the hypnotic effect of etomidate is fairly short.

Table 2 shows the effect of etomidate on SEP latency and amplitude in neurological normal patients. The ages of the two groups of patients were 50 ± 3 and 52 ± 3 years, respectively. Height was 67 ± 1 inches in the patients who received 0.4mg/kg etomidate and 67 ± 2 inches in the patients who received 0.1mg/kg etomidate. Neither dose of etomidate altered the latency of N14 recorded over the second cervical ver-

Table 1 - SEP monitoring parameters

Equipment	Nicolet Med 80
Stimulus electrode	Sterile 23 gauge needles near a median nerve
Stimulus intensity	Motor threshold (1.9 - 3.9mamp)
Stimulus frequency	5.9/sec; 256-512 stimuli in awake patients, 128 stimuli in anesthetized patients
Stimulus duration	200μsec
Ground	Extremity
Sensitivity	100μv full scale
Recording electrodes	Cup electrodes attached with collodium and filled with electrode gel; impedance <3kohm
Electrode location	Contralateral scalp (C3' or C4'), FPZ and over second cervical vertebra (SC2); C3' - FPZ or C4' - FPZ and SC2 - FPZ monitored simultaneously
Bandpass filters	15-1500Hz
Waves assessed	N14, initial negative wave over second cervical vertebra; about 14msec N20, initial negative scalp wave about 20msec P23 positive wave which follows N20; about 23msec N20-P23, amplitude from peak of N20 to trough of P23 N = negative; P = positive plus nominal latency

tebra. Both doses of etomidate increased latency of both N20 (about 1msec) and P23 (about 2msec). The latencies appeared to return to control within 5 minutes of injection. Amplitude was increased within 1 minute of injection in the patients who received 0.4mg/kg etomidate (230%) and amplitude was declining although still above control 5 minutes following injection (180%). The increase in amplitude was less following 0.1mg/kg etomidate (maximum of 50% increase at 1 minute post injection).

Fig. 5 shows the time course of wave augmentation following etomidate (0.1mg/kg, iv bolus) in a patient anesthetized with isoflurane (1.0% in oxygen). The waves prior to etomidate injection which were obtained approximately 5 minutes apart were reproducible. A rapid and sustained increase in amplitude occurred following etomidate injection. Fig. 6 shows the effect of bolus etomidate administration in neurologically normal patients receiving 0.4mg/kg (upper panel) and 0.1mg/kg (lower panel). In both groups of patients, the time to peak effect as well as the magnitude of augmentation was variable.

TABLE 2 – Effect of 0.1 mg/kg and 0.4 mg/kg Etomidate on SEP Latency and Amplitude

TIME (min)	0	1	2	3	4	5
ETOMIDATE (0.4 mg/kg N=12)						
N14 (msec)	13.9 + 0.4	14.0 + 0.5	14.2 + 0.5	14.1 + 0.5	14.1 + 0.5	14.0 + 0.5
N20 (msec)	19.8 + 0.5	21.0 + 0.5*	21.1 + 0.6*	21.0 + 0.5*	21.0 + 0.6*	20.7 + 0.6*
P23 (msec)	24.5 + 0.8	26.0 + 0.7*	25.3 + 0.6	25.5 + 0.6	25.6 + 0.8*	25.3 + 0.9
N20-P23 (uvolt)	1.6 + 0.2	3.7 + 0.9*	3.7 + 0.9*	3.1 + 0.6*	2.9 + 0.7*	2.9 + 0.8*
ETOMIDATE (0.1 mg/kg N=12)						
N14 (msec)	14.6 + 0.5	14.8 + 0.4	14.8 + 0.5	14.8 + 0.4	14.9 + 0.6	15.0 + 0.5
N20 (msec)	21.7 + 0.6+	22.7 + 0.7+*	22.8 + 0.8+*	22.6 + 0.7+*	22.1 + 0.6+*	22.5 + 0.6+*
P23 (msec)	26.8 + 1.0+	28.3 + 0.9+*	28.3 + 1.1+*	28.2 + 1.1+*	27.5 + 1.0+	27.8 + 1.1+
N20-P23 (uvolt)	0.8 + 0.2+	1.2 + 0.2+*	1.0 + 0.2+	1.1 + 0.2+*	1.0 + 0.2+	1.0 + 0.2+

* P<.05 – Compared to time 0
+ P<.05 – Compared to 0.4 mg/kg etomidate

Discussion

In this report, we demonstrate that etomidate increases the amplitude of scalp recorded SEP waves in patients with severe cervical stenosis. Because scalp waves were augmented despite spinal stenosis which prevented normal waves in the awake patient, our previous speculation that this effect is rostral to the spinal cord is supported (13). Additionally, the efficacy of etomidate in allowing SEP monitoring in patients with cervical stenosis is demonstrated. The previously reported depression of SEP by anesthetic gases would have probably further depressed the waves and interfered with intraoperative monitoring.

The role of anesthetic management in intraoperative SEP monitoring is poorly defined. This lack of definition arises because of the heterogeneous population thought to benefit from such monitoring. The extremes are neurologically normal patients (Harrington Rod placement) and neurologically abnormal patients such as the two patients presented in this work. Patients with severe cervical stenosis may present difficulties in intraoperative monitoring because of decreases in spinal transmission and suppression of scalp recorded waves due to conventional anesthesia (12, 16, 18). Despite these limitations, monitoring may be important because spinal cord injury may occur due to spinal hypoperfusion or position related ischemia. These impaired patients are at high risk of additional neurological injury early in anesthesia and surgery (for example, positioning). Because the time necessary to produce a wave suitable for evaluation is inversely proportional to the signal to noise ratio, improvement of the signal (wave amplitude) will decrease the time to diagnosis of waveform change. Thus using etomidate allows more rapid patient assessment following a potential change in neurological status.

In both patients with cervical stenosis, scalp recorded waves were augmented 1-2 minutes after bolus administration of etomidate, and the effect appeared to decrease within 5 minutes of administration. To verify this time course following bolus injection, we assessed the time course of the drug effect in patients without neurological abnormalities of the upper extremity. In all patients receiving etomidate (n = 24), wave augmentation occurred, but as shown in Fig. 6 the amount of wave augmentation varied as did the time to peak effect. In all patients, the effect appeared to decrease by 5 minutes following administration.

Etomidate is a satisfactory anesthetic drug for use during neurosurgical procedures. Like barbiturates, etomidate decreases cerebral blood flow and cerebral metabolism (17). Because of the short length of action it must be administered by continuous infusion for long operations (5, 19). Etomidate may be used as an alternative to barbiturates to induce anesthesia, because of a lesser degree of hypotension, particularly in elderly patients (28). The sleep time following etomidate is similar to thiopental when equivalent doses are given (19).

We have no additional information concerning the relationship between the etomidate dose and the magnitude of SEP wave augmentation since only two doses of etomidate, 0.4mg/kg and 0.1mg/kg were studied. The larger dose is sufficient to induce anesthesia quickly, while the smaller dose is sufficient to supplement inhalational anesthesia without producing hypotension. Therefore, we have evidence of the time course following bolus injection, but no indication of dose response.

Of concern are reports that etomidate depresses adrenocortical function (6). Etomidate causes the adrenocortical response to anesthesia and surgery to be depressed, beginning as soon as one hour following drug administration (6). The responsivity of the adrenocortical response seems to return rapidly following short term administration of etomidate. Since the patients in this study received steroids to minimize post-operative edema, this depression would be of no clinical importance.

In conclusion, etomidate administration should be considered in patients who have small scalp recorded SEP waves due to spinal stenosis. The drug can be administered by a continuous infusion to augment the wave amplitude until the risk of neural injury has passed.

References

1. Baines, D.B.; Whittle, I.R.; Chaseling, R.W.; Overton, J.H.; Johnston, I.H.: Effect of halothane on spinal somatosensory evoked potentials in sheep. Br. J. Anaesth., 57: 896-899, 1985.
2. Criado, A.; Maseda, J.; Navarro, E.; Escarpa, A.; Avello, F.: Induction of anaesthesia with etomidate: Haemodynamic study of 36 patients. Br. J. Anesth., 52: 803-805, 1980.
3. Drummond, J.C.; Todd, M.M.; Sang, H.: The effect high dose sodium thiopental on brain stem auditory and median nerve somatosensory evoked responses in humans. Anesthesiology, 63: 249-252, 1985.
4. Engler, G.L.; Spieholz, N.I.; Bernhard, W.N.; Danziger, F.; Merkin, H.; Wolff, T.: Somatosensory evoked potentials during Harrington instrumentation for scoliosis. J. Bone Joint Surg., 60: 528-532, 1978.
5. Fragen, R.J.; Avram, M.J.; Henthorn, T.K.; Caldwell, N.J.: A pharmacokinetically designed etomidate infusion regimen for hypnosis. Anesth. Analg., 62: 654-660, 1983.
6. Fragen, R.J.; Shanks, C.A.; Molteni, A.; Avram, M.J.: Effects of etomidate on hormonal responses to surgical stress. Anesthesiology, 61: 652-656, 1984.
7. Ghoneim, M.M.; Yamada, T.: Etomidate: A clinical and electroencephalographic comparison with thiopental. Anesth. Analg., 56: 479-485, 1977.
8. Giese, J.L.; Stockham, R.J.; Stanley, T.H.; Pace, N.L.; Nelissen, R.H.: Etomidate versus thiopental for induction of anesthesia. Anesth. Analg., 64: 871-876, 1985.
9. Horrigan, R.W.; Moyers, J.R.; Johnson, B.H.; Eger, E.I.; Margolis, A.; Goldsmith, S.: Etomidate vs. thiopental with and without fentanyl -- A comparative study of awakening in man. Anesthesiology, 52: 362-364, 1980.
10. Lamont, R.L.; Wasson, S.I.; Green, M.A.: Spinal cord monitoring during spinal surgery using somatosensory spinal evoked potentials. J. Pediatr. Ortho., 3: 31-36, 1983.
11. Larson, S.J.: Somatosensory evoked potentials in lumbar stenosis. Surg. Gynecol. Obstet., 157: 191-196, 1983.
12. McPherson, R.W.; Mahla, M.; Johnson, R.; Traystman, R.J.: Effects of enflurane, isoflurane and nitrous oxide on somatosensory evoked potentials during fentanyl anesthesia. Anesthesiology, 62: 626-633, 1985.
13. McPherson, R.W.; Sell, B.; Traystman, R.J.: Effects of thiopental, fentanyl, and etomidate on upper extremity somatosensory evoked potentials in humans. Anesthesiology, 65: 584-590, 1986.
14. Moss, E.; Powell, D.; Gibson, R.M.; McDowell, D.G.: Effect of etomidate on intracranial pressure and cerebral perfusion pressure. Br. J. Anaesth., 51: 347-352, 1979.
15. Pathak, K.S.; Brown, R.H.; Cascorbi, H.F.; Nash, C.L., Jr.: Effects of fentanyl and morphine on intraoperative somatosensory cortical evoked potentials. Anesth. Analg., 63: 833-837, 1984.
16. Peterson, D.O.; Drummond, J.C.; Todd, M.M.: Effects of halothane, enflurane, isoflurane, and nitrous oxide upon somatosensory evoked potentials in man. Anesthesiology, 65: 35-40, 1986.
17. Renou, A.M.; Vernhiet, J.; Marcrez, P.; Constant, P.; Billerey, J.; Khadaroo, M.Y.; Caille, J.M.: Cerebral blood flow and metabolism during etomidate anaesthesia in man. Br. J. Anaesth., 50: 1047-1051, 1978.
18. Samra, S.K.; Vanderzant, C.W.; Domer, P.A.; Sackellares, J.C.: Differential effects of isoflurane on human median nerve somatosensory evoked potentials. Anesthesiology, 66: 29-35, 1987.
19. Sear, J.W.; Walters, F.J.M.; Wilkins, D.G.; Willatts, S.M.: Etomidate by infusion for neuroanesthesia. Anaesthesia, 39: 12-18, 1984.
20. Symon, L.; Wang, A.D.; Costa, E.; Silva, I.E.; Gentile, F.: Perioperative use of somatosensory evoked response in aneurysm surgery.. J. Neurosurg., 60: 269-275, 1984.

Brain and Spinal Cord Monitoring by Multispatial and Multimodal Evoked Potentials during Aortic Surgery

Y. Maruyama;[*] K. Shimoji; H. Fujioka; T. Takada; H. Endoh

Summary

The most serious complications of aortic surgery are ischemic spinal cord and/or brain dysfunctions caused by an aortic clamp or emboli (15). As neurological signs and symptoms of ischemic lesions by aortic clamping are masked during anesthesia, an alternative measure should be undertaken for monitoring the brain and spinal cord functions. Although several neurophysiological techniques have been successfully used to detect early signs of CNS dysfunction due to carotid surgery (10, 17, 31, 34), there are some difficulties in monitoring CNS functions in aortic surgery. First, it is hard to define the anticipated sites of ischemia preoperatively, since there are considerable anatomical variations of arterial outflows to the spinal cord (7). Second, although the skin surface recording of spinal cord potential (SCP) has been attempted (32), it is often hard to reproduce the potential and also takes a considerable amount of time to average the response.

Therefore, we have developed a new method of monitoring the spinal cord and brain functions during aortic surgery. We report herein that mul-

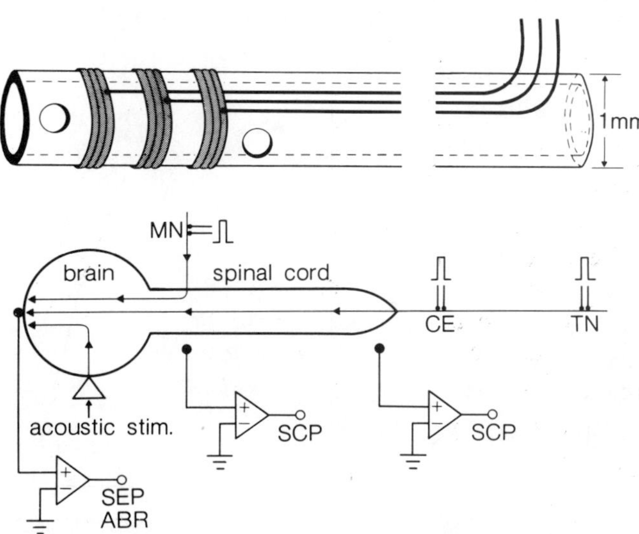

Fig. 1. Schematic drawings of an epidural catheter electrode (top) and the experimental arrangements (bottom). MN = median nerve; CE = cauda equina; TN = tibial nerve; SEP = somatosensory evoked potential; SCP = spinal cord potential; ABR = auditory brainstem response.

* Department of Anesthesiology, Niigata University School of Medicine, 1-757 Asahi-Machi, Niigata 951, Japan

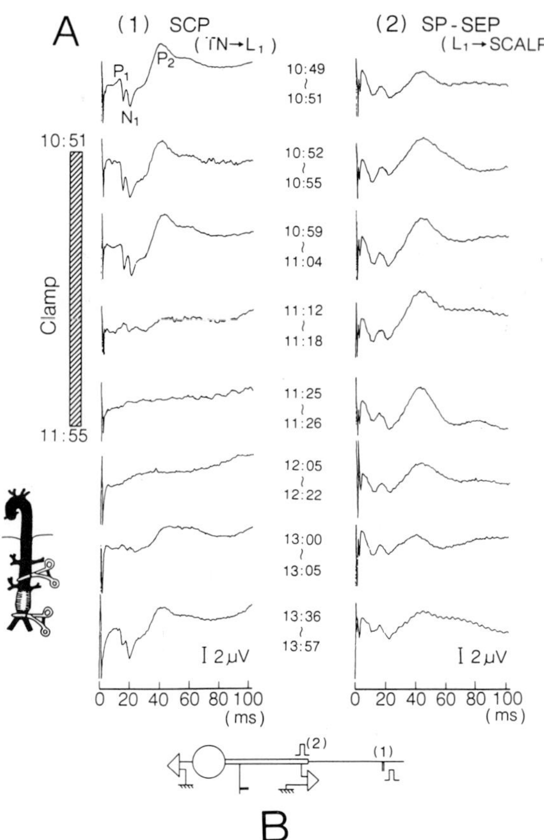

Fig. 2. (A). Specimen records of the SCP (from the epidural space at L1 vertebral level) in a patient (case 3 in Table 1) who underwent a Y-graft. Note that the SCP disappeared completely without any substantial change in the SEP by cross-clamping of the abdominal artery. Positivity of the PES potential to the skin surface (reference) produces an upward deflection in this and all subsequent recordings.

Fig. 2. (B). Graphic representation of amplitude changes in each component of the SCP and SEP by cross-clamping of the abdominal artery (shaded bar).

timodal and/or multi-spatial recordings of evoked potentials, using a specially designed epidural catheter electrode, might be of value for this purpose. A preliminary account has been made elsewhere (18, 30).

Methods - Patients and Anesthesia

Ten patients aged between 24 and 72 years were investigated. Each patient had given informed consent to our program approved by the institutional review board of this university hospital. Disease, operation, method of circulatory assist, and ischemic episode with temperature are listed in Table 1. All patients were free of neurological deficit. Meperidine, 1mg/kg, and atropine, 0.5mg, were used as premedication. Before induction of anesthesia, all electrodes for recording and stimulation were placed at the appropriate sites. A large-bore intravenous catheter was inserted into the cephalic vein for administration of the drugs and for infusion of an electrolyte solution (lactated Ringer). Anesthesia was induced with thiamylal, 2-5mg/kg. The trachea was intubated following pancuronium, 0.1mg/kg, and the lungs ventilated with 60% nitrous oxide in oxygen. The ventilation volume was adjusted to keep end-tidal carbon dioxide tension constant (35-45 torr) (monitor with an infrared analyzer). A few minutes after the induction of anesthesia, fentanyl, 50-75μg/kg (six patients), or morphine, 2.5mg/kg (two patients), was injected intravenously in three to five incremental doses. The remaining patients (cases 4 and 10) underwent enflurane (1.5-1.75%) anesthesia (Table 1).

Radial artery and balloon-tipped pulmonary artery catheters were inserted. Monitoring of EEG, EKG, heart rate, systemic arterial pressure, pulmonary arterial pressure, cardiac output, arterial blood gases, and rectal and esophageal temperatures was established. Hypertension and hypotension during the procedure were treated by intravenous nitroglycerin and dopamine, respectively.

Recording of evoked potentials

For recording somatosensory evoked potentials (SEPs), the P3 and P4 standard EEG electrode positions based on the international 10-20 system were used, with the reference electrode placed at the earlobe. Surface electrodes, Ag-AgCl, 10mm in diameter, were used for recording the evoked potentials from the scalp in response to stimulation of the median nerve (MN) at the wrist, the tibial nerve (TN) at the popliteal fossa, terminal cone at L1 vertebral level or cauda equina (CE) at L4 vertebral level from the posterior epidural space (PES). Strength for peripheral nerve stimulation was adjusted to a level just below the pain threshold. The body ground for all electrical activities was led from the surface electrode (1x1cm) attached to the shoulder.

The technique for recording the evoked potentials from the epidural space has been described previously (27, 29). A pair of catheter electrodes were introduced into the PES at L1 through L4, and, at the C6 level before the start of anesthesia in one subject. The electrodes were stainless steel wires, 50μm in diameter, attached to the tip of a polyethylene tube (Fig. 1). The reference needle electrodes (230μm in diameter and 10mm in length) were inserted into the skin close to the supraspinal ligament of the same spinal levels.

Since the SCP was often small and easily contaminated by electrocardiogram artifacts, it was obtained at intervals between the T and P waves of the electrocardiogram. This was accomplished in the following manner. The electrocardiogram was displayed on the screen of an oscilloscope with the sweep triggered by every other QRS complex (28). A pulse was generated at the start of each sweep with a variable delay and was fed into the stimulator as a trigger. Thus, the stimulus was delivered after every other QRS complex of the electrocardiogram. The SCPs, which were monitored on other channels of the oscilloscope, were led to a computer (ATAC 2300) for averaging. The computer was triggered by the stimulating pulse. The average response (n = 25-50) were photographed or plotted by an X-Y plotter. At the same time, all electrical activities were continuously recorded on an ink-writing polygraph (Nihon Kohden RM-150 M). The time constant used was 2.0 seconds for both the SCP and electrocardiogram. Stimulus parameters used were square waves, 0.1ms in duration, 15-30 (for peripheral nerves) and 2-10mA (for epidural stimulation) in peak current.

Before recording, the approximate position of the electrodes in the PES was assessed by applying weak electrical pulses through these recording electrodes. If an electrode was situated laterally in the PES, the stimulation caused slight twitching of the ipsilateral muscles of the same segment; if it was located in the midsagittal plane, the stimulation caused weak bilateral muscle twitching. In addition, if the electrode was situated properly in the PES, a stimulus intensity as low as 1.5-4.5V (1.9-5.6mA) produced small visible segmental muscle twitching (20, 26, 28). These procedures did not cause the subjects any discomfort. If the electrode slipped out of the PES and was

A

Case7 Dissecting Aneurysm (I)

(Arch Replacement)

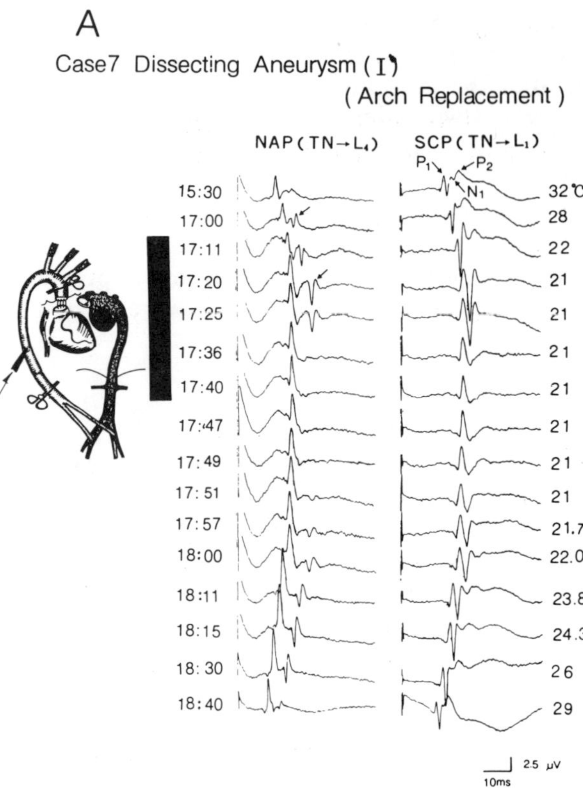

Fig. 3 (A). Simultaneous recording of nerve action potential (NAP) recorded from the PES at L4 vertebral level and spinal cord potential (SCP) recorded from the PES at L1 vertebral level in a patient with dissecting aneurysm (case 7). Both the NAP and spike potential (P1) of the SCP were prolonged in latency and increased in amplitude by the decrease in body temperature. After the arch-replacement was performed, clamping was carried out on the artificial vessel for manipulation of the aneurysm, as shown in the schematic drawing. By this clamp, latencies of both the NAP and P1 were prolonged with increases in their amplitudes. The N1 and P1 waves of the SCP disappeared after a transient facilitation and reappeared gradually after declamping. A spike potential marked by an arrow in the NAP trace indicates a reflex potential.

(B). Graphical representation of the sequential changes in the P1 latency and in amplitudes of both the N1 and P2 with rectal temperature. During the clamp, rectal temperature remained at 21°C and the P1 latency showed almost the same value throughout. On the other hand, amplitudes of the N1 and P2 waves rapidly decreased after transient facilitation by the clamp.

located in the adjacent muscles, segmental muscle twitches could hardly be elicited even by increasing the stimulus intensity, and evoked responses could barely be recorded with this electrode. The location of the recording electrodes in the PES was roughly estimated by these procedures (20, 26, 28). Thus, when the

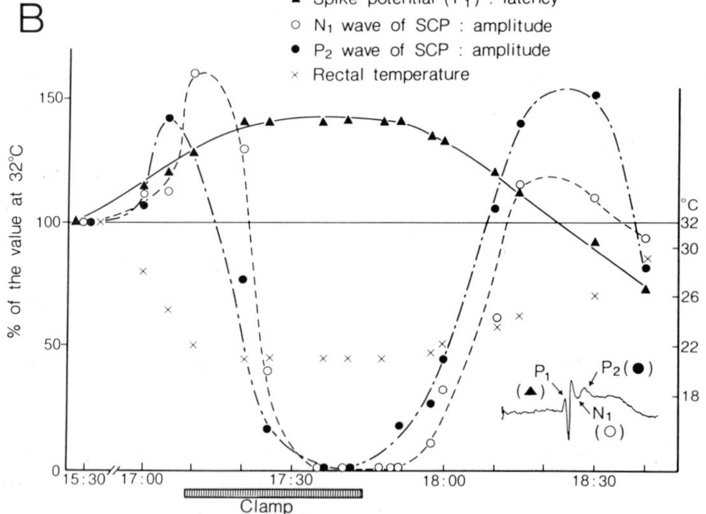

electrode was considered to slip out of the PES, it was repositioned properly in the PES (26, 27). The x-ray examinations at the end of each recording strongly supported the usefulness of this method for estimating the electrode position, and the related segment was roughly estimated by Pernkopf's atlas (13).

The SCPs recorded from the PES are divided into three patterns: Those evoked by segmental, ascending and descending volleys (20, 26, 28-30). The segmentally evoked

SCP consists of an initially positive spike (P1), reflecting the afferent volley along the roots, followed by slow negative wave (N1) believed to be synchronized activity of dorsal horn neurons, and slow positive wave (P2) assumed to reflect primary afferent depolarization (20, 26). For instance, the radial, median or ulnar nerve stimulation produces the segmental SCP in the PES at C5 through T1 vertebral level (cervical enlargement); and tibial nerve or cauda equina (CE) stimulation evokes the segmental SCP in the PES at T9 through L1 vertebral level (lumbar enlargement) (20, 26, 28-30). The waveform characteristics of these potential changes produced in the PES are just the same as those directly recorded from the cord surface (21). Mostly, the SCPs recorded from the PES at L1 vertebral level (case 3-8, 10) in response to tibial nerve stimulation and those recorded at C6 vertebral level (case 8) elicited by afferent volleys were subjected to this study (Table 1). Nerve action potential (NAP) was recorded from the PES at L4 vertebral level (20).

Using a specially designed catheter electrode which had two side holes near its tip (Fig. 1), an aspiration test was carried out frequently to ensure that the tip of the catheter had not eroded into a blood vessel. The catheter electrode was gently introduced to a minimum depth (no more than 5cm). Frequent flushing with saline and aspiration were undertaken during and after the operation. Neurological examination was also carried out after the operation.

The MN-SEP and TN-SEP involve neuronal activities through the median nerve - cervical spinal cord to the brain and those through the tibial nerve - total length of spinal cord to the brain, respectively. Therefore, to detect an ischemic injury confined to the brainstem, the auditory brainstem response (ABR) was also recorded in five subjects (Fig. 1). For recording the ABR, clicks, having a 100μs duration and an intensity of 100dB NHL (normal hearing level), were delivered through bilateral shielded earphones at 20Hz and the averaged responses (n=2000) were recorded from Cz with reference to the earlobes. The ABR and SEPs with or without SCP were alternatingly recorded throughout the operation. In all recordings, upward deflection was set to be positive.

The peak latency and peak-to-peak amplitude of the early components (around 20ms in latency) were chosen for analysis in the SEP [Fig. 2 (A)]. Component V of the ABR was selected for analysis due to its stable reproducibility (9).

Results

Change in the SCP occurred in three patients with cross-clamping of the aorta (cases 3, 7 and 10 in Table 1).

The segmentally evoked SCP was completely abolished by clamping of the abdominal aorta in case 3 (Fig. 2) and 10. However, simultaneous recording of the SEP from the scalp produced by epidural stimulation at the L1 level through the same electrodes used to record the SCP revealed that the disappearance of the segmental SCP was not caused by ischemia in the lumbar spinal cord, but by that in the peripheral nerve.

In case 7, the scalp-recorded SCP (produced by epidural stimulation), segmental SEP and nerve action potential (NAP) were all decreased by aortic clamping (Fig. 3). Responding to these changes, further procedures were abandoned and the aortic clamp was released immediately after the disappearance of the SCP. There were also no postoperative sequelae in this case.

Changes in the waveforms of the evoked potentials were noticed in six of the ten patients; in three of them (cases 1, 2, 9), they were produced by the short-term circulatory arrest and in the other three (cases 3, 7, 10) were due to the aortic clamping. Table 2 shows the summary of changes in amplitudes of the evoked potentials recorded in the three patients subjected to clamping of the aorta with or without circulatory as-

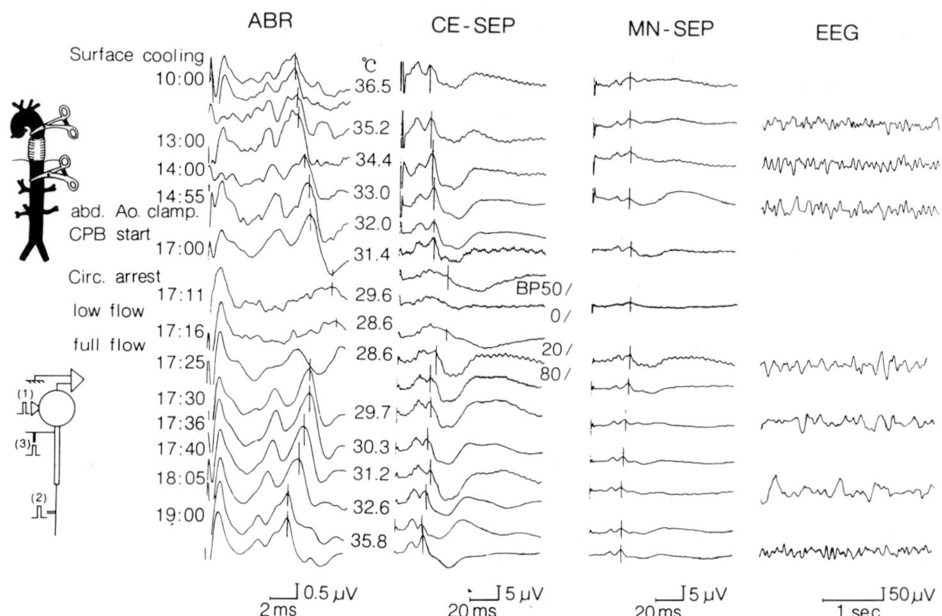

Fig. 4 (A). Simultaneous and sequential records of ABR, CE-SEP and MN-SEP with EEG in case 1. Vertical lines are given to show clearly the peak latencies of component V of ABR and early negative components of SEPs (N20 in CE-SEP; N25 in MN-SEP). Time and esophageal temperature are shown at the left and right sides of the ABR records, respectively. CZ (ABR) or P3 (SEP) to the ear-lobe with relative positivity of CZ or P3 producing an upward deflection.

(B). Changes in latency and amplitude of ABR (component V), CE-SEP and MN-SEP by hypother-mia, abdominal aortic cross-clamping (A.C.) and circulatory arrest (C.A.) during aortic surgery.

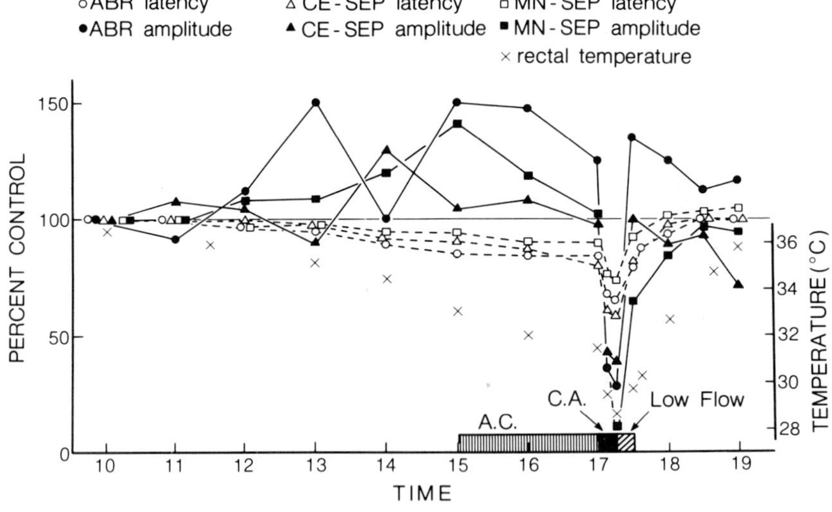

sist. Upon recirculation, the changes in the evoked potentials were completely reversed during the operation in all six cases, and there were no postoperative neurological abnormalities.

Fig. 4 shows the specimen records of the ABR, cauda equina (CE)-SEP and median nerve (MN)-SEP (A) in case 1, and the sequential representation of changes in their latencies and amplitudes (B). Latencies of all evoked potentials were prolonged in parallel to the decrease in body temperature [Figs. 3, 4 (A)]. Amplitudes of all evoked potentials showed a considerable variablity among the subjects in response to the decrease in body temperature (Figs. 3, 4). The ABR, however, disappeared at around 20°C of rectal temperature, while the SEP was still obtained at the same body temperature (cases 2, 9).

Ischemia caused by total circulatory arrest produced an abrupt prolongation of latencies and a decrease of amplitudes in all scalp-recorded evoked potentials.

Discussion

The present study has shown that systemic ischemia produces a simultaneous amplitude decrement with latency prolongation in all evoked potentials from the scalp in response to acoustic and somatosensory stimulations, whereas local ischemia caused by aortic clamping induces an unparalleled change in the evoked potential recorded from the involved area (Fig. 2).

If ischemia was present in the spinal cord at the T1 through T4 or L1 level, believed to be the most vulnerable site to ischemia (3, 7, 15), by clamping of the descending aorta in the patient shown in Fig. 4, the cauda equina (CE)-SEP, which traverses through the CE, the entire length of the spinal cord, brainstem and thalamus to the cerebral cortex, must have decreased in amplitude and prolonged in latency sparing the MN-SEP and ABR. This is because the MN-SEP travels through the nerve, cervical spinal cord, brainstem, and thalamus to the cerebral cortex without passing the T1 through T4 and L1 segments of the cord. The ABR also should not be changed because its tract does not contain the spinal cord, thus serving as a reference potential for diag-

---

Table 1. Diseases and anesthetic methods in ten patients subjected to the study.

Patients:	9 males, 1 female
Age:	24 - 72
Disease:	Descending aortic aneurysm (5) Dissecting aortic aneurysm (2) Abdominal aortic aneurysm (1) Coarctation of aorta (1) Pseudoaneurysm (1)
Anesthesia:	High dose fentanyl (6) High dose morphine (2) N₂O/enflurane (2)

nosis. Even by a simultaneous clamping of the descending aorta, there was a parallel change in all evoked potentials with gradual development of hypothermia (Fig. 4).

Further, in the patient shown in Fig. 2, if the cord function had been monitored only by the tibial nerve-SEP, there could be misjudgment. Placement of the electrodes at both below (the epidural electrode at L1 through L2 vertebral level) and above (the scalp electrode) the anticipated ischemic site (the lumbosacral enlargement in the zone of Adamkiewicz's artery) along the sensory path should be a proper arrangement for accurate monitoring of the cord function in this particular case.

Previous studies have shown that the ABR (9, 12, 22) and SEP (4, 33) components increase in latency with lowered temperature. Two possible mechanisms have been thought to account for the increase in latency as a function of decreasing temperature; firstly, synaptic delay, and secondly, decreased conduction velocity along the axons (4). Studies in animal neural pathways suggest that latency increases result largely from increased synaptic delay compared to the decrease in conduction velocity (9, 12, 22). Transsynaptic excitation has been shown to fail at a temperature of between 10 and 20°C with minimal disturbance to the velocity and amplitude of axonal conduction (2, 6). Parallel prolongations of the latencies of all evoked potentials with a progressive decrease in temperature, in spite of differences in stimulation modality and stimulation site in the present study, might have been caused by this characteristic feature of synaptic transmission and axonal conduction affected by temperature.

The effects of hypothermia on the evoked potential amplitudes have been controversial. Amplitude increases (22) as well as decreases (5, 14) of evoked potentials have been reported. It is possible that the rate of cooling and/or the degree of hypothermia may be responsible for these controversial effects of temperature on the evoked potential amplitudes. Rossi and Britt (22) have demonstrated in their well-controlled study in cats that amplitude increase occurred at between 37 and 32°C, whereas

Table 2. Ischemic regions in 3 patients who underwent aortic cross-clamp, diagnosed by evoked potential monitoring.

	Stimulation Modality and Site	Evoked Potential and Recording Place	Change in Evoked Potential	Ischemic Area Diagnosed
Case 3	TN stim →	SCP (L_1)	disappeared	peripheral nerve
	L_1 stim →	SEP	no change	
Case 7	TN stim →	NAP (L_4)	decreased slightly	lumbosacral spinal cord
	TN stim →	SCP (L_1)	disappeared	
	L_1 stim →	SEP	decreased	
Case 10	TN stim →	SCP (L_1)	disappeared	peripheral nerve
	L_1 stim →	SEP	no change	
	Acoustic stim →	ABR	no change	

the amplitudes of all five components in the ABR decreased at approximately a linear rate below 32°C. In the present study, the temperature in the neural tissue was presumed to be higher than that read by the esophageal temperature. This might cause the variable amplitude changes during hypothermia down to 30°C. With profound hypothermia, however, all evoked responses should be suppressed. The critical temperature at which evoked potentials cease to be activated might depend upon their characteristics. For instance, the ABR disappeared at a temperature of around 20°C (cases 2, 9), while the SEP was still clearly recorded in the present study (Table 1).

There is ample evidence that changes in the evoked potential reflect those in oxygen delivery to neural tissue, and failure of return of the evoked potential to the prehypoxic level is associated with pernament neurologic changes. Ischemic changes in the evoked potentials have been assigned to be latency increase and amplitude decrease (11, 23-25, 35) with sometimes a transient excitation (11, 33). Many SEP studies have aimed at the demonstration of supratentorial lesions in patients with cerebral dysfunction due to ischemia (24, 25, 35). These studies, however, were limited to the recording of SEPs from the scalp in response to the sensory stimulation with a single modality and site. In such patients who undergo clamping of the aorta, acute ischemia can develop not only in the brain, but in the spinal cord, particularly at the T1 through T4 or L1 level (7, 15). Intensive sequential monitoring of the evoked potentials of multimodalities and multispaces as attempted in the present study might give the earliest signs and sites of ischemia.

Although greater debate exists about the source of the early complex (N20, P25) of the MN-SEP, it is suggested either to be the primary somatosensory cortex (1), the thalamocortical radiation (8) or the thalamus (9). Thus, when changes in the early complexes of the SEPs do occur without any change in the ABR, it may indicate that the ischemic lesion is present either below the level of the brainstem (cervical spinal cord in the MN-SEP and a certain level of spinal cord in the TN-SEP or CE-SEP) or beyond the structure (thalamus to cortex). When a parallel change is demonstrated in all these evoked potentials during surgery, it may reveal a systemic effect. Indeed, all evoked potentials (the ABR, MN-SEP and/or CE-SEP) were simultaneously abolished or attenuated during the total circulatory arrest as in case 1.

In conclusion, our first preliminary experience suggests that intraoperative monitoring of the multimodal and multispatial evoked potentials during aortic surgery is a useful technique to detect early CNS dysfunction and also to define the site of a lesion caused by ischemia.

Summary

To detect brain and spinal cord ischemia produced by aortic cross-clamping, the somatosensory evoked potentials from the skull and spinal epidural space, and auditory brainstem responses, were recorded simultaneously and sequentially in patients during aortic surgery.

All evoked potentials were prolonged lineally in latency and varied in amplitude as the body temperature decreased. Systemic ischemia caused by total circulatory arrest produced an abrupt latency prolongation and amplitude decrease in all modality responses. Local ischemia by aortic clamping demonstrated unparalleled changes in evoked potentials, which made it possible to detect the site of ischemia.

Thus, recording of multimodal and multispatial evoked potentials along the sensory tract might provide a more accurate monitoring of an ischemic lesion and its site in the brain and/or spinal cord during aortic surgery.

Acknowledgement

This work was supported in part by a grant-in-aid (#59870051) from the Japanese Ministry of Education, Science and Culture.

References

1. Arezzo, J.; Legatt, A.D.; Vaughan, H.G.: Topography and intracranial sources of somatosensory evoked potentials in the monkey. 1. Early components. Electroenceph. Clin. Neurophysiol., 46: 155-172, 1979.
2. Benita, M.; Conde, H.: Effects of local cooling upon conduction and synaptic transmission. Brain Res., 36: 133-151, 1972.
3. Bromage, P.R.: Epidural Analgesia. Philadelphia: W.B. Saunders Co., 1978, pp. 231-232.
4. Brooks, V.B.: Studies of brain function by reversible local cooling. In: R.H. Adrian et al. (eds), Review of Physiology, Biochemistry and Pharmacology. New York: Springer-Verlag, 1983, Vol. 95, pp. 200-109.
5. Brown, M.C.; Smith, D.I.; Nuttall, A.L.: The temperature dependency of neural and hair cell responses evoked by high frequencies. J. Acoust. Soc. Amer., 73: 1662-1670, 1983.
6. Budnick, B.; McKeown, K.L.; Wiederhold, W.C.: Hypothermia-induced changes in rat short latency somtosensory evoked potentials. Electroenceph. Clin. Neurophysiol., 51: 19-31, 1981.
7. Carpenter, M.B.; Sutin, J.: Human Neuroanatomy, 8th ed. Baltimore: Williams & Wilkins, 1983, pp. 707-741.
8. Celesia, G.G.: Somatosensory evoked potentials recorded directly from human thalamus and Sm 1 cortical area. Arch. Neurol., 36: 399-405, 1979.
9. Chiappa, K.H.: Evoked Potentials in Clinical Medicine. New York: Raven Press, 1983, pp. 203-250.
10. DeBakey, M.E.; Crawford, E.S.; Cooley, D.A.; Morris, G.C., Jr.; Garrett, H.E.: Cerebral arterial insufficiency. One- to 11-year results following arterial reconstructive operation. Ann. Surg., 161: 921-945, 1965.
11. De Weerd, A.W.; Looijenga, A.; Veldhuizen, R.J.; van Huffelen, A.C.: Somatosensory evoked potentials in minor cerebral ischaemia: Diagnostic significance and changes in serial records. Electroenceph. Clin. Neurophysiol., 62: 45-55, 1985.
12. Doyle, W.J.; Fria, T.J.: The effects of hypothermia on the latencies of the auditory brainstem response (ABR) in the rhesus monkey. Electroenceph. Clin. Neurophysiol., 60: 258-266, 1985.
13. Ferner, H.: Eduard Pernkopf Atlas der Topographischen und Angewandten Anatomie des Menschen. Munich: Urban & Schwarzenberg, 1963.
14. Kaga, K.; Takiguchi, T.; Myokai, K.; Shiode, A.: Effects of deep hypothermia and circulatory arrest on the auditory brainstem responses. Arch. Otorhinolaryngol., 225: 199-205, 1979.
15. Lake, C.L.: Cardiovascular Anesthesia. New York: Springer-Verlag, 1984, pp. 383-409.
16. Lehmkuhl, D.; Dimitrijevic, M.R.; Renouf, F.: Electrophysiological characteristics of lumbosacral evoked potentials in patients with established spinal cord injury. Electroenceph. Clin. Neurophysiol., 59: 142-155, 1984.
17. Markand, O.N.; Dilley, R.S.; Moorthy, S.S.; Warren, C., Jr.: Monitoring of somatosensory evoked responses during carotid endarterectomy. Arch. Neurol., 41: 375-378, 1984.
18. Maruyama, Y.; Fujioka, H.; Shimoji, K.: Monitoring of cerebrospinal function during cardiovascular surgery. Jap. J. EEG EMG, 13: 46, 1985.
19. Maruyama, Y.; Shimizu, H.; Fujioka, H.; Shimoji, K.; Takahashi, H.; Honma, T.; Yasukawa, K.; Nakamura, T.: Spinal cord function monitoring by spinal cord potentials during spine and spinal surgery. In: S. Homma; T. Tamaki (eds), Fundamentals and Clinical Application of Spinal Cord Monitoring. Tokyo: Saikon Publishing Co., 1984, pp. 191-201.
20. Maruyama, Y.; Shimoji, K.; Shimizu, H.; Kuribayashi, H.; Fujioka, H.: Human spinal cord potentials evoked by different sources of stimulation and conduction velocities along the cord. J. Neurophysiol., 48: 1098-1107, 1982.
21. Nashold, B.S., Jr.; Ovelmen-Levitt, J.; Sharpe, R.; Higgins, A.C.: Intraoperative evoked potentials recorded in man directly from dorsal roots and spinal cord. J. Neurosurg., 62: 680-693, 1985.
22. Rossi, G.T.; Britt, R.H.: Effects of hypothermia on the cat brainstem auditory evoked response. Electroenceph. Clin. Neurophysiol., 57: 143-155, 1984.
23. Schramm, J.; Hashizume, K.; Fukushima, T.; Takahashi, H.: Experimental spinal cord injury produced by slow, graded compression: Alterations of cortical and spinal cord potentials. J. Neurosurg., 50: 48-57, 1979.
24. Shahani, M.; Bharucha, E.P.; Capadia, G.D.: Comparative study of early and late somatosensory evoked potentials in patients with hemiplegia and/or hemianesthesia. In: C. Barber (ed.), Evoked Potentials. Lancaster, MTP Press, 1980, pp. 465-474.

25. Shibasaki, H.; Yamashita, Y.; Tsuji, S.: Somatosensory evoked potentials. Diagnostic criteria and abnormalities in cerebral lesions. J. Neurol. Sci., 34: 427-439, 1977.

26. Shimizu, H.; Shimoji, K.; Maruyama, Y.; Matsuki, M.; Kuribayashi, H.; Fujioka, H.: Human spinal cord potentials produced in lumbosacral enlargement by descending volleys. J. Neurophysiol., 48: 1108-1120, 1982.

27. Shimoji, K.; Higashi, H.; Kano, T.: Epidural recording of spinal electrogram in man. Electroenceph. Clin. Neurophysiol., 30: 236-239, 1971.

28. Shimoji, K.; Ito, Y.; Ohama, K.; Sawa, T.; Ikezono, E.: Presynaptic inhibition in man during anesthesia and sleep. Anesthesiology, 43: 388-391, 1975.

29. Shimoji, K.; Kano, T.; Higashi, H.; Morioka, T.; Henschel, E.O.: Evoked spinal electrograms recorded from epidural space in man. J. Appl. Physiol., 33: 468-471, 1972.

30. Shimoji, K.; Maruyama, Y.; Shimizu, H.; Fujioka, H.; Taga, K.: Spinal cord monitoring - a review of current techniques and knowledge. In: J. Schramm; S.J. Jones (eds), Spinal Cord Monitoring. Berlin: Springer-Verlag, 1985, pp. 16-28.

31. Sundt, T.M., Jr.; Sharbrough, F.W.; Piepgras, D.G.; Kearns, T.P.; Messick, J.M., Jr.; O'Fallon, W.M.: Correlation of cerebral blood flow and electroencephalographic changes during carotid endarterectomy. Mayo. Clin. Proc., 56: 533-543, 1981.

32. Takaki, O.; Oku, S.; Kishi, Y.; Kobori, M.; Okumura, F.: Intraoperative monitoring of somatosensory evoked potentials in surgery of the aortic aneurysm. Jpn. J. Anesthesiol., 33 (suppl.): S219, 1984.

33. Taylor, M.J.; Borrett, D.S.; Coles, J.C.: The effects of profound hypothermia on the cervical SEP in humans: Evidence of dual generators. Electroenceph. Clin. Neurophysiol., 62: 184-192, 1985.

34. Thompson, J.E.: Complications of carotid endarterectomy and their prevention. World J. Surg., 3: 155-165, 1979.

35. Tsumoto, T.; Hirose, N.; Nonaka, S.; Takahashi, M.: Cerebrovascular disease: Changes in somatosensory evoked potentials associated with unilateral lesions. Electroenceph. Clin. Neurophysiol., 35: 463-473, 1973.

Anesthesia Influence on Recording: Summary

A. Koht[*]

Introduction

The advancements in various medical fields enhanced spinal surgery and increased the volume of more aggressive procedures. These operations have been associated with altered spinal function and different methods have been proposed to minimize such risk. Among these are local anesthesia, wake-up test, and more recently evoked potentials. Local anesthesia for spinal surgery is not practical and the wake-up test is limited to the time of testing. Somatosensory evoked potentials (SSEP), a safe, noninvasive, easy, and continuous technique for assessing the functional integrity of the spinal cord, offers a valuable alternative. However, many factors such as anesthesia, pharmacology, physiology, and technical errors can limit the value of evoked potentials. Nonsurgical factors account for 40% of SSEP changes and 25% of these changes are related to anesthesia (16). The yield of surgical monitoring is enhanced by the reduction of nonsurgical factors. The appreciation of anesthetic effects, comprehension of the triangle of anesthesia, and better utilization of such information decrease nonsurgical factors and increase the value of SSEP monitoring. A stable level of anesthesia and hemodynamics facilitates the differential diagnosis of SSEP changes. Intravenous anesthesia supplemented with low concentration of volatile agents offers that level of stability and may be an ideal technique for SSEP monitoring.

In this chapter the general characteristics of SSEP changes related to anesthesia are discussed and outlined for anesthesia with minimal effect on evoked potentials enclosed.

Characteristics of SSEP changes related to anesthesia

Such changes, usually bilateral, follow the administration or change in concentration of existing agent and manifest itself at higher cortical more than subcortical or spinal responses. In fact the effects of anesthesia are farther weakened when responses are recorded directly from the spinal cord or the epidural space. The selective effect of anesthetic agents at a higher cortical rather than subcortical or spinal levels reinforce the value of SSEP recording from spinal leads. Utilizing recording sites at the spinal cord, epidural space, or the emphases on subcortical evoked potentials avoids much of the anesthetic effects and increases the sensitivity of SSEP monitoring. These responses are less effected by fluctuations in anesthetic, physiological, and pharmacological factors.

---

* Director of Neurosurgical Anaesthesia, Northwestern University, 303 E. Chicago Avenue, Chicago, IL 60611

Inhalation agents affect amplitude more than latency, tend to be slow, dose dependent, follow the anesthetic partial pressure, and in time could abolish the SSEP signals (20, 24). Fat solubility, inhaled anesthetic concentration, and cardiac output alter the way inhalation agents affect SSEP recordings.

Contrary to the slow effect of the inhalation agents, intravenous drugs have an abrupt effect after a bolus injection (12, 15, 19, 22, 30). These changes are dose related at a lower slope than inhalation agents and usually stabilize within 20 minutes. The period of stabilization relates to the drug and the dose being used. After this period signals gradually return toward normal when the serum levels decline.

In general, anesthesia shifts the baseline values of the preoperative evoked potential curve to the right and also moves the amplitude to a new smaller baseline. Such effect occurs after the induction of anesthesia and continues throughout the rest of the operation. This fact should not be missed during intraoperative monitoring.

Nonanesthetic drugs, such as sodium nitroprusside, behave similar to intravenous drugs. However, their influence is related to their systemic action rather than a direct effect on the nervous system.

The effect of premedication

Drugs such as atropine, scopolamine, morphine, meperidine, fentanyl, diazepam, and droperidol have minimal or no effects on evoked potentials (9, 10, 25). Both meperidine and droperidol were studied as intravenous injections. Although intravenous route is not the usual way for premedication, it is still helpful to understand their effect. In this study meperidine enhanced the amplitude of cortical recording and increased the absolute and interpeak latencies. However, the effects of droperidol on cortical evoked potentials were not consistent. In a recent article, Loughnan had described the effects of intravenously injected fentanyl and diazepam in awake patients (17). The dose for fentanyl was 200μ -001Dg and for diazepam 20mg. Both drugs showed little effect on SSEP in the hour following injection. In general, premedication tended to decrease apprehension and smooth the induction of anesthesia. By doing so they decrease muscle tension and patient's apprehension and may have a positive effect on evoked potentials. Such relaxation minimizes muscle twitches, artifact, and produces better wave forms and saves time during data averaging. Such sedation has been known and utilized for a long time in the diagnostic laboratory (4).

The effect of induction agents

Induction utilizes either intravenous or inhalation agents; however, most patients other than children are induced by intravenous route. Due to the shift of evoked potential curves under anesthesia, it is essential to re-establish the baseline as soon as possible within the first half hour.

Thiopental

This is the most widely used drug for the induction of anesthesia. It has a short half life and works fast. The effect of thiopental on SSEP has been studied in regular induction doses (3-5mg/kg) (12, 19, 22) and in higher doses that produce coma in man (6) and in animals (30). Induction doses of thiopental did not effect amplitude or latency of cervical evoked potentials but slightly increased the latency of cortical responses. However, these changes were short lived and stabilized within half an hour. The high dose used in humans (77.5mg/kg) showed a dose related changes in latencies and amplitudes of various evoked responses and increase in central conduction time. However, signals were still useful to monitor at a dose surpassing twice the dose of thiopental needed to produced isoelectric EEG (22.5mg/kg). The higher doses (20-180mg/kg) used in animal experiments showed a dose dependent effect at high cortical responses but a lesser effect at lower brain stem and spinal responses. Central conduction time (CCT) was not significantly altered in this study. In conclusion, thiopental is a suitable drug for the induction and maintenance of anesthesia. During induction, if thiopental is used in conjunction of high dose narcotics, it is essential to lower thiopental's dose to 1/8-1/4 of the usual dose to avoid hypotension. After induction the doses of thiopental used for amnesia range from 1.5 - 2mg/kg/h, preferably by infusion technique. In either situation, induction or maintenance, the direct effects on evoked potentials are minimal.

Etomidate

Recent studies of the effects of etomidate on SSEP showed increases in the amplitude of the cortical response (N20 and higher) by as much as 150-400%, and decreases of the cervical response (14, 19). This effect is not related to the vehicle used for injection (19). Latency of cortical but not cervical responses were prolonged. Such changes in SSEP's were in contrast to the decrease in amplitude of early and late cortical responses seen in auditory evoked potentials (31). The latency of both high cortical SSEP's and middle to late auditory responses increased while no changes occurred at the brain stem level. These studies of etomidate and SSEP suggest to withhold etomidate during a critical stage of the operation. Such action minimizes the confusion caused by either the rise or decline in amplitude [see this volume] (more details in chapter D 26 by Dr. McPherson). On the other hand, it may be useful in patients with very small signals. A continuous infusion may be used to augment such small signals. However, such hypothesis needs further research.

Narcotic agents, morphine, fentanyl, and sufentanil

Fentanyl effect on evoked potentials has been studied in a variety of doses (7, 11, 16, 17, 19, 22, 23, 25). Pathak et al. compared the intermittent and continuous infusion of morphine and fentanyl (22). Doses for fentanyl were 50-100µg as intermittent injections and 1.5-2.5µg/kg/h for infusion. Doses of morphine were 5-10mg injections and 150-250µg/kg/h infusions. This study showed a decrease in amplitude and an increase in latency of the responses at the higher cortical more than the subcortical levels. Changes were dose dependent and occurred more with intermittent than infusion techniques. McPherson and his coworkers found 25µg/kg fentanyl slightly increase latency and decrease amplitude of cortical responses but spare subcortical and spinal potentials (19). A higher dose of fentanyl (75µg/kg) in open heart surgery did not affect either the amplitude or latency of SSEP responses when corrected for temperature (11). In auditory evoked potentials 25-50µg/kg fentanyl did not have an effect on either absent or minimal and did not interfere with the usage of SSEP monitoring.

We studied sufentanil at an induction dose of 5µg/kg (equal to 50-75µg fentanyl) in fifteen unpremedicated patients (15). Our findings consisted of a slight increase in latency of the cortical response stabilized within 20 minutes and was not associated with changes in cervical potentials. Amplitude was reduced in 13 patients and increased in two. Changes in SSEP, due to sufentanil, are most likely to occur early following moderate bolus delivery, and subsequently remain stable.

It is clear that a variety of narcotics at different induction doses can be used during SSEP monitoring without altering the value of this monitor.

Midazolam

The effects of midazolam on SSEP is not widely known and the only study available to me at this time is in German. It shows a decrease in amplitude and an increase in

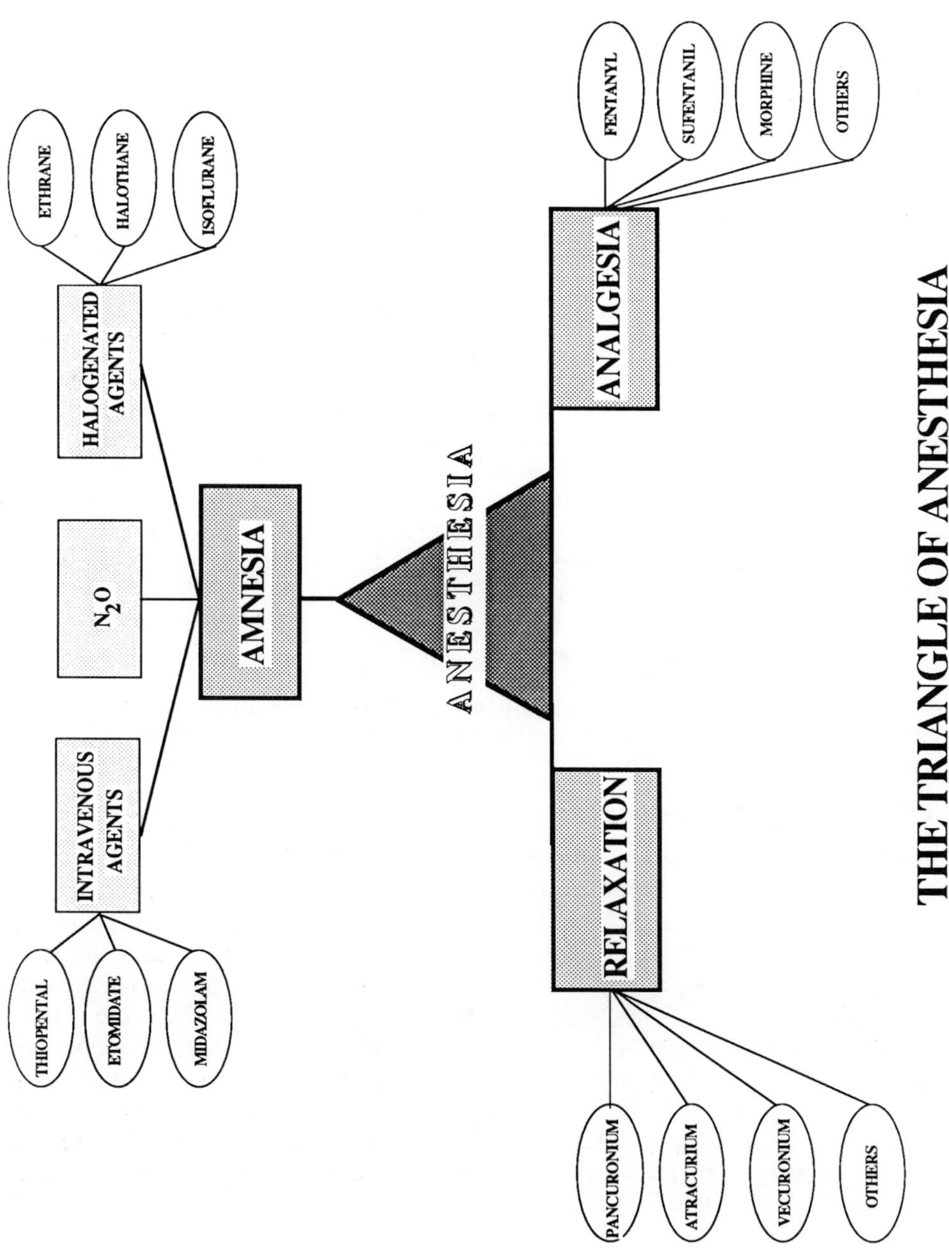
ETHRANE
HALOTHANE
ISOFLURANE
HALOGENATED AGENTS
N2O
THIOPENTAL
ETOMIDATE
MIDAZOLAM
INTRAVENOUS AGENTS
AMNESIA
ANESTHESIA
FENTANYL
SUFENTANIL
MORPHINE
OTHERS
ANALGESIA
RELAXATION
PANCURONIUM
ATRACURIUM
VECURONIUM
OTHERS
THE TRIANGLE OF ANESTHESIA

latency of the high cortical responses after the induction dose of midazolam (14). However, these changes did not limit the value of this test.

Diazepam

There is a lack of studies evaluating the SSEP changes related to the induction dose of diazepam. However, there are two reports pointing to the negative effect of 10mg and 20mg doses of diazepam when injected intravenously (4, 17). Our own experience, in anesthetized patients, hints to a depressing effect on amplitude and a slight increase in latency of cortical evoked potentials with no effect on cervical responses. The two different observations may be related to the conscious level or drug interaction. Our patients received, half an hour before the injection of 10mg diazepam, a bolus induction dose of sufentanil equal to 5μg/kg and were asleep at that time.

Muscle relaxants

These drugs do not exert direct effect but may have an indirect impact. First, the use of muscle relaxants during induction and before enlisting muscle twitches makes it difficult to obtain the optimal stimulating site. Stimulation intensity decreases according to the distance of the stimulating electrodes from the nerve being stimulated and such change in intensity may effect evoked potentials (33). Second, during light stages of anesthesia muscle twitches interject a sizable amount of artifact and render SSEP data averaging a difficult matter (16). The injection of muscle relaxants will decrease such artifacts and enhance the acquisition of evoked potentials (13). Such effects are common to all muscle relaxants; however, it is logical that a long acting agent or a continuous infusion may be a better choice.

Inhalation agents for induction of anesthesia

This method requires higher concentration of the inhalation agents which can decrease SSEP signals significantly to the point of losing the identifiable signals (see maintenance).

The effect of maintenance agents

Anesthesia implies relaxation, analgesia, and amnesia. Different components of this triangle are furnished by different drugs (Fig. 1). Relaxation was usually achieved by one of several muscle relaxants. Analgesia and amnesia may be achieved either by inhalation agents alone, at high concentration, or by a combination of narcotic and amnesic drugs. The use of high dose narcotics, to achieve complete analgesia, is superior to other agents because of the better preservation of SSEP signals. The same is true for the use of muscle relaxants to achieve relaxation instead of depending on deep levels of inhalation agents. The third part of the anesthesia triangle is amnesia and it can be achieved by one of several methods. First, the use of intravenous agents such as thiopental, etomidate, etc., as intermittent or infusion modes. Second is the use of halogenated agents. Third, the use of nitrous oxide in 60-70% concentration. The effect of intravenous agents have been discussed earlier so only inhalation agents will be discussed here.

Halothane

The common view from early reports is to avoid inhalation agents, especially halothane. Such statements were qualitative in nature rather than quantitative. Recently, Salzman et al. had studied the utility of combining halothane anesthesia and evoked potential monitoring (24). They studied the use of SSEP monitoring in 116 patients undergoing surgery for spinal fusion and demonstrated the possible utility of halothane in low concentration. In their study 0.5% halothane slightly increased the latency of the SSEP peaks and variably decreased the amplitude. This effect did not change significantly during the entire period thereafter. During the combined use of low concentration halothane 0.5% and nitrous oxide 66%, they were able to monitor successfully 96% of their patients. This is in agreement with our study of 395 spinal cord patients where we had 96.5% success rate using, for amnesia, thiopental infusion or one of the inhalation agents (16). However, increasing halothane concentration increased changes and decreased the success rate in obtaining recognizable and reproducible wave forms. This study, and our own experience, support the utility of lower concentration of halothane combined with a narcotic base anesthesia during SSEP monitoring to achieve sleep and amnesia.

Ethrane

Ethrane with and without the combination of nitrous oxide has been studied in humans (20, 32). In both situations ethrane effected evoked potentials in dose related fashion with more prominent effects when used in conjunction with nitrous oxide. The amplitude of cortical evoked potentials decreased and the latency increased. However, these changes at 0.5% ethrane were less than the effects of 50% nitrous oxide. Both studies demonstrated the compatibility of ethrane with evoked potential monitoring if used in low concentration. The range for ethrane compatibility with SSEP's if used alone is 0.5-1% and is less than 0.5% if used in conjunction with nitrous oxide.

Isoflurane

This agent has similar effects on evoked potentials as halothane and ethrane (20, 26). A dose dependent increase in latency and a decrease in amplitude with a more selective effect on the cortical responses rather than the cervical. However, a low concentration of isoflurane less than 0.5% exert a minimal effect on SSEP, provide amnesia, and is suitable with intraoperative monitoring (for more details see chapter by Dr. Samra in this volume). Also, it exerts a dose dependent increase in CCT in normothermic patients (11).

Nitrous oxide

This agent has been recommended as part of the nitrous oxide narcotic technique on the assumption of its lack of effect on evoked potentials. However, recently few papers have addressed this point (1, 20, 27, 28). In these studies nitrous oxide displayed a depressant effect by decreasing the amplitude and increasing the latency of the cortical waves. Nitrous oxide in 50% concentration has more effect on latency and amplitude than 0.25-1% isoflurane or ethrane (20). In addition, nitrous oxide in the concentrations usually used, 65-70%, could easily lead to hyposemia, ischemia, and changes in evoked potentials.

Intravenous agents for the maintenance of anesthesia

During maintenance of anesthesia intravenous agents are used in two and maybe the three angles of the anesthesia triangle. It is needed for analgesia, relaxation, and may be for amnesia. The agents used are the same as discussed during induction but the doses are smaller and as a result their effect on SSEP is lesser. Such agents may be used either as infusion or intermittent injections (7).

The effect of adjunct drugs

Several drugs may be used during the course of anesthesia for one reason or another. The knowledge of their effects on evoked potentials is essential during the course of monitoring.

Cardiovascular drugs

Nitroprusside, trimethaphan, and hydralazine may indirectly effect SSEP signals by altering blood pressure. A drop in blood pressure below a critical point is known to change SSEP signals (5, 8). Effects from calcium channel blockers are not known. However, my experience with Nimodipine (during my sabbatical year at the University of Erlangen-Nurenberg, West Germany) indicates the lack of recognizable effect of (0.3mg/kg/h) either amplitude or latency of SSEP signals.

Other drugs

Antibiotics are not known to have an effect on SSEP. Phenytoin when used in neurosurgical patients and head trauma is reported not to have any effect on SSEP or on CCT (4). Triiodothyronine used in hypothyroid patients increased the amplitude of early SSEP component and this effect was blocked by propranolol (29). Mannitol, steriod, and furosemide have not been specifically addressed. However, it is conceivable to exert an indirect effect related to their influence on intracranial pressure (ICP). Intracranial hypertension that exerts some changes on SSEP and drugs that lower ICP may alter evoked potentials in patients with raised intracranial pressure. Our clinical experience, at this time, does not support such a hypothesis.

The effect of adjunct techniques used during anesthesia

During the course of surgery several techniques are used in conjunction with anesthesia. Such techniques may have an effect on the SSEP monitoring and if neglected can lead to false results.

Hypothermia

Lowering body temperature has a linear depressing effect on latency, amplitude, and CCT (11, 18). All responses are affected bilaterally with more significant changes at higher cortical than both subcortical or spinal levels. The CCT (N20-N14) increased exponentially with decreasing temperature $\{CCT_{te} = 1.066^{(37-te)}$ where te = esophageal temperature$\}$ (11). Also, spinal conduction time (SCT) (N13-N10) increased during hypothermia but less steeply than CCT $\{SCT_{te} = 1.047^{37-te)}\}$ 37-te). Hypothermia was studied also in middle latency auditory evoked potentials. Such responses were obtainable as long as the temperature was above 23°C and the main changes in the potentials were in latency. Amplitude was not uniformly effected (13). Rewarming has a similar but opposite effect. Localized hypothermia such as arm or surgical site cooling can affect evoked potentials (16).

Hyperventilation

Aggressive hyperventilation can lead to low CO_2 and to a decrease in cerebral blood flow (CBF). A decrease in CBF below 20ml/100gm/min is associated with changes in evoked potentials (2). The effects of hyperventilation and hypoventilation in association with ethrane have been studied in humans and cats during the visual and auditory evoked potentials (3). In this study, 2.5-3.7% ethrane increased the amplitude of cortical evoked potentials and this rise in amplitude was further increased when CO_2

was lowered by hyperventilation. With hypoventilation, CO_2 increased to the sixties and the amplitude of cortical evoked potential decreased. We do not know at this time if such effect is persistent with other drugs or with SSEP modality.

Hemodilution

A decrease in hemoglobin, if not accompanied by an increase in cardiac output, can lead to a decrease in oxygen delivering capacities and tissue ischemia which leads to SSEP changes (21). Restoring blood volume and hemoglobin can recover the evoked potentials.

Deliberate hypotension

A drop of blood pressure below certain critical level is associated with a drop in CBF below the ischemia threshold. This drop below ischemia threshold is associated with changes in SSEP (2, 5). Changes in SSEP due to deliberate hypotension under influence of hypothermia are not known. However, in middle latency evoked potentials, these effects are intact. In another word the effects of deliberate hypotension on SSEP under different degrees of hypothermia are the same as normothermic patients (13). The use of deliberate hypotension, though it is safe during neurosurgical and spinal cord surgery, may become safer if used in conjunction with evoked potential monitoring (8). However, such statement of good correlation between SSEP changes during deliberate hypotension and the neurological outcome is not supported by others (5).

Anesthetic techniques for SSEP monitoring

Three techniques are available at this time to provide anesthesia during SSEP monitoring (7). Total intravenous anesthesia, nitrous narcotic, and narcotic with halogenated agents. All three techniques require good premedication, preferably agents with amnesic effect such as scopolamine. The choice of the anesthetic technique is a personal preference and depends on the familiarity with the different agents. Any of the three techniques can safely be used if the anesthesiologist is familiar with the bases of evoked potential monitoring and the anesthetic effects (16). This fact should be a shared information between all members of the monitoring team. The bases for all three techniques are narcotic agents combined with muscle relaxants and supplemented with amnesic agent. The three techniques utilize the same two angles of the anesthesia triangle, namely analgesia and muscle relaxation, and differ at the third angle only. Amnesia is the point of difference and this can be accomplished by one of three methods. Before discussing the three techniques we should mention the relative benign effect of narcotics and the potential additive effects of many agents.

Total intravenous anesthesia

The base for this technique is a moderate to high dose narcotic supplemented by muscle relaxants and hypnotic drug such as thiopental, etomidate, midazolam, etc., given as a continuous infusion or intermittent injections. Advantages to this technique are the use of agents with minimal effects on SSEP. Disadvantages include the heavy usage of infusion sets and potential disconnection in these lines, higher cost, and also this is the technique with the least familiarity to the anesthesiologist. Blood pressure is not well controlled and requires vasodilator to be controlled. In our experience, this technique, due to technical difficulties, had a higher incidence of intraoperative recall than the other two techniques.

Narcotic nitrous oxide

This implies the use of moderate to high doses of a narcotic, either as an infusion or intermittent injections supplemented by 60-70% nitrous oxide. As it is in the previous technique, blood pressure may fluctuate and require medical management. In addition, the margin of safety for oxygen delivery is decreased due to the high concentration of nitrous oxide used. Lastly, nitrous oxide carries a stronger effect on cortical evoked potentials than the agents used in the other two techniques.

Narcotic halogenated agents

As it is in the other two techniques, narcotics are the bases for anesthesia supplemented by low concentration halogenated volatile agents such as isoflurane, ethrane, and halothane. All three agents are suitable for SSEP monitoring if used in low concentration (0.5-0.75 MAC). The blood pressure is better controlled in this technique than in the other two, and the rate of recall is much less. Actually, our own experience in over two hundred cases and that of McPherson indicate the total absence of recall. Because isoflurane has advantages over both halothane and ethrane in neurosurgical patients, it may be a better choice in such patients. Advantages to this technique is higher oxygen delivery than the nitrous narcotic technique, more cardiovascular stability than the other two techniques, less infusion setup than the total intravenous anesthesia, and it is easier to grasp by most anesthesiologists.

Conclusion

The anesthesiologist's better understanding of the effects of anesthetic agents, drugs, and techniques on evoked potentials enhances SSEP monitoring. This basic knowledge is essential to the success of SSEP monitoring and it is more important than the fight over which technique to use or not use. An informed knowledgeable anesthesiologist is able to provide anesthesia in any technique during SSEP monitoring without altering the value of this test. However, an uninformed anesthesiologist will fail to provide the optimal anesthesia for a successful evoked potentials monitoring. The general shared information, good dialogue, and understandings of the SSEP problems among all members of the operating room team are invaluable for the success of intraoperative monitoring.

References

1. Benedetti, C.; Chapman, C.R.; Colpitts, Y.H.; Chen, A.C.: Effects of nitrous oxide concentration on event related potentials during painful tooth stimulation. Anesthesiology, 56: 360-364, 1982.
2. Branston, N.M.; Symon, L.; Crockard, H.A.; Pazstor, E.: Relationship between the cortical evoked potential and local cortical blood flow following acute middle cerebral artery occlusion in baboon. J. Exp. Neurol., 45: 195-208, 1974.
3. Burchill, K.J.; Stackard, J.J.; Myer, R.R.; Bickford, R.G.: Visual and auditory evoked responses during enflurane anesthesia in man and cats. Electroenceph. Clin. Neurophysiol., 39: 434, 1975.
4. Chiappa, K.H.: Evoked Potentials in Clinical Medicine. Raven Press, pp. 248-249, 1985.
5. Dong, W.K.; Bledsoe, S.W.; Eng, D.Y.; Heavner, J.E.: Profound arterial hypotension in dogs: Brain electrical activity and organ integrity. Anesthesiology, 58: 61-71, 1983.
6. Drummondd, J.C.; Todd, M.M.; Sang, H.: The effect of high dose sodium thiopental on brain stem auditory and median nerve somatosensory evoked responses in humans. Anesthesiology, 63: 249-254, 1985.
7. Frazier, W.T.; Odom, S.H.; Biggs, B.D.: Anesthetic technique for spinal cord monitoring. In: Spinal Cord Monitoring. J. Schramm; S.J. Jones (eds). Berlin-Heidelberg-New York, Springer, pp. 69-81, 1985.
8. Grundy, B.L.; Nash, C.L.; Brown, R.H.: Arterial pressure manipulation alters spinal cord function during correction of scoliosis. Anesthesiology, 54: 249-253, 1981.
9. Grundy, B.L.; Brown, R.H.; Clifton, P.C.: Effects of droperidol on somatosensory cortical evoked potentials. Electroenceph. Clin. Neurophysiol., 50: 158, 1980.

10. Grundy, B.L., Brown, R.H.: Meperidine enhances somatosensory cortical evoked potentials. Electroenceph. Clin. Neurophysiuol., 50: 177, 1980.

11. Hume, A.; Durkin, M.: Central and spinal somatosensory conduction times during hypothermic cardiopulmonary bypass and some observation on the effects of fentanyl and isoflurane anesthesia. EEG Clin. Neurophysiolog., 65: 46-58, 1986.

12. Kimovec, M.A.; Serpico, L.C.; Sloan, T.B.; Koht, A.; Carlvin, A.O.: Effects of thiopental on cerebral evoked potentials. Anesth. Analg., 65: 580, 1986.

13. Kileny, P.; Dobson, D.; Gelfand, E.T.: Middle-latency auditory evoked potentials during open heart surgery with hypothermia. Electroenceph. Clin. Neurophysiol., 55: 268-276, 1983.

14. Kochs, E.; Treede, R.D.; Schulte An Esch, J.: Evozierte potentiale unter narkoseeinleitung mit midazolam in kombination mit etomidat oder thiopental. In: Benzodiazepine in Ansthesie und Intensivmedizin. J. Schulte am Esch (ed.), pp. 153-170, 1986.

15. Koht, A.; Kimovec, M.A.; Sloan, T.B.; Carlvin, A.O.: The effects of sufentanil on median nerve somatosensory evoked potentials. Anesth. Analg., 65: 581, 1986.

16. Koht, A.; Sloan, T.; Ronai, A.; Toleikis, R.: Intraoperative deterioration of evoked potentials during spinal surgery. In: Spinal Cord Monitoring. J. Schramm; S.J. Jones (eds), Springer-Verlag Berlin Heidelberg, pp. 161-166, 1985.

17. Loughnan, B.L.; Sebel, P.S.; Thomas, D.; Rutherfoord, C.F.; Rogers, H.: Evoked potentials following diazepam or fentanyl. Anaesthesia, 42: 195-198, 1987.

18. Markand, O.N.; Warren, C.H.; Moorthy, S.S.; Stoelting, R.; King, R.D.: Monitoring of multimodality evoked potentials during open heart surgery under hypothermia. EEG Clin. Neurol., 59: 432-440, 1984.

19. McPherson, R.W.; Sell, B.; Traystman, R.J.: Effects of thiopental, fentanyl, and etomidate on upper extremity somatosensory evoked potentials in humans. Anesthesiology, 65: 584-589, 1986.

20. McPherson, R.W.; Mahla, M.; Johnson, R.; Traystman, R.J.: Effects of enflurane, isoflurane and nitrous oxide on somatosensory evoked potentials during fentanyl anesthesia. Anesthesiology, 62: 626-633, 1985.

21. Nago, S.; Roccaforte, P.; Moody, R.A.: The effects of isovolemic hemodilution and reinfusion of packed erythrocytes on somatosensory and visual evoked potentials. J. Surg. Res., 25: 530-537, 1978.

22. Pathak, K.S.; Brown, R.H.; Cascorbi, H.F.; Nash, C.L.: Effects of fentanyl and morphine on intraoperative somatosensory cortical evoked potentials. Anesth. Analg., 63: 833-837, 1984.

23. Pathak, K.S.; Brown, R.H.; Nash, C.L., Jr.; Cascorbi, H.F.: Continuous opioid infusion for scoliosis fusion surgery. Anesthesiology, 62: 841-845, 1983.

24. Salzman, S.K.; Beckman, A.L.; Marks, H.G.; Bunnell, W.P.; MacEwen: Effects of halothane on intraoperative scalp-recorded somatosensory evoked potentials to posterior tibial nerve stimulation in man. EEG Clin. Neurophysiolog., 65: 36-45, 1986.

25. Samra, S.K.; Lilly, D.J.; Rush, N.L.; Kirsh, M.M.: Fentanyl anesthesia and human brain stem auditory evoked potentials. Anesthesiology, 61: 261-265, 1984.

26. Sebel, P.S.; Ingram, D.A.; Flynn, P.J.; Rutherfoord, C.F.; Rogers, H.: Evoked potentials during isoflurane anaesthesia. Br. J. Anaesth., 58: 580-585, 1986.

27. Sebel, P.S.; Flynn, P.J.; Ingram, D.A.: Effects of nitrous oxide on visual auditory and somatosensory evoked potentials. Br. J. Anaesth., 56: 1403-1407, 1984.

28. Sloan, T.B.; Koht, A.: Depression of cortical somatosensory evoked potentials by nitrous oxide. Br. J. Anaesth., 57: 849-852, 1985.

29. Straumanis, J.J.; Shagass, C.: Electrophysiological effects of triiodothyronine and propranolol. Psychopharmacologia, Berlin, 46: 283-288, 1976.

30. Sutton, L.N.; Frewen, T.; Marsh, R.; Jaggi, J.; Bruce, D.: The effects of deep barbiturate coma on multimodality evoked potentials. J. Neurosurg., 57: 178-185, 1982.

31. Thornton, C.; Heneghan, C.P.H.; Navaratnarajah, M.; Bateman, P.E.; Jones, J.G.: Effects of etomidate on the auditory evoked responses in man. Br. J. Anaesth., 57: 554-561, 1985.

32. Thurner, F.; Schramm, J.; Romstck, J.: Effects of fentanyl and enflurane on cortical and subcortical SEP during general anesthesia in man. In: Spinal Cord Monitoring. J. Schramm; S.J. Jones (eds), Berlin-Heidelberg-New York, Springer, pp. 82-89, 1985.

33. Tsuji, S.; Lueders, H.; Dudley, S.; Dinner, D.S.; Lesser, R.P.; Klem, G.: Effect of stimulus intensity on subcortical and cortical somatosensory evoked potentials by posterior tibial nerve stimulation. Electroenceph. Clin. Neurophysiol., 59: 229-237, 1984.

Operative Data

Criteria for Detection and Pathological Significance of Response Decrement during Spinal Cord Monitoring

S.J. Jones;[*] L. Howard; F. Shawkat

The intervention criteria for spinal cord monitoring (i.e., the changes in various response parameters believed to represent a significant defect of conduction and therefore calling for corrective action) have without exception been established empirically in the light of previous experience with the particular monitoring and surgical techniques employed. Retrospective analysis of a sufficient number of cases will eventually yield quantitative information concerning the incidence and severity of neurological impairment associated with a given degree and duration of response decrement, but there is general agreement that more "scientifically" based criteria are desirable. In this respect there are at least 4 questions to be answered: 1) what degree of response decrement constitutes a definite change? 2) what is the clinical significance of a given decrement? 3) over what period is the decrement likely to be fully or partially reversible? 4) how effectively can detection followed by reversal (or even attempted reversal) of a decrement reduce the likelihood and severity of clinical impairment? This paper will attempt to answer question 1) in the context of epidural spinal SEP monitoring of neurologically normal scoliosis patients, will argue that 2) and 3) are essentially unanswerable in a human context but may be approached by animal experiments, and that 4) can only be addressed in humans by retrospective examination of previous cases.

Recordings of spinal SEPs made at various times during surgery were used to define the inherent variability of the response in quantitative terms and so establish statistical criteria for what constitutes a "definite" decrement. Ten patients were selected, in whom both the course of surgery and the outcome were judged to be uneventful. Data obtained from 7 other cases were rejected on the following grounds: In 2 the position of the electrode tip was changed; in 1 insufficient recordings were obtained at certain "key" stages of surgery; in 1 there was a qualitative change in the waveform after tightening of sublaminar wires; in 1 there was a transient flattening of the response with a known precipitating factor; in 2 there were transient flattening during the course of a difficult and lengthy procedure. In none, however, was there any significant postoperative defect. The 10 selected cases were therefore considered to represent the "normal" variability of the response.

Details of the surgical procedure differed somewhat between patients but the 4 major stages during which SEPs were recorded were defined as follows:

Stage 1 - immediately following insertion of the epidural electrode after exposure of vertebrae.

---

[*] Medical Research Council and Royal National Orthopaedic Hospital, Queen Square, London WC1N 3BG, England

Stage 2 - distraction with outrigger, insertion of sublaminar wires and/or compression rodding.

Stage 3 - distraction with Harrington rod and tightening of wires.

Stage 4 - closure just prior to insertion of final sutures.

The time elapsed between each stage was on average approximately 30 minutes. At each stage at least 4 responses were recorded with alternate stimulation of the posterior tibial (peroneal) nerve at either knee. The stimulus was sufficient to produce contraction of the calf muscles, the amplifier settings were 200-2000Hz, the recording configuration was monopolar with reference to paraspinal muscles, the digitization frequency was 16.7 points/msec and approximately 500 responses were averaged at a rate of 20/sec.

For each group of recordings obtained at each stage of the operation, means and standard deviations were calculated of 5 parameters (Fig. 1): 1) Component I amplitude; 2) Component II amplitude; 3) Component III amplitude; 4) Overall latency; 5) Overall duration. Group means were then calculated across all 10 patients, and the means of the intrasubject standard deviations were also calculated.

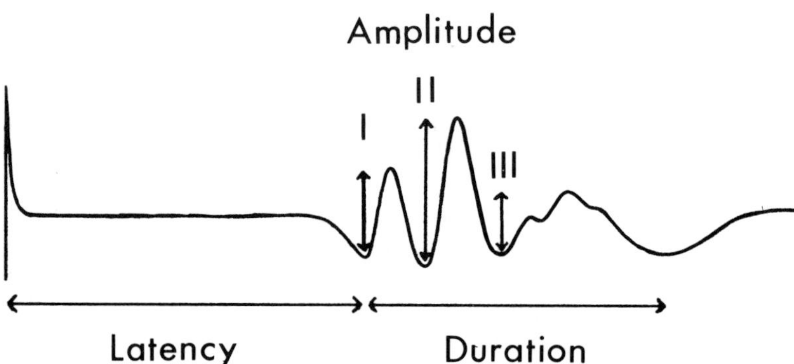

Fig. 1. Stereotypical waveform of spinal SEP, indicating parameters assessed for variability.

A conservative criterion for a response decrement was adopted, of 3 standard deviations below (amplitude, duration) or above (latency, duration) the mean (representing a probability of chance occurrence of less than 0.5%). This was calculated as a percentage of the mean of each response parameter (Figs. 2-5). On average, in the initial stage of monitoring a drop in the amplitude of Component I by more than 39% (left) or 27% (right) could on this criterion be regarded as "definite" (Fig. 2). These figures varied somewhat throughout the 4 stages but apparently not in any consistent fashion (although in Stage 3 the criterion for a "definite" decrement was below 25% on both sides). No significant change in the mean amplitude of Component I was detected at any time, although there was a slight numerical decrease through Stages 1-4.

Similar findings were obtained with respect to Component II amplitude (Fig. 3). The criterion for a "definite" decrement ranged from 27% to 39% reduction, with no apparent relation to the stage of surgery. There was no significant change in mean amplitude over time, although numerically the values showed a slight, progressive reduction. For Component III the criterion for a "definite" change ranged from 35% to 60% (Fig. 4), on account of its mean amplitude being somewhat smaller than that of Components I and II but its standard deviation similar. The mean amplitudes of Component III were numerically largest in Stage 4 for both left and right sided stimuli, but the trend was not statistically significant.

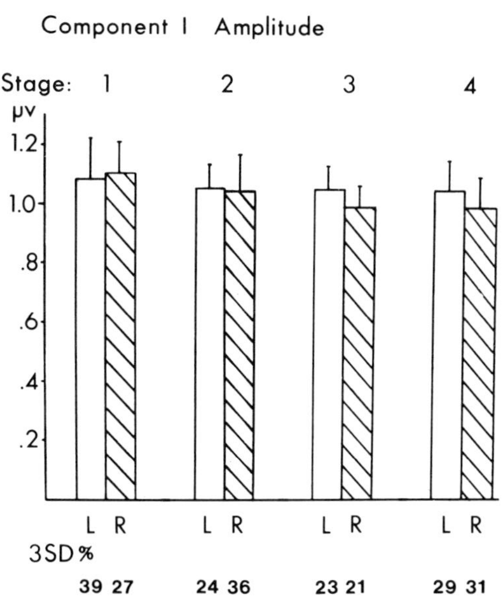

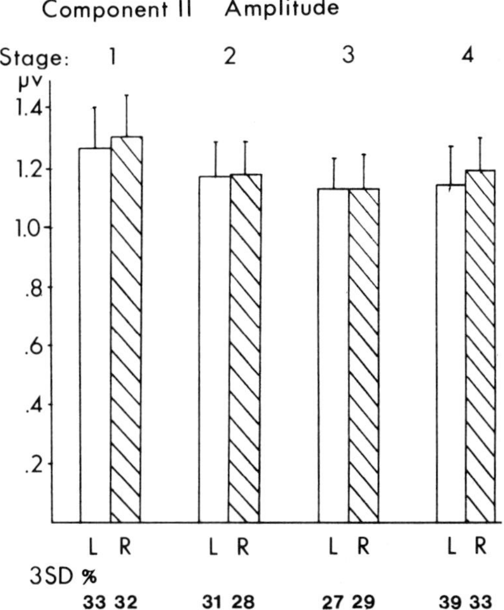

Fig. 2. Mean amplitude of Component I (left and right sided stimulation) across 10 patients at 4 stages of surgery, plus means of the intrasubject standard deviations. The criterion for a "definite" decrement was defined as an amplitude reduction of more than 3 standard deviations, which is calculated as a mean percentage of the amplitude at each stage.

Fig. 3. Mean amplitudes and mean intrasubject standard deviations of Component II, plus criteria for a "definite" decrement.

Latency values were highly consistent between sequentially recorded responses, with the result that a "definite" change could be defined as one exceeding only 2.5-3.8% (Fig. 5). Latency was the only parameter to show a significant trend over time, the increase being greatest between Stages 1 and 2. This was considered likely to be due to cooling of the spine and did not appear to be of any pathological significance. No significant trend was observed in the duration of the response, for which a "definite" change would be one exceeding 11-18%. In practice, however, neither the latency nor the duration of the spinal SEP are found to be useful monitoring indices.

It is thus possible to define the magnitude of changes which, if trivial causes can be excluded, may with a low probability of error be regarded as "definite". This, however, does not take into account the possibility of a change in the morphology of the response, which is sometimes seen even when the location of the recording electrode is apparently not altered. Also, whether or not the response can be said to have "definitely" deteriorated says nothing about the clinical significance of such changes, which was the second question raised in the opening paragraph. A definite decrement will usually (it is hoped) provoke the surgeons to take corrective action, and if this is successful the extent of neurological impairment associated with a permanent response decrement will not be discovered. Response decrements occasionally persist in spite of all restora-

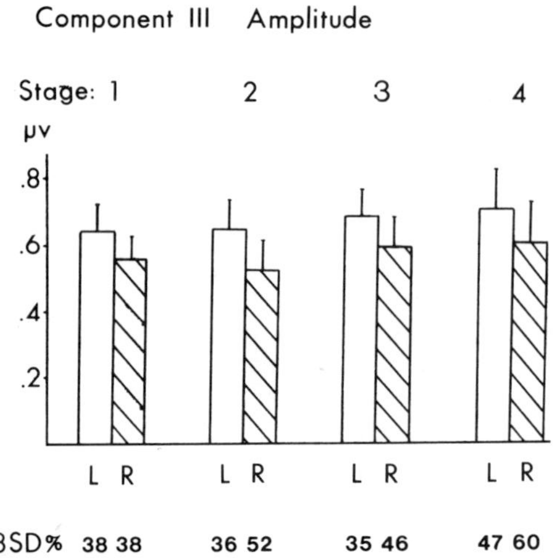

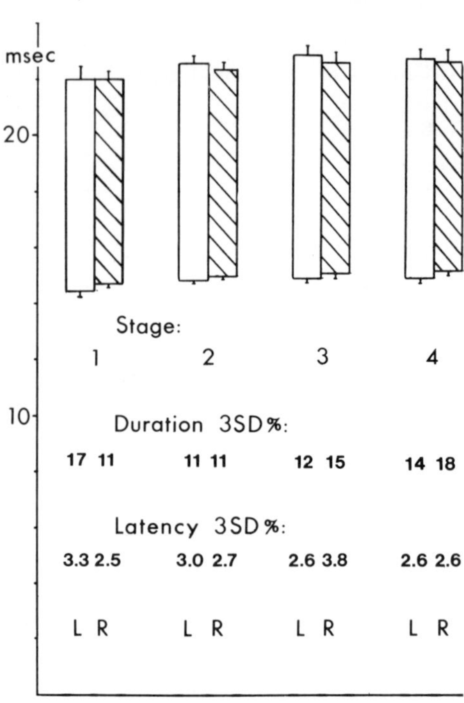

Fig. 4. Mean amplitudes and mean intrasubject standard deviations of Component III, plus criteria for a "definite" decrement.

Fig. 5. Mean latency (bottom of bar) and duration (top-bottom) of the spinal SEP at each stage of surgery, plus means of the intrasubject standard deviations and criteria for a "definite" decrement.

tive attempts, and temporary decrements may also result in minor neurological symptoms (see below), but even in specialist scoliosis centres such events are too infrequent to enable absolute rules governing the relationship between response decrement and neurological impairment to be determined. The issue may be partially resolved by animal experiments, but not fully answered since there are no adequate animal models of the spinal deformities which represent the "norm" in the human monitoring situation.

There is no doubt that many "definite" SEP decrements occurring during surgery are fully reversible in the sense that the response apparently returns entirely to normal, and that consequent neurological deficits may be absent or at least temporary and too insignificant to be noticed in the immediate postoperative period. For obvious reasons, however, it is not possible to determine experimentally the period over which conduction defects are completely or partially reversible. Animal studies may be of some assistance, but are of limited value on account of species differences and the absence of complicating factors such as skeletal deformities which are the norm in human patients. As with the previous question, therefore, it is inevitable that certain assumptions will continue to be made concerning the reversibility of cord conduction defects detected electrophysiologically.

The final question asks whether or not the early detection of conduction failure, reversible or irreversible, is of real value in preventing iatrogenic neurological defects. A controlled study, with the evidence of the monitor presented to or withheld from the surgeons in equal numbers of cases receiving identical surgical treatment, would require participation of extremely large numbers of patients on account of the low incidence of complications even in unmonitored cases. Furthermore, such a controlled trial is generally felt to be unethical although the reasons have yet to be clearly stated. In the writer's opinion, the fundamental difference between the monitoring question and a drug trial in which 50% of patients receive treatment and 50% placebo is that in the latter the anticipated consequences are a) benefit from the drug or b) no change, with the added possibility that the untreated patients may opt for real treatment at the end of the trial. In spinal cord monitoring the corresponding consequences are a) no change and b) possibly disastrous impairment of function, with little or no prospect of subsequent treatment and recovery. For this reason, no patient should be asked to consent to participate in a controlled monitoring trial.

In order to assess the value of monitoring, therefore, one has recourse to retrospective analysis. The data presented in Table 1 was compiled from 410 patients undergoing posterior fusion with Harrington instrumentation (Jones et al., 1985), very few of whom received the "wake-up" test. The incidence of paraplegia or severe neurological impairment was zero, which when compared with the world-wide incidence of such problems is clearly a favourable indication. This cannot be regarded as conclusive, however, on account of the wide range of techniques employed and the great variation in expertise between different centres. Neither can the 410 cases be realistically compared with the preceding 410 in the same centre, since over the period concerned (more than 10 years) there have been enormous developments in surgical and anaesthetic techniques. A number of useful criteria can be defined, however, based solely on the findings in the monitored patients.

The overall incidence of mild postoperative neurological symptoms was 2.2% (9/410, Table I), ranging from enhanced reflexes to temporary incontinence and paraesthesiae in the lower extremities. In the majority of patients with unremarkable outcome no "definite" spinal SEP changes were observed, and in a further larger proportion transient amplitude changes of 30-50% were associated with no significant neurological impairment. For patients in whom there was a transient or sustained decrement of more than 60%, however, the incidence of mild postoperative problems was 28% (7/25), representing a thirteen-fold increase over the incidence in the patient group as a whole. It is therefore reasonable to conclude (although it is scientifically unrigorous to extrapolate from past cases to present ones in whom different surgical techniques might be used) that at the time the response is seen to be attenuated by more than 60% there is already a greater than 1 in 4 probability of complications, even assuming that some restorative steps will be taken. According to the effectiveness of such steps in reversing the decrement the risk of impairment is reduced or increased; among those in whom the decrement was partially or completely reversed the incidence of problems was 18% (3/17) while among those with irreversible decrements the incidence was 50% (4/8). One of the two patients who had mild postoperative symptoms in spite of no significant SEP decrement had signs of a sacral root lesion, which may have been missed on account of the tibial nerve volley entering the cord at a higher segmental level.

With such statistical evidence it is possible to provide the surgeon with quantitative information regarding the morbidity associated with any given degree of response decrement. Moreover, when the criterion for a significantly increased risk of postoperative impairment (60% amplitude reduction, affecting all components) comfortably exceeds that for a "definite" SEP decrement (conservatively, 40% reduction of components I and II), there exits a "safety zone" within which restorative steps may be taken in the knowledge that, if no further deterioration is seen, no serious consequences are likely to be incurred. This provides an opportunity for the person responsible for the monitor to check fully for technical defects and for the surgeon to contemplate whether to continue with surgery or to moderate his intended procedures.

A further question which is often raised concerns the incidence of so-called "false positives", that is to say decrements which go beyond the criteria empirically determined as representing an increased likelihood of iatrogenic impairment but not subsequently found to result in any detectable defect. The incidence of false positives can be reduced by setting the intervention criterion at a higher level, but only with the effect of reducing the incidence of correctly detected defects in similar proportion. Some techniques are likely to be inherently associated with higher or lower false positive rates and it is the author's opinion that cortical SEP monitoring will inevitably produce higher false positive rates unless fully effective measures are taken to limit spurious fluctuations due to changes in anaesthetic level and blood pressure. In contrast to the spinal SEP technique, therefore, it is unlikely that confidence in the reliability of a cortical SEP monitor will ever be sufficiently great to be able to dispense entirely with the "wake-up" test.

To argue from a different standpoint, it is debatable whether a response decrement which exceeds the criterion for a "definite" change should ever be regarded as an unwanted or "false" indication. If technical factors can be discounted the only possible conclusion is that a real defect of conduction has occurred, and that the amplitude loss represents a reduction in the number of axons contributing to the compound sensory nerve volley in the spinal cord. Although statistically speaking there is still a favourable chance that the decrement will not prove to be of clinical importance, the surgeon should still wish to be informed that an adverse physiological effect has resulted from his actions.

Table I. Incidence of postoperative neurological deficits compared with intraoperative spinal SEP decrement in 410 patients undergoing posterior fusion with Harrington instrumentation (Jones et al., 1985).

Neurological deficit	Severe	Mild	None
Overall (N = 410)	0 (0%)	9 (2.2%)	401 (97.8%)
Technical failure (N = 17)	0 (0%)	0 (0%)	17 (100%)
Amplitude Decrement:			
<60% (N = 485)	0 (0%)	2 (0.5%)	408 (99.5%)
>60% (N = 25)	0 (0%)	7 (28%)	18 (72%)
>60% transient (N = 17)	0 (0%)	3 (18%)	14 (82%)
>60% permanent (N = 8)	0 (0%)	4 (50%)	4 (50%)

Reference

Jones, S.J.; Carter, L.; Edgar, M.A.; Morley, T.; Ransford, A.O.; Webb, P.: Experience of epidural spinal cord monitoring in 410 cases. In: J. Schramm; S.J. Jones (eds.): Spinal Cord Monitoring. Springer-Verlag Berlin Heidelburg, 1985.

Real-Time Intraoperative Monitoring during Neurosurgical and Neuroradiological Procedures

R.J. Chabot;[*] E.R. John; L.S. Prichep

Introduction

As neurosurgical and neuroradiological techniques and equipment have advanced, the ability to preserve the functional integrity of neurologically compromised central nervous system structures has become more of a reality. Neurophysiological monitoring systems have been shown to be quite useful in this regard. Brainstem auditory and somatosensory evoked potentials (BAEP's and BSEP's) have been utilized to monitor lateral and medial brainstem function (1, 2, 3) during posterior fossa procedures. Subcortical and cortical somatosensory evoked potentials have been monitored during surgical and radiological procedures for aneurysms, AVM's, and tumors (4, 5, 6, 7). Visual evoked potentials (VEP's) have been monitored during procedures where visual system structures are deemed to be at risk (AVM's, pituitary and sella turcica tumors) (8, 9, 10). Finally, 7th and 8th cranial nerve function has been monitored with brainstem auditory and facial evoked potentials (11, 12). In fact, a recent review of the literature uncovered over 200 monitoring papers published over the past 10 years.

Although it is generally agreed that intraoperative evoked potential monitoring can be quite useful, several critical issues remain unresolved. These issues are interrelated and revolve around a critical shortcoming of all currently reported monitoring systems. Stable, reliable evoked potentials require averaging and thus, the presentation of relatively large numbers of stimuli (BAEP, n = 2048; SEP's, n = 256; VEP, n = 256, etc.). The relatively low signal/noise ratio which occurs in the operating room environment (compromised central nervous systems structures and high electrical background noise) adds to this problem. EP update time in current systems tends to be quite long (2-5 minutes) and real-time feedback to the surgeon is not possible. This lack of moment-to-moment information about the effects of specific surgical and neuroradiological maneuvers compromises the utility of intraoperative monitoring. In addition, EP quality is often quite poor and variable intraoperatively, and it can be difficult to obtain reliable quantitative measures of EP features over time. Thus, workers within this field have difficulty defining what constitutes a critical EP change that signals impending CNS compromise and warrants immediate intervention.

Within this paper we will present recently developed intraoperative EP monitoring techniques which alleviate several of the previously mentioned problems. Technical advances allow stable, reliable EPs to be collected and updated in real-time providing

[*] Psychiatry Department, Brain Research Laboratory, NYU Medical Center, 550 First Avenue, New York, NY 10016

immediate feedback about surgical maneuvers. The signal/noise ratio is increased via on-line automatic artifact rejection and by signal enhancement via optimal digital filtering (13, 14). This allows EPs to be collected with minimal stimulus presentations which in conjunction with the use of overlapping running subaverages of EPs allows real-time feedback. Finally, automatic peak detection using self-normed statistical peak confidence intervals allows critical latency changes to be easily identified. Specific details of this monitoring system as well as examples of the effects of various surgical and neuroradiological maneuvers will be presented.

Methods

The Brain State Analyzer (BSA) contains an intraoperative EP monitoring system which incorporates several technological features not contained in traditional EP averaging systems. Specific details about each feature as well as their role in facilitating intraoperative monitoring will be presented.

1. Automatic artifact rejection. The output of the analog to digital convertor is continuously monitored on-line. The user defines an artifact threshold voltage level and any samples containing values beyond this threshold value are automatically excluded from the current EP. This threshold value can be changed as warranted by operating room conditions. Artifact rejection increases EP signal/noise ratios by excluding noise contamination from EP averages.

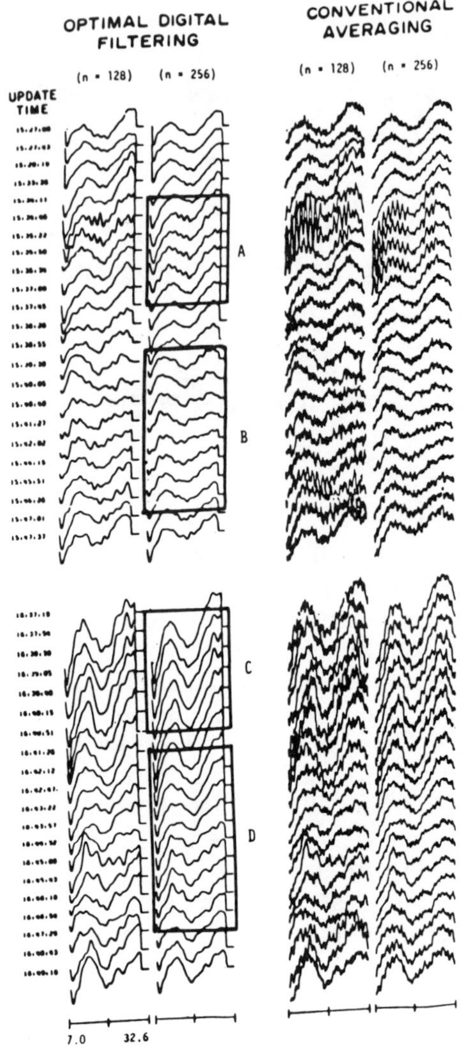

Fig. 1. Compressed evoked potential array plot of BSEP's obtained during the resection of a spinal cord tumor.

2. Signal enhancement - optimal digital filtering. Digital data are collected for 32 "groups" of; a) background electrophysiological and environmental "noise", no stimulus presented and b) "signal plus noise", appropriate stimulus (click, flash, or electrical pulse) delivered. For brainstem somatosensory and auditory EPs each 'group' represents the average of 64 sampling epochs of either 12.8 or 25.6msec. For cortical somatosensory or visual EPs each "group" represents the average of 8 sampling epochs of either 256 or 512msec. Spectral analysis of each of the 32 groups in 'noise' and 'signal' plus 'noise' conditions is then calculated using a 256 point Fast Fourier Transform. The amplitude and phase of each spectral component across the bandwidth of the amplifier is obtained for each group. Phase variance, a measure of the synchroniza-

tion of each spectral component with the onset of data acquisition in each sample (noise and signal) is then calculated. Spectral components which describe an EP will display the same phase whereas, those describing background noise will have variable phase. Phase synchrony (1 minus phase variance) is then displayed for each spectral component for the set of 32 groups of noise samples and the 32 groups of evoked potentials. Those portions of the averaged wave shapes which are phase-locked to data acquisition, and are likely to reflect the evoked response, will be described by spectral components with high phase synchrony. Contaminating background noise will not be phase-locked to data acquisition and will have low phase synchrony. The operator can then examine phase synchrony diagrams of 'noise' and 'signal' plus 'noise' baseline conditions and can identify those frequency domains which contain signal information and exclude noise. An 'optimal' digital filter can then be selected for future monitoring based on this information. This filter can be changed or updated should background noise change (i.e., microscope brought into case) or should the physiological characteristics of the patient change (i.e., anesthesia change or surgical decompression). The major consequence of the use of this 'optimal filter' is to increase the signal/noise ratio by filtering out background noise.

3. Automatic peak detection - calculation of 95% peak latency confidence intervals. "Optimal" digital filtering enhances the EP signal to such an extent that automatic peak detection becomes feasible within each 'group' or subaverage collected. This is accomplished within the BSA by locating the zero crossing of the first derivative. All peaks located in the grand EP average (32 groups) are displayed and the user selects any number of peaks for subsequent analysis (believed to reflect integrity of CNS structure being monitored). The mean and standard deviation of each selected peak latency is then calculated across the 32 baseline groups. The standard error of estimation and 95% confidence intervals can then be calculated for each peak for any sample size. A sample size providing reliable 95% confidence intervals is then selected for future monitoring. This mini/max decision requires operator judgement based upon the need for fast updates with the realization that reliability decreases as smaller sample sizes are used. The calculation of 95% confidence intervals provides a statistical criteria for determining what constitutes a critical EP latency change using a previously determined and often updated self-normed baseline.

4. Running subaverages. The final advanced feature is concerned with how EP's are displayed during monitoring. The baseline grand EP average is displayed along with any alternate baseline EP chosen. This is followed on the monitoring screen by three running subaverage EP's. Each running subaverage is based upon the sample size selected during 95% confidence interval selection. The first and second subaverage EP overlap by 50% and the second and third overlap by 50%. The first and third subaverage EP do not overlap. As a new "group" (8 stimulus presentations for cortical EP's and 64 for brainstem EP's) is added to the first subaverage, the oldest group is dropped out and added to the second subaverage. This pattern is repeated within the second and third subaverage EP's. The consequence of using these running subaverages is that EP changes can be evaluated in real-time or the amount of time needed to collect as little as one group (i.e., BAEP = 64 clicks at 27.6/sec).

Subjects

Over the past 4 years we have utilized the above system during over 400 neurosurgical and 200 neuroradiological procedures. The neurosurgical procedures include: Posterior fossa tumors; sella turcica and pituitary tumors; posterior fossa/cerebellar/midbrain AVM's and aneurysms; cranial nerve decompressions; and spinal cord tumors. Neuroradiological procedures involving balloon occlusion and embolization of brainstem, cerebellar, midbrain and cortical vessels for AVM's and aneurysms have

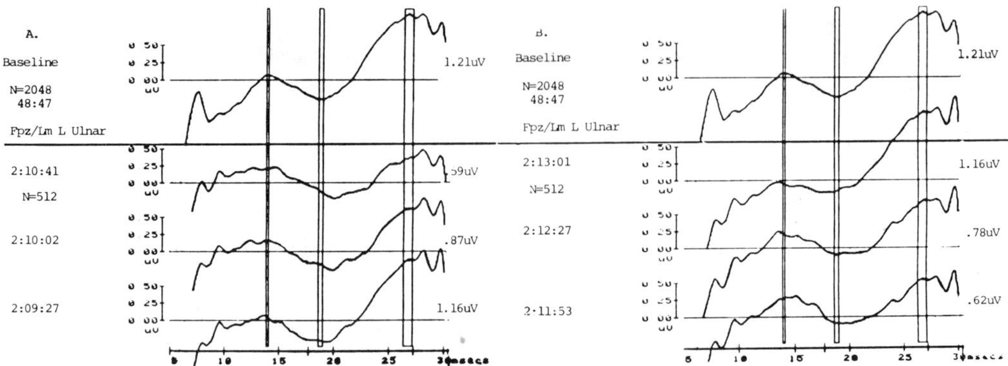

Fig. 2. BSEP's collected during the neuroembolization of the superior cerebellar artery in a 7-year-old with a cerebellar AVM.
A. BSEP changes due to restriction of blood flow to brainstem.
B. BSEP returns to baseline with resumption of normal flow.

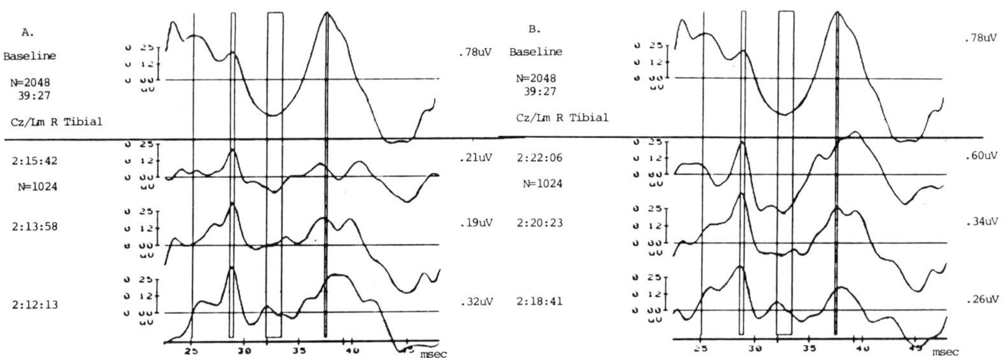

Fig. 3. BSEP's collected during clipping of feeder vessel to left parietal AVM.
A. Decrease in thalamo-cortical projection peaks during temporary vessel occlusion.
B. Gradual increase in thalamo-cortical peaks after clip removal.

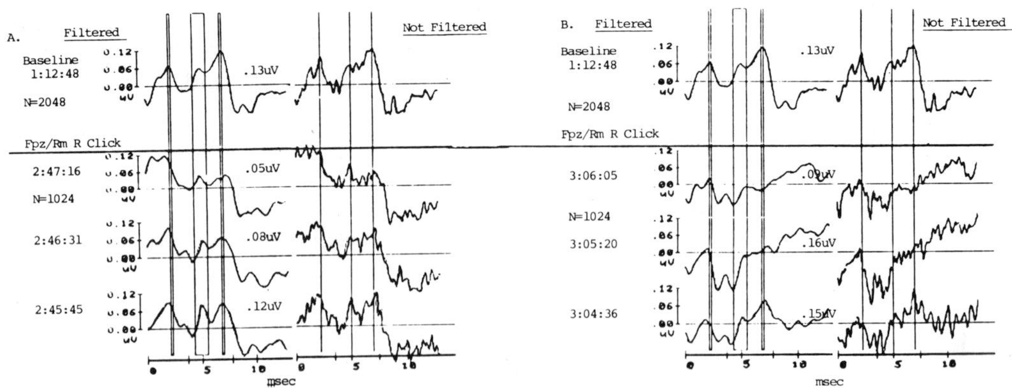

Fig. 4. BAEP changes during removal of an acoustic neuroma from an intact 8th cranial nerve.
A. Decreases in wave P during tumor resection.
B. Loss of waves PIV and PV when blood supply to 8th nerve is compromised.

also been monitored. Selected examples chosen to represent the major features of this system as well as demonstrating the effects of various surgical and vascular manipulations will now be presented.

Results

A. Use of optimal digital filter. Optimal digital filtering proved to be quite useful within the surgical environment for two interrelated reasons both leading to increased signal/noise ratios. First, digital filtering based upon "noise" and "signal plus noise" baseline computations effectively allowed most background environmental noise to be filtered out of averaged EP's. Secondly, the digital filters for most neurosurgical patients which optimized EP wave shapes were quite different from those found adequate when monitoring EP's under similar anesthetic management in neurologically normal patients. The upper frequency limit chosen was usually lower in neurosurgical patients reflecting slowing and morphological changes in their EP wave shapes. It was also found necessary to change digital filters during surgical procedures as patient and environmental conditions changed. For example, during 6 randomly selected posterior fossa procedures utilizing BAEP monitoring digital filters were redefined 20 times. Lower frequency limits varied from 30 to 312Hz, while upper limits varied from 468 to 1054Hz.

B. Stability and sensitivity of EP's after filtering. The stability and sensitivity of optimally filtered EP's is illustrated in the compressed evoked potential array (CEPA) of BSEP's presented in Fig. 1. These BSEP's were collected during the resection of a spinal cord tumor. Successive sliding window BSEP wave shapes, using sample sizes of 128 or 256 stimulus presentations with or without optimal digital filtering, are graphically presented. This figure illustrates several important features of our monitoring system. First, stable reproducible brainstem EP's can be recorded within the intraoperative environment utilizing minimal numbers of stimulus presentations (256 in this case). Second, a comparison of the filtered and conventionally averaged EP's shows the effectiveness of digital filtering within the operating room. Clearly, it would be impossible to make clinical decisions based upon the unfiltered EP wave shapes. Third, BSEP update times closely approximated real-time, averaging about 35 seconds. This update time represents the upper limit and is slower than usual for our system due to our utilization of reduced stimulation rates to prevent fatigue in this abnormal spinal cord and because of artifacting due to environmental noise. It should be noted that this background noise virtually obliterates the conventionally averaged BSEP's, whereas the optimally filtered wave shapes remain uncontaminated. Fourth, this figure illustrates the sensitivity of filtered BSEP's to surgical manipulations. Time period A represents BSEP changes (decrease in P15 amplitude and flattening of wave shape between P15 and N19) during initial tumor resection. Time period B illustrates the effects of cavitron. These include immediate decreased amplitude of the positivity following N19 with little decrement in P15 or N19 amplitude. BSEP stability during a pause in tumor resection can be seen in time period C. The rapid effects of heat generated by a laser burst are seen in time period D. P15 is relatively unaffected, whereas N19 decreases rapidly followed by a decrement in the positivity following N19. Although the above changes can be more quickly observed in the smaller sample size column (N = 128), we would probably monitor with a sample size of 256 since the EP wave shape is more stable and less influenced by momentary disturbances.

The dissociations between changes in P15 and N19 described above were observed in numerous spinal cord operations during resection within the intramedullary substance. Although this dissociation is not invariable, it occurs with enough frequency to raise reservations about the notion that successive BSEP peaks are generated by the sequential activation of different anatomical structures. Our observations with real-time

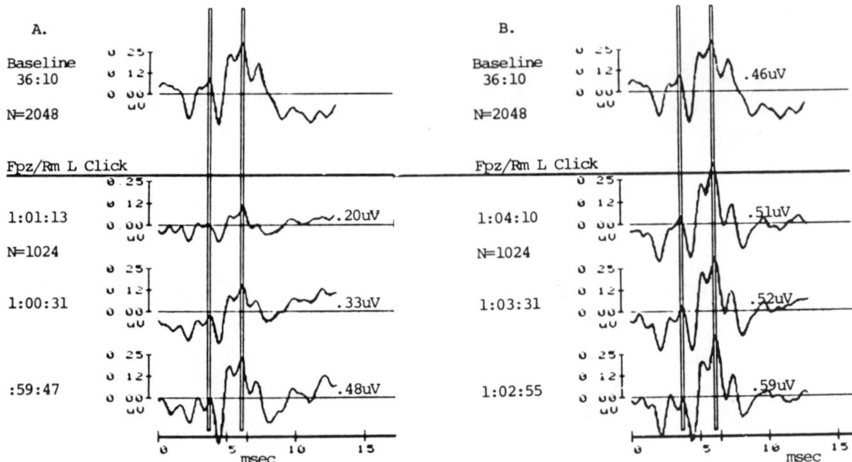

Fig. 5. Contralateral BAEP changes during brainstem retraction for a posterior fossa tumor.
A. Decrease in wave PV amplitude shortly after retractor slips.
B. Recovery of PV after retractor removal.

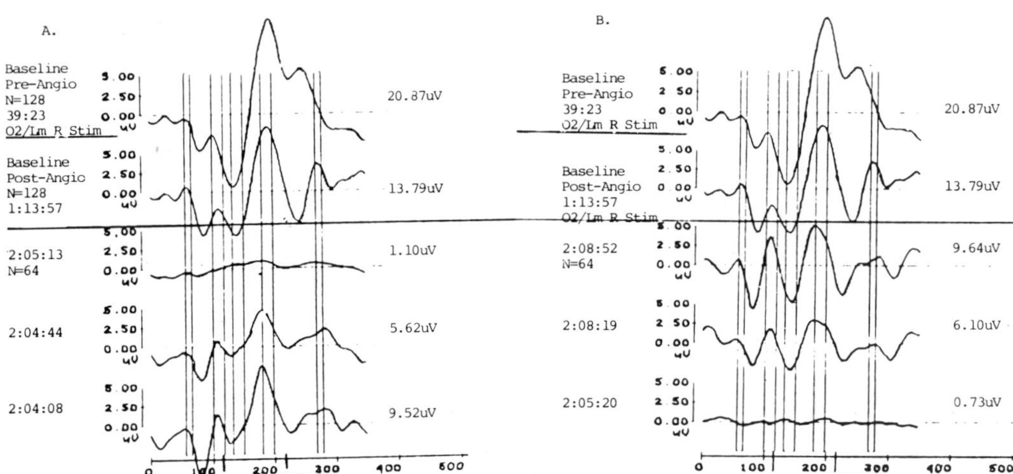

Fig. 6. VEP changes during neuroembolization of a parietal/occipital AVM.
A. Rapid decrease in P100 and P200 when basilar artery is temporarily occluded.
B. Rapid recovery after catheter advanced beyond basilar artery.

Fig. 7. VEP changes during removal of a meningioma from the optic nerve.
A. Decrease in P100 amplitude during tumor resection.
B. Recovery during pause in tumor resection.

updates suggest that P15 and N19 represent conduction at different velocities from multiple pathways within the spinal cord. These pathways can be affected separately or in combination by surgical maneuvers.

C. Intraoperative BSEP monitoring. BSEP's were monitored during surgical and neuroradiological procedures involving the spinal cord as well as during procedures involving brainstem and cortical structures. P15 is believed to reflect the function of the medial lemniscus, whereas N19 represents outflow to the thalamus with the following positivity representing thalamo-cortical projections. BSEP's were quite useful during tumor resections and vascular procedures involving the posterior fossa, cerebellum, and somatosensory or motor cortex areas. Thalamo-cortical projection time was found to be sensitive to impending ischemic events during aneurysm surgery. Figs. 2A and 2B illustrate the sensitivity of BSEP's to impending brainstem and cortical damage. BSEP wave shapes collected during the neuroembolization of a cerebellar AVM in a 7-year-old female are presented. The top wave shapes shown in A and B represent our initial baseline. The following three wave shapes under each baseline represent running sub-averaged wave shapes collected during embolization. Note that N19-P27 amplitude decreased dramatically over a period of less than 1 minute (Fig. 2A). This occurred when a balloon catheter slipped during embolization of the superior cerebellar artery causing blood flow to be cut off to the brainstem. The catheter was rapidly moved and the BSEP returned to baseline as shown in the wave shape in Fig. 2B. Note, that in this and all figures that follow, the third nonbaseline wave shape occurs before the second waveshape in time and that the second occurs before the first waveshape. Fig. 3 represents BSEP changes during surgical and neuroradiological treatment of a left parietal AVM which was located near the motor strip and had two feeder vessels off the left anterior cerebral artery. The top waveshapes in Figs. A and B show our baseline BSEP after right tibial nerve stimulation. The three waveshapes in Fig. 3A show decreases in the peaks representing thalamo-cortical projection after a feeder vessel collapsed after clipping. The three waveshapes in Fig. 3B show a slow temporary return toward baseline although the initial baseline amplitude was never reached. This 11-year-old female awoke with weakness of the right leg which resolved.

D. Intraoperative BAEP monitoring. BAEP's were monitored during posterior fossa and cerebellar surgical and neuroradiological procedures. BAEP's were useful in preventing damage to the auditory system as well as to brainstem structures. Figs. 4 and 5 present examples which illustrate the above statements. Fig. 4A presents BAEP waveshapes collected during the removal of an acoustic neuroma from an intact 8th cranial nerve. The top waveshape in A and B is the baseline measure. The three waveshapes in A show changes in P5 during tumor resection from the 8th nerve. The waveshapes to the right of the filtered waveshapes illustrate the difficulty of interpreting changes from the same waveshapes with digital filters set at the level most commonly utilized by other monitoring systems (50-3KHz). The three waveshapes in B show the rapid loss of waves P4 and P5 when the 8th nerve was lost due to the inadvertent cutting of a blood vessel which supported 8th nerve function. Again, the unfiltered waveshapes on the right illustrate the importance of optimal filtering. Fig. 5 shows reversible BAEP changes which occurred after a retractor slipped (without surgeon awareness) deeper into brain structures during removal of a petrous meningioma. The top waveshapes in Figs. A and B represent baseline measures. Fig. 5A shows decreases in wave V amplitude with retractor slippage and Fig. 5B shows recovery after retractor removal.

E. Intraoperative VEP monitoring. VEP's to flash stimulation proved to be reliable indicators of impending damage to the optic nerve, visual pathways and visual cortex during surgical and neuroradiological procedures involving these structures. Figs. 6 and 7 present examples of typical VEP changes. Fig. 6 shows dramatic reversible decreases in P100 and P200 amplitude during the neuroembolization of a parietal/occipital AVM. This dramatic decrease occurred when the catheter reached the base of the basilar artery. The waveshapes in Fig. 6A show rapid decreases in amplitude and those in Fig. 6B recovery after the catheter was advanced. Note this figure also presents two baseline VEP waveshapes and illustrates the effect of a right vertebral artery angiogram between the collection of the two baselines. Fig. 7 illustrates VEP changes occurring during surgical removal of a meningioma from the optic nerve in a 74-year-old female. Waveshapes in Fig. 7A show the loss of P100 during tumor removal with recovery shown in Fig. 7B when resection was discontinued.

Discussion

The combined utilization of automatic artifacting, optimal digital filtering, statistical criteria for quantifying EP parameter change, and running subaverage EP's allowed us to provide real-time feedback to neurosurgeons and neuroradiologists within the intraoperative environment. Reliable, stable evoked potentials could be obtained even from patients with compromised central nervous system structures. The ability to recalculate baseline EP's and adjust digital filters whenever significant surgical (decompression, ischemia, edema, etc.) medical (anesthesia, temperature, electrolyte balance changes) or environmental (background noise changes due to cavitron, laser, ultrasound introduction) changes occurred allowed us to continually monitor under conditions which reflected the current status of the patient within the current environment. The major consequence of this reoptimization is that reliable EP's could be continually obtained with minimal numbers of stimulus presentations. This allowed feedback about the physiological effects of surgical manipulations to become available within a real-time framework throughout the duration of the surgical or radiological procedure.

This real-time feedback has several immediate benefits from both a medical and research viewpoint:

A. Medical benefits

1. Constant feedback about the effects of surgical manipulations on neural tissue can be provided;

2. During tumor resection EP changes or the lack of changes can help the surgeon distinguish healthy functional tissue from nonviable tissue and can aid in identifying critical structures (i.e., localize optic or auditory nerve in tumor mass);

3. EP changes can be used to monitor the physiological consequences of temporary vessel occlusions and can aid neuroradiologists during embolization procedures (Fig. 2);

4. The influences of cavitron and laser use on adjacent neural structures can be identified;

5. The influence of supposedly innocuous events can be identified (i.e., Fig. 6 baselines show the effects of an angiogram on VEP's).

B. Research benefits

Information central to resolving the following research questions can bed obtained when real-time feedback is available.

1. What is the relationship between specific EP parameter changes and the development of specific neurological deficits? For example, which peaks, if any, associated with cortical and subcortical somatosensory EP's reflect the function of the motor pathways?;

2. What degree and duration of change in EP parameter measures signify a critical event requiring immediate surgical intervention?;

3. What are the quantitative relationships between EP parameter changes and intraoperative physiological events such as anesthetic effects, temperature changes, etc.?

Although the above lists are incomplete at best, they do provide criteria for evaluating the effect that real-time intraoperative EP monitoring can have in the future. Current neurosurgical and neuroradiological techniques may be modified as a result of the information provided by such monitoring. Procedures may be continued beyond what was believed possible because no significant EP changes are noted or they may be discontinued because increasingly dramatic EP changes are occurring. Certainly, much future research must be conducted with real-time feedback, but the early results suggest that the incidence of central nervous system dysfunction during neurosurgical and neuroradiological procedures can be reduced.

References

1. Hashimoto, I.; Ishiyama, Y.; Totsuka, G.; Mizutani, H.: Monitoring brainstem function during posterior fossa surgery with brainstem auditory evoked potentials. In: C. Barber (ed.): Evoked Potentials. Baltimore, University Park Press, 1980; 377-390.
2. Raudzens, P.A.; Shetter, A.G.: Intraoperative monitoring of brainstem auditory evoked potentials. Neurosurg., 1982; 57: 341-348.
3. Piatt, J.H.; Radtke, R.A.; Erwin, C.W.: Limitations of brainstem auditory evoked potentials for intraoperative monitoring during a posterior fossa operation: Case report and technical note. Neurosurg., 1985; 16: 818-821.
4. Carter, L.P.; Raudzens, P.A.; Ginaes, C.; Crowell, R.M.: Somatosensory evoked potentials and cortical blood flow during craniotomy for vascular disease. Neurosurg., 1984; 15: 22-28.
5. Tamaki, T.; Takano, H.; Takakuwa, K.: Spinal cord monitoring: Basic principles and experimental aspects. Central Nervous System Trauma, 1985; 2: 137-149.
6. Nuwer, M.R.; Dawson, E.: Intraoperative evoked potential monitoring of the spinal cord: Enhanced stability of cortical recordings. Electroenceph. Clin. Neurophysiol., 1984; 59: 318-327.
7. Berenstein, A.; Young, W.; Ransohoff, J.; Benjamin, V.; Merkin, H.: Somatosensory evoked potentials during spinal angiography and therapeutic transvascular embolization. J. Neurosurg., 1984; 60: 777-785.
8. Feinsod, M.; Selhorst, J.B.; Hoyt, W.F.; Wilson, C.B.: Monitoring optic nerve function during craniotomy. J. Neurosurg., 1976; 44: 29-31.
9. Albright, A.L.; Sclabassi, R.J.: Cavitron ultrasonic surgical aspirator and visual evoked potential monitoring for chiasmal gliomas in children. J. Neurosurg., 1985; 63: 138-140.
10. Costa, E.; Silva, I.; Wang, A.D.; Symon, L.: The application of flash visual evoked potentials during operations on the anterior visual pathways. Neurol. Res., 1985; 7: 11-16.
11. Prichep, L.S.; John, E.R.; Ransohoff, J.; Cohen, N.; Benjamin, V.; Ahn, H.: Real-time intraoperative monitoring of cranial nerves VII and VIII during posterior fossa surgery. In: C. Morocutti; P.A. Rizzo (eds.): Evoked Potentials: Neurophysiological And Clinical Aspects. Elsevier Science Publishers, 1985; 193-202.
12. Ojemann, R.G.; Levine, R.A.; Montgomery, W.M.; McGaffigan, P.: Use of intraoperative auditory evoked potentials to preserve hearing in unilateral acoustic neuroma removal. J. Neurosurg., 1984; 61: 938-948.
13. Fridman, J.; John, E.R.; Bergelson, M.; Kaiser, J.B.; Baird, H.W.: Application of digital filtering and automatic peak detection to brainstem auditory evoked potentials. Electroenceph. Clin. Neurophysiol., 1982; 53: 405-416.
14. John, E.R.; Baird, H.; Fridman, J.; Bergelson, M.: Normative values for brainstem auditory evoked potentials obtained by digital filtering and automatic peak detection. Electroenceph. Clin. Neurophysiol., 1983; 54: 153-160.

Clinical Study of Spinal Cord Evoked Potentials

H. Baba;[*] K. Tomita; S. Umeda; N. Kawahara; S. Nagata; S. Nomura; H. Yugami

Summary

The author's experience with ascending SEPs during 36 cases of spinal surgery is reported. Qualified SEP was obtained in 10 of 12 who underwent anterior spinal surgery and in 19 of 24 with posterior spine surgery. In anterior spinal surgery, 4 cases (40%) showed amplitude attenuations more than 30% of baseline amplitude, and in posterior spinal surgery, more than 30% amplitude attenuation was seen in 5 cases (26%) with laminoplasty. There were two cases which showed more than 50% amplitude attenuation; however, postoperative results were not catastrophic (54).

Introduction

Monitoring of spinal cord function using various electrodiagnostic procedures has rapidly increased (7), and hence, objective measurement of neurological deterioration has become commonplace, (2, 3, 6). Many papers are based on experimental data concerned with changes of spinal cord evoked potentials (SEP) in various types of spinal cord injury (3); however, interpretation of potential changes is not always easy. In 1982, the authors started using epidural ascending SEP monitoring to detect injured spinal cord segments and to detect early spinal cord damage during neurological surgery. The findings of mapped and intraoperatively monitored SEP in this relatively small series has been reported elsewhere (1). In this report, our more recent findings concerning ascending SEP changes in bipolar or monopolar recording mode and intraoperative changes are described.

The essential characteristics of the SEP are as follows: Stability in the acquisition of evoked responses, reproducibility and consistency, identification of the disappearance of each peak and its localization, abnormalities in the amplitude and/or latency and those degrees, and the relationship of SEP changes and neurological status postoperatively. Based on these characteristics, it is possible to make neurophysiological predictions with some confidence. In authors' previous report, it was reported that both amplitude increase and decrease has occurred in injured spinal cord segments confirmed with alternative diagnosis of myelography; positive-going killed end potentials have been elicited in about one third of cases with moderate to severe myelopathy. When the bipolar recording mode is used between the active anode and the cathode, the SEP amplitude represents the potential differences set up between two active electrodes. Thus, a very localized lesion longitudinally, or a ventral lesion may not bring any remarkable amplitude changes in ascending SEPs recorded from the dorsal epidural surface. In clinical settings, almost one half of the cases with myelopathy

[*] Department of Orthopaedic Surgery, School of Medicine, Kanazawa University, Kanazawa 920, Japan

showed amplitude abnormalities in the first and/or second components, however, there were increased false-negatives in the other half.

In this study, we describe our experience with ascending SEP monitoring during spinal surgery, and its utility in making level-specific diagnosis of cervical and/or thoracic myelopathy or cauda equina lesions.

Clinical material and methods

Surgical ascending SEP monitoring was performed on 36 patients who underwent neurosurgical procedures during the past three years. In more than three hundred patients treated operatively during this period, we monitored only high-risk cases considered to be at risk of intraoperative insult to the spinal cord. The ages ranged from 13 to 73 years; 25 were males, 11 were females. Six patients had cervical spondylotic myelopathy (CSM) with more than three vertebrae involved, 11 had ossification of the posterior longitudinal ligament (OPLL), 5 had fracture-dislocation of the thoracolumbar spine, and 14 had spinal cord tumors or other conditions. There were 12

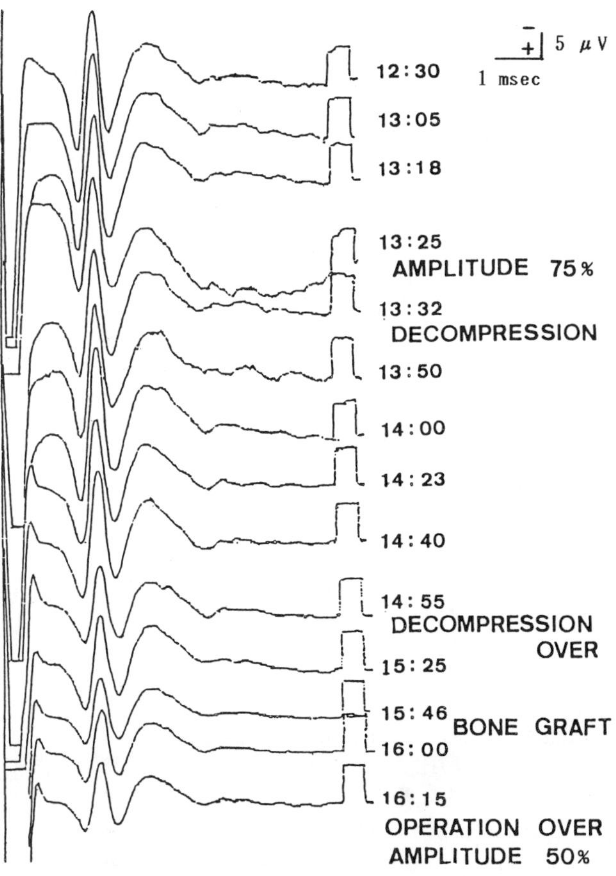

Fig. 1. Intraoperative ascending SEP recorded epidurally at C4–C5 from a case who underwent anterior subtotal spondylectomy, resection of OPLL and strut bone grafting from C2 to C6. The amplitude attenuation of N1 during decompressive procedure (floating and resection of OPLL) was significant.

cases who underwent anterior decompression with fusion, 19 posterior laminectomy or laminoplasty, and 7 removal of tumor or drainage of the syrinx. In the anterior spinal surgery cases which required at least two or three subtotal spondylectomy, SEP was monitored when removing compressive lesions of osteophytes or OPLL, and in posterior spinal surgery, when lifting the entire laminae or exposing tumor from the spinal cord. In three cases with cauda equina tumor, in addition to monitoring ascending SEP from the conus medullaris, peripheral M-wave responses and the evoked S-waves were monitored to identify nerve distribution.

The techniques employed in this study have been described previously. In the majority of the cases, both recording and stimulating electrodes were introduced before anesthetic induction, but some nervous patients required electrode placement after anesthetic induction. None of the patients showed any complications associated with electrode introduction to the dorsal epidural surface. The recording electrode was placed about 3 vertebrae rostral to the level of spinal surgery, and the stimulating electrode at the thoracolumbar junction or cauda equina. We used a DISA 1500 digital

EMG system because it satisfactorily amplified exceedingly small current levels in the spinal cord. Signals were fed through a DISA 15C04 preamplifier and averaged on a DISA 15G21 averager. The duration of the rectangular stimulus was 0.2ms and frequency was 20 pulses per second. The low and high frequency filters were at 10 and 10k Hz, respectively, and the analysis time was 10 to 50ms.

Results - Ascending SEP in anterior spinal surgery

Anterior decompression and fusion was performed in 6 cases with cervical involvement and in 6 cases with thoracic or thoracolumbar lesions. Serious complications associated with the surgical procedures were not seen in any cases. Extensive bleeding during surgery occurred unexpectedly in two cases with cervical OPLL and in one with thoracic OPLL. Three cases out of 12 (25%) did not recover well, and two of those three were cases which had experienced intraoperative bleeding. Qualified SEP were recorded in 10 cases; the exceptions were two young females with severe paraplegia due to thoracolumbar spine deformities. In these two cases, we measured extremely reduced afferent volleys travelling up the cauda equina, which had seriously degenerated. This was surmised to be respon-

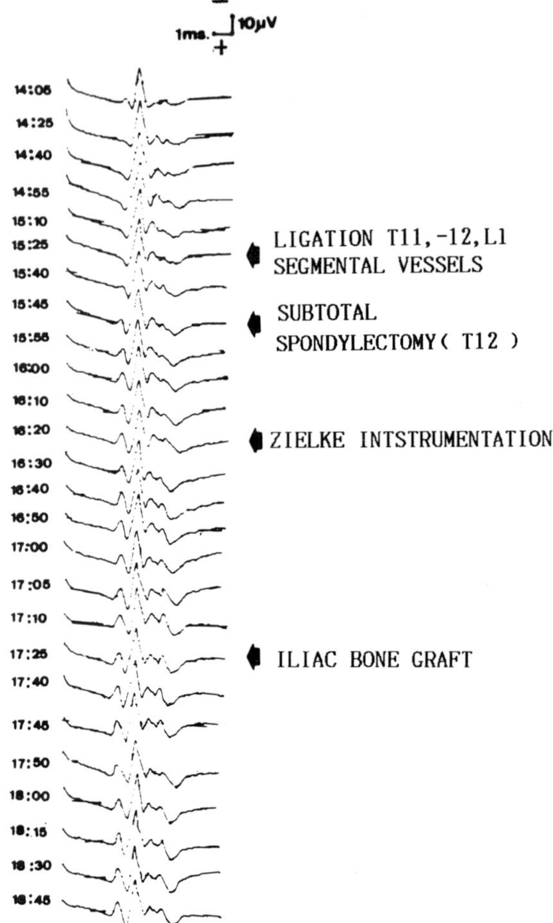

Fig. 2. Serial SEP traces recorded from a 51-year-old female, sustaining T12 bursting fracture from fall, who underwent anterior decompression and fusion followed by modified Zielke instrumentation. A small negative-positive wave which precedes N1 clearly appeared after the anterior decompression. This patient was completely relieved of the radicular pain in her left leg.

sible for the failure in recording SEP from the lower thoracic spinal cord. In the 10 cases where good SEPs were acquired, two cases who underwent anterior resection of the elongated OPLL did show more than 50% amplitude attenuation during surgical manipulation, and 30% transient amplitude attenuation occurred in two other cases. The first two cases did not show marked neurological recovery. Fig. 1 illustrates the serial responses recorded from a 48-year-old male who underwent subtotal vertebrectomy and resection of OPLL from C2 to C6. The first negativity, which was termed "N1" according to our nomenclature, significantly reduced in amplitude when floating and resection of OPLL was completed. Intravenous bolus injection of steroids (Decadron) was repeated but the amplitude remained reduced. Of particular interest were the thoracolumbar responses shown in Fig. 2, recorded from 51-year-old female presenting with unilateral radiating pain due to conus medullaris irritation from a T12 burst fracture. There was a very small negative and positive component preceding the large negativity (N1), which increased in amplitude after subtotal spondylectomy of T12 to decompress the conus medullaris, and remained stable thereafter. The slow positivity, which appeared following ligation of the left segmental vessels of T11, T12, and L1 gradually enlarged during bone grafting. Amplitude and latency of N1 did not show any alteration throughout the surgery.

Results - Ascending SEP in posterior spinal surgery

There were 24 cases involved in this monitoring study; 19 cases with laminectomy or laminoplasty, 3 with Harrington Instrumentation, and 2 with cauda equina tumor. Good quality recordings of SEP were obtained in 19 cases; the exceptions were 3 with thoracic spinal cord neoplasms and 2 with thoracolumbar fracture dislocation. In 19 cases, more than 30% attenuation of the amplitude of the major components of SEP (N1 or N2) occurred; Five cases received decompressive laminoplasty and in four of these, the

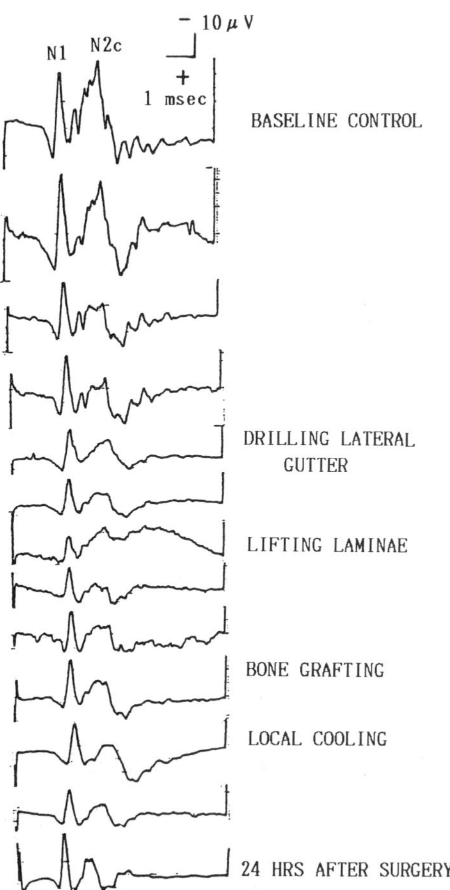

Fig. 3. Ascending SEP recorded from 72-year-old male with C2 through T2 OPLL myelopathy. The baseline N2 complex (N2c) appeared stable during laminoplasty. N1 and N2c were reduced in amplitude transiently when the laminae were lifted.

amplitude decreased up to half of the baseline amplitude and remained there. In the remaining case, the SEP disappeared permanently following transient amplitude augmentation associated with severe hypotension and fatal bleeding. Of the other four cases, three showed good neurological outcomes. There was no relationship between neurological results and the amplitude attenuation. Two of 13 cases with posterior decompression showed significant increment of the amplitude following laminar manipulation. Assessment of latency changes after decompression of the cord was difficult indeed, and moreover, did not show any valid changes with some exception. Fig. 3 illustrates typical responses recorded from a 72-year-old male with C2 through T2 OPLL myelopathy. Laminoplasty of C2 through C6 and laminectomy from C7 to T3 were performed. The second polyphasic negativity in the baseline control was stable 24 hours after surgery, and transient amplitude attenuation of the first upward negativity (N1) was shown during en bloc lifting the lamina backward. His postoperative neurological result was good in the JOA Assessment scoring system. Latencies of both negativities shortened 24 hours after the surgery.

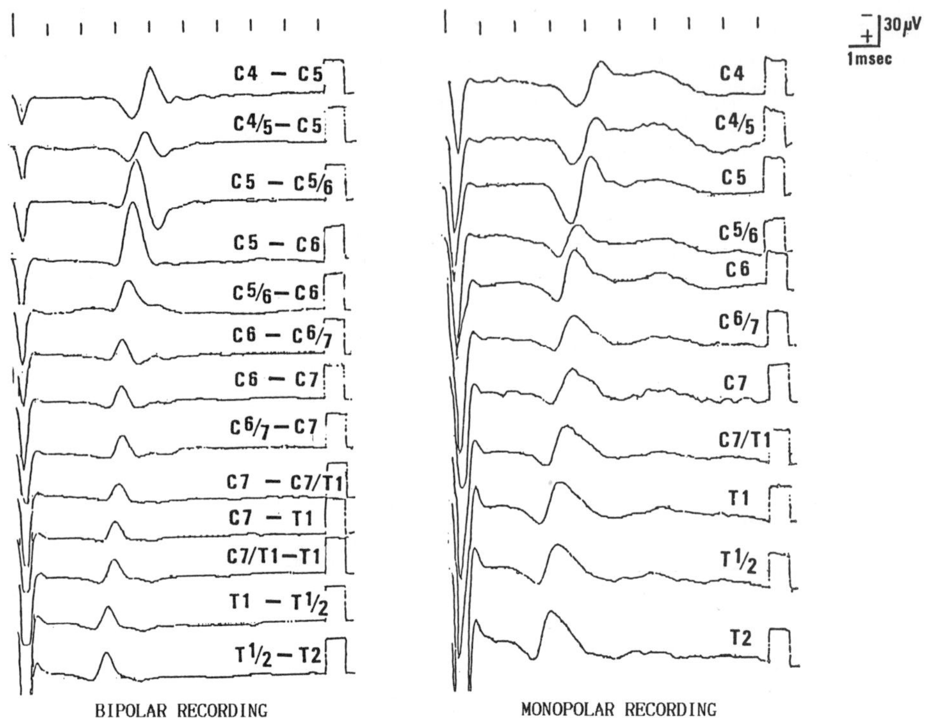

Fig. 4. Difference of the mapped SEP between bipolar and monopolar recording, recorded from 38-year-old male with cervical myelopathy. A large positive-going killed end potential appeared at C5 in monopolar recording.

Discussion

Among various physiological tests available, the authors have utilized invasive methods of somatosensory evoked potentials (SSEP) in the treatment of cervical myelopathy. Although there have been many papers describing the usefulness of these potentials (4, 6), we find that in detecting early insult to the spinal cord, they are much less stable and technically more difficult to use intraoperatively. In author's experience with a comparatively small group of cases, ascending SEP monitoring offered very qualified responses to lower spinal cord or cauda equina stimulation.

In seven cases (19%), acquisition of the ascending SEP was not possible; two with severe paraplegia from congenital spine deformities, three with neoplasms, and two acute phase of fracture dislocation. In reviewing these cases, diminished afferent volley generating SEP was thought to be the main reason of the failure. The ascending SEP is mediated predominantly via the posterior column and partially via lateral spinothalamic pathways. Direct injury to these somatosensory pathways may result in significant changes in SEP; however, pathology within the lateral corticospinal tracts or the anterior horns without any involvement of the posterior columns may not bring any change in the ascending SEP. It may be safely documented that surgical intervention directly to the dorsal spinal cord is well monitored by the ascending SEP, but surgery to the anterior or anterolateral pathology causing pyramidal tract involvements may not be monitored. For elucidation of this phenomenon, see Levy et al. (5). This further offers an explanation for the number of false-negative results as shown in the previous mapping study (1).

There exists another item of importance to be noted: in bipolar recording, evoked responses may be of different configurations from those of monopolar recording. Fig. 4

illustrates differences in mapped SEPs between the two recording modes. In bipolar recording, there was amplitude increase of the large negative wave followed by a small positive wave between C5 and C5 through C6; however, a large positive-going killed end potential made the precise diagnosis at C5. Based on our experience with this case, the authors utilized both of the recording modes in the recent series.

Ascending SEP monitoring epidurally is a useful method for detecting spinal cord dysfunction; however, because it predominantly reflects somatosensory afferent pathways and because of variability in recording modes, great care must be taken in the neurophysiological interpretation of these readings.

References

1. Baba, H.; Shima, I.; Tomita, K. et al.: Clinical usefulness of spinal cord evoked potentials. J. Schramm; S.J. Jones (eds), Spinal Cord Monitoring. Berlin Heidelber: Springer-Verlag, 1985, pp. 245-249.
2. Bunch, W.H.; Scarff, T.B.; Besser, M. et al.: Spinal cord monitoring. J. Bone Joint Surg. (Am.), 65: 707-710, 1983.
3. Dinner, D.S; Luders, H.; Lesser, R.P. et al.: Intraoperative spinal somatosensory evoked potential monitoring. J. Neurosurg., 65: 807-814, 1986.
4. Hahn, J.F.; Lesser, R.; Klem, G. et al.: Simple technique for monitoring intraoperative spinal cord function. Neurosurg., 9: 692-695, 1981.
5. Levy, W.J.; York, D.H.; McCaffrey, M. et al.: Motor evoked potentials from transcranial stimulation of the motor cortex in humans. Neurosurg., 15: 287-302, 1984.
6. Macon, J.B.; Poletti, C.E.; Sweet, W.H. et al.: Conducted somatosensory evoked potentials during spinal surgery. Part 2: Clinical applications. J. Neurosurg., 57: 354-359, 1982.
7. Whittle, I.R.; Johnston, I.H.; Besser, M. et al.: Recording of somatosensory evoked potentials for intraoperative spinal cord monitoring. J. Neurosurg., 64: 601-612, 1986.

True and False Positive Amplitude Attenuations during Cortical Evoked Potential Spinal Cord Monitoring

R. C. More; M. R. Nuwer;[*] E. G. Dawson

Introduction

Spinal cord impairment is a well recognized but uncommon complication during surgical correction of spinal deformities. Somatosensory evoked potential (EP) monitoring has been used to detect spinal cord impairment intraoperatively, so that immediate measures can be taken to minimize the residual effects. EP amplitude decrease is the main factor used for monitoring. Amplitude attenuation of greater than 50% is generally regarded as a significant reason to worry about neurologic complications. However, EP amplitude can be decreased transiently without significant postoperative neurologic sequelae. It would be clinically useful to ascertain the incidence at which varying degrees and durations of amplitude attenuation are observed, and what relationship these changes have to clinical sequelae.

We review here our experience with EP monitoring, with specific attention to those patients who had amplitude attenuations in their EPs intraoperatively.

Methods

Between 8/81 and 2/86, 138 operations were performed at the University of California, Los Angeles, Medical Center by one experienced surgeon (E.C.D.) for the correction of spinal deformities. All used somatosensory Ep monitoring under the direction of one of us (M.R.N.), using techniques previously devised specifically to reduce random background noise and variability (Nuwer and Dawson, 1984a, 1984b, and in this volume).

The diagnosis in these patients was scoliosis in 132, kyphosis in five and lordosis in one. There were 110 females and 28 males, with ages ranging from 11-56 years. Harrington rods were used in 134 and Luque rods in four.

A Nicolet Pathfinder II averager was used to record EPs. Baseline EP testing was performed often on the afternoon prior to operation, to ascertain the presence of recordable potentials and to allow scalp measurement and electrode placement. Gold disc electrodes were glued to the scalp with collodion and left in place until the postoperative recovery period. Electrodes were placed at a minimum of five locations, including Fz, Cz, Pz, E1 and E2 according to the 10% extension of the International EEG ten-twenty system (Chatrian, 1985; Nuwer, 1986). Electrodes were reglued at the

---

* UCLA Department of Neurology, Reed Neurological Research Center, 710 Westwood Plaza, Los Angeles, CA 90024

beginning of the operation. In most cases two channels of recording were obtained simultaneously, Cz-Pz, and E1-E2, with Fz used as a ground.

Bare platinum EEG needle electrodes were placed subdermally for peroneal nerve stimulation at the knee or posterior tibial nerve stimulation at the ankle. Usually right and left sided nerves were separately stimulated, and the best EP was used during most of the monitoring. Bilateral simultaneous stimulation was avoided. The stimuli consisted of square-wave pulses, usually of at least 20mA intensity (twice motor threshold), duration 0.25msec, stimulation rate 5.1 per second, sample size 200-600 trials per average, allowing a new EP every 30-120 seconds. Filters were set at 20-3,000Hz.

Most patients were given balanced anesthesia consisting of nitrous oxide and narcotic (in most cases fentanyl). Halogenated inhalation agents were usually avoided but isoflurane was used in 6 cases because of asthma or other medical conditions. Nondepolarizing muscle blockade was given to some patients.

Results

Of 138 patients in the series, the EP signal was unobtainable in the operating room under anesthesia in 8 patients, including 5/6 cases in which isoflurane was used (and other patients with pre-existing neurological disorders).

The remaining 130 patients had stable, reproducible EPs at the beginning of the operation. Of these 117 had stable EPs the entire case, and 13 had amplitude attenuations of >35% at some point during the case.

Nine patients had amplitude attenuations at some point before distraction of the spine. We reviewed the anesthetic records of these patients in detail and found a period of hypotension to <80 torr (mm Hg) was associated with the attenuation in two patients, and a change in the level of nitrous oxide anesthesia was associated in one patient. In all three patients, the EPs returned to baseline immediately after these factors were corrected. Six patients had EP attenuation before distraction for unknown reasons. Of these 6, three recovered within a few minutes, two were stable at the lower amplitude, and one was lost altogether.

Four patients had EP amplitude attenuations of >35% at the time of distraction of the spine. Of these, 3/4 recovered to baseline within 15 minutes. The fourth patient did not have recovery of the EP, and is the only patient in our series with new postoperative neurological impairment. This case deserves discussion in more detail.

This latter patient was a 19-year-old male with congenital insensitivity to pain who developed a Charcot-related kyphotic deformity of 75 degrees at L1-L2 region of his spine, which corrected to 40 degrees on hyperextension. T11-L3 double Harrington rod instrumentation and posterior spinal fusion with iliac crest bone graft was performed. The EPs were stable early in the procedure with the average early positive EP peak amplitude of 0.30μV with a latency of 35msec. After distraction the amplitude decreased to 0.15-0.20μV and the latency was unchanged. Slowly, over the next few minutes the amplitude returned to 0.30μV. Towards the end of the case the amplitude again decreased to 0.15-0.20μV, and remained there until completion of the case (Fig. 1).

Postoperatively the patient had normal motor function, but within six hours he developed acute lower extremity motor deterioration. He was returned emergently to the operating room for rod removal. Exploration of the cord revealed no obvious lesions. He was left with a complete cord syndrome at L1 which improved over days to moderate motor paresis (distal worse than proximal) without sensory loss. This was felt to be compatible with a central cord syndrome of vascular origin. Repeat EPs were normal at eight days postoperative. At fourteen months postoperative he had normal proximal motor strength and 3/5 distal motor strength. Shortly thereafter he suffered an acute neurological deterioration, underwent an anterior decompression and

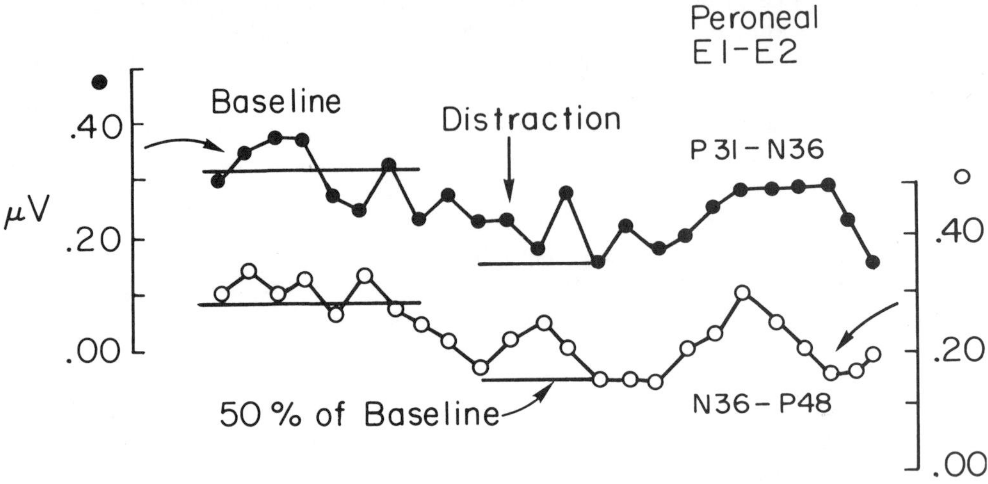

Fig. 1. The EP amplitudes dropped to 50% of baseline after distraction during monitoring of this patient's Harrington rod procedure. The amplitude attenuation lessened transiently and then worsened again at the end of the case. The patient awoke moving both lower extremities well, but developed substantial weakness within 6 hours postoperatively. The amplitudes shown are for the N36 vertex EP, measured to the preceding P31 and also to a succeeding P48 (Lesser et al., 1986).

stabilization, and has since returned to his baseline status. He ambulates with the assistance of bilateral knee-ankle-foot orthoses.

Discussion

Of 130 patients, we had twelve patients with false positive amplitude attenuations of >35% intraoperatively, without postoperative neurological impairment. Of these, 9/12 occurred before distraction of the spine at a time when the spinal cord was not in any apparent jeopardy. The other 3/12 episodes of amplitude attenuation occurred at the time of distraction.

Amplitude attenuations at the time of distraction were a rare event, occurring in only four patients out of 130. Interestingly, of these four patients, the only one without amplitude recovery by the end of the operation was also the one with postoperative impairment. Clearly, any attenuation at distraction needs to be carefully assessed. If there is not recovery, the surgeon must consider relieving the distraction on the spine immediately. When we initially started monitoring we considered a 50% amplitude decrease to be the criterion for raising an alarm. The present data suggest that a 35% amplitude decrease should be considered reason to begin worrying, especially if such a loss is persistent (>15min) and not associated with hypotension or changes in level of anesthesia. By our original criteria we considered the case presented here as a "false negative", and discussed it as such briefly in previous report on such phenomena (Lesser, 1985). By a 35% criterion, this case would be considered a "true positive" monitoring alarm.

In our case of postoperative spinal cord impairment, the wake-up tests would not have been helpful. The patient had awoken with intact neurological function, and it is almost certain that he would have been intact during a wake-up test.

Our patient with postoperative paraplegia also reinforces several known lessons. The first is that patients with kyphosis or those with pre-existing neurological deficits are at high risk for new neurological complications (MacEwen et al., 1975). Second, EPs may err at predicting motor outcome in the chronic setting (e.g., eight day post-onset or much later) despite their accuracy in the acute setting (McGarry et al., 1984; Rowed et al., 1978; York et al., 1983).

Our preferred stimulating and recording parameters were chosen only after a careful, systematic study of the effects of these choices on random, unwanted background noise and variability (Nuwer and Dawson, 1984a, 1984b, and in this volume). Our preferred and recommended settings minimize such unwanted variability. Other stimulus and recording techniques studied were associated with greater amounts of the random, unwanted background noise and variability. Monitoring teams who use EP stimulating and recording settings different from ours may well incur worse rates of transient and persistent false positive events.

Conclusions

Cortical evoked potential spinal cord monitoring can be accomplished while limiting false-positive (false-alarm) events to less than 10% of cases. A criterion for alarm of >35% amplitude loss seems appropriate, and more sensitive for detecting true-positive events than the widely used 50% criterion. Among our 130 patients, persistent changes (>15 minutes) in the EP amplitude were only seen in 4 cases (3%), and in only one patient at the time of distraction. This patient was the one in our group with postoperative neurological impairment.

References

1. Lesser, R.P.; Raudzens, P.; Luders, H.; Nuwer, M.R.; Goldie, W.D.; Morris, H.H., III; Dinner, D.S.; Klem, G.; Hahn, J.F.; Shetter, A.G.; Ginsburg, H.H.; Burd, A.R. (1986): Postoperative neurological deficits may occur despite unchanged intraoperative somatosensory evoked potentials. Ann. Neurol. 19: 22-25.
2. McGarry J.; Friedgood, D.L.; Woolsey, R.; Horenstein, S.; Johnson, C.: Somatosensory evoked potentials in spinal cord injury. Surg. Neurol. 22: 341-343, 1984.
3. MacEwen, G.D.; Bunnel, W.P.; Sriram, K.: Acute neurological complications in the treatment of scoliosis: A report of the Scoliosis Research Society. J. Bone Joint Surg. 57A: 404-408, 1975.
4. Nuwer, M.R.; Dawson, E.C.: Studies of the sites, rates and filters that best eliminate background noise and variability during cortical evoked potential spinal cord monitoring. (In this volume).
5. Nuwer, M.R. (1986): Evoked potential monitoring in the operating room. Raven, New York.
6. Nuwer, M.R.; Dawson, D. (1984a): Intraoperative evoked potential monitoring of the spinal cord: Enhanced stability of cortical recordings. Electroencephalogr. Clin. Neurophysiol. 59: 318-327.
7. Nuwer, M.R.; Dawson, E. (1984b): Intraoperative evoked potential monitoring of the spinal cord: A restricted filter, scalp method during Harrington instrumentation for scoliosis. Clin. Orthop. 183: 42-50.
8. Rowed, D.W.; McLean, J.A.G.; Tator, C.H.: Somatosensory evoked potentials in acute spinal cord injury: Prognostic value. Surg. Neurol. 9: 203-210, 1978.
9. York, D.H.; Watts, C.; Raffensberger, M.; Spagnolia, T.; Joyce, C.: Utilization of somatosensory evoked cortical potentials in spinal cord injury: Prognostic limitations. Spine 8: 832-839, 1983.

Cortical and Spinal Intraoperative Recordings in Uneventful Monitoring and in Cases with Neurologic Changes

E. Watanabe;[*] J. Schramm; J. Romstöck

Introduction

Spinal cord monitoring using epidural electrodes during an operation was initially applied in patients undergoing correction of spinal deformities (3). Because the response obtained by this technique is definitely large, it was noticed by several authors that the stability is considerably high in the intrathecal recording compared with conventional SEP with skin recording (4, 5, 6, 7, 13). This method was also used in neurosurgical cases (1, 2, 5, 10, 16) and several differences were found between deformity cases principally without cord lesions and neurosurgical cases with definite cord lesions (5, 8, 9, 11, 15). The advantages of spinal recording include quick availability and higher stability even in most patients with cord lesions. The advantages of cortical recording, on the other hand, include that it can be easily applied and that it can cover the whole course of the operative procedure including preoperative and postoperative control recordings. These aspects were discussed in a recent review article (9). The value of simultaneous multi-level recordings has been discussed (2, 4). In this report we present our results of scalp and spinal recordings and compare these two techniques referring to significant potential change and detection of postsurgical neurological outcome.

Patients and methods

During the last two years, we have monitored 63 cases. Recordings were usually done from scalp and intra or epidural space after the stimulation of median nerve or peroneal nerve. We analyzed 40 of them concerning the effects of recording and stimulation sites. The effect of various stimulation and recording sites have been presented in detail elsewhere (12, 14). The diagnoses were: 12 meningiomas, 2 neurinomas, 3 syringomyelias, 3 metastases, 3 thoracic discs and miscellaneous other lesions. No spinal deformity or spinal trauma cases were included. There were 14 lesions in the cervical region, 18 in thoracic region and 7 in the lumbar region.

Scalp responses were recorded with platinum-iridium plate electrodes from the scalp at Cz-Fz for peroneal nerve and C3/4-Fz for median nerve stimulation. Intrathecal potentials were recorded with platinum tipped wire electrode (1.2mm; diameter) placed in epidural or subdural space in the operating field above and below the lesion.

* Neurochirurgische Klinik der Universitt Erlangen-Nrnberg, Schwabachanlage 6, D-8520 Erlangen, Federal Republic of Germany

Potentials were averaged with a NICOLET CA-1000/2000 Evoked Potential System with floppy disk storage.

Input filters were set at 30-3KHz bandpass. Amplification was 25-50μV full scale sensitivity with automatic noise rejection circuit. Peripheral nerves were stimulated transcutaneously on the median nerve at the wrist or the peroneal nerve at the knee at 5.3Hz with bipolar plate electrodes. Constant current square wave (0.2-0.4ms) pulses were used with supramaximal nerve activation. 200-400 runs were used per average.

In most of these cases, we used more than one stimulation and recording mode in the same patient. In the operating theater, we usually started with scalp recordings and as soon as the dura was exposed we placed a pair of epidural recording electrodes above and below the lesion (10).

Results

Relation with postoperative neurology

Fig. 1 shows a summary of relationships between intraoperative potential changes and postoperative neurological changes. Each star represents one patient, and on the right column, the neurological outcome is shown, on the left column the intraoperative potential changes are shown. We divided our patients into four categories according to potential and neurology changes. First with no potential and no neurological changes; second, potential deterioration; third, with neurological deterioration; and fourth, potential improvement. These categories have some overlap, as is shown in Fig. 1. The last three categories will be discussed in this report as they are helpful in assessing usefulness.

Category 1: Intraoperative potential deterioration

The potential deteriorated in 4 cases during surgery (Table 1). In 3 of the cases, scalp recordings showed deterioration and postoperative neurology also showed deterioration (correct detection). The remaining one patient showed no neurological change after surgery in spite of definite deterioration of intraoperative potentials. This could be called a false positive case. Fig. 2 shows the tracing of the potentials in a patient with correct detection from the group who underwent a thoracic disk operation.

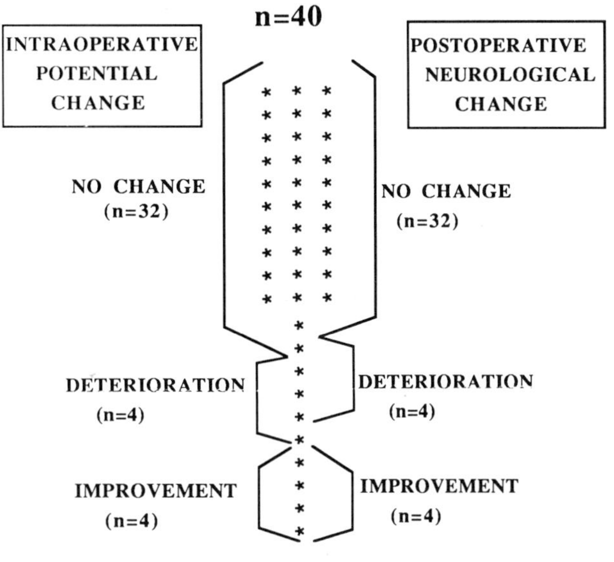

Fig. 1. Summary of intraoperative potential changes (left column) and postoperative neurological changes (right column). Asterisk indicate one case. Improvement of potential means either the increase of amplitude or the decrease of latency of N20 in cortical SEP or negative peak of spinal potential. Deterioration means the decrease of amplitude or the increase of latency. Amplitude change was taken as significant when the change was more than 50%. For latency, change more than 10% was taken as significant.

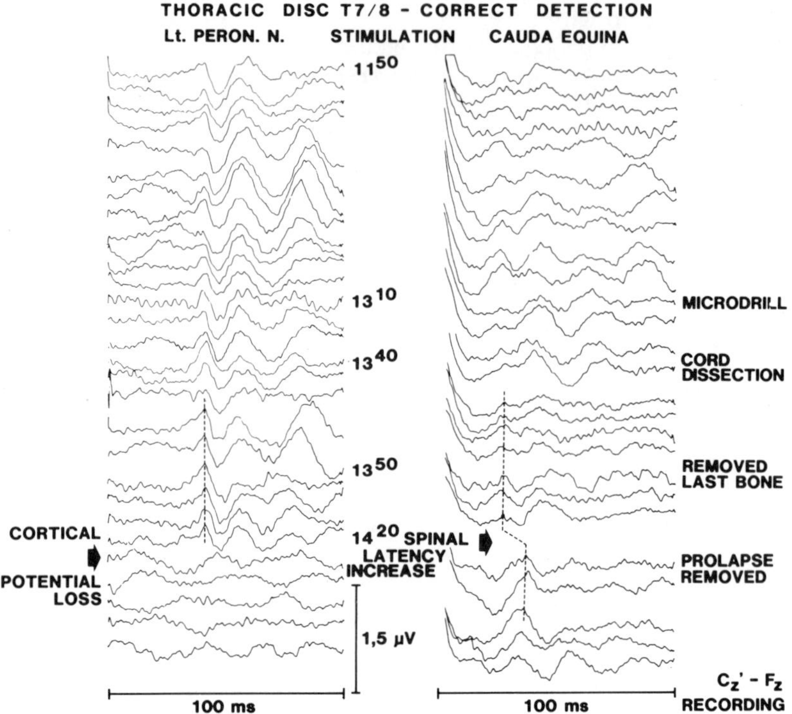

Fig. 2. Cortical recording after cauda versus peroneal nerve stimulation in a case with T7/8 thoracic disc operation. Potential after peroneal nerve stimulation showed a sudden amplitude decrease when the thoracic disc space. Note at the same moment that the potential after cauda equina stimulation showed latency elongation rather than amplitude decrease. This patient showed a persistent worsening of neurologic status after surgery. Positivity downward.

The scalp potential after peroneal nerve stimulation showed a sudden potential loss when the thoracic disc prolapse was removed from the spinal canal anterolaterally and pulled back into the disc space. The scalp potentials after cauda equina stimulation showed different alterations, that is N20 showed only a sudden increase in latency but no decrease in amplitude. This clear discrepancy might be explained by the fact that with cauda equina stimulation, more sensory neurons are activated as compared to unilateral peroneal nerve stimulation. This patient showed an increase in existing paraparesis after the operation which did not recover within one year.

In spinal potentials, only one out of four cases showed correct detection. From the remaining three cases (Table 1), one showed no potential from the beginning and in another case, no spinal recording was done. In the last case, as the scalp potential showed amplitude and latency change during laminectomy, the spinal potential was already abnormal from the beginning of recording. Fig. 3 shows the potential of this case with a thoracic meningioma after peroneal nerve stimulation. Scalp potential was monitored from the beginning of surgery and disappeared during laminectomy probably due to the mechanical influence from the dissection. Although the scalp potential recovered within 30 minutes, this particular phase was not covered by the period of intrathecal spinal recording which was started after laminectomy. The spinal potential was already definitely abnormal showing no typical wave form from the beginning. After operation this patient showed weak paresis in one leg which recovered in one week. This case shows a dissociation between the cortical potential and a longer lasting

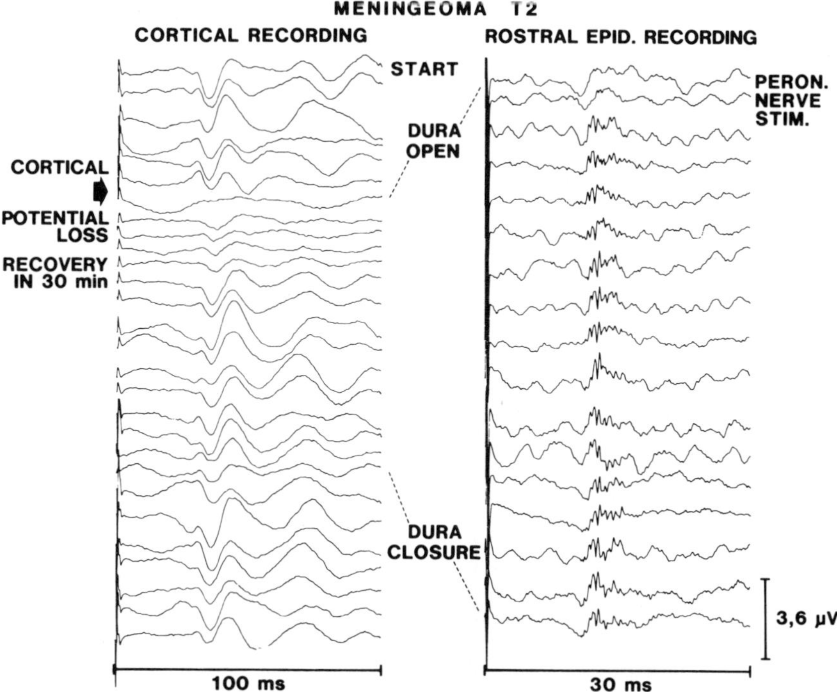

Fig. 3. Cortical recording (left column) and spinal epidural recording (right column) after peroneal nerve stimulation. During laminectomy the cortical potential decreased in amplitude suggesting mechanical influence from laminectomy. The spinal recording was started just after laminectomy and at this moment the potential gradually improved its shape but the patient developed weak paresis in one leg which recovered in one week. Positivity downward.

alteration of the spinal potential which, however, was not lost. This case documents a good example of the disadvantage of intrathecal spinal monitoring particularly concerning the events before or during laminectomy. The remaining two cases of spinal potentials failed to discover the change because spinal potentials were not obtained. The one case, where spinal potentials were unobtainable from the beginning, illustrates the significance of the effect of the lesion.

Category 2: With neurological deterioration

The next group with neurological deterioration after surgery is listed in Table 2. This group includes nearly the same patients as the preceding category. Scalp potential showed correct detection in 3 of 4 cases. In another case, scalp potential showed no change despite the postsurgical deterioration in paraparesis (false negative). Spinal recording, however, never showed correct detection. The reasons for that are no recording (n = 1), no potentials (n = 1), and changes during laminectomy (n = 1). The latter case is the same case that is shown in Table 1. In this group spinal recording again shows a poor result regarding detection mainly because of technical reasons (2 of 3 cases).

Category 3: With potential improvement

Potential improved during surgery in four cases. As is shown in Table 3, scalp potential showed correct detection in 3 cases out of 4. In one case, however, spinal potential from above the lesion showed improvement during the tumor removal. Unfortunately in the remaining two cases, no spinal monitoring was done during surgery.

Discussion

In epi- or intradural monitoring during spinal cord surgery with a laminectomy, the invasiveness of this method creates no further problems additional to the surgical insults (10, 11). Our results show higher rates of useful and stable recording in spinal monitoring compared to scalp monitoring (12). This tendency might be caused by the existence of the spinal cord lesion especially in our series which consists of neurosurgical patients with cord compression and neurological defect. In consequence, direct spinal monitoring technique might fit better for neurosurgical cases.

Whittle (15) reported two monitoring cases during spinal neurosurgery. From this experience, he recommended the use of intrathecal monitoring because of the high amplitude enables fewer repetition and more stable recording.

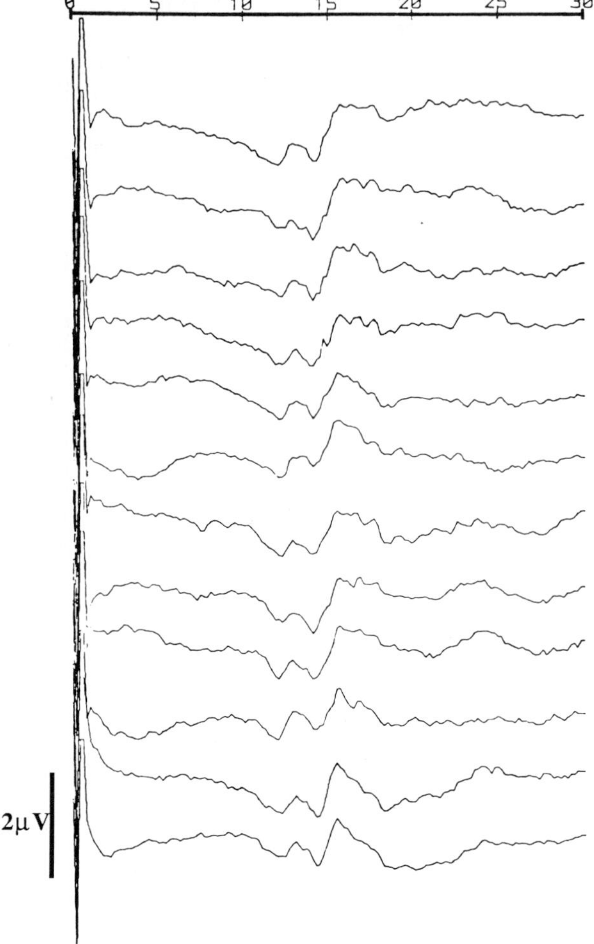

Fig. 4. Spinal epidural recording after peroneal nerve stimulation on a patient with a thoracic meningioma (T12). Note that during tumor extirpation the shape of the negative wave became sharper. Cortical potential did not change. After the operation the patient showed improvement of hypesthesia of lower extremities. Positivity downward.

Britt et al., (1) used multicontact electrodes for intrathecal monitoring in neurosurgical cases and found that bipolar recording using either two of these contact points allows lesser artifact than conventional monopolar recording. Dinner et al., (2) monitored 220 patients with spinal lesions and compared the recording quality of surface and interspinous electrodes. The stability and reproducibility of potentials were similar in 54% of cases, interspinous electrode provided better recording in only 9% of cases. The results of interspinous recording are worse than our results using intrathecal recording (14). This may be due to the higher amplitude acquired with intrathecal recording. In their study, 5 cases with mass lesion in the spinal cord showed some change either in potential or neurology. Three of them showed false negative, one case showed false positive and the remaining case showed correct detection. The existence of more false negative cases demonstrates the need for more precise analysis of the waves as we will discuss later.

If we consider our results concerning detectability of postoperative neurological outcome, the rate of correct detection was higher in the scalp recording than in the spinal recording. Since the number of cases with perioperative potential changes or neurological changes is still small, and unfortunately there were several cases in which spinal recording was not properly performed for various reasons, it is too early to draw final conclusions as to the differential value of these methods. Further monitoring cases completely covered by sufficient spinal recordings and with intraoperative potential changes or postoperative neurological changes are needed. We would like to discuss some of the problems in spinal monitoring to be solved in the future. As is depicted in Fig. 4, one of the drawbacks of spinal recording is the fact that dysfunction of the cord could sometimes occur during laminectomy which would not be detected using our recording protocol of intraspinal monitoring. It might be overcome by beginning spinal cord monitoring by introducing the electrode into epiligamental space between the lamina before the laminectomy is started. For that reason we have changed our protocol and now we place small electrodes between the lamina before taking the bone away. Another problem is coming from the very fact that the spinal cord potential is significantly more stable than the scalp potential (8). In other words, it was to be discussed whether the spinal SEP is not too sensitive.

Certainly it may be concluded that criteria for evoked potential indicating impending danger to the cord should be quite different for spinal SEP as compared to cortical SEP. We are usually concentrating only on the latency and amplitude changes in the decision of "potential change." We have several cases, however, in which we feel a necessity for using further criteria including the wave form. In Fig. 4 a case of meningioma at 12th thoracic level is shown. During extirpation of the tumor mass, the spinal cord potential evoked by posterior tibial nerve stimulation gradually changed its shape, i.e., up-going leg of the first negative peak became steeper and the following positive peak became more prominent. The pre-existing hypesthesia in lower extremities improved after the operation. But conventional parameters such as latency or amplitude of the negative peak did not show significant change predicting no neurological improvement. This case and the case in Fig. 3 where spinal SEP were not so normal seem to indicate a necessity of adopting some "wave form parameters" in the evaluation of spinal SEP.

Conclusion

Cortical SEP demonstrated good reliability in detecting postoperative neurological change. The first glance conclusion regarding the reliability of spinal SEP recordings, however, seem to be wrong as there were just not enough well documented cases with obtainable spinal SEP throughout the surgical procedure. Therefore this matter will have to be taken up again when a large case number with intraoperative potential changes and/or postoperative neurological changes has been collected.

Table 1 - Intraoperative potential deterioration (n = 4)

Cortical Potential:
Correct detection   3
False Positive   1

Spinal Potential:
Correct detection   1
-- change before electrode 1
insertion
-- no potential   1
-- no recording   1

Cases with intraoperative potential deterioration. The cases which showed postoperative neurological deterioration as was expected from intraoperative potential deterioration were counted as correct detection. The cases which showed no postoperative neurological deterioration were counted as false positive. In one case which is assigned as "no recording," spinal recording was not done due to a technical problem.

Table 2 - Postoperative neurological deterioration (n = 4)

Cortical Potential:
Correct detection   3
- false negative   1

Spinal Potential:
Correct detection   0
- no recording   1
- no potential   1
- false negative   1
- change before electrode 1
insertion

Cases with postoperative neurological deterioration. The cases which showed postoperative neurological deterioration as was expected from intraoperative potential deterioration were counted as correct detection. The cases which showed no intraoperative potential deterioration were counted as false negative. In a case which is assigned as "no recording," spinal recording was not done due to a technical problem. In a case indicated as "no potential," no potential was obtained in spinal cord recording due to the effect of the lesion.

Table 3 - Intraoperative potential improvement (n = 4)

Cortical Potential:
Correct detection   3
False negative   1

Spinal Potential:
Correct detection   2
- no recording   2

Cases with intraoperative potential improvement. The cases which showed postoperative neurological improvement as was expected from intraoperative potential improvement were counted as correct detection. In one case which is assigned as "false negative," intraoperative cortical potential showed no change despite postoperative neurological improvement. In this case, however, spinal potential showed improvement as is presented in Fig. 4. In cases which are assigned as "no recording," spinal recording was not done due to technical problems.

References

1. Britt, R.H.; Ryan, T.P.: Use of a flexible epidural stimulating electrode for intraoperative monitoring of spinal somatosensory evoked potentials. Spine, 11: 348-351, 1986.
2. Dinner, D.S.; L ders H; Lesser, R.P.; Morris, H.H.; Barnett, G.; Klem, G.: Intraoperative spinal somatosensory evoked potential monitoring. J. Neurosurg., 65: 807-814, 1986.
3. Jones, S.J.; Carter, I.; Edgar, M.A.; Morley, T.; Ransford, A.O.; Webb, P.J.: Experience of epidural spinal cord monitoring in 410 cases. In: J. Schramm; S.J. Jones (eds.): Spinal Cord Monitoring. Springer, Berlin, Heidelberg, New York, 215-220, 1985.
4. Maccabee, P.J.; Levine, D.B.; Pinkhasov, E.I.; Cracco, R.Q.; Tsairis, P.: Evoked potentials recorded from scalp and spinous processes during spinal column surgery. Electroencephalogr. Clin. Neurophysiol., 56: 569-582, 1983.
5. Macon, J.B.; Poletti, C.E.; Sweet, W.H.; Ojemann, R.G.; Zervas, N.: Conducted somatosensory evoked potentials during spinal surgery. Part 2: Clinical Applications. J. Neurosurg., 57: 354-359, 1982.
6. Maruyama, Y.; Shimizu, H..; Fujioka, H., et al.: Spinal cord function monitoring by spinal cord potentials during spine and spinal surgery. In: S. Homma; T. Tamaki (eds.): Fundamentals and Clinical Application of Spinal Cord Monitoring. Tokyo, Saikon, 191-202, 1984.
7. Ohmi, Y.; Tohno, S.; Harata, S.; Nakano, K.: Spinal cord monitoring using evoked potentials recorded from epidural space. In: S. Homma; T. Tamaki (eds.). Tokyo, Saikon, 203-210, 1984.
8. Romstöck, J.; Watanabe, E.; Schramm, J.: Variability of spinal epidural SEP from below and above spinal cord lesion -- the significance of the lesion. This volume.
9. Schramm, J.: Spinal cord monitoring: Current status and new developments. CNS Trauma, 2: 207-227, 1985.
10. Schramm, J.: Intraoperative spinal cord monitoring. Adv. Neurosurg., 14: 17-21, 1986.
11. Schramm, J.; Romstöck, J.; Thurner, F.; Fahlbusch, R.: Variance of latencies and amplitudes in SEP monitored during operations with and without cord manipulation. In: J. Schramm; S.J. Jones (eds.): Spinal Cord Monitoring. Springer, Berlin, Heidelberg, New York, 186-196, 1985.
12. Schramm, J.; Romstöck, J.; Watanabe, E.: Cortical versus spinal recordings in intraoperative monitoring of space occupying spinal lesions. To be published In: C. Barber; T. Blum (eds.): Evoked Potential III. Boston, Butterworthes.
13. Takano, H.; Tamaki, T.; Noguchi, T.; Takakura, K.: Comparison of spinal cord evoked potentials elicited by spinal cord and peripheral nerve stimulation. In: J. Schramm; S.J. Jones (eds.): Spinal Cord Monitoring. Springer, Berlin, Heidelberg, New York, 29-34, 1985..
14. Watanabe, E.; Schramm, J.; Romstck, J..: Intraoperative monitoring of cortical and spinal potentials using different stimulation sites. To be published In: R. Villani; B. Grundy (eds.): Evoked Potentials: Intraoperative and ICU Monitoring. Springer, Berlin, Heidelberg, New York, 1987..
15. Whittle, I.R.; Johnston, I.H.; Besser, M.: Spinal cord monitoring during surgery by direct recording of somatosensory evoked potentials. J. Neurosurg., 60: 440-443, 1984.
16. Whittle, I.R.; Johnston, I.H.; Besser, M.: Intraoperative recording of cortical somatosensory evoked potential as a method of spinal cord monitoring during spinal surgery. Aust. N.Z.J. Surg., 56: 309-317, 1986.

Direct Recording of Spinal Evoked Potentials to Peripheral Nerve Stimulation by a Specially Modified Electrode

N. K. Nainzadeh;[*] M. G. Neuwirth; R. Bernstein; L. S. Cohen

Introduction

Intraoperative spinal cord monitoring of somatosensory evoked potentials (SEP) has become widely accepted in the past decade, and is becoming a standard of care in major spinal centers. Various techniques now exist for continuous monitoring of spinal cord function by cortical SEP (16, 18, 21), vertebral bone recording (17), epidural recording (1, 11, 13, 23), and intraspinal ligament placement of recording needle electrode (9). Additional studies have also examined SEP pickup at cervical, thoracic, and lumbosacral levels of the spine (20). Stimulation sites vary, but are either unilateral or bilateral peripheral nerve and/or centrally in the cauda equina (12) or thoracic spine (23). Of these techniques, the simplest and least invasive is cortical somatosensory evoked potentials. The origin of cortical somatosensory evoked potentials is thought to be the pyramidal cells of the cerebral cortex (26). It is affected by various factors, such as inhalation of halogenated anesthetic agents (21), the stimulation rate of peripheral nerve (19), drugs (diazepam, haloperidol) (8), and hypotension (1, 4). The technique of recording spinal evoked potentials by placement of a Kirschner wire in the spinous process (17) or placement of a needle in the intraspinal ligament (9) usually will permit reliable monitoring of spinal cord function in the low thoracic and lumbar region, but in the high thoracic and cervical spine, signals are too small to permit reliable interpretation of changes in responses. Placing a needle (11) or a pair of wire (13) electrodes in the epidural space causes concern about dural laceration.

To overcome this difficulty, we modified the posterior column stimulator electrode (Neuromed Unistem 2) and used it for epidural recording. In a series of 44 patients undergoing posterior spinal fusion with instrumentation, both cortical and epidural responses were recorded. Changes in latency, amplitude, stability, and reproducibility of the cortical and epidural responses were compared.

Methodology

Forty-four (44) patients undergoing posterior spinal fusion with Harrington and/or Luque rod instrumentation were studied. The diagnoses were as follows: Idiopathic scoliosis, 37; spinal fracture, 3; metastatic lesion of spinal vertebrae, 2; osteomyelitis of spine, 1; and spondylolisthesis, 1. The ages ranged from 10-72 years with a mean of

---

[*] Department of Rehabilitation Medicine, Hospital for Joint Diseases Orthopaedic Institute, 301 E. 17th Street, New York NY 10003

17.5. There were 37 females and 7 males. All patients were anesthetized with Surital and intubated using succinylcholine. They were maintained with Fentanyl, nitrous oxide and oxygen, and a muscle relaxant. Inhalation agents were not used in order to maintain cortical responses.

Induced hypotension to help control blood loss was achieved with either sodium nitroprusside or nitroglycerine administered by means of a controlled drip. A level of 60mm Hg as a mean arterial pressure was usually used.

Monitoring consisted of the electrocardiogram, body temperature by means of an esophageal probe, arterial blood pressure through an indwelling radial artery catheter, esophageal stethoscope, urinary output, and arterial blood gases.

Blood loss was determined by measuring the volume of blood in the suction and the weighing of sponges and laparotomy pads.

All fluid and blood administered was warmed by means of blood warmers.

Simultaneous square wave electrical pulse stimuli were applied transcutaneously with a plastic block fitted with two stainless steel discs 2cm apart and a pair of strip-type electrodes filled by conductive gel and secured by tape at the ankle on the medial malleolus and anterior aspect of the ankle between two malleoli to the posterior tibial and superficial peroneal nerves respectively, with the following parameter: Duration of 0.3 milliseconds and intensity of 25mA were used for both cortical and epidural. Stimulation rate of 2.9 Hz was used for cortical and 19.9 Hz for epidural. Cortical responses were recorded 2cm behind the Cz point of 10/20 international EEG system with Fz reference and A1 or A2 as a ground. These responses were amplified over a band width of 32-160Hz and a duration of 200 milliseconds. 200 epochs were averaged twice to insure consistency. Baseline cortical responses were recorded 24-48 hours preoperatively to have a baseline. Another tracing was taken following induction of anesthesia. These were kept for comparison with further recording. Every 2 or 3 minutes a new recording was made. Amplitude and latency of peaks were compared with the first tracing, right and left legs were stimulated alternatively. This was continued until the time of wound closure.

Epidural responses were recorded by specially modified and custom designed dorsal column stimulator electrodes, "Neuromed Unistem 2." These electrodes are pure platinum specially modified for 2.5cm spacing on center. Each is 4mm long, 1.2mm in diameter with a surface of $7.536mm^2$ and is connected to a special alloy wire with polyethylene insulating sheets encasing the entire length of the connecting wires. The proximal electrode is used for pickup and the distal for reference (bipolar recording) (Fig. 1). The electrode is placed by the surgeon one interspace cephalod to the superior segment to be instrumented. A midline thoracic laminotomy is performed using a Leksell rongeur and a small-angled Kerrison rongeur; then the electrode is passed manually into the epidural space for a distance of about 2 inches. It is then sutured in place at the proximal end of the skin incision and further anchored with steri drape. No force is used to pass the electrode. If it will not pass easily, it is removed entirely and the size of the laminotomy increased. A ground electrode is taped to the greater trochanter. The spinal evoked responses were amplified over a bandwidth of 200-2000Hz, with a duration of 50 milliseconds. 400-800 epochs were averaged twice. The right and left legs were stimulated alternatively. The first tracing was kept for comparison as a baseline, and a new tracing was made every 2-3 minutes. Amplitude and latency of peaks were compared with the first tracing, and this was continued until the time of wound closure when the epidural electrode was removed.

Results

The epidural electrode was easy to insert and the leads were unobstructive to the surgeon (Fig. 2). There was no evidence of any immediate epidural hemorrhage or any

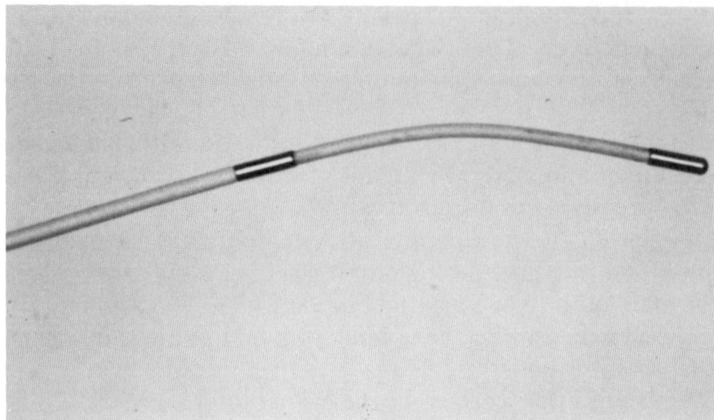

Fig. 1. Epidural recording electrode modified "Neuromed Unistem 2."

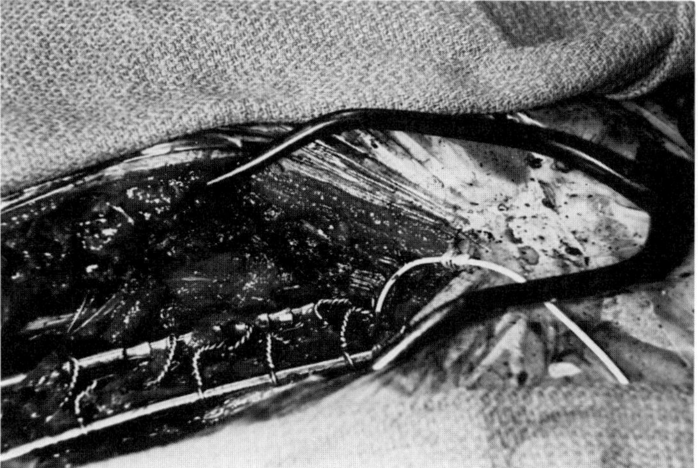

Fig. 2. Epidural electrode in place. Located in midline proximal to Harrington and Luque rods.

other short or long-term problems with a 36-month follow-up since the first case. In a few cases, the epidural electrode was displaced by the suction tube and was replaced without any complication. In 3 of 37 cases of idiopathic scoliosis with no preoperative neurological ·abnormality, epidural responses were not obtained due to our technical error (stimulation intensity mistakenly reduced to 5 milliamps). In one case of metastatic carcinoma of the spine with severe neurological abnormality of the lower extremities, no reliable response was obtained. Therefore, we excluded these four cases from our study group. In the remaining 40 cases (80 extremities), we were able to obtain constant reproducible cortical and epidural responses for each patient throughout the course of the surgical procedure. The initial epidural evoked potential (EEP) following electrode placement was used as the control pattern during the entire course of operation. The latency of the first negative peak (N1) was used as a control for each patient (Fig. 3). This latency varied from patient to patient depending on the level of the pickup electrode and height of the patient. The spinal wave form was a series of negative, positive deflection with a duration of 5.8-6.5 milliseconds (average 6.2 mil-

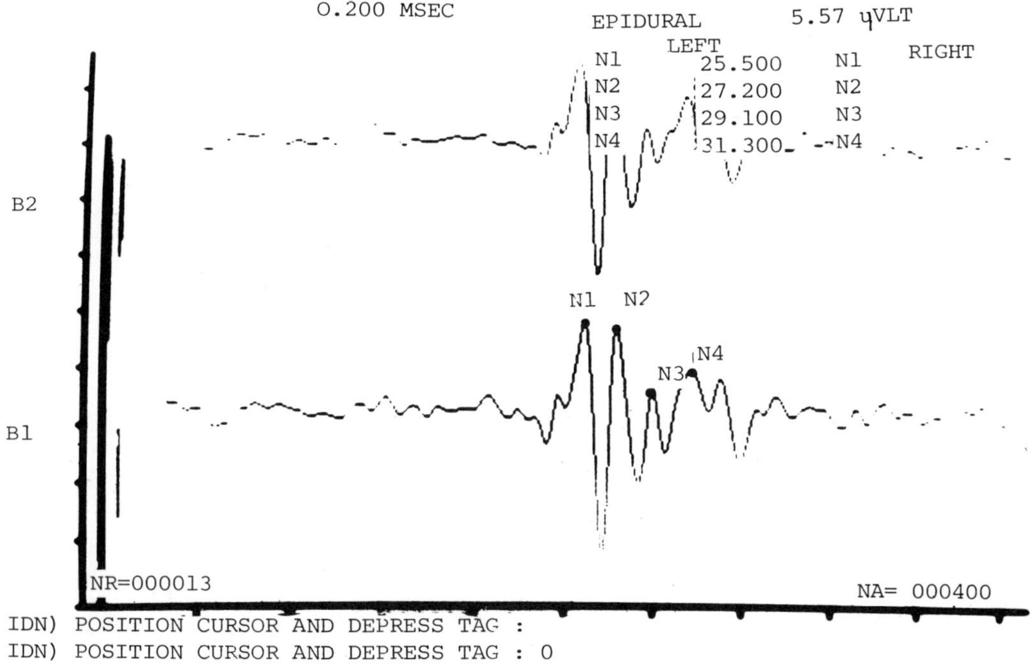

Fig. 3 & Fig. 4. Epidural responses: B1 = initial epidural evoked potential (EEP). B2 configuration of EEP and different wave form components latency measurement.

liseconds). In 40 cases (80 extremities), the wave form could be resolved into at least three components (Fig. 4): The first, a triphasic (positive/negative/positive) wave resembling conventional nerve action potentials (N1); the second, a fairly sharp negative (N2) wave; the third, a broader negative (N3) wave; occasionally, a fourth negative wave could be discerned (N4). The amplitude of the response measured from the deepest through the highest peak (usually the second positive to second negative) was between 0.12μv to 2.3μv with a mean of 1.7μv. If the position of the electrode remained undisturbed, the wave form was fairly constant for that individual, although a slight latency increase was noticed as the operation progressed but was not more than 0.8 milliseconds with an average of 0.53 milliseconds. There was a slight side-to-side latency difference of 0.1-0.5 milliseconds and, in four cases as high as 0.8 milliseconds, with an average of 0.46 milliseconds. The amplitude of the first negative peak potentials (N1) fluctuated by less than 30% from the control value in 39% of cases, and between 40-50% in 52% of patients. In three cases, there was an increased amplitude of up to 75%. In one case, there was unilateral decreased amplitude of 76%. This was in the case of a 32 year old female with a diagnosis of idiopathic scoliosis with right thoracic/left lumbar curvature. During the first distraction, the amplitude of N1 (first negative) wave of the left side decreased to 76% of the initial amplitude, and this returned to predistraction value in less than 20 minutes. Right side amplitude remained unchanged. In this patient, latency did not increase (Fig. 5). Amplitude of first negative (N1) compared to N2, N3, and N4 did not change significantly in high thoracic when compared to low thoracic and high lumbar (Fig. 6), but in low lumbar N1 amplitude was much larger than N2 and N3 (Fig. 6). Although interesting, these preliminary results require some

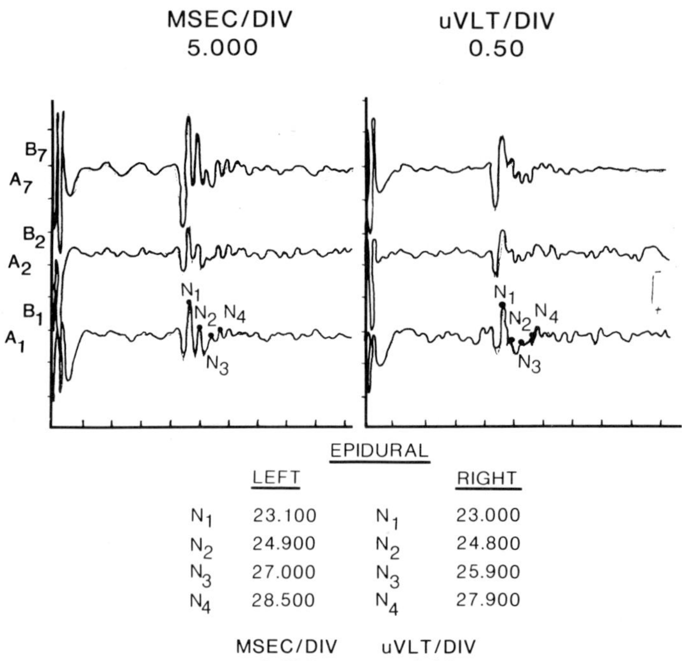

Fig. 5. Changes in EEP amplitude on the left side immediately following distraction. A7 = initial EEP wave form A2 = following distraction. A1 = several minutes later amplitude improved B7, B1, B2 right side recording showed no significant changes.

Fig. 6. Shows configuration of EEP responses at different level of the spine. Note except at L4 level that N1 is markedly larger than other components, the remaining levels show no significant differences in amplitude.

caution before definite conclusions are drawn, since the number of patients investigated in low thoracic and lumbar are small (Table 1).

CSEP responses showed a single initial positive potential with peak latency of 40 milliseconds followed by a large negative, then positive, and negative peaks of 49, 59, and 71 milliseconds respectively (Fig. 7). Latency of responses within the individual cases showed increase of less than 3 milliseconds and amplitude change was not more than 50%, provided all the factors affecting them, such as inhalation anesthetic agents, hypotension (systolic pressure of < 60mm Hg; hypercarbia, PCO_2 more than 42torr), hypothermia (core temperature of less than 35°C) and a stimulation rate of peripheral nerve of more than 3 per second were avoided.

In one case with core temperature of 32.2°C, the cortical response's latency of P1 (P40) increased by more than 5 milliseconds, no changes were noted in the latency and amplitude of epidural responses (N1). This occurred during the time of Harrington rod

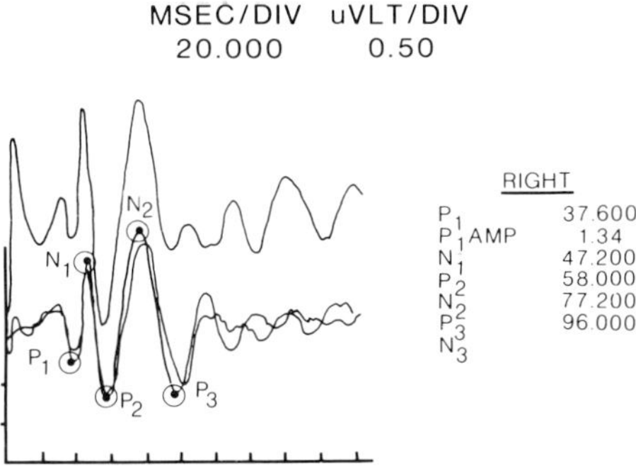

	MEAN	SD	N	RANGE
P_1	40.0	2.00	80	36.0 – 43.4
N_1	50.0	3.10	80	44.6 – 60.6
N_2	60.8	4.80	80	52.0 – 78.6
P_2	71.6	8.50	80	64.0 – 94.0

Fig. 7. Recorded 2 centimeter posterior to Cz and Fz reference in response to simultaneous electrical stimulation of the superficial peroneal and posterior tibial nerves at the ankle. Intraoperative latencies: Subject anesthesized (age 10-72).

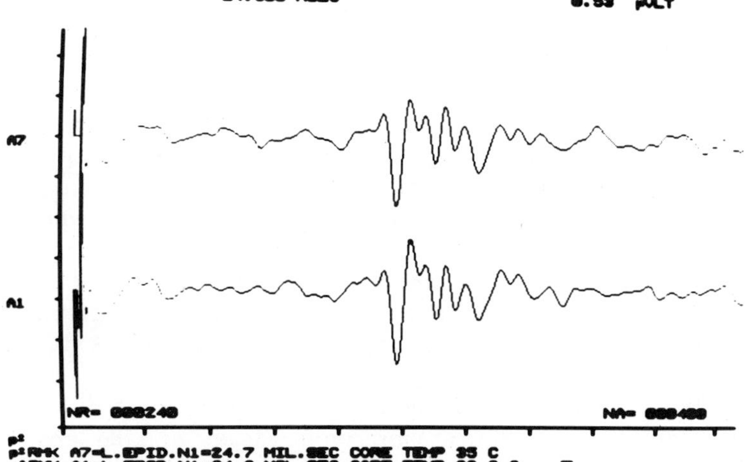

Fig. 8. A7 = initial EEP temperature 35° Centigrade. A1 = core temperature is reduced to 33.2° Centigrade. Note there is no changes in amplitude and latency of EEP evoked response. B2 = core temperature 35° Centigrade. B1 = core temperature 33.6° Centigrade. Note there is an increase latency of more than 6 milliseconds.

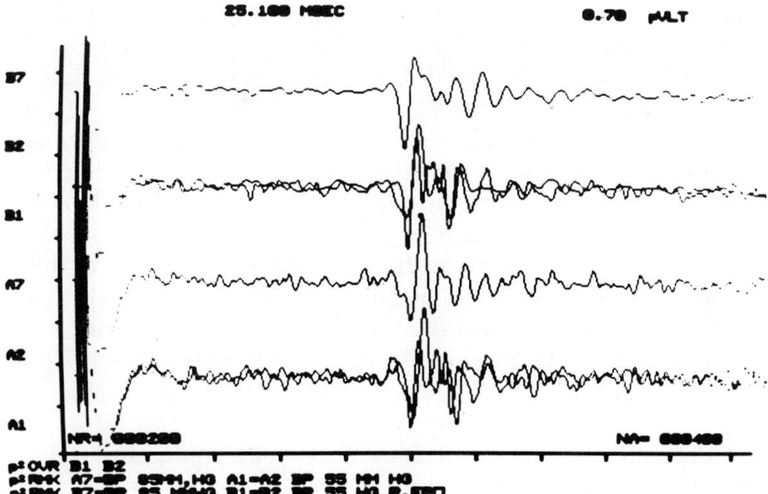

Fig. 9. Effect of hypotension on EEP. A7 = BP: 85mm Hg. A1 = BP 55mm Hg. (left side). Note no changes in amplitude and latency of EEP occurred. B7 = BP 85mm Hg. B1 = BP 55mm Hg (right side).

instrumentation (Fig. 8). With hypotensive anesthesia and a systolic pressure of 55mm Hg, there were no changes in the latency and amplitude of EEP, when CSEP showed significant changes in latency and amplitude with systolic pressure of less than 80mm Hg during the spinal distraction period (Fig. 9). With a use of halogenated inhalation anesthetic agents, there was no change in amplitude and latency of EEP when CSEPs were completely abolished (Fig. 10).

EEPs in this study were found to be more stable with less variability in latency and amplitude values than CSEP. They also allow for greater degree of flexibility in the choice of anesthetic technique since inhalation agents, which affect cortical evoked responses but not epidural evoked responses, can be used.

Discussion

Intraoperative monitoring of the spinal cord during spinal surgery allows for greater safety with less risk to the patient than the wake-up test (25). CSEP was the first method used, but is known to be affected by the following: 1) Inhalation halogenated anesthetic agents; 2) hypotension (1, 4); 3) hypothermia (core temperature less than 35°C) (5); 4) stimulation presentation rates (more than 3/second) (19); and 5) hypercarbia (PCO_2 greater than 42torr).

Data obtained from animal studies suggests that conducted SEPs recorded from the dorsal spinal cord are the result of the conduction in stimulation of the dorsal column ipsilateral to the peripheral nerve stimulation (13). EEPs appear to be generated from the posterior and posterior lateral column of the spinal cord (13, 25). Tsuyama has determined that the initial spike wave (N1) represents the dorsal lateral column and the polyphasic wave (N2, N3, N4) represents the dorsal column (22, 24) and, according to Tamaki and Tsuyama, the polyphasic component is the part that can be affected by anoxia, anesthesia, and the rate of stimulation (22, 23). We had no episodes of hypoxia and therefore cannot comment on its effect on EEP, but stimulation rates up to 20/second and halogenated inhalation anesthetic agents Fentanyl and nitrous oxide did not alter these responses. Whitle cited similar experiences as well (27). According to Bradshaw (1), the early positive-negative-positive complex is generated in the dorsal roots. He supports this hypothesis with the following data.

In his series, this complex was best observed when electrodes were located in the upper lumbar level, placed lateral to the midline at thoracolumar levels, and the second component (N2, N3, N4) complex of negative potentials presumably reflecting postmyoptic activity at root entry level of the spinal cord.

In the study by Jones et al., (11) of epidural, vertebral, and cortical responses, EEP recording was found to be more useful because of greater stability, less susceptibility to anesthesia or fluctuation in blood pressure. EEPs were also recorded with faster stimulation rates and found to have increased amplitude (10, 11). Jones et al., (11), in their study of epidural recording with peripheral nerve stimulation, found the following advantages: 1) Safety factor of only one electrode near the cord; 2) convenience for surgeon; 3) more natural afferent volley; and 4) fewer problems with directly conducted artifacts. Bradshaw et al., (1) in their study of epidural, cortical, and vertebral bone responses found vertebral bone responses to be reproducible, but had a 50% reduction in amplitude compared to epidural. Bradshaw felt epidural recording demonstrated lateralization of the EEP abnormality which correlated with motor function. EEP showed partial recovery with the easing of traction. He also felt that EEP response showed neurological deficit earlier than would have been demonstrated by the wake-up test (1).

Hahn et al., (9) studied alternate stimulation sites, such as cauda equina instead of peripheral nerve. In cases where peripheral nerve stimulation is not possible. It is a vi-

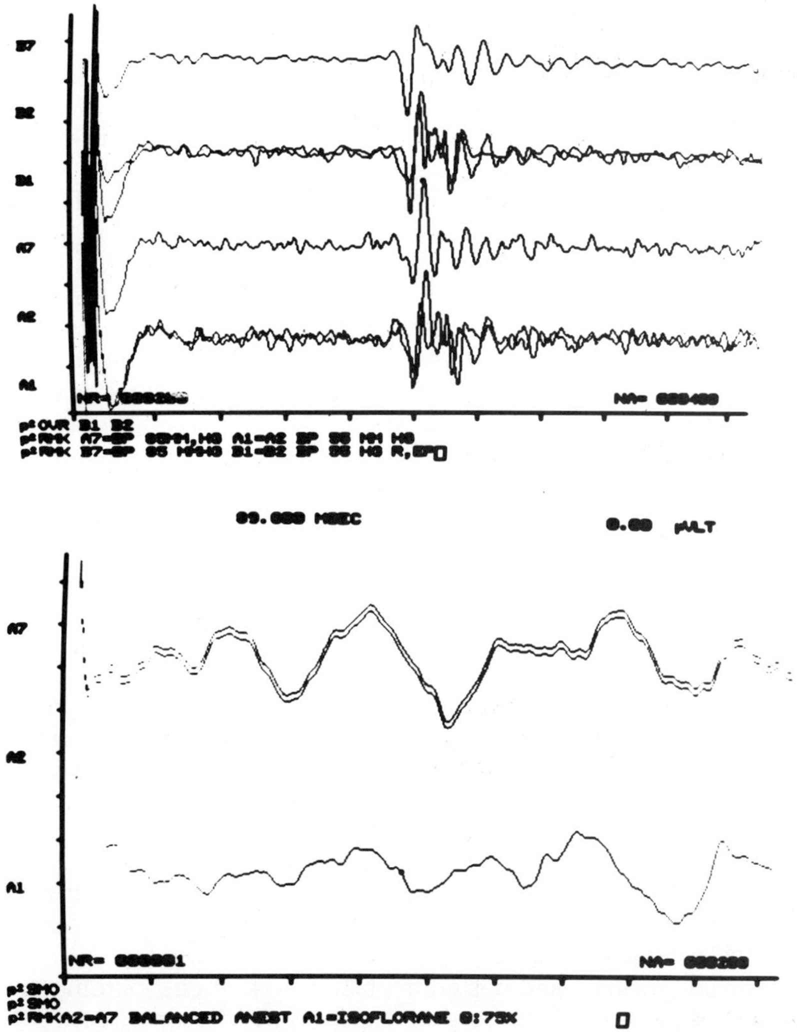

Fig. 10. A7 = A2 CSEP prior to use of inhalation agents. A1 = following use of inhalation agent. B7 = B2, B1 right side, A7, A2, A1, prior and following use of inhalation agents. Note no change in EEP amplitude or latency.

able alternative. The risks include infection, cerebellar or tentorial herniation, hemorrhage, and as a late complication, arachnoiditis (11).

In our series of 40 patients changes in the first peak (N1) in each case from start to the end of procedure did not exceed more than 0.8 milliseconds (average of 0.53 milliseconds) with no postoperative neurological abnormality and we feel this amount of change in latency is acceptable. Changes of more than 0.8 milliseconds may indicate a problem in spinal cord conductivity. Amplitude changes of more than 50% was considered abnormal.

Our study is a preliminary report on the use of EEP using a modified custom made electrode. The electrodes used have the following advantages:

1. They are not needles and are safe to use without risk of complications.

2. They are placed during surgery, so risk of infection is low.

3. With both pickup and reference electrode at the epidural space, EMG artifact from paraspinal muscles is eliminated (bipolar recording) (13).

4. They provide consistency of latency measurement from unilateral lower extremity, peripheral nerve stimulation and allows for lateralization activation from ipsilateral peripheral nerve stimulation (13).

5. Responses have not been shown to be affected by the factors that affect CSEP (11, 13, 21, 27).

6. There is a rapid analysis time (one minute for each two tracings) allowing for a closer monitoring of spinal function.

We have no intention of promoting this technique over CSEP's for all cases of spinal surgery. During the past several years of spinal cord monitoring, we were able to obtain stable reproducible and reliable CSEP responses with no evidence of false positive or negative recordings. In a retrospective study of 157 cases of spinal cord monitoring (14, 15) we found changes in latency of P1 (P40) of more than 3 milliseconds and amplitude decrease of more than 50% to be abnormal. Brown et al., cited similar findings as well (3). The use of CSEP is the easiest way of monitoring when epidural recording either is not applicable or not desired by the surgeon; CSEP could be used, but this requires that rigid attention must be paid to create a constant operating room environment suitable for CSEP. Once this is done, extraneous causes of wave form alterations are eliminated and any changes which do occur can be directly attributable to spinal cord trauma. Epidural recordings have been shown to have greater stability and less variability in latency of responses. This method allows for greater ease of reading, less change in latency and amplitude and greater flexibility in anesthetic technique and less sensitivity to hypotension and, therefore, increased accuracy in diagnosing intraoperative problems. We feel that epidural recording with peripheral nerve stimulation sites offers the best available method, to date, to monitor and evaluate intraoperative spinal cord function during posterior spinal surgery. Further study is necessary to monitor any other possible complication or limitation in this method than we have already mentioned.

Summary

In a series of 44 patients undergoing posterior spinal fusion with instrumentation, both cortical and epidural recordings were done to compare changes in latency and amplitude of the evoked responses.

The superficial peroneal and posterior tibial nerves of one extremity were stimulated simultaneously at the ankle. Cortical responses were recorded 2cm behind the Cz position of the 10/20 international EEG system, and epidural responses were recorded by a specially modified posterior column stimulator electrode (Neuromed Unistem 2) with a 2.5 CMS between active and reference electrode (bipolar recording). The electrodes are pure platinum, each 4mm long and 1.2mm in diameter. They were placed in epidural space one level above the spinal correction. The epidural responses showed a greater degree of stability with little alteration in latency as compared to cortical responses (throughout a procedure, latency changes ranged 2.3-3.0 milliseconds in cortical and 0.2-0.8 milliseconds in epidural recordings). It was found that epidural responses being less affected by factors known to modify cortical responses (e.g., inhalation anesthetic agents) and provide a more reliable functional intraoperative monitoring of the spinal cord. The modified electrode is easy to place and maintain in the operating field, providing clear, reproducible responses.

Table 1

Level breakdown of epidural electrode placement

T1 level	1 case
T2 level	2 cases

T3 level	15 cases
T4 level	9 cases
T5 level	4 cases
T7 level	1 case
T8 level	2 cases
T9 level	4 cases
L1 level	1 case
L4 level	1 case
Total:	40 cases

References

1. Bradshaw, K.; Webb, J.K.; Fraser, A.M.: Clinical evaluation in spinal cord monitoring in scoliosis surgery. Spine, Vol. 9, 6: 636-643, 1984.
2. Brown, R.H.; Nash, C.L., Jr.: Current status of spinal cord monitoring. Spine, 4: 466-470, 1979.
3. Brown, R.H.; Nash, C.L., Jr.; Berraile, J.A.; Maddio: Cortical evoked potentials monitoring, a system for intraoperative monitoring of spinal cord function. Spine, 9: 256-261, 1984.
4. Bunegrin, L.; Albin, M.D.; Helsel, P.; Herrera, R.: Cerebral blood flow and the evoked responses. J. Cerebral Blood Flow Metabolism, S-226-227, 1981.
5. Desmedt, J.E.: Somatosensory cerebral evoked potential in man: Handbook of electroencephal. In: W.A. Cobb (ed.), Cl. Neurophysiol., Vol. 9, pp. 55, Elsevier, Amsterdam, 1979.
6. Engler, G.L.; Spielholtz, N.I.; Berneard, W.N.; Donsinger, F.; Meskin, H.; Wolf, T.: Somatosensory evoked potentials during Harrington instrumentation for scoliosis. J. Bone Joint Surg., 60.A: 528-532, 1978.
7. Grundy, B.L.; Heros, R.C.; Teng, A.S.; Doyle, E.: Intraoperative hypoxia detected by evoked potentials monitoring. Anesth. Analg. (Clevel), 60: 437-439, 1981.
8. Grundy, B.L.: Monitoring of sensory evoked potentials during neurosurgical operations: Methods and applications. Neurosurg., Vol. 11, 4: 556-575, 1982.
9. Hahn, J.F.; Lesser, R.; Klem, G.; Lueders, H.: Simple techniques for monitoring intraoperative spinal cord function. Neurosurg., 9: 692-695, 1981.
10. Jones, S.J.; Edgar, M.A.; Ransford, A.O.: Sensory conduction in the human spinal cord: Epidural recordings made during scoliosis surgery. J. Neurol. Neurosurg. & Physchiat., 45: 446-451, 1982.
11. Jones, S.J.; Edgar, M.A.; Ransford, A.O.; Thomas, N.P.: A system for the electrophysiological monitoring of the spinal cord during operation for scoliosis. J. Bone Joint Surg., Vol. 65, B, 2: 134-139, 1983.
12. Lueders, H.; Hahn, J.; Gurd, A.; Tsuji, S.; Dinner, O.; Lesser, R.; Klern, G.: Surgical monitoring of spinal cord function: Cauda equina stimulation technique. Neurosurg., Vol 11, 4: 482-485, 1982.
13. Macon, J.B.; Poletti, C.E.: Conducted somatosensory evoked potentials during spinal surgery: Part 1 and part 2. J. Neurosurg., 57: 349-359, 1982.
14. Nainzadeh, N.K.; Lane, M.E.; Graham, J.J.; Neuwirth, M.; Bernstein, R.: Somatosensory evoked potentials (SSEP) obtained by simultaneous stimulation of the superficial peroneal and posterior tibial nerves as indicator of spinal cord function during spinal surgery. Electroencephalogr. Cl. Neurophysio., Vol. 56, 3: P.S. 140-141, 1983.
15. Nainzadeh, N.K.; Lane, M.E.; Graham, J.J.; Neuwirth, M.; Bernstein, R.: Somatosensory evoked potentials -- Intraoperative indicators of spinal cord function: A clinical analysis of 157 surgical spinal procedure. Orthop. Transaction (J. Bone Joint Surg.), Vol. 8, 1: 160-161, 1984.
16. Nash, C.L., Jr.; Loring, R.A.; Schatzinger, L.A.; Brown, R.H.: Spinal cord monitoring during treatment of spine. Clin. Ortho. Related Research, July-August, 126: 100-105, 1977.
17. Nordwal, A.; Anelgoard, J.; Harada, Y.; Volencia, P.; McNeil, D.R.; Brown, J.C.: Spinal cord monitoring using evoked potentials recorded from feline vertebral bone. Spine, Vol. 4, 6: 486-494, 1979.
18. Nuwer, M.R.; Dawson, E.C.: Intraoperative evoked potentials monitoring of the spinal cord: Enhanced stability of cortical recording. Electroenceph. Neurophysiol., 59: 318-327, 1984.
19. Pratt, H.; Politoske, D.; Starr, A.: Mechanically and electrically evoked somatosensory evoked potentials in humans: Effects of stimulus presentation rate. Electroencephal. Cl. Neurophysio., 48: 312-317, 1980.
20. Sherwood, A.: Characteristics of somatosensory evoked potentials recorded over the spinal cord and brain of man. IEEE Transaction on Biomed. Engineer, Vol. BME-28, 7: 481-487, 1981.
21. Spielholtz, N.I.; Benjamin, M.V.; Engler, G.L.; Ransohoff, J.: Somatosensory evoked potentials during decompression and stabilization of the spine: Methods and findings. Spine, Vol. 4, 6: 500-505, 1979.

22. Tamaki, T.: Basic analysis of evoked spinal cord potentials elicited by direct spinal cord stimulation. In: C.L. Nash (ed): Proceedings of the Spinal Cord Monitoring Workshop. Data Acquisition and Analysis. Cleveland, Case Western, Reserve University, pp. 17-24, 1979.
23. Tamaki, T.; Noguchi, T.; Takano, H.; Tsuji, H.; Nakagaway, T.; Imai, K.; Inoue, S.: Spinal cord monitoring as a clinical utilization of the spinal evoked potentials. Clin. Ortho . Related Research, 184: 58-64, 1984.
24. Tsuyama, N.; Tsuzumi, N.; Kurokawa, T.; Imai, T.: Clinical application of spinal cord action potential measurement. Int. Orthop. (S. JCOT), 2: 39, 1978.
25. Vauzelle, C.; Stagnara, P.E.; Jouvinoux, P.: Functional monitoring of the spinal cord activity during spinal surgery. Clin. Orthop., 93: 173-178, 1973.
26. Walter, W.G.: Evoked response general. In: Van Leeawen vs. Lopes da Silva FH, Kamp A (ed.): Handbook of Electroencephalog. and Cl. Neurophysio.: Evoked Responses. Amsterdam, Elsevier Scientific Publishing Co., Vol. 8A, pp. 20-32, 1975.
27. Whittle, I.R.; Johnston, I.H.; Besser, M.; Taylor, T.K.F.; Overton, J.: Intraoperative spinal cord monitoring during surgery for scoliosis using somatosensory evoked potentials. Aust. NZ, J. Surg., 5-4: 553-557, 1984.

Intraoperative Somatosensory Evoked Potential Monitoring: The Rochester Experience

R. Q. Knight;[*] D. P. K. Chan; D. N. Smith; J. R. Devanny; K. V. Jackman

Introduction

Recent developments in spinal surgery have produced remarkable advances in the management of spinal trauma and deformity. Orthopaedic surgeons today have become more aggressive in their approach to these problems as well. A valuable technique, that has often protected both the patient and surgeon from catastrophic complications, is the monitoring of somatosensory evoked potentials (3, 5, 6). With the recent explosion in computer technology, the once cumbersome task of evaluating the electrical activity of the brain and spinal cord has become "user friendly."

The focus of this presentation is derived from data on forty-eight intraoperative cases involving SEP monitoring. This study, by the Department of Orthopaedic Surgery at the University of Rochester, was conducted from January 1, 1985, to June 1, 1986. The patient population included thirty females and twelve males with an average age of 34.1 years, ranging from 17 months to 71 years.

Our interest in the uses of SEP monitoring was stimulated by our knowledge of the spinal cord's sensitivity to ischemia and the evoked potential's sensitivity in detecting such ischemic changes (2, 4, 7). Our colleagues, in Neurology, have used the summation of electrical potentials produced by cortical activity for decades as the EEG. The canine

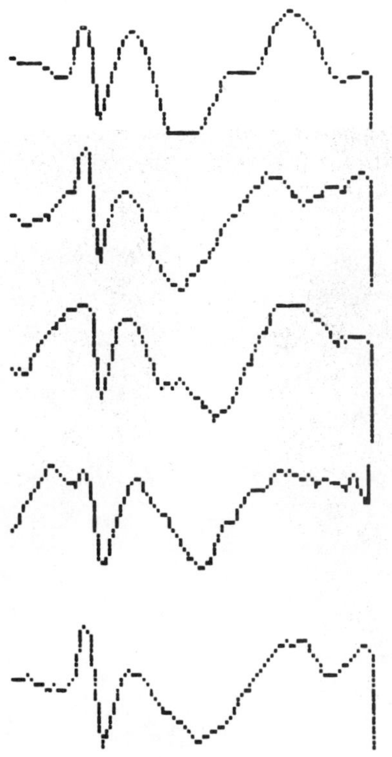

Fig. 1. Normal SSEPs produced by the Siegen Neuroscope. Upward deflections are positive and downward deflections negative.

* New York Medical College, Westchester County, 503 Grasslands Road, Valhalla, NY 10595

model for spinal cord injury has been used with repeated success in demonstrating various modes of stimulation and reception for evoked potentials, however, the precise meaning of the potentials often remains an enigma (1, 7).

Anatomically, the spinal cord possesses a segmental blood supply. Radicular vessels which branch from the aorta primarily, anastomose at the level of the intervertebral foramen to produce the singular anterior and dual posterior spinal arteries. With the advent of more frequent anterior spinal surgery, the disruption of this segmental blood supply is a valid concern. Similarly, posterior procedures that include distraction instrumentation should not be undertaken lightly, especially in the elderly patient.

The world of evoked potential monitoring for spinal surgery is largely divided into two categories, but the choices available within those categories are numerous.

Evoked potential testing

Stimulation sites	Monitoring sites
Somatosensory	Cortical
Spinal	Spinal
	Epidural
	Bony
	Monopolar
	Bipolar

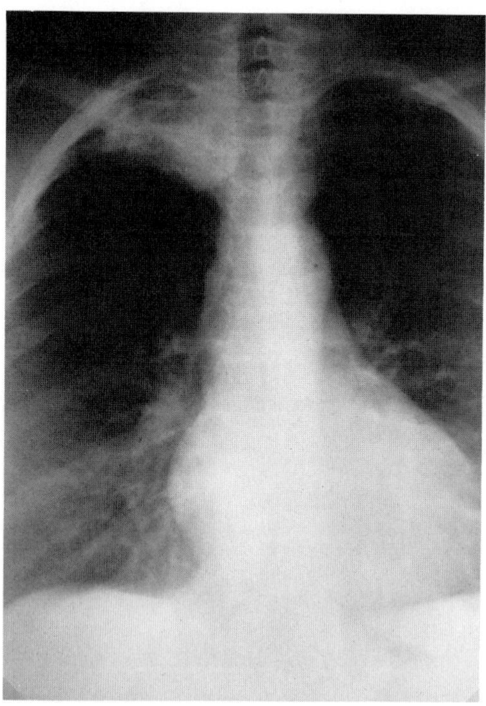

Fig. 2. A-P chest radiograph demonstrating blastic lesion in right upper lung field, preoperatively.

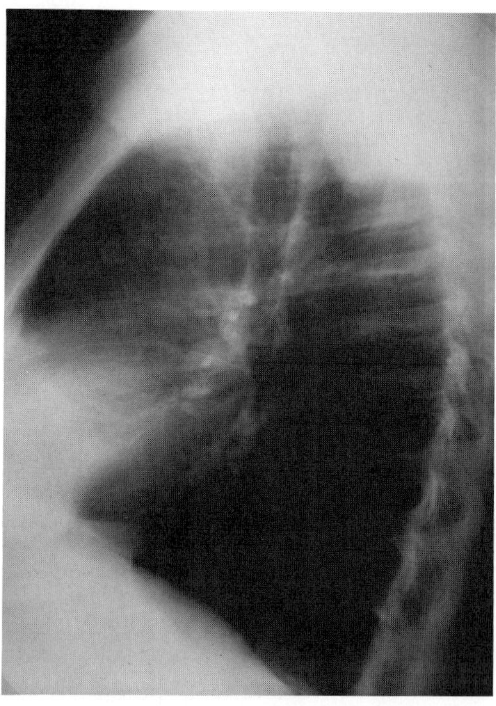

Fig. 3. Lateral radiograph of right sided thoracic lesion, preoperatively.

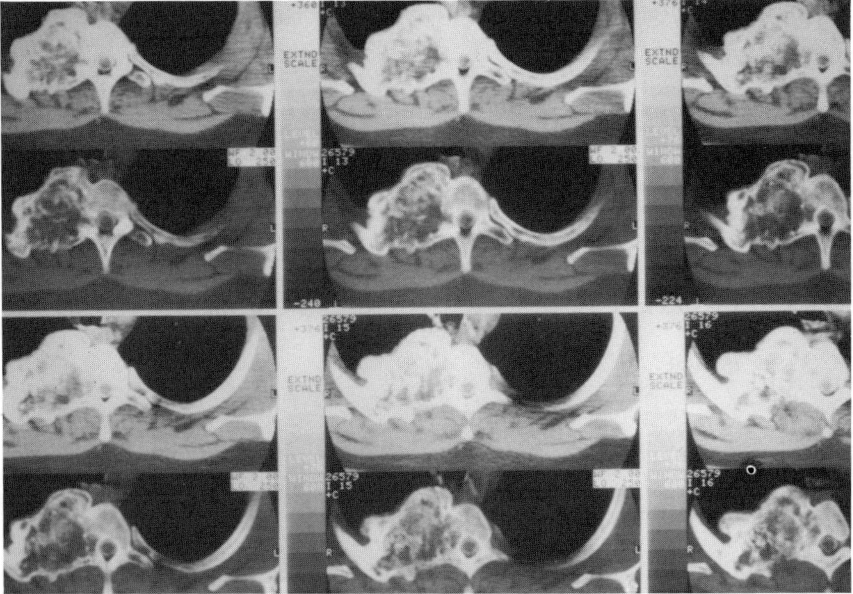

Fig. 4. C.T. scan of thoracic spine at T3 through T4 level. Note the osteoblastic lesion involving the body, transverse process and rib at the affected levels.

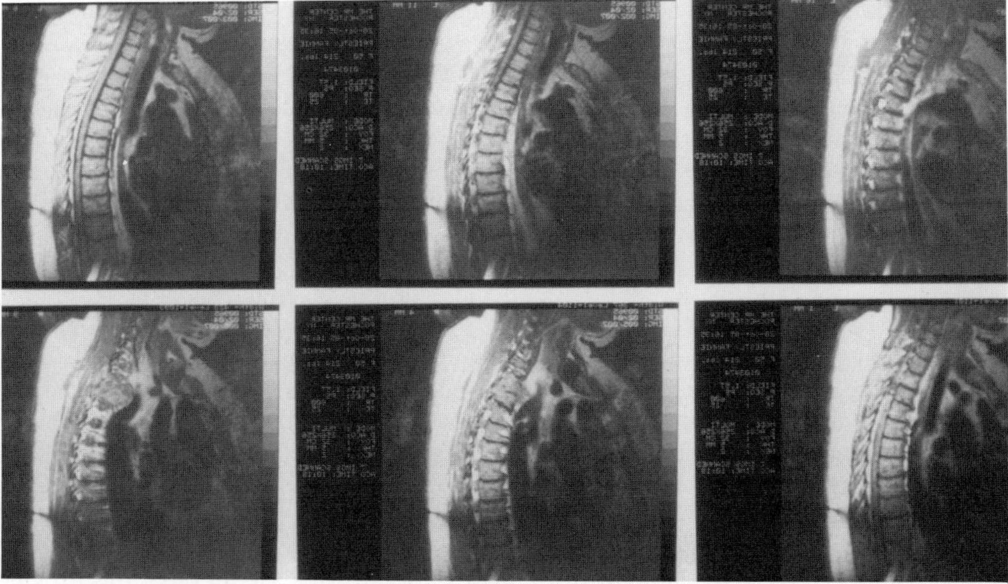

Fig. 5. M.R.I. of the thoracolumbar spine. Note the multi-level involvement.

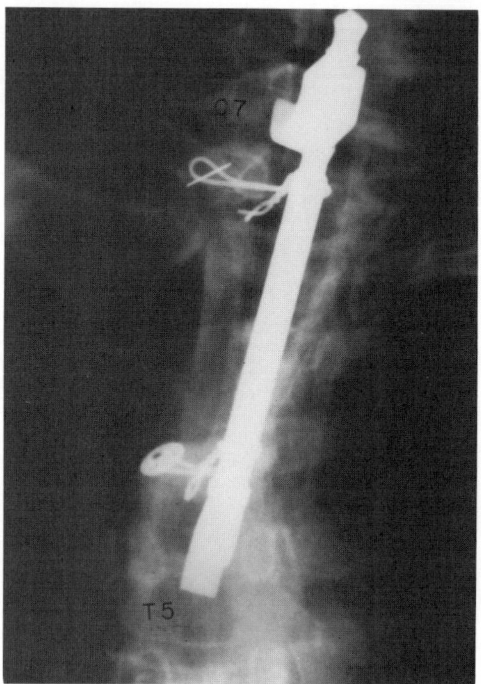
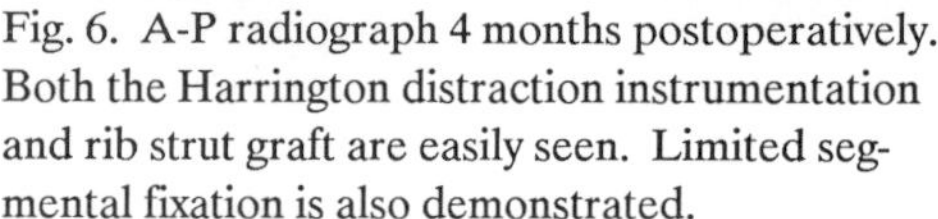
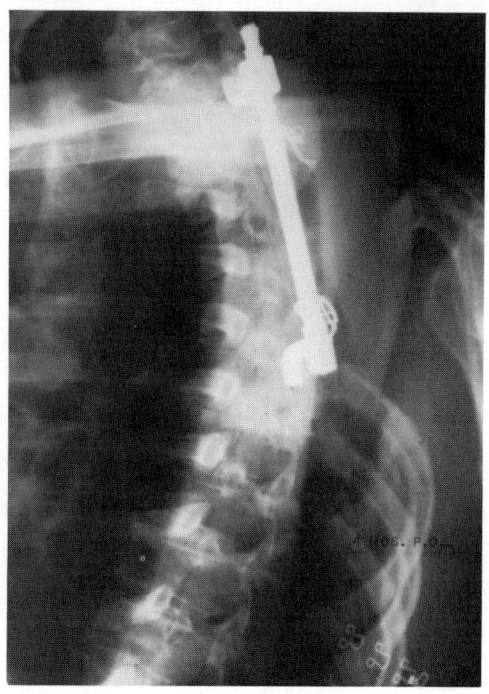

Fig. 6. A-P radiograph 4 months postoperatively. Both the Harrington distraction instrumentation and rib strut graft are easily seen. Limited segmental fixation is also demonstrated.

Fig. 7. Lateral radiograph 4 months postoperatively.

The versatility of SEP monitoring allows for its adaptation to most clinical situations. Both systems have their proponents and opponents. The bottom line is that if used appropriately, either system can be effective and a combination of the two systems is often useful. The peripheral-cortical mode is criticized for more frequent medication interference, especially halogenated inhalation agents, while the spinal-spinal mode is criticized for being invasive, albeit minimally. The quality of the recordings is largely a technical matter and is based on the secure placement of electrodes and the sophistication of the computer program. The data generated here was produced by peripheral stimulation of the posterior tibial nerve at the medial malleolus bilaterally and cortical summation at FZ and CZ, using a Siegen Neuroscope. This program tabulated the response to two hundred consecutive peripheral stimuli. We had the capacity to vary the number, duration, and intensity of the stimulus (Fig. 1). The studies were initiated with an intensity of 14 to 16 milliamps and a duration of 200 microseconds. Computer printouts are then available in both monitor and hardcopy for interpretation by the physicians involved. In the elderly patient, whose neural pathways may be muted by the aging process, a stimulus of greater intensity and duration is usually required. The contrary is often true in the pediatric population, whose superimposed EEG may obscure the underlying SEP. Intraoperative SEPs are designed to replace the Stagnara wake-up test. In our experience, normal triple phase SEPs were sufficient to preclude intraoperative wake-up.

Our study has used triple phase monitoring to complete one test. The patients are studied preoperatively, intraoperatively and postoperatively. We feel that this provides additional valuable information in assessing significant changes in the intraoperative phase of recording. The case load was heavily weighed with orthopaedic (43) monitorings, although the neurosurgical (3) and vascular (2) services were also involved. Orthopaedically, the operative experience was quite varied and included both spinal deformity and trauma.

Orthopaedic operative experience

<u>Scoliosis (34)</u>   <u>Trauma/Deformity (9)</u>
Adult (18)   Kyphosis (6)
Idiopathic (10)   Fracture (2)
Neuromuscular (5) Tumor (1)
Congenital (1)

Anterior procedures (17)
Posterior procedures (26)

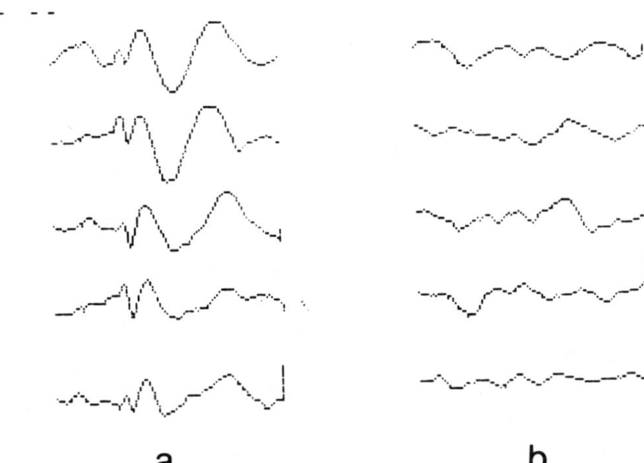

a   b

Fig. 8. Each tracing depicted here is the product of 200 individual stimuli. They are computed and displayed every 2 minutes. The frequency, intensity, and duration are constant in these tracings. (A). Normal preoperative tracing. (B). Abnormal intraoperative tracing. Stimuli intensity was increased in (B).

The logistics of providing SEP monitoring did not add significantly to the perioperative or operative time. This technique, however, does require a reasonable degree of communication and cooperation between the monitoring team, i.e., the surgeon, the anesthesiologist, and in our situation, the neurologist. In our hands the aberrations from medication induced artifacts were minimal, affecting only one out of the forty-eight monitorings (2.1%). Three patients demonstrated abnormal SEPs during all three phases of monitoring. The abnormal studies were correlated clinically with neurologic deficits secondary to spinal cord injury. Eight patients (16.7%) experienced intraoperative SEP changes with two cases resulting in persistent postoperative abnormality.

Causes of intraoperative SEP changes

Fluctuations in blood pressure (3)
Cord irritation (3)
Halogenated anesthetic (1)
Electrical interference (1)

In the cases of persistent postoperative changes, one patient was a C7 quadriplegic and no alteration was noted in his postoperative neurologic examination. However, the second patient was a healthy fifty-five year old female who was paraplegic postoperatively. The intraoperative insult to the cord was thought to be direct contusion in both cases. During the first case an interbody fusion was performed at C6 through C7, and with placement of the tricortical bone plug, a drop in amplitude was noted on his SEPs.

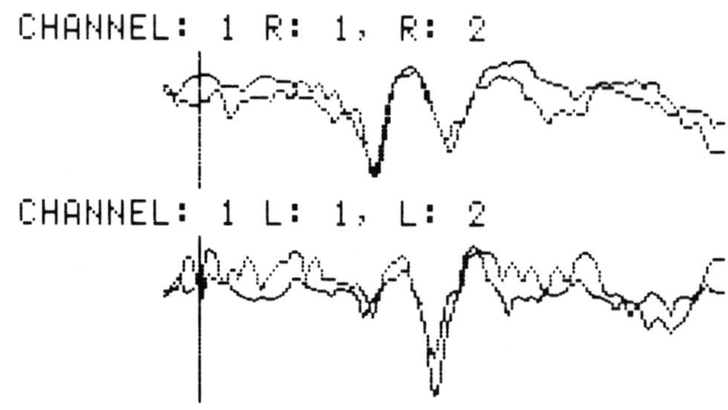

Fig. 9. Two months status post intraoperative loss of SEPs and postoperative paraplegia. Both wave forms are abnormal due to the increased latency. The presence of SEPs suggest neural recovery which was correlated clinically.

The situation involved the resection of a locally aggressive osteochrondrama of T3 (Figs. 2, 3, 4, 5). Following exposure of the lesion from a combined anterior-posterior approach, two-thirds of the T3 vertebra required resection including a portion of T2 and T4. The bodies were resected from a posterior to anterior direction, using a osteotome (Figs. 6, 7). Although the pressure applied to the cord appeared minimal at the time, following resection her SEPs were noted to have diminished amplitude and increased latency (Fig. 8). During the postoperative period serial monitorings were used to assess the return of her neurologic function (Fig. 9).

The data complied from this study demonstrates that cortical evoked somatosensory potential monitoring has, in our experience, been an effective means of assessing neurologic function in the intraoperative setting. Of the forty-eight triple phase studies reported here, forty-four (91.7%) were normal postoperatively and consequently so were the neurologic examinations. The four remaining patients (8.3%) had abnormal neurologic function. Although the information rendered is not always clear cut, our experience with SEP monitoring has been a positive one. The adverse affects of anesthetic agents have not been a major problem. SEPs can be utilized to monitor a patient's neurologic recovery and normal triple phase SEPs negate the necessity for intraoperative wake-up testing.

References

1. Cohen, A.R.; Young, W.; Ransohoff, J.: Intraspinal localization of the somatosensory evoked potential. Neurosurg, Vol 9, No. 2, pp. 157-162, August, 1981.
2. Coles, J.G. et al.: Intraoperative detection of spinal cord ischemia using somatosensory cortical evoked potentials during thoracic aortic occlusion. Ann Thoracic Surg, Vol. 34, No. 3, pp. 299-306, September, 1982.
3. Keim, H.A. et al.: Somatosensory evoked potentials as an aid in the diagnosis and intraoperative management of spinal stenosis. Spine, Vol. 10, No. 4, pp. 338-344, 1985.
4. Laschinger, J.C.: Detection and prevention of intraoperative spinal cord ischemia after cross-clamping of the thoracic aorta: Use of somatosensory evoked potentials. Surgery, Vol. 92, No. 6, pp. 1109-1116, December, 1982.
5. Nash, C.L. et al.: Spinal cord monitoring during operative treatment of the spine. Clinical Orthopaedics and Related Research, No. 126, pp. 100-105, 1977.
6. Perot, P.L.: The clinical use of somtasensory evoked potentials in spinal cord injury. Clin Neurosurg, Vol. 20, pp. 367-381, 1972.
7. Satomi, K.; Nishimoto, G.I.: Comparison of evoked spinal potentials by stimulation of the sciatic nerve and the spinal cord. Spine, Vol. 10, No. 10, pp. 884-890, 1985.

Somatosensory Evoked Potentials (SEP) Intraoperative Monitoring during Cranial Vertebral Compression and Instability

D. E. McDonnell;[*] H. F. Flanigin; H. G. Sullivan

Abstract

Four patients suffering from advanced myelopathy with severe quadriparesis from mechanical compression of the cervical medullary neuro-axis and instability of the cranial vertebral junction underwent staged surgical decompression and osseous stabilization. These patients were in particular jeopardy for exacerbation of their neurologic impairment due to the anterior compression, posterior encroachment, and instability.

SEP monitoring was performed on these patients continuously through each procedure. The SEPs were generated from median nerve stimulation bilaterally, and were recorded at CZ-C3 and CZ-C4. When the recorded evoked potentials changed, the surgeon was warned, and the surgical maneuvers were noted. These events and their lateralization were noted as they occurred. These changes were subsequently analyzed and collated for each operative procedure. The number of events which occurred during the four anterior decompression procedures ranged from three (3) to sixty (60) events for a total of eighty (80) events. The number of events which occurred during the four posterior fusion procedures ranged from two (2) to fifteen (15) events, for a total of forty (40) events. All four (4) patients subsequently exhibited considerable functional improvement over preoperative neurologic function with a follow-up period ranging from nine (9) months to two (2) years. SEP monitoring contributed to these improved functional results by improving the safety of the surgical decompression.

Introduction

Mechanical compression and instability at the craniovertebral junction produces an insidious and relentless myelopathy which, when left unchecked, will eventually lead to quadriplegia and death from respiratory failure (24). The cause may be congenital in the form of a developmental anomaly or acquired from inflammatory, neoplastic, or traumatic destruction of supporting elements. The treatment of such lesions is a multi-staged surgical decompression of the cervical medullary junction of the neuro-axis and stabilization of the craniovertebral junction (25, 33). There is a great hazard for aggravating the injury and deficit by such surgical manipulation (29).

We present four patients with advanced myelopathy due to cranial vertebral compression and instability whose somatosensory evoked potentials (SEP) were monitored during each of their spinal procedures (7, 11, 14). Extraneous conditions that would

---

[*] Section of Neurological Surgery, Medical College of Georgia, Augusta, GA 30912

otherwise alter SEPs such as type of anesthesia, body temperature, irrigation solution temperature, and drugs used were carefully controlled (15, 16, 17, 18, 28).

Alterations in SEPs measured during the procedures were due to actual surgical manipulation (1, 4). The SEP changes were stored on electronic disc and clinical notations made at each incident of change. Each incident of SEP change was later reviewed, analyzed, and tabulated to determine how such information affected the surgery and what untoward effects if any were due to a specific episode of manipulation. The results of this analysis and conclusions are herewith presented.

All four patients suffered from osseous compression and instability at the craniovertebral junction and had an element of basilar impression or a spinal canal diameter narrowed to less than 10mm anterior-posterior at C1-CZ. Normally that diameter is greater than 20mm (13, 27). Likewise all four patients were incapacitated by advanced myelopathy. The cause of this functional disturbance was due to focal reduction of cross sectional area of both the neural axis-cylinders of the medulla and cord as well as the vasculature supplying them (4, 12). The lesions affecting these patients were all chronic so that the posterior spinal ligaments were hypertrophic and densely adherent to the adjacent dura (25).

These patients were in serious jeopardy from 1) anterior compression, 2) posterior encroachment, and 3) instability (24, 25, 26, 33).

The rationale for treatment was surgical correction of the pathological tension, distortion, and compression of the neuro-axis so as to re-establish neural conductivity and to renew regional blood flow. A multistaged treatment plan used in these patients to accomplish this is outlined as follows (25):

A. Attempted reduction by cervical traction, with demonstrated fixation.
B. Tracheostomy/feeding gastrostomy or jejunostomy.
C. Anterior resection of encroaching structures (bone and ligaments).
D. Immobilization in craniocervical traction between stages.
E. Occipito C1-CZ-C3 rib graft fusion by posterior approach.
F. Halo brace immobilization for six (6) months.

Because these lesions had been chronic, persisting from one (1) to twelve (12) years, the malalignment could not be reduced by skull traction. They were likewise debilitated because of the advanced stage and chronicity of their myelopathy, and adequate nutrition had to be maintained during the immediate postsurgical convalescent period. Those who did not have gastrostomy and tracheostomy before the anterior resection required these procedures afterward under more trying circumstances. After airway and nutrition routes were established, definitive treatment could be implemented. This was initiated by anterior resection of the compressive elements, either via the trans-oral (3, 13, 30) or the trans-cervical route (32). For this decompression to be complete, the exposed anterior aspect of the dural sac must bulge and pulsate into the excision exposure (25, 30).

The posterior longitudinal, tectorial, and alar ligaments were found to be thickened and tenaciously adherent to the adjacent dura. The dens and lateral masses were, by definition, encroaching deeply into the canal anteriorly, and became more mobile as their resection progressed. Even slight movement of these structures transmitted forces to the underlying spinal cord; this was reflected by changes in the SEP's. The safety of such surgical maneuvers was enhanced by using microsurgical technique under magnification from the operating microscope (27), a high speed drill, and a C-arm fluoroscope (30). The anterior rim of foramen magnum, dens, and rostral 3/4ths of C-2 body and adjacent posterior ligaments were resected, so that the dura bulged and pulsated into the resection site.

Postoperative position and instability was controlled between stages by skull traction with the patient supine and in mild extension.

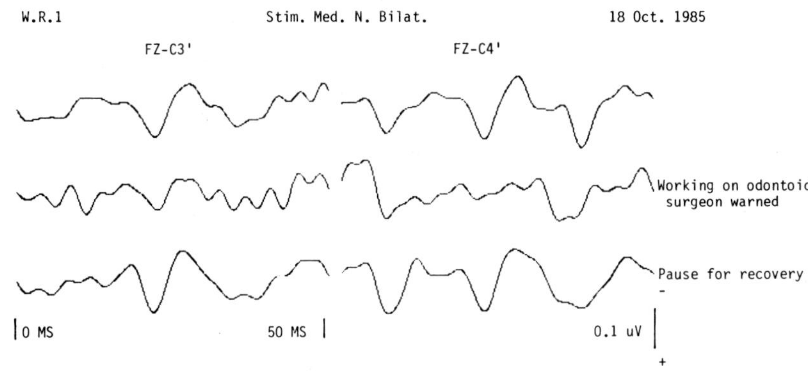

Fig. 1. Patient 3. Decreased cortical evoked responses while working on odontoid with recovery.

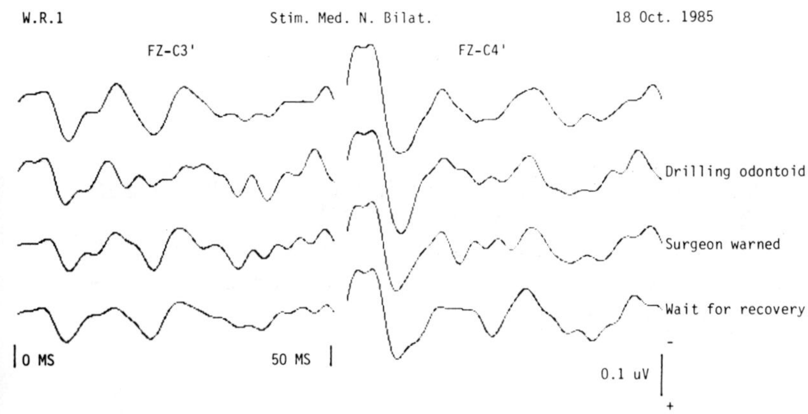

Fig. 2. Patient 3. Decreased cortical evoked responses while working on odontoid with recovery.

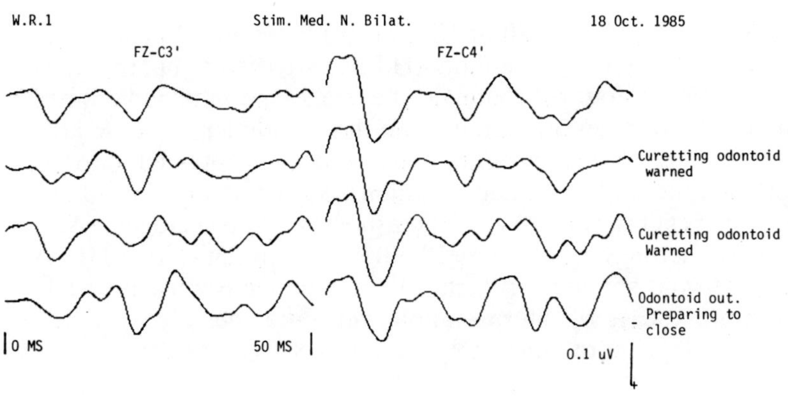

Fig. 3. Patient 3. Decreased cortical evoked responses while working on odontoid with recovery.

After recovery from the anterior decompression, usually one or two weeks, the posterior stabilization procedure was performed. This involved incorporation of the occiput with the posterior elements of C1-CZ and C-3. The preferred graft material was autogenous bone, either from rib or iliac crest. The authors' preference is rib graft secured to the laminae and posterior occiput with doubled twisted #28 wire.

Immediately following surgery, the patient was placed in a halo orthotic device for external stabilization. Each patient was so maintained for six (6) months when osseous incorporation of the grafts was achieved (25, 26, 33).

Intraoperative evoked potential monitoring

Intraoperative monitoring of somatosensory evoked potentials (SEP) reflects electrophysiologic evidence of neuro-axial functional integrity (1, 4, 6, 7, 10, 11, 12, 15, 28, 31). Alterations in response latencies and amplitude serves as a warning that structural and functional changes are occurring during surgical manipulation (1, 4, 7, 10, 11, 12, 15, 28, 31). Even minor changes are significant, i.e., the neural tissues are already severely compromised (14, 16, 18, 19, 29).

Methods and instrumentation

For intraoperative somatosensory evoked response monitoring during operations at the medullospinal junction the following technique is used.

Stimulation.

Bilateral stimulation of the median nerve at the wrist is carried out using needle electrodes (*Grass Instruments Co., Quincy, MA), with a minimal interelectrode distance of 3.0cm, the cathode being proximal (8, 9, 16, 17, 18, 22, 23).

Either the Grass (*) G10DECMA two channel stimulator or the Nicolet (#) SM300 stimulator is used to deliver an isolated stimulus of 200µs duration at 4.7-7.7 per second. The current is less than 20mA (22).

Recording.

Spiral fetal scalp monitoring electrodes are used for two channel recording, referring C3' and C4' to CZ or Fz (20, 21). Impedance is below 5,000ohms.

Averaging.

Averaging is done on a Nicolet (#) Med 80 or Nicolet (#) Pathfinder I computer. In either case, programs written by one of the authors (H.F.F.) are used to operate the computer. Continuous acquisition of four (4) channels of averages are obtained on the Med 80, and two (2) channels of averages are obtained on the Pathfinder I. These are designed to sequence in plotting so that a continuous record is obtained during monitoring, and appropriate time and comment notations can be made, including lateralization. The sweep time is 50ms, with 100 to 500 responses being averaged. Artifact rejection is used. Filter settings range from 10-30Hz LFF to 250-1KHz HFF. Sweep delay may be used to avoid stimulus artifact. We had previously used digital smoothing frequently to improve ease of interpretation, but more recently we have found that digital filtering provides a more satisfactory record (5, 8, 9, 17, 18, 20).

Interpretation.

Interpretation of the recordings has required almost constant observation. The patients served as their own controls, since alterations in function were already present (16, 18). A significant increase in latency or a decrease in amplitude of more than 50%

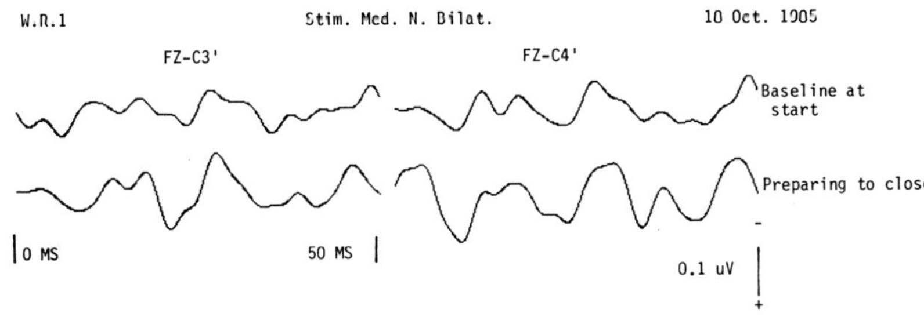

Fig. 4. Patient 3. Comparison of cortical evoked responses at start and end of anterior decompression.

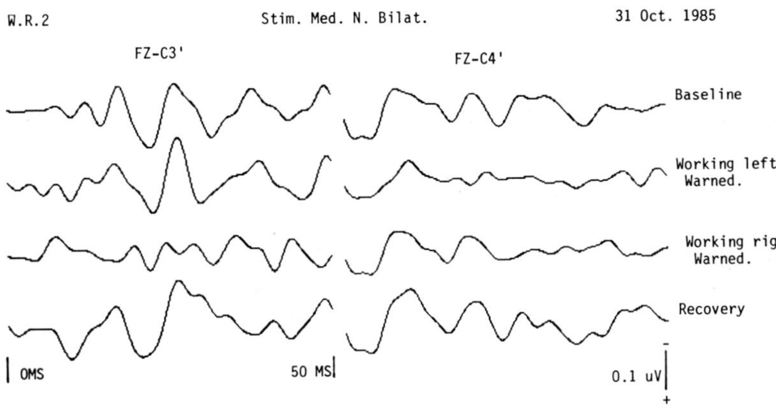

Fig. 5. Patient 3. Shifting decrease in cortical evoked response opposite side of dissection of laminae.

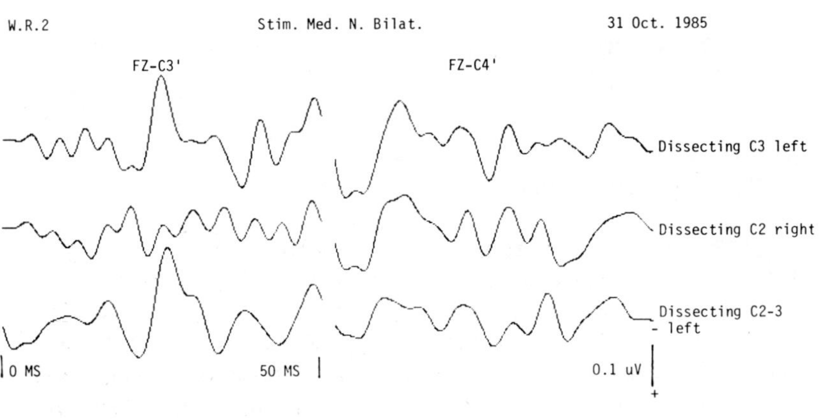

Fig. 6. Patient 3. Decreased cortical response on left during dissection under right C2 lamina.

(fifty percent) was reported to the surgeon as a warning including lateralization (14). A frequent dialogue between surgeon and monitor permitted meaningful correlation between steps in the surgery and the observed evoked response events (16).

Clinical presentations

All four (4) patients presented with advanced myelopathy. Three (3) were bed-ridden and totally incapacitated for self-care, and one could only walk a few steps. They all suffered from neural compression at the cervical medullary junction which was aggravated by craniovertebral instability (Table 1).

Patient #1. M.B. is a 34-year-old black female with a seven year history of progressive quadriparesis and a spastic neurogenic bladder. She had urinary incontinence and wore a diaper. She was incapable of self care and spent most of her time in bed, although she could stand independently and could walk short distances with assistance. Examination revealed her to have a web neck and low hair line over her posterior neck. She stood with a stooped posture. Her grip strength was 50% of normal. Her joint sensation was present in all limbs. She was hyperreflexic throughout, and her plantar responses were extensor bilaterally. Her gait was spastic, unsteady, and required assistance.

Her diagnosis was platybasia and basilar impression with assimilation of C-1 with the occiput. Additionally there was an "Os Odontoideum" with anterior translation of C-1-Occiput on C-2 associated with telescoping of C-2 into the spinal canal. This deformity was not reducible by skull traction.

Surgical procedures.

Oct. 26, 1984 - Tracheostomy/Gastrostomy.
Nov. 12, 1984 - Transoral resection of C-1 and C2; Cervical traction.
Dec. 7, 1984 - Posterior fusion Occiput to C-1-2-3; Halo brace application.

Complications.

Atelectasis of the right lung and pneumonia; mental confusion and agitation; all cleared.

Follow-up.

Sept., 1986 - Walks with cane; controls urine; combs hair; capable of some self-care.

SEP warnings.

Nov. 12, 1984 - Five (5) separate events.
Dec. 7, 1984 - Eleven (11) separate events.

Patient #2. M.A. is a 45-year-old black female with a twelve (12) year history of sever rheumatoid arthritis and eighteen (18) month history of neck pain and spastic quadriparesis, which had advanced to where she was bed-ridden. Examination revealed her to have marked weakness in both upper and lower limbs. Weak grip prevented her from feeding herself. She had unsustained antigravity strength in both upper and lower limbs. Her joint sensation was intact. She had advanced rheumatoid joint deformities in all of her limbs.

Her diagnosis was rheumatoid erosion of the odontoid process and the adjacent support ligaments with anterior translation of C-1 on C-2, associated with telescoping of C-2 into the spinal canal resulting in basilar impression (26).

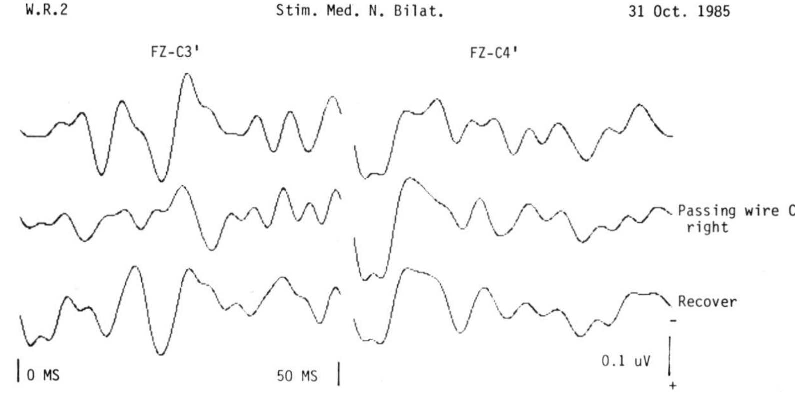

Fig. 7. Patient 3. Decreased left cortical evoked response during passage of wires under right C4 lamina with recovery.

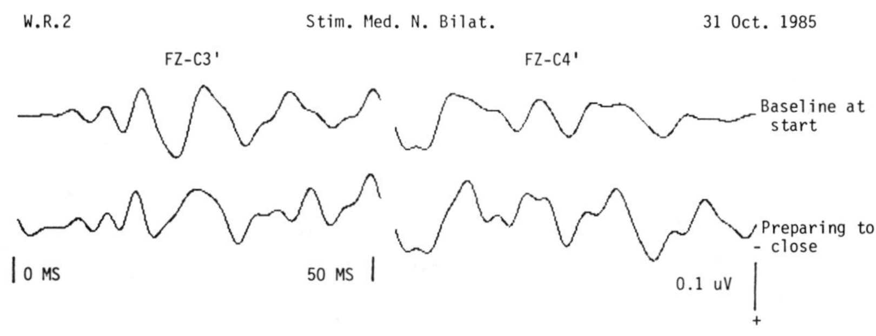

Fig. 8. Patient 3. Comparison of evoked responses at start and end of posterior wiring and fusion.

Surgical procedures.

Mar. 27, 1985 - Tracheostomy/Gastrostomy.
Apr. 1, 1985 - Transoral resection of C-1 and C-2; maintained in skull traction.
Apr. 18, 1985 - Posterior fusion Occiput to C-1-2-3; Halo brace application.

Complications.

None

Follow-up.

May 2, 1986 - Walks 250 feet with a walker; feeds herself well; persistent clonus and Hoffman sign.

SEP warnings.

Apr. 1, 1985 - Three (3) separate events.
Apr. 18, 1985 - Two (2) separate events.

Patient #3. W.R. is a 60-year-old white alcoholic female with a nine (9) month history of headache, neck pain, and progressive quadriparesis. This began when she refused treatment for a type III fracture of C-2 after a fall down a flight of stairs. Initially the fracture was nondisplaced and she was neurologically intact. She was lost to follow-up until returned by her son because she gradually had become severely quadriparetic, bed-ridden and unable to care for herself. She had sustained several falls prior to admission to the hospital. Examination revealed her to have a torticollis to the right. There was a partial Brown-Sequard Syndrome on the left. she was able to overcome gravity in the lower limbs, with the right side being stronger. Her joint sensation as well as sphincter control were intact. There was hyperreflexia throughout and her plantar response was absent on the right and extensor on the left.

Her diagnosis was a chronic type III odontoid fracture with a nonreducible (7) mm anterior dislocation seven of odontoid C-1 on body of C-2. There was also a segmentation defect of C-2 and C-3, a Klippel-Feil deformity. There was a heavy, fibrotic, retroodontoid pannus-like scar seen by magnetic resonance as a soft tissue density.

Surgical procedures.

Oct. 18, 1985 - Transcervical resection odontoid and C-2 while preserving most of C-1; maintained in cervical traction.
Oct. 31, 1985 - Posterior cervical fusion C-1-2-3-4; Halo brace application.

Complications.

CSF cutaneous fistula from cervical wound; organic mental syndrome with confusion and agitation; pneumothorax after harvesting rib graft requiring a tube thoracostomy.

Follow-up.

April, 1986 - Walks independently; cares for herself; neurologically intact.

SEP warnings.

Oct. 18, 1985 - Twelve (12) separate events.
Oct. 31, 1985 - Twelve (12) separate events.

Patient #4. D.D. is a 31-year-old disabled white male with a four (4) year history of progressive stiffness and clumsiness which eventually caused him to be fired from his job as a baker. He was able to walk independently but was generally spastic and hyperreflexic. His plantar responses were extensor bilaterally, with spontaneous clonus. There was a partial Brown-Sequard Syndrome on the left. His sphincter control was intact.

His diagnosis was os odontoideum with cranial setting and basilar impression.

Surgical procedures.

Feb. 3, 1986 - Transcervical resection of os odontoideum, C-1, and rostral 2/3 C-2.
Feb. 5, 1986 - Tracheostomy/Gastrostomy
Feb. 11, 1986 - Posterior fusion Occiput to C-1-2-3; Halo brace application.

Complications.

Bradycardia; hyponatremia - mental confusion; pneumonitis; malnutrition - gingivitis; left arm weakness increased.

Follow-up.

Aug. 22, 1986 - Good mental spirits; walks independently with spastic gait capable of self care; feeds himself; has voluntary sphincter control; left arm weakness persists.

SEP warnings.

Feb. 3, 1986 - Sixty (60) separate events.
Feb. 11, 1986 - Fifteen (15) separate events.

Safe surgical manipulation of the mechanically distorted neural axial cervical medullary junction presents a technical challenge best met by microsurgical technique and continuous SEP monitoring. SEP suppression occurred with the drilling or curetting away the compressing bony structures, dissecting the pannus and thickened ligaments, manipulating the loose odontoid, and irrigating with cold saline solution (Fig. 1). The anesthetic agent also affected the SEP's. These changes included both prolongation of latencies and diminution of wave amplitude. These changes usually reverted back to baseline values after a few minutes when manipulation was stopped, except as noted (2, 23) (Figs. 2 and 3).

SEP improvement in the form of shortened latencies and higher amplitudes occurred intraoperatively with completion of the anterior decompression. For this to occur the dura had to be bulging and pulsating into the decompression site (Fig. 4). During posterior wiring and fusion, dissecting and passing wires under the laminae carried a definite risk (Figs. 5, 6, 7). As in the first stage the SEP parameters tended to improve with completion of the posterior fusion when the posterior arch of C-1 was tied into the fusion thus reducing the anterior translation of C-1 on C-2 (Fig. 8). Clinical outcome, however, cannot necessarily be related to or predicted by SEP being unchanged or improved (19, 31).

Conclusions

Surgical manipulations of compromised neural tissue can affect its electrophysiology and possible integrity more frequently than is usually realized. Attention to SEP warnings allow the surgeon to desist from the activity, resulting in recovery of the SEPs usually to baseline (2, 6, 19). SEP monitoring also encourages a gentleness of surgical technique thus reducing inadvertent tissue distortion. We believe that this improves long-term functional results even in the severely compromised patient with complex compressive myelopathy at the cervical medullary junction from anterior compression and cranial cervical instability.

References

1. Allen, A.; Starr, A.; Nudleman, K.: Assessment of sensory function in the operating room utilizing cerebral evoked potentials: A study of fifty-six surgically anesthetized patients. Clin. Neurosurg. 28: 457-481, 1981.
2. Allison, T.: Recovery functions of somatosensory evoked responses in man. Electroencephalogr. Clin. Neurophysiol. 14: 331-343, 1962.
3. Apuzzo, M.L.J.; Weiss, M.H.; Heiden, J.: Transoral exposure of the atlantoaxial region. Neurosurgery 3: 301-207, 1978.
4. Bohlman, H.H.; Bahniuk, E.; Field, G.; Raskulineoz, G.: Spinal cord monitoring of experimental incomplete cervical spinal cord injury. Spine 6: 428-436, 1981.
5. Boston, J.R.; Ainslie, P.J.: Effects of analog and digital filtering on brainstem auditory evoked potentials. Electroencephalogr. Clin. Neurophysiol. 48: 361-364, 1980.
6. Brodkey, J.S.; Richards, D.E.; Blasingame, J.P.; Nalson, F.E.: Reversible spinal cord trauma in cats: Additive effects of direct pressure and isonemia. J. Neurosurg. 37: 591-593, 1972.
7. Brown, R.H.; Nash, C.L.: Current status of spinal cord monitoring. Spine 4: 466-470, 1979.
8. Chiappa, K.H.; Ropper, A.H.: Evoked potentials in clinical medicine. Part 1. New Engl. J. Med. 306: 1140-1150, 1982.

9. Chiappa, K.H.; Ropper, A.H.: Evoked potentials in clinical medicine. Part 2. New Engl. J. Med. 306: 1205-1211, 1982.

10. Croft, T.J.; Brodkey, J.S.; Nulsen, F.E.: Reversible spinal cord trauma: A model for electrical monitoring of spinal cord function. J. Neurosurg. 36: 402-406, 1972.

11. Cusick, J.F.; Mykleburst, J.B.; Larson, S.J.; Sances, A., Jr.: Spinal cord evaluation by cortical evoked responses. Archives Neurol. 36: 140-143, 1979.

12. Dolan, E.J.; Transfeldt, E.E.; Tator, C.H., et al.: The effect of spinal distraction on regional spinal cord blood flow in cats. J. Neurosurg. 53: 756-764, 1980.

13. Dolan, K.D.: Cervicobasilar relationships. Radiol. Clin. North Am. 15: 155-166, 1977.

14. Dorfman, L.J.; Perkash, I.; Bosley, T.M.; Cummins, K.L.: Use of cerebral evoked potentials to evaluate spinal somatosensory function in patients with traumatic and surgical myelophathies. J. Neurosurg. 52: 654-660, 1980.

15. Grundy, B.L.: Intraoperative monitoring of sensory evoked potentials. Anesthesiology 58: 72, 1983.

16. Grundy, B.L.: Monitoring of sensory evoked potentials during neurosurgical operations: Methods and applications. Neurosurgery 11: 556-575, 1982.

17. Grundy, B.L.: Electrophysiologic monitoring: EEG and evoked potentials. In: P. Newfield; J. Cottrell (eds.). Manual of Neuroanesthesia. Boston, Little Brown, pp. 28-59, 1983.

18. Grundy, B.L.: Evoked potential monitoring. In: C.D. Blitt (ed.). Monitoring in Anesthesia and Critical Care Medicine. Churchill Livingston, New York, 345-411, 1985.

19. Grundy, B.L.; Nelson, P.B.; Doyle, E.; Procopio, P.T.: Intraoperative loss of somatosensory evoked potentials predicts loss of spinal cord function. Anesthesiology 57: 321, 1982.

20. Jasper, H.H.: The ten twenty electrode system of the International Federation. Electroencephalogr. Clin. Neurophysiol. 10: 371-375, 1958.

21. Jones, S.J.: Short latency potentials recorded from the neck and scalp following median nerve stimulation in man. Electroencephalogr. Clin. Neurophysiol. 43: 853-863, 1977.

22. Lesser, R.P.; Koehle, R.; Lueders, H.: Effect of stimulus intensity on short latency somatosensory evoked potentials evoked by median nerve stimulation. Electroencephalogr. Clin. Neurophysiol. 47: 377-382, 1979.

23. Lesser, R.P.; Lueders, H.; Hahn, J.; Klem, G.: Early somatosensory potentials evoked by median nerve stimulation: Intraoperative monitoring. Neurology (NY) 31: 1519-1523, 1981.

24. List, C.F.: Neurologic syndromes accompanying developmental anomalies of the occipital bone, atlas and axis. Archives Neurol. Psychiatry 45: 577-616, 1941.

25. Menezes, A.H.; VanGilder, J.C.; Graf, C.J.; McDonnell, D.E.: Craniocervical abnormalities - A comprehensive surgical approach. J. Neurosurg. 53: 444-455, 1980.

26. Menezes, A.H.; Van Gilder, J.C.; Clark, C.R., El-Khoury, G.: Odontoid upward migration in rheumatoid arthritis - An analysis of 45 patients with "cranial settling". J. Neurosurg. 63: 500-509, 1985.

27. de Oliveira, E.; Rhoton, A.L., Jr.; Peace, D.: Microsurgical anatomy of the region of the foramen magnum. Surgical Neurol. 24: 293-352, 1985.

28. Raudzens, P.A.: Intraoperative monitoring of evoked potentials. Ann. N.Y. Acad. Sci. 388: 308, 1982.

29. Schramm, J.; Hashizume, K.; Fukushime, T.; Takahashi, H.: Experimental spinal cord injury produced by slow, graded compression: Alterations of cortical and spinal evoked potentials. J. Neurosurg. 50: 48-57, 1979.

30. Spetzler, R.F.; Selman, W.R.; Nash, C.L., Jr.; Brown, R.H.: Transoral microsurgical odontoid resection and spinal cord monitoring. Spine 4: 506-510, 1979.

31. Spielholz, N.I.; Benjamin, M.V.; Engler, G.L.; Ransohoff, J.: Somatosensory evoked potentials during decompression and stabilization of the spine - Methods and findings. Spine 4: 500-505, 1979.

32. Stevenson, G.C.; Stoney, R.J.; Perkins, R.K.; Adams, J.E.: A transcervical transclival approach to the ventral surface of the brainstem for removal of a clivus chordoma. J. Neurosurg. 24: 544-551, 1966.

33. VanGilder, J.C.; Menezes, A.H.: Craniovertebral abnormalities and their treatment. In: H.H. Schmidek; W.H. Sweet (eds.): Operative Neurosurgical Techniques-Indications, Methods and Results. Grune & Stratton, New York, 1221-1235, 1982.

Variability of Epidural SEP from Below and Above Spinal Cord Lesions - The Effect of the Lesion on Spinal SEP

J. Romstöck;[*] E. Watanbe; J. Schramm

Introduction

Clinical neurophysiologists are familiar with the phenomenon of variability in their measurements. Variability is, however, the enemy to a short-cut definition of alarm criteria for monitoring, as changes may be slight and gradual, but nevertheless important. Continuous intraoperative SEP recordings are contaminated by the use of electrical tools, by the surgeon's manipulation during dissection and by changes in the patient's spinal cord function. It is the source, the degree and the interaction of the spontaneous fluctuations which are crucial for the definition of alarm criteria in intraoperative monitoring. Some studies have been devoted to reliability of alarm criteria, spontaneous variability (6, 19) and improvement of variability (15). Concerning spinal SEP it has been suggested that the waveform may be of importance for the detection of abnormality (17, 18, 21). Therefore, apart from the usual criteria "peak amplitude and latency," an additional criterion, namely "wave area," would be of interest. In patients undergoing neurosurgical operations, contrary to orthopedic cases, some pathological alteration in evoked potentials can be expected preoperatively and right from the beginning of the monitoring session. Thus in neurosurgery, even in patients without postoperative neurological worsening, a higher degree of fluctuation will be expected in successively recorded SEP. As this effects the reliability of alarm criteria and makes comparative studies of individual patients more difficult, the neurosurgeon must be interested in the influence of the lesion and his operative maneuvers on the evoked potential.

The purpose of this paper is to investigate the influence of the patient's pathology on the spinal evoked response under the given conditions of a neurosurgical operation and what maximum variability can be expected in event-free monitoring sessions.

Patients and methods

28 patients, separated into two groups, were examined using epidural recording technique.

Group 1: 11 patients with cervical lesions and median nerve stimulation.

Group 2: 17 patients with thoracic or conus medullaris lesions and peroneal nerve stimulation.

Patients in both groups were operated on intra or extramedullary space occupying lesions such as meningioma (1 cervical/8 thoracolumbar), neurinoma (3 cervical), syrin-

---

* Neurochirurgische Klinik der Universität Erlangen-Nürnberg, Schwabachanlage 6, D-8520 Erlangen, Federal Republic of Germany

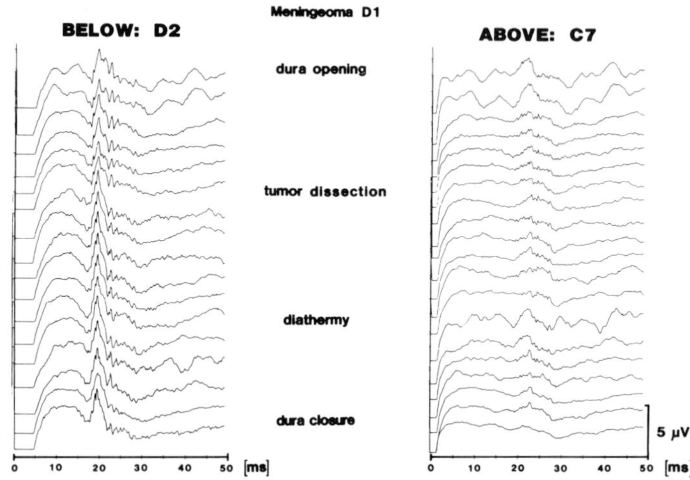

Fig. 1. Original tracings recorded continuously during dissection of a meningioma at D1. Please note that the much smaller potential from above (right) is easily affected by surgical manipulation (diathermy). Although single tracings show poor quality, no complete potential loss was observed. The 37 year old patient, preoperatively suffering from spastic paraparesis showed no postoperative worsening and recovered within 3 weeks.

Fig. 2. Epidural SEP of different quality. The difficulty of proper peak determination in the right tracing (N1) makes latency measurement variable within approximately 3 uSec (= 10% of analysis time). The dotted area is defined by an auxiliary line reaching from P1 through P2 and calculated by microcomputer aid. Area measurement does not consider only one peak maximum, but takes into account each single subpeak contributing to the complete volley potential.

gomyelia (5 cervical/1 thoracic), metastasis (2 thoracic), and others (2 cervical/6 thoracic). None of the patients showed complete loss of either sensory or motor functions. No case exhibited permanent worsening of neurological findings postoperatively.

All measurements were taken using a Nicolet CA 1000/2000 clinical averager linked to a microcomputer and stored on floppy discs for off-line data evaluation.

Stimulation

Frequency 5.1 - 7.1Hz. Constant current pulses of 20mA, square wave stimulus duration 0.3µs Median nerve stimulation at the wrist, peroneal nerve stimulation at the popliteal fossa near the fibula, using bipolar disc electrodes.

Recording

Analysis time of 30µs for median nerve stimulation, 50µs for peroneal nerve stimulation. Bandpass 30 - 3000Hz. 200 - 400 repetitions generating one waveform. Amplification of 25-50 µV, using automatic artifact rejection. Epidural recordings were performed using a pair of platinum tipped wire electrodes (1.2mm) in a frontal

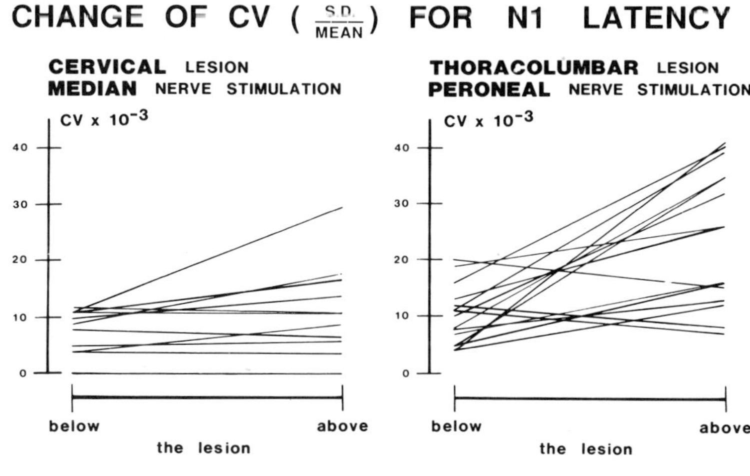

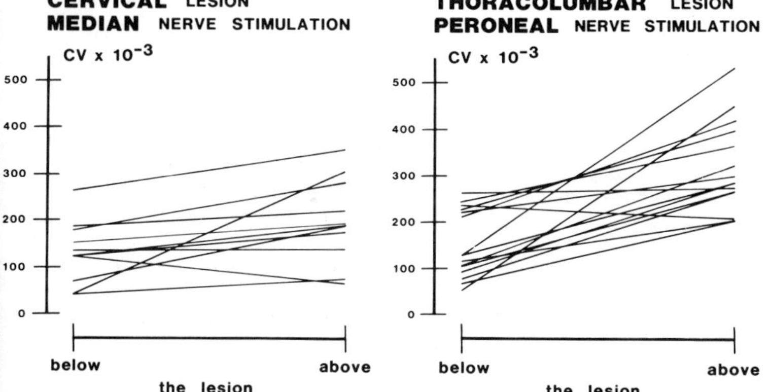

Fig. 3. Variability of N1 through latency and P1 through N1 through amplitude in epidural SEP following median nerve (left) and peroneal nerve stimulation (right). Each line represents one patient, displaying the difference of variability (expressed by the coefficient of variability = CV) between recordings from below and above the lesion. The diagrams to the left show very similar variation between below and above, i.e., there is hardly an effect of the lesion on median nerve SEP. Potentials following peroneal nerve stimulation (right) show statistically higher variability in recordings having passed the cord lesion. In all diagrams it can be seen that, in neurosurgical monitoring, we may be confronted with completely different fluctuation and quality even in patients with comparable stimulation and recording sites. Please note the different scales for CV values of latency and amplitude.

reference montage. Following laminectomy, direct recordings were picked up continuously from caudal and rostral epidural space adjacent to the edges of dura incision (Fig. 1, Fig. 4). All settings of stimulation and recording parameters were kept stable; anesthesia was carried out using neurolept-analgesia avoiding halothane and bolus drug application. Great care was taken to prevent dislocation of the recording electrodes. Due to the variable time span the dura was exposed to, the total number of epidural tracings ranged from 5 to 25 per operation.

Data analysis

Epidural responses were evaluated by measuring latencies of the first and the last reproducible positive (P1, P2) and the major negative deflection (N1) and the interpeak amplitude P1 through N1 of each single tracing. To simplify measurements, peaks were determined following the assumption that epidural potentials correspond to the classical triphasic shape of volley potentials; thus minor subpeaks were ignored. Area measurement was defined by drawing an auxiliary line from the first (P1) to the last (P2) positivity, calculating the area lying within the potential's trace and the additional line by microcomputer aid (Fig. 2).

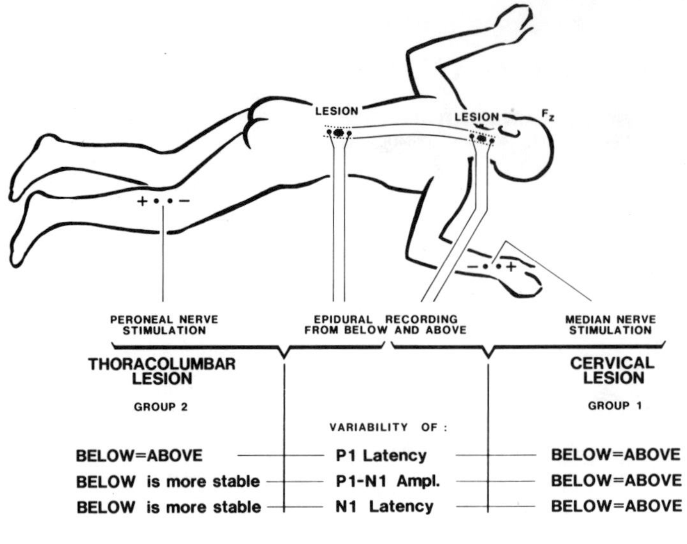

Fig. 4. Synopsis of methods and results.

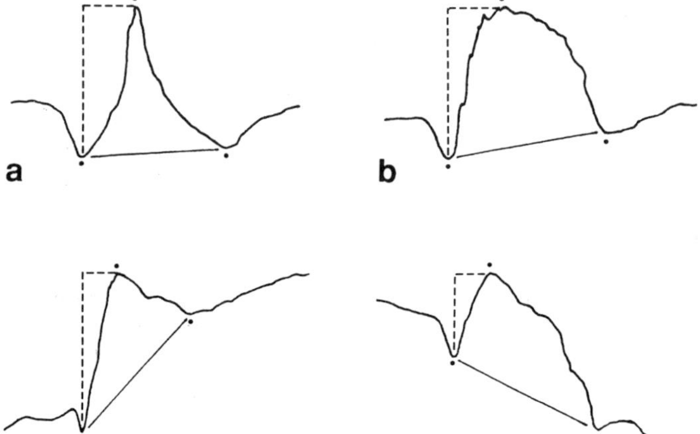

Fig. 5a, 5b. Different wave area in two schematic tracings of spinal SEP with identical N1 and P1 latency and amplitude. These differences would not be detected by conventional latency and amplitude evaluation.

Fig. 5c, 5d. Identical wave area in two schematic tracings of spinal SEP with different amplitude caused by baseline shift. These amplitude changes would be noted in conventional evaluation, although the amount of electrical energy would be similar.

For each monitoring case from latencies, amplitude, and area, the mean, standard deviation, and the coefficient of variability were calculated from all tracings. In this way the variability of all recordings from above and below was compared. The coefficient of variability (CV) is the ratio of standard deviation and mean value obtained from a set of values. It gives an idea of fluctuation in a series of measurements, whatever the absolute values of the data may be. Using the t-test for paired data the differences between CV values of recordings from above and below the lesion were tested (22).

To check the maximum percentage of amplitude and latency change within one monitoring case the difference between the lowest and the highest value of a recording sequence was noted.

Results

Cervical lesions with median nerve stimulation

In this group none of the patients showed an amplitude change of more than 50% or a latency change of more than 10% on evaluating the raw data. Fluctuations of latency, amplitude, and wave area from below and above the lesion were very closely related to each other within the same monitoring case. Within the whole group variability of the rostral and caudal potentials did not differ significantly ($p > 0.05$) (Fig. 3, left).

Thoracolumbar lesions with peroneal nerve stimulation

In group 2 recordings from above the lesion, the recording site of interest for monitoring, exhibited much greater variability. Although none of the patients suffered additional postoperative neurological deficits, 5 cases (30%) showed maximum amplitude changes of more than 50%. In 3 patients (18%) combined amplitude change of more than 50% and latency change for N1 of more than 10% was observed (Fig. 3, right).

Comparing variability of rostral and caudal recordings for all patients in this group, as expressed by the CV value, a significant higher fluctuation was revealed above the lesion for N1 through latency, P1 through N1 through amplitude and wave area (p < 0.01). For P1 through latency this effect could not be observed.

Wave area

In cervical lesions no difference between recording from above and below the lesion was found as expressed by the CV value. In thoracic lesions wave area fluctuated significantly more above the lesion than below, as measured with the CV value (p < 0.01).

This means that, for whatever reason, in thoracic lesions with peroneal nerve stimulation variability from above is higher not only in comparison to the cervical median nerve group, but also with respect to potentials from below the thoracic lesion. In cervical lesions, however, no significant differences between above and below the lesion were observed. A summary of our findings is displayed in Fig. 4.

Discussion

The concept of spinal cord monitoring is based on the assumption that recordings will worsen (as compared to the base line obtained prior to surgical manipulation) in the case of a monitoring event during operative maneuvers. In clinical practice, however, a more differentiated look at the paradigm of SEP monitoring has to be made. It has been pointed out that the presence of a cord lesion is of great importance for signal quality and thus monitoring reliability (18). Most scoliosis patients show preoperative cortical SEP of very good quality which a priori provide a favorable signal-to-noise ratio intraoperatively. In the case of spinal cord distraction (Luque, Harrington rods) a slow gradual effect is exercised on all cord fibers. Years of experience in orthopedic monitoring showed primarily good correlation between postoperative neurological status and intraoperative potential change (5, 7, 9, 13, 14, 16, 23, 28). The arbitrary definition of 50% amplitude decrease and 10% latency increase proved to be suitable to define periods of danger using cortical recordings, although false negative cases have been reported (4). If cord distraction is released within these alarm criteria, there is a good chance for the spinal cord to recover and for the patient to retain good neurological function postoperatively.

In neurosurgical monitoring, however, we will be confronted with completely different basic fluctuations, depending on the severity of cord dysfunction. In many patients with space occupying lesions cortical potentials are unobtainable or of unsatisfactory quality (18). Epidural recording technique, which is known to provide stable peaks, high amplitudes and which is less easily affected by anesthesia or blood pressure, may help in these cases (8, 11, 12, 13, 20, 25, 26). Thus, we have been interested in studying epidural SEP with respect to variability and adverse effects of the lesion. Our main interest was directed to the question of criteria for abnormal spinal SEP being redefined, as has been suggested previously (19, 24).

Cervical epidural SEP following median nerve stimulation proved to be very stable showing similar variability below and above the lesion site. This may be caused by more cases with less deficit in the cervical group, but other reasons seem more important: The greater number of fibers from median nerve, the smaller temporal dispersion of the traveling potential and possibly stationary generators in the cervical cord.

As our results indicate, there is a remarkable difference of quality between thoracic SEP from below and above the lesion at any thoracic level. In part this might be explained by the fact that with higher thoracic levels lower amplitudes are encountered due to temporal dispersion of the volley potential (1, 27). This effect is superimposed by the considerable influence of the cord lesion itself. During spinal cord surgery effects acting at the level of the lesion may only affect a part of fibers passing through. Mechanical alterations during dissection, changes in the regional blood flow and local edema provoke recovering and worsening effects on nerve tissue simultaneously. Therefore, the short time functional changes, seen in successively recorded SEP, may represent these influences or they may just reflect the natural variation seen in any neurophysiological recording. Exact time courses between fiber destruction and obvious SEP changes in man cannot be given in spite of many animal models investigating acute and chronic spinal cord compression (2, 3, 10, 17).

Evaluation of wave area was included because we felt that measurement of only one or two major peak latencies and amplitudes might not be sufficient to describe the total amount of electrical activity traveling along the cord or arising from stationary generators (Fig. 5). This approach may be understood as testing another criterion in description of intraoperative spinal SEP changes. To bypass the effect of temporal dispersion of a volley potential and to express the total electrical energy of nerve tissue at a given moment, area measurement may be useful. In the case of unfavorable signal-to-noise ratio, as in spinal tumors, uncertainty from correct peak evaluation can be reduced. It is surprising that our results did not show more stability in area measurement than for latencies or amplitudes, although the most common effects of noise (overlying sine waves, baseline shifting . . .) should zero out. This gives good support to the fact that it is indeed the lesion induced noise which causes SEP variability in thoracic recordings. Why the lesions have a more pronounced effect in thoracic locations has been discussed above. Future clinical investigations must show if epidural recordings in neurosurgical monitoring are superior to cortical SEP and if area measurement or other wave form parameters are able to explain more about the spinal cord.

Conclusion

Amplitude decrease of 50% and latency increase of 10% may be useful as intervention criteria in cervical levels with median nerve stimulation, but seem unsuitable in thoracolumbar lesions with peroneal nerve stimulation.

Alarm criteria vary for different recording, stimulation, and lesion sites. They have to be adapted to each particular case by critically observing the quality and fluctuation of the first few baseline recordings. On-line trend plotting of data is useful for correct detection of monitoring events.

Epidural potentials seem to need a more detailed evaluation than commonly suggested for cortical SEP, i.e., other criteria than amplitude and latency seem necessary.

References

1. Cracco, R.Q.: Spinal evoked response: Peripheral nerve stimulation in man. Electroencephalogr. Clin. Neurophysiol., 35: 379-386, 1978.
2. Cracco, R.Q.; Evans, B.: Spinal evoked potential in the cat: Effects of asphyxia, strychnine, cord section and compression. Electroencephalogr. Clin. Neurophysiol., 44: 187-201, 1978.
3. Cusick, J.F.; Myklebust, J.; Larson, S.J.; Sances, A.: Spinal evoked potentials in the primate: Neural substrate. J. Neurosurg., 49: 551-557, 1978.
4. Dinner, D.S.; Luders, H.; Lesser, R.P.; Morris, H.H.; Barnett, G.; Klem, G.: Intraoperative spinal somatosensory evoked potential monitoring. J. Neurosurg., 65: 807-814, 1986.
5. Engler, G.L.; Spielholz, N.I.; Berhard, W.N.; Danziger, F., et al.: Somatosensory evoked potentials during Harrington instrumentation for scoliosis. J. Bone and Joint Surg., 60-A: 528-532, 1978.

6. Gonzalez, E.G.; Hajdu, M.; Keim, H.; Brand, L.: Quantification of intraoperative somatosensory evoked potentials. Arch. Phys. Med. Rehabil., 65: 721-725, 1984.

7. Grundy, B.L.: Monitoring of sensory evoked potentials during neurosurgical operations: Methods and applications. Neurosurg., 11: 556-575, 1982.

8. Jones, S.J.; Carter, L.; Edgar, M.A.; Morley, T.; Ransford, A.O.; Webb, P.J.: Experience of epidural spinal cord monitoring in 410 cases. In: J. Schramm; S.J. Jones (eds.): Spinal Cord Monitoring. Springer, Berlin, Heidelberg, New York, Tokyo, pp. 215-220, 1985.

9. Koht, A.; Sloan, T.; Ronai, A.; Toleikis, J.R.: Intraoperative deterioration of evoked potentials during spinal surgery. In: J. Schramm; S.J. Jones (eds.): Spinal Cord Monitoring. Springer, Berlin, Heidelberg, New York, Tokyo, pp. 161-166, 1985.

10. Larson, S.J.; Walsh, P.R.; Sances, A.; Cusick, J.F.; Hemmy, D.C.; Mahler, H.: Evoked potentials in experimental myelopathy. Spine, 5: 299-302, 1980.

11. Lesser, R.P.; Luders, H.; Hahn, J.; Klem, G.: Early somatosensory potentials evoked by median nerve stimulation: Intraoperative monitoring. Neurology, 31: 1519-1523, 1981.

12. Luders, H.; Gurd, A.; Hahn, J.; Andrich, J.; Weiker, G.; Klem, G.: A new technique for intraoperative monitoring of spinal cord function. Spine, 7: 110-115, 1982.

13. Maccabee, P.J.; Levine, D.B.; Pinkhasov, E.I.; Cracco, R.Q., et al.: Evoked potentials recorded from scalp and spinous processes during spinal column surgery. Electroencephalogr. Clin. Neurophysiol., 56: 569-582, 1983.

14. Nordwall, A.; Axelgaard, J.; Harada, Y.; Valencia, P.; McNeal, D.R.; Brown, J.C.: Spinal cord monitoring using evoked potentials recorded from feline vertebral bone. Spine, 4: 486-494, 1979.

15. Nuwer, M.R.; Dawson, E.G.: Somatosensory Evoked Potential Monitoring: Measurement Of Variability. In: R.H. Nodar; C. Barber (eds.): Evoked Potentials II. Butterworth, Boston, pp. 510-513, 1984.

16. Raudzens, P.A.: Intraoperative monitoring of evoked potentials. Ann. N.Y. Acad. Sci., 388: 308-326, 1982.

17. Schramm, J.; Krause, R.; Shigeno, T.; Brock, M.: Experimental investigation on the spinal cord evoked injury potential. J. Neurosurg., 59: 485-492, 1983.

18. Schramm, J.; Romstöck, J.: Cortical versus spinal recordings in intraoperative monitoring of space-occupying spinal lesions. In: C. Barber; T. Blum (eds.): Evoked Potentials III. Butterworth, Boston, 1987.

19. Schramm, J.; Romstöck, J.; Thurner, F.; Fahlbusch, R.: Variance of latency and amplitude in SEP monitored during spinal operations with and without cord manipulation. In: J. Schramm; S.J. Jones (eds.): Spinal Cord Monitoring. Springer, Berlin, Heidelberg, New York, Tokyo, pp. 186-196, 1985.

20. Schramm, J.; Watanabe, E.; Romstöck, J.: Cortical and spinal intraoperative recordings in uneventful monitoring and in cases with neurological changes (this volume), 1987.

21. Shimoji, K.; Matsuki, M.; Shimizu, H.: Wave-form characteristics and spatial distribution of evoked spinal electrogram in man. J. Neurosurg., 46: 304-313, 1977.

22. Sokal, R.R.; Braumann, C.A.: Significance tests for coefficients of variation profiles. Syst. Zool., 29: 50-66, 1980.

23. Spielholz, N.I.; Benjamin, M.V.; Engler, G.L.; Ransohoff, J.: The prevention of iatrogenic spinal cord damage. Int. Orthop., 4: 500-505, 1981.

24. Tamaki, T.; Tsuji, H.; Inoue, S.I.; Kobayashi, H.: The prevention of iatrogenic spinal cord damage. Int. Orthop., 4: 313-317, 1981.

25. Watanabe, E.; Schramm, J.; Romstöck, J.: Intraoperative monitoring of cortical and spinal potentials using different stimulation sites. To be published in: R. Villani; B. Grundy (eds.): Evoked Potentials: Intraoperative and ICU Monitoring. Springer, Berlin, Heidelberg, New York, Tokyo, 1987.

26. Whittle, I.R.; Johnston, I.H.; Besser, M.: Recording of spinal somatosensory evoked potentials for intraoperative spinal cord monitoring. J. Neurosurg., 64: 601-612, 1986.

27. Wood, C.C.; Allison, T.: Interpretation of evoked potentials: A neurophysiological perspective. Canad. J. Psychol/Rev. Canad. Psychol., 35(2): 113-135, 1981.

28. Worth, R.M.; Markand, O.N.; DeRosa, P.; Warren, C.: Intraoperative somatosensory evoked response monitoring during spinal cord surgery. In: J. Courjon; ;J. Maugiere; M. Revol (eds.): Advances In Neurology, Vol. 32: Clinical Applications of Evoked Potentials In Neurology. Raven Press, New York, pp. 367-374, 1982.

Acute Evoked Potential Changes in Operative Treatment: A Summary

J. Schramm[*]

Introduction

Intraoperative changes in evoked potentials (EP) monitored during spinal surgery are plentiful and it has been pointed out correctly in More's paper that "it would be clinically useful to ascertain the incidence at which varying degrees and durations of amplitude attenuation are observed and what relationship these changes have to clinical sequelae" (16). As has been pointed out in a previous review (22), the reasons for variability of intraoperatively obtained EP's and the influences of different anesthetic agents are numerous. In that same review the author has pointed out that ". . . the clinical relevance of evoked potential changes during monitoring needs further clarification . . ." This paper will try to summarize the different experiences and interpretations of intraoperative potential changes with particular regard to their clinical usefulness. While this is particularly important pertaining to the definition of intervention criteria, I will also take the liberty of discussing, in a more general fashion, the relationship between commonly used concepts, paradigms, and conclusions.

A Problem Of Definition

If this paper is to deal with "acute changes" these acute changes must be seen within the framework outlined in Figure 1. It is necessary to recall the parameters within which acute changes are commonly considered. Usually amplitude and latency are used. Other components mentioned have been wave configuration (9, 19) and repetitive impulse conduction (5, 13). One also realizes that "an acute change" can only be defined if the baseline value against which this change is measured is known and in which time frame and what amount of change has to be found to be considered a significant change. Knowing the literature we soon realized that we still have to define what makes an acute change an "event" in terms of monitoring.

So, if we are talking about "acute changes" we have to ask 1) Changes compared to what? (i.e., the definition of a baseline), 2) How acute are these changes? (i.e., the definition of a time frame), 3) The parameters observed for an acute change, and 4) What makes a change an event?

In Figure 1 other expressions are also used: Stable recordings, events, warning criteria and neurological change. Jones et al., in their paper (9) have given a beautiful definition of intervention criteria in the clinical setting. The various definitions of intervention criteria have been reviewed some time ago (22).

---

* Neurochirurgische Klinik der Universität Erlangen-Nürnberg, Schwabachanlage 6, D-8520 Erlangen, Federal Republic of Germany

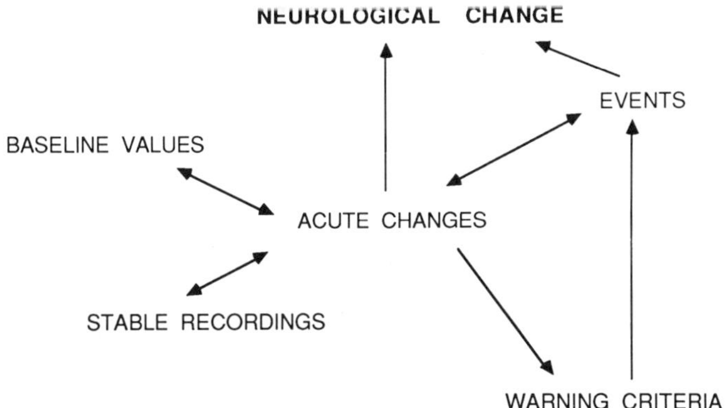

Figure 1: Framework of criteria related to acute changes in intraoperative monitoring of evoked potentials.

Changes Compared To What - The Baseline

Most authors use the recordings obtained before and after induction of anesthesia as baseline values. In a narrower sense only the traces immediately recorded before an "event" may be used as baselines. In any case the baseline value usually consists of a sample of recordings and therefore a definition of baseline values usually falls back onto statistical terms. Baseline values may vary considerably or may be rather stable. Therefore, a number of authors have tried to evaluate the stability of intraoperative recordings in order to develop a more precise idea of the variability of baseline values.

Regarding stability this may be defined using the usual mathematical or statistical criteria, i.e., defining the range of normal variation within two or three standard deviations around the mean. A purely empirical approach is also justified where the range of fluctuation (of latencies or amplitudes) is described in a sub-population of the monitored patients where nothing happened in the sense of intraoperative surgical event or postoperative neurological deterioration. Another way to define stability of evoked potentials is to mathematically express normal variability by using coefficients of variability as has been proposed by our group (24). Romstck and co-workers in their paper discussed several other papers which have been devoted to the study of spontaneous variability and the degrees of variations (19). Their paper tries to investigate the influence of the patient's pathology on the spinal evoked response. The presence of a cord lesion is of great importance for signal quality and monitoring reliability. The incidence of unobtainable EP or potential of unsatisfactory quality is much higher in neurosurgery cases (25). Conversely, the incidence of unobtainable recordings is low in series which include only orthopedic cases (9, 16) or only a small proportion of cases with impaired cord function (10, 17). The composition of the patient population influences in this way the precise definition of a baseline value and makes it harder to recognize "an acute change." Therefore, if one talks about reliability of monitoring, one has to strictly separate cases with normal spinal cord function from those with already impaired spinal cord function. This separation unfortunately is hardly ever done by the authors. There are, of course, large series available in the meantime where only orthopedic or only neurosurgery cases have been monitored.

The baseline values are also influenced by the type of monitoring used. It is generally accepted that spinal potentials recorded at bone or epidural level are more stable than cortical potentials. In other terms, changes observed in spinal recordings may be much smaller and still be of significance compared to changes in cortical recordings. Thus, intervention criteria should differ for spinal and cortical recordings.

In a paper by Chabot and colleagues (2) an interesting approach to the definition of baseline has been applied together with some other features of data manipulation. A new baseline value is obtained frequently and rapidly by adding a new small subset of averages and leaving out the oldest subset. It is too early to evaluate the value of this technique, especially as their software has not been available to other groups yet. In summary, it should be remembered that the "baseline" is not a uniform and homogenous value. It is influenced by the mode of recording (cortical versus spinal) and by the patient population and it remains doubtful whether a purely mathematical approach will be the solution (6, 24, 25). Still, a mathematical approach could be useful if the averaging equipment is supplemented by software giving on-the-spot analysis of the development of mean values and standard deviations of previously defined peak amplitudes and latencies.

Other Parameters Observed For An Acute Change

Our previous report described that there may be statistically significant variations in amplitude of cortical EP in a fair number of neurological normal cases (24). This indicates that a conventional analysis of the range of normality alone is no solution. The paper by Dinner et al., (4) has demonstrated clinically that amplitude and latency changes alone do not always avoid false-positive and false-negative monitoring. Therefore, it has been suggested that the morphology of the response may be of importance (9, 19) and a particular case is mentioned in the paper by Baba et al., (1). The importance of wave form analysis in the experimental set up has been pointed out previously (23).

Another way of reducing the ambiguity in the assessment of acute changes is to do a multi-level recording (4, 13, 14, 25). Despite the well known advantages and disadvantages of cortical recordings (18, 22) it increases the likelihood of excluding anesthesiologic or physiological factors responsible for acute change.

Another attempt of introducing other criteria apart from amplitude and latency is made by Ryan and Britt (20); Chabot et al., (2); and Dill et al., (3). The measurements of conduction velocity as used by Macon and Poletti (15), and Ryan and Britt (20) needs further assessment and may, of course, be used only in epidural recordings above and below the lesion. There is a useful discussion of the value of CV measurements in Ryan and Britt's paper outlining the pitfalls associated with this approach (20). Another way of testing the spinal cord was by using paired stimuli (5, 8, 21).

What Makes A Change Significant?

Changes in amplitude and latency of evoked potentials occur all the time and even in what we consider to be extremely stable EP's there are continuous minute changes (e.g., epidural EP's below the lesion before the start of manipulation in a technically good case). Quite obviously this is not the type of change anybody thinks of. But not even the larger changes, often seen with amplitudes varying up to 30, 40, or 50% of baseline values (8, 16, 17, 18, 24, 25) are necessarily significant. All these changes are seen rather frequently and only become important when their relationship to the clinical outcome is taken into consideration. As has been pointed out by Jones et al., (9), the only way to look at the relevance of the degree of amplitude and latency changes is retrospective analysis. Most authors of this chapter have submitted their cases to a retrospective analysis and it has become clear that amplitude attenuation of 30 or 35% for cortical EP hardly ever is of clinical significance. The incidence of pathological postoperative changes increases once the amplitude changes more than 50% and it quadruples when the amplitude loss is more than 60% (9). Dinner et al., (4) have expressed it similarly but numerically less precise from their large series of 220 cases: A marked change in the SEP's indicated a high chance of developing a neurological deficit and, if there was not change, the chance of neurological postoperative deficit was extremely low. It is the author's opinion that a more precise description of the relationship between baseline, acute change and clinical significance of that change will not be achievable in the foreseeable future.

What Makes The Change An Event?

A change in any of the above mentioned parameters, even a significant change, is not necessarily an "event" in terms of monitoring neurological function. Even if the significance of an acute change is given, applying either mathematical or empirical criteria (e.g., an amplitude loss of more than 60%), even then it is not necessarily an event as has been nicely demonstrated by Dinner et al., (4); and Lesser et al., (12).

We should speak of an "event" in terms of monitoring if the patient develops some neurologic sequelae. Other authors have used the term event to describe peculiarities, mishaps, or significant steps in the surgical procedure. Therefore, two types of events may be associated with acute changes of any evoked potential parameters: Surgical and clinical events (11).

An acute change to be noteworthy must therefore be associated with either a surgical event or a clinical event. This relationship between the clinical changes and the evoked potential changes has been mentioned before (22). It must also be pointed out that a clinical deterioration is not easily classified as being significant or not. In some papers paresthesia in the legs after scoliosis surgery have been qualified as neurologic deterioration, whereas many neurosurgeons would not necessarily consider this to be a remarkable neurologic deterioration. Here again, the different patient populations play a role. In neurosurgical patients a typical danger of surgery is paraplegia, and paresthesia would not necessarily be considered a severe disturbance by neurosurgeons. The point of view of a young patient with scoliosis, however, might be quite different. It should therefore be stated precisely in all papers what type of neurologic change was encountered.

Concerning latency changes these are obviously not so important and Jones et al., even go so far as to declare them not useful monitoring indices in spinal recordings (9). In the same paper, however, they nicely demonstrate how the degree of changes may be defined as definite and at the same time that a definite change in amplitude may be quite variable for different wave components. This is one way to take a more differentiated approach to the analysis of evoked potentials.

The limits of normal baseline value variations which have been used as warning criteria in the past, do overlap with the values of stable recordings and both groups of values are contained in all acute changes (Fig. 2). The discrimination between a change and an event remains therefore difficult. At present it seems likely that only retrospective analysis will allow correlations between intraoperative evoked potential changes and postoperative neurologic sequelae. Whether new approaches in the monitoring

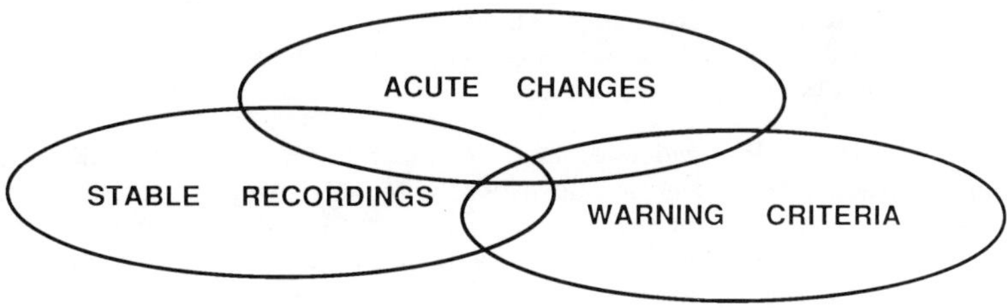

Figure 2: Overlap as an obstacle in judging on acute intraoperative evoked potentials.

procedure (2), the examination of other parameters, or possibly the use of new software will improve reliability of predicting postoperative neurologic outcome in monitoring of the spinal cord, remains open.

Summary

It is obvious that the borders of normal or acceptable acute changes defined by purely mathematical methods are not necessarily clinically meaningful.

Acute changes become meaningful by the association with a surgical or clinical event. Currently the association of a clinical event with potential changes may only be detected in retrospective analysis.

The reliability of monitoring judged by the incidence of false-positives (acute changes not associated with clinical changes) should be judged separately for patients with normal spinal cord function and for patients with impaired spinal cord function.

Acute changes, the acceptable limits of acute changes, and consequently the "intervention criteria," may be quite different depending on the mode of stimulation and the mode of recording.

The evaluation of acute changes in parameters other than just latency and amplitude is not yet finished. Other parameters, such as waveform or reaction to double-stimuli testing supplemented by modern software, deserve attention.

Intervention criteria should be defined for each modality differently: Cortical and spinal recordings, epidural or peripheral stimulation, normal or diseased spinal cords.

An update of modern semi-automatic averaging with modern computer software seems necessary.

Bibliography

1. Baba, H.; Tomita, K.; Umeda, S. et al.: Clinical study of spinal cord evoked potentials. This volume.
2. Chabot, R.J.; John, E.R.; Prichep, L.S.: Real-time intraoperative monitoring during neurosurgical and neuroradiological procedures. This volume
3. Dill, R.J.; Lam, C.F.; Katz, S.: Feature enhancement techniques for detection of spinal cord injury. Pattern Recognition, 1976; 8: 163-172.
4. Dinner, D.S.; Lueders, H.; Lesser, R.P. et al.: Intraoperative spinal somatosensory evoked potential monitoring. J. Neurosurg., 1986; 65: 807-814.
5. Gerhard, H.; Wurzer, K.; Demmer, G. et al.: Single and double stimuli SEP examinations in spinal tumours and in multiple sclerosis. In: Schramm, J.; Jones, S.J. (eds.). Spinal Cord Monitoring, Berlin-Heidelberg-New York, 1985; pp. 308-315.
6. Gonzalez, E.G.; Hajdu, M.; Keim, H. et al.: Quantification of intraoperative somatosensory evoked potential. Arch. Phys. Med. Rehab., 1984; 65: 721-725.
7. Jones, S.J.; Edgar, M.A.; Ransford, A.D.: Sensory nerve conduction in the human spinal cord: Epidural recordings made during scoliosis surgery. J. Neurol. Neurosurg. Psychiatr., 1982; 45: 446-451.
8. Jones S.J.; Edgar, L.; Morley, M.A. et al.: Experience of epidural spinal cord monitoring in 410 cases. In: Schramm, J.; Jones S.J. (eds.). Spinal Cord Monitoring, Berlin-Heidelberg-New York, 1985; pp. 215-220.
9. Jones, S.J.; Howard, L.; Shawkat, F.: Criteria for detection and pathological significance of response decrement during spinal cord monitoring. This volume.
10. Knight, R.Q.; Chan, P.K.D.; Smith, D.N. et al.: Intraoperative somatosensory evoked potential monitoring: The Rochester experience. This volume
11. Koht, A.; Sloan, T.; Ronai, A. et al.: Intraoperative deterioration of evoked potentials during spinal surgery. In: Schramm, J.; Jones, S.J. (eds.). Spinal Cord Monitoring, Berlin-Heidelberg-New York, 1985; pp. 161-166.
12. Lesser, R.P.; Raudzens, P.; Lueders, H. et al.: Postoperative neurological deficits may occur despite unchanged intraoperative somatosensory evoked potentials. Ann. Neurol., 1986; 19: 22-25.
13. Lueders, H.; Lesser, R.; Gurd, A. et al.: Recovery functions of spinal cord and subcortical somatosensory evoked potentials to posterior tibial nerve stimulation: Intrasurgical recordings. Brain Research, 1984; 309: 27-34.

14. Maccabee, P.J.; Levine, D.B.; Pinkhasov, E.I. et al.: Evoked potentials recorded from scalp and spinous processes during spinal column surgery. Electroenceph. Clin. Neurophysiol., 1983; 56: 569-582.

15. Macon, J.B.; Poletti, C.E.; Sweet, W.H. et al.: Conducted somatosensory evoked potentials during spinal surgery. Part 2: Clinical applications. J. Neurosurg., 1982; 57: 354-359.

16. More, R.C.; Nuwer, M.R.; Dawson, E.G.: True and false positive amplitude attenuations during cortical evoked potential spinal cord monitoring. This volume.

17. Nainzadeh, N.K.; Neuwirth, M.G.; Bernstein, R. et al.: Direct recording of spinal evoked potentials to peripheral nerve stimulation by a specially modified electrode. This volume.

18. Nuwer, M.R.; Dawson, E.C.: Intraoperative evoked potential monitoring of the spinal cord. A restricted filter, scalp method during Harrington instrumentation for scoliosis. Clin. Ortho., 1984; 183: 42-50.

19. Romstöck, J.; Watanabe, E.; Schramm, J.: Variability of epidural SEP from below and above spinal cord lesions - the effect of the lesion on spinal SEP. This volume.

20. Ryan, T.P.; Britt, R.H.: Spinal and cortical somatosensory evoked potential monitoring during corrective spinal surgery with 108 patients. Spine, 1986; 11, 4: 352-361.

21. Sherwood, A.M.: Somatosensory evoked potentials recorded in the epidural space in man. Proceedings of the Engineering in Medicine and Biology Society on Engineering and Computing in Health Care; September, 1982, Philadelphia, Pennsylvania. New York, IEEE Press: 50-55, cited in 20.

22. Schramm, J.: Spinal Cord Monitoring: Current status and developments. CNS, Vol. 2, 1985; 3: 207-227.

23. Schramm, J.; Krause, R.; Shigeno, T. et al.: Experimental investigation on the spinal cord evoked injury potential. J. Neurosurg., 1983; 59: 485-492.

24. Schramm, J.; Romstöck, J.; Thurner, F. et al.: Variance of latencies and amplitudes in SEP monitoring during operation with and without cord manipulation. In: Schramm, J.; Jones, S.J. (eds.). Spinal Cord Monitoring, Berlin-Heidelberg-New York, 1985; pp. 186-196.

25. Schramm, J.; Romstöck, E.; Watanabe, E.: Cortical versus spinal recordings in intraoperative monitoring of space occupying lesion. To be published in: Barber, C.; Blum, T. (eds.). Evoked Potentials III. Butterworths, New York-London.

26. Verma, N.P.; Peters, G.M.; Jacobs, L.A. et al.: An assessment of the variability of early scalp-components of the somatosensory evoked response in uncomplicated, unshunted carotid endarterectomy. Clin. Electroenceph., 1985; 16: 157-160.

Continuous/Chronic Changes In Evoked Potentials

Somatosensory Evoked Potentials in Chronic Spinal Cord Injury: An Update

W. Young;[*] D. Mollin

The first of this series of symposia on spinal cord monitoring held in St. Louis in 1979 (Nash & Brown, 1980) emphasized cortical somatosensory evoked potentials (SEP). Questions at that time centered on issues such as: "What should we name the different SEP components?", "Where are their sources and sinks located?", "How should the evoked potentials be done?" By the First International Symposium in Tokyo, Japan in 1982 (Homma & Tamaki, 1984), investigators were beginning to present clinical series with patients numbering in the hundreds. Discussion revolved around the far field and spinal cord recordings of somatosensory evoked potentials. By the Second International Symposium held in Erlangen, Germany in 1984 (Schramm & Jones, 1985), the techniques of far field and spinal cord recordings of SEP had become well established and presentations centered around interpretation and creation of standards. In the recently held Third International Symposium, speakers emphasized further refinements of SEP methods and the newly developed motor evoked potential in different clinical disorders. Although still young, spinal cord monitoring is beginning to mature.

Two problems, however, continue to delay full acceptance of spinal cord monitoring into the mainstream of clinical practice to a level comparable with more established techniques; i.e., electrocardiography (EKG), electroencephalography (EEG), and electromyography (EMG). First, the reliability of SEP in certain neurological disorders is still being questioned. Second, the physiological and morphological bases of SEP changes are not yet well understood. Despite experience with SEP in thousands of patients, there are still currently no accepted standards for interpreting SEP changes. In this article we shall discuss these two problems, some of the progress that has been made, and the future directions of research necessary to resolve these issues.

Reliability and usefulness of SEP

Much ado has been made over the reliability of SEP in different clinical situations, particularly the occasional failure of SEP results to correlate with neurological findings. Here, we shall argue that SEP and neurological examinations yield overlapping sets of information concerning neurological deficits. Discrepancies between the two are not only acceptable but perhaps desirable. SEPs should not duplicate but rather add to the information provided by neurological examinations. SEP should be judged on whether it contributes to clinical decisions, not necessarily how well it correlates with neurological findings. We should move away from the issue of false correlations to the more important issue of how best to utilize SEPs in different clinical situations.

---

[*] Department of Neurosurgery, New York University Medical Center, 550 First Avenue, New York, NY 10016

Neurological vs. SEP findings

The golden standard against which investigators always compare SEP is the neurological examination. In recent years, several reports of false correlations between SEPs and neurological findings have appeared (York et al., 1983; McGarry et al., 1984; Hahn & Latchaw, 1984; Ginsburg et al., 1985; Lesser et al., 1986). These reports have typically emphasized the failure of SEPs to correlate with motor findings. For example, Ginsburg et al., (1985) presented a patient who suffered significant motor deficits postoperatively but failed to show SEP changes during the operation, concluding with a warning that intraoperative SEPs do not always reflect motor function.

SEPs test the somatosensory pathways activated from a specific peripheral stimulation site (usually a peripheral nerve). Unfortunately, neurological findings selected for comparison with SEPs seldom correspond to the nervous structures and pathways tested by the particular SEP protocol used. For example, most investigators do not restrict the comparison to the specific dermatomes activated by the particular SEP protocol used. Many investigators compare SEP with motor function. Some do not apply appropriate controls to rule out general decreases in cortical excitability (Young, 1981; Young, 1986b). It is a wonder that false correlation rates are not higher than the 10-30% reported.

Neurological examinations constitute but one of many descriptive approaches toward the study of the nervous system. The observation that neurological and neurophysiological findings do not correlate in all cases is hardly surprising considering the dissimilarities of stimuli and output in the two approaches. Neurological examinations document subjective sensation based on receptor activated and frequency coded sensory input into the nervous system. In contrast, SEPs represent the responses to gross electrical stimulation of peripheral nerves, producing synchronous volleys of action potentials (Gardner et al., 1985). Injured nervous structures, in particular, respond differently to these two types of stimuli. Finally, SEPs represent electrical signals from the brain whereas subjective sensations reported by the patient represent behavior.

Neurophysiological tests should not be judged solely on how well they correlate with neurological findings. Neurological examinations, like neurophysiological tests are fallible. For example, if a patient has no detectable neurological function, it does not necessary indicate that the patient has no residual axons capable of conducting across the lesion site. Studies by Dimitrijevic et al., (1983, 1984) have amply documented the presence of "subclinical" function in so-called "complete" spinal cord injured patients. There will be some situations where neurological examinations will provide the definitive information and other situations where neurophysiological testing will be better. Ideally, the two approaches should complement each other.

Spinal cord injury

At NYU Medical Center, we have monitored SEP in > 500 spinal injured patients in the past 10 years. Some of the results have been reported (Young, 1982, 1985; Young et al., 1982; Flamm et al., 1985). In general, because spinal injured patients have such abnormal SEP (only 36% of the patients have recognizable early components), comparisons of specific component amplitudes and latencies with values from normal populations are difficult and often impossible. Using a subjective scoring approach which categorized SEP findings on a scale of 0-3 (0 = absent, 1 = trace, 2 = definite, 3 = near normal), we found that SEPs more consistently predicted eventual sensory recovery than the neurological examination during the first few days after injury (Young, 1982).

Rates of false correlations were low during the acute phase of spinal cord injury. At the extreme ends of the scale, patient with no SEP on either leg on admission had a

95% probability of no clinically detectable sensory function in the legs. The 5% of cases where some function was detected, most typically had sensation in dermatomes outside of those innervated by the posterior tibialis nerve, the stimulation site in our standard SEP protocol. Patients admitted with a definite SEP on either leg had a 97% probability of having some sensory preservation in the lower limbs. Actually, in several cases where the neurological examination suggested complete sensory loss while SEPs indicated preservation, re-examination of the patient revealed the presence of sensation.

In chronic spinal cord injury, however, false correlation rates between SEP scores and sensory function were on the order of 20-30%. A majority of these false correlations involved bizarre long latency waves on the SEP which often (17% of the patients) are present in patients with severe sensory losses. Some cases can be attributed to poor technique. Excluding these cases, we had a combined false negative and false positive rate of ~ 10% in chronic spinal injured patients. Approximately 1 out of 10 patients had either a definite SEP with minimal sensory function in the legs or no SEP in the presence of some sensation in the legs. We have never seen normal SEPs in a patient with complete loss of motor or sensory function. Conversely, we have yet to see a patient with no SEP response and intact sensory function.

SEPs have proven to be invaluable in several clinical situations; in uncooperative, unconscious, or malingering patients. Neurological examinations are often indecisive in such cases and SEP can provide critical information for clinical decisions. SEPs also serve as a check on the neurological examinations. Neurophysiological testing may uncover subclinical residual function (Dimitrijevic et al, 1984) or distal lesions marked by a more proximal primary lesion (Dimitrijevic et al, 1982). Finally, SEPs have become; an integral part of our protocol for follow-up of treatments in patients (Flamm et al., 1985; Young, in press). Because of the time and expense involved in having neurologists examine patients, long-term and frequent follow-ups are often omitted. Neurophysiological testing, by contrast, can be carried out by technicians and at weekly or even daily intervals.

Intraoperative monitoring

Although intraoperative SEP monitoring is usually considered to reflect acute spinal cord injury, the vast majority of patients undergoing operations already have some lesion in their spinal cord. Therefore, intraoperative spinal cord monitoring often represents a rather specialized situation, the possible imposition of an acute injury upon a chronically injured spinal cord. One of the most frequent mistakes made by clinicians is their tendency to apply criteria developed for normal SEP to intraoperative monitoring of abnormal SEPs.

Monitoring patients with compromised spinal cord function differs from monitoring normal patients in the following respects. First, SEPs in patients with pre-existing lesions are often much more sensitive to anesthesia than SEPs in normal patients; i.e., those undergoing correction of scoliosis. Fig. 1 shows an example of a patient with cortical SEPs that were unusually sensitive to low level halothane anesthesia (0.5%). Second, we have also found that if preoperative SEPs are small, variable, and delayed, they tend to be less reliable reflections of functional changes in the operating room.

In the past 10 years, we have monitored >500 cases of spinal cord procedures at the NYU Bellevue Medical Center (Young & Berenstein, 1985). These include cases of neuroradiological procedures where individual arteries of the spinal cord were occluded or embolized, neurosurgical removal of tumors and disks, orthopedic correction of unstable spinal cord fractures, and cardiovascular surgical repair of dissecting thoracic aortic aneurysms. A recent review of 100 consecutive cases studied between June, 1985 to January, 1987 (Young & Mollin, in press) revealed that 20% could not be

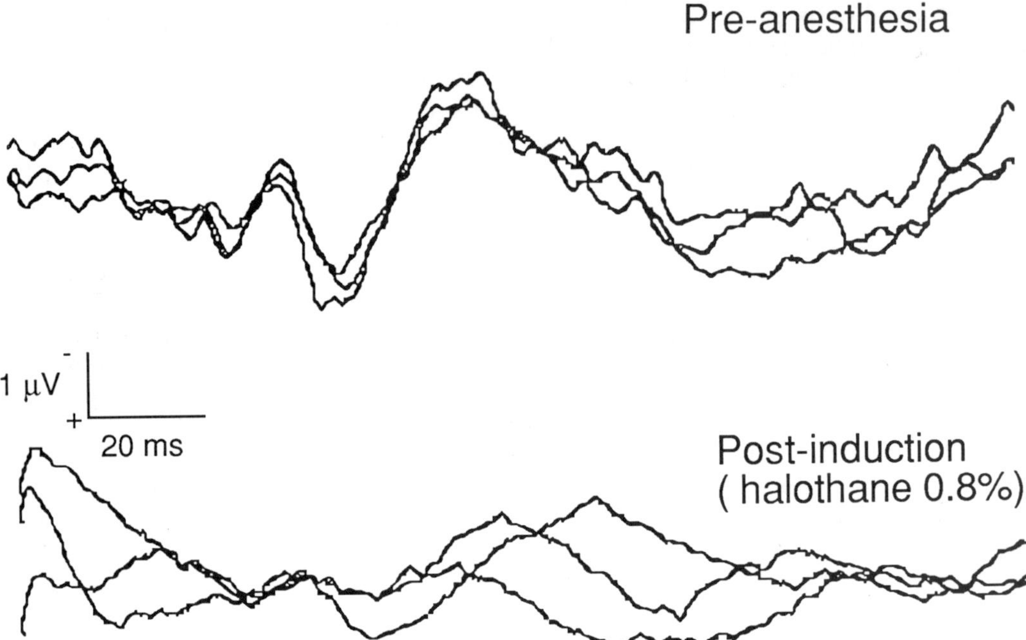

Fig. 1. Sensitivity of SEP to halogenated anesthesia in a patient with an incomplete T12 through L1 unstable fracture. The top set of traces show cortical SEP recorded from the patient before induction of anesthesia. Induction of anesthesia with 0.8% halothane severely depressed SEP amplitude. All traces represent averages of 100 responses activated from unilateral posterior tibial nerve simulation, analysis time of 200mSec, negative is up, stimulation frequency is 2.3Hz, and scalp recording sites are on the Fz and C3'.

monitored because their SEPs were of such low amplitude or variable after induction of anesthesia that the responses could not be interpreted. This contrasts with <2% of cases monitored for routine correction of scoliosis.

Intraoperative monitoring of spinal cord angiography and embolization procedures (1, 29) is possibly the most striking example of the usefulness of SEP in clinical use. During procedures where the anterior spinal artery (ASA) is occluded or filled with radiographic contrast dyes, we have consistently observed rapid decreases in SEP amplitudes (29). SEP monitoring during selective spinal angiography has speeded up the procedures, halving the time required. The incidence of post-procedure complications has been significantly reduced from >20% to <1%. Although SEPs were used to follow patients after embolization procedures, the neurological improvements were sufficiently striking that SEP played little role in establishing this observation. Thus, in this clinical situation, the issue of false negative and false positive correlations between SEP changes and postoperative neurological deficits was moot.

SEP monitoring has not had as great an impact on neurosurgical cases for the following reasons. In the majority of the cases where SEP changes were encountered, the surgical procedure proceeded despite the SEP changes. For example, when the surgery was for intraspinal tumor removal, the clinical decisions seldom depended on SEP findings. Unlike spinal angiography and embolization procedures where temporary vascular occlusions are used to test the outcome of a given maneuver, similar provocative tests are not available for surgical manipulations of the spinal cord. Once damage has been done with a surgical instrument, the SEP simply documented the change. Although a neurosurgeon may choose to stop the procedure and wait for recovery of SEP, such delays may pose some risk to the patient.

Feedback and provocative tests

Several investigators have emphasized the need for SEP techniques that provide more rapid feedback of information. The presumption is that the earlier the feedback, the more likely the surgeon can stop the action causing the damage. Current SEP averaging methods and recording paradigms require about a minute for completion of the data acquisition. Reduction of the feedback time to a few seconds may increase the usefulness of evoked potentials. However, because brain and spinal cord tissues can be irreversibly damaged after only 1-2 minutes of ischemia, the rapid reaction time required may not be practical. In some cases, such as during removal of tissues, the action is not reversible.

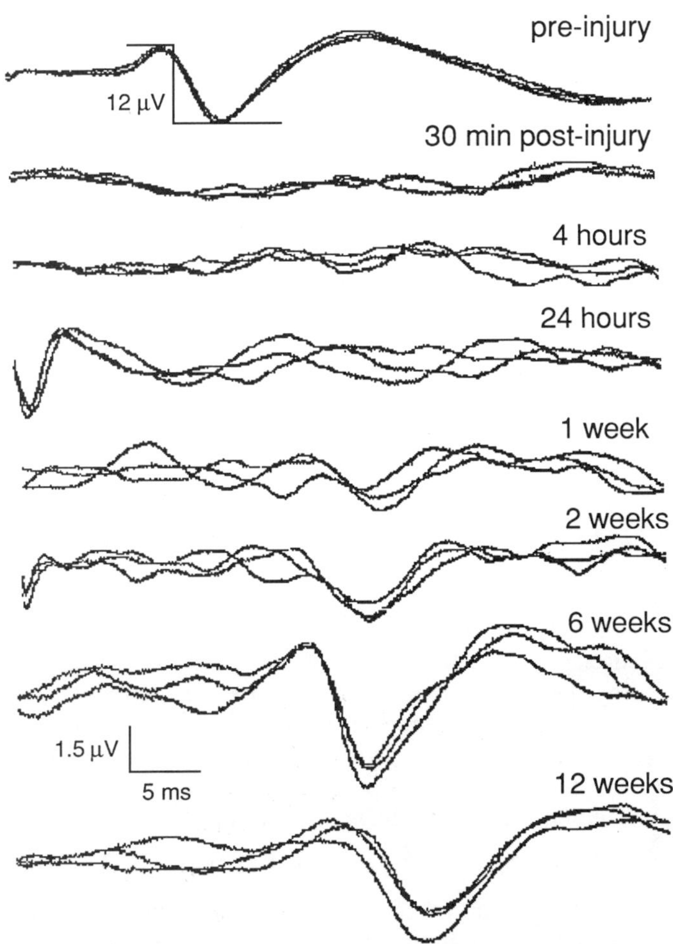

Fig. 2. Cortical somatosensory evoked potential changes in cat after a 10gm weight dropped 20cm onto rigidly supported thoracic spinal cord (T8). The traces represent cortical potentials recorded with epidural screw electrodes, one placed at the vertex and the other overlying somatosensory cortex contralateral to the stimulated side. 100 responses activated by stimulation of the sciatic nerve at 2.3Hz were averaged for each trace. Note the difference in the amplitude scaling of first pre-injury traces and the subsequent traces. This injury generally produces a 90% loss of axons at the lesion site at 3 months after injury.

An alternative and perhaps better approach is to use "provocative" tests, like the one developed in transvascular embolism (1, 29). The principle of a provocative test is to subject the tissue to a reversible insult and use SEP to monitor the effects of this maneuver prior to carrying out an irreversible step. For example, temporary vascular occlusion prior to the embolism or sacrifice of the vessel may provide critical information concerning the potential risks of the step. However, in cases where tissues are being removed, a different type of test may be necessary. For example, one approach may be to cool or apply local anesthetic to the tissue to be removed.

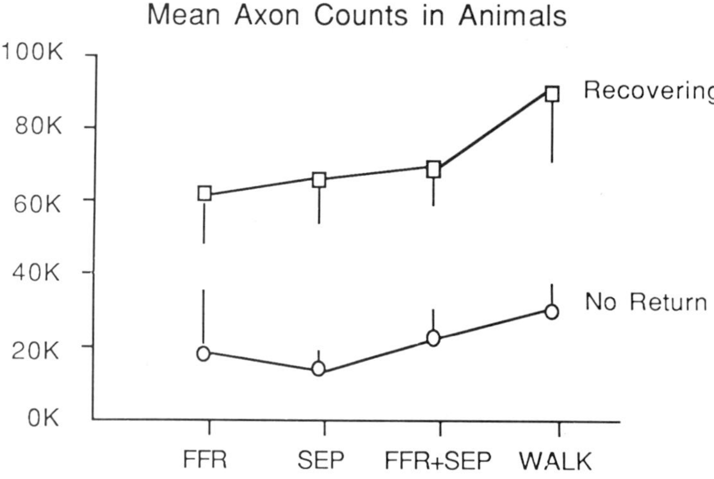

Fig. 3. Relationship of mean axon counts from 10 recovering and 10 paralyzed cats subjected to a variety of injuries and weight drop contusions of thoracic spinal cords. The squares and circles represent mean values while the bars indicate standard errors of means. The mean axon counts in animals that recovered somatosensory evoked potentials (SEP), free fall responses (FFR), both SEP and FFR (SEP + FFR), and walking (WALK) are given.

The usefulness of spinal cord monitoring depends on how clinicians use the information. Neurosurgeons and orthopedic surgeons may be able to gain critical information by stressing the spinal cord. For example, an important question that often arises in spinal cord injury care is whether the spinal cord is at risk to further injury due to an unstable fracture. SEP may provide valuable information during flexion extension x-ray studies of the spinal cord. Likewise, SEP monitoring may be useful during external realignment and traction of the spinal column.

Physiological and morphological bases of SEP changes

Losses of evoked potentials can have many causes besides loss of the axons mediating the response. In humans, the causes are often difficult to establish due to the increasing difficulties of obtaining autopsy confirmations of lesions. For a better understanding of SEP changes we must turn to animal studies. These studies have recently shown that SEP losses in acute spinal cord injury are likely to be due to extracellular ionic derangements (31, 32, 35, 36) and possibly blood flow (34). In chronic spinal cord injury, Blight (1984) has found suggestive evidence that demyelination plays a major role in the failure of axons to conduct across the lesion site. The excitability of proximal and distal structures will influence the amplitude of evoked potentials. Finally, the procedure of obtaining the SEP by averaging repetitive stimuli will introduce several variables that influence axonal conduction; i.e., axonal fatigue due to the repetitive stimulation and the presence of inhibitory feedback and feed forward loops in the ascending sensory pathways. These must be considered when interpreting evoked potential changes.

Patterns of SEP loss and recovery in spinal cord injury

A distinct pattern of SEP loss and recovery occurs in injured animal spinal cords. Cortical responses are immediately lost after a contusive injury to the spinal cord (27, 38). In treated or moderately injured spinal cord, SEPs often recovered within 1-2 hours (26, 38). With severe injuries, the SEPs either do not recover at all or recovered only transiently at 1-2 hours and then were lost a second time at 3-4 hours after injury.

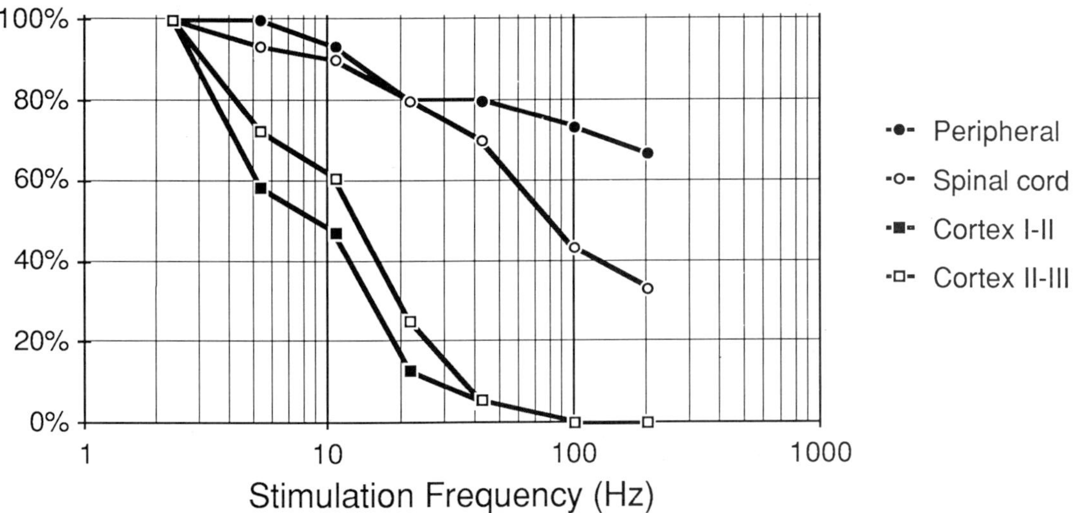

Fig. 4. Fatigue of SEP components with increasing stimulus frequencies. The somatosensory potentials were activated by sciatic nerve stimulation at frequencies shown in the abscissa. The peripheral nerve component (Peripheral) recorded bipolarly between an electrode placed on the lumbosacral spinal cord and a knee reference. The spinal component (Spinal Cord) was recorded epidurally from the cervical spinal cord. The cortical components were recorded bipolarly from the screw electrodes placed epidurally at the vertex and over the somatosensory cortex; yielding positive-negative-positive peaks (respectively Cortical I, Cortical II, and Cortical III). The peak to peak amplitudes are shown (Cortex I-II, Cortex II-III). All amplitudes are given as percentages of the control amplitude at stimulation frequency of 2.3Hz.

These early recoveries of SEP can be correlated with extracellular ionic derangements at the contusion site (see below). SEPs may recover a second time at 18-36 hours and may again be lost at 48-56 hours (unpublished). The causes of the longer term changes in evoked potential conduction across the lesion site are not well understood.

Fig. 2 illustrates the cortical SEP changes that occur in a cat injured with a 10gm weight dropped 20cm onto the spinal cord with rigid clamping of the vertebral column at the impact site. This particular injury model allows the survival of approximately 5% of the axons at the lesion site at 3 months after injury (2, 3, 5). The SEPs were recorded from epidural screws placed at the vertex and over the somatosensory cortex. Activated from posterior tibial nerve in the hind limbs, 100 responses were averaged. Note the similarity in appearance of these potentials with the human cortical SEPs. The major difference is the latency of the cortical components.

Prior to injury, the initial negative-positive complex of the SEPs had a peak-to-peak amplitude of 12μV. Note that the amplitude scale for the rest of the traces is different and is shown on the lower left. Contusion caused an immediate loss of SEP, as shown in the SEP obtained at 30 minutes and 4 and 24 hours after injury. The early positive-negative deflection seen at 24 hours after injury is not a cortical potential and may be related to a far field injury potential. At 2 weeks, the SEP began to recover and eventually reached 30-40% of pre-injury amplitudes by 6 weeks. However, at 12 weeks the SEPs had diminished in amplitudes as well as longer latencies.

The morphological and physiological bases underlying some of the SEP changes in chronic spinal cord injury are not well understood. What do SEP changes represent in terms of morphology and physiology of the structure generating the response? In humans, this question cannot be answered with certainty because of the difficulty of obtaining well preserved pathological material and invasive neurophysiological recordings from brain and spinal cord structures. In animals, this question has not received the attention that it merits. For example, although SEP have been used to monitor spinal cord injury in many studies, there have been few studies which have systematically correlated SEP and numbers of axons crossing spinal injury sites.

Axon counts and morphometry in injured spinal cords

In recent studies of cat spinal cords, we collected data on the axon counts and morphometry at the lesion center (2, 3), correlating these with cortical SEPs and free fall response (FFR). The latter are stereotyped muscle responses mediated by vestibulospinal and reticulospinal pathways (15). The total number of axons crossing the lesion center and their radial locations were documented in 20 animals at 3-6 months after a contusion injury. Normally, cats have approximately 500,000 axons that are countable on light microscope in the mid thoracic spinal cord. After a 20gm weight is dropped 20cm onto the thoracic spinal cord, the number of axons passing through the lesion site is dramatically reduced to 30,000-50,000, a >90% loss of axons. With more severe injuries where, for example, a 13gm weight dropped 20cm onto the spinal cord while the vertebral column is rigidly fixed, total axon counts typically are <5% of normal (5).

The axons that survive at the lesion center tend to be concentrated in the rim of tissue adjacent to the dural surface, decreasing exponentially in density with distance from the dural surface. Usually, about 90% of the surviving axons at the lesion site are located within 300-400μ of the dural surface. In addition to this depth related loss of axons, there is also a selective loss of larger axons, unrelated to depth. Many of the axons are demyelinated or dysmyelinated (4) with physiological evidence of temperature sensitive conduction failure that is attributable to the demyelination.

SEPs can recover in spinal injured animals with surprisingly low axon counts. Fig. 3 summarizes these data. The mean total axon counts at the lesion sites of 10 recovering and 10 paralyzed animals at 12 weeks after injury are shown with standard errors of means. The cats that had SEP recovery had a mean axon count of 64,000. In animals that did not recover SEP, the mean axon count was about 18,000. The surviving axons in the different parts of the spinal cord are fairly evenly distributed between the dorsal and ventral halves of the cord. The mean axon counts of animals with FFRs were similar to those that had SEP. Walking appeared to require more axons with those animals recovering locomotion having an average of ~90,000 axons and those that did not 38,000 axons.

Some animals with <50,000 axons crossing the injury site recovered SEP. However, 25,000-30,000 axons appeared to be a lower limit. Few animals with <30,000 axons recovered SEP, FFR, or walking while most animals with >50,000 recovered one or more of these three different tests of function. Note that if we consider only ascending axons, the difference is much smaller. Probably less 25,000 ascending axons or about 5% of the original axon population of the spinal cord may be able to support recovery of SEP in cats. In adult humans, 5% of axons in spinal cords would be about 1 million.

Demyelination

Demyelination may play a role in determining the loss and recovery of SEP in chronic spinal cord injury. Although our data suggests a certain minimum number of axons that is required for SEPs, there was significant overlap in the axon counts of animals that did and did not recover SEP, FFR, and walking. Some animals with >30,000 axons could not walk while others with <30,000 would walk. This of course

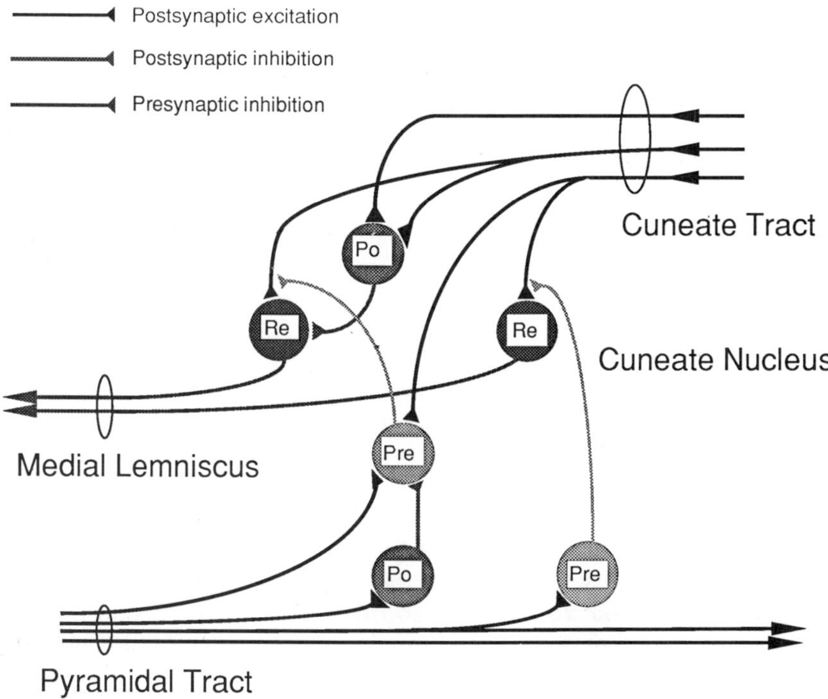

Fig. 5. Inhibition in the dorsal column nuclei. A schematic diagram of the circuit of presynaptic and postsynaptic inhibition in the cuneate nucleus illustrates the complex inhibition mechanisms in the somatosensory pathway. Pyramidal tract axons terminate on neurons (Po) which inhibit the cuneate afferents terminating on medial lemniscal relay neurons (Re) both postsynaptically (via Po) and presynaptically (via Pre). Cuneate afferents activate the medial lemniscal relay neurons postsynaptically and also exert recurrent presynaptic inhibition (via Pre) or postsynaptic inhibition (via Po). Thus, repetitive stimulation of cuneate afferents and postsynaptic inhibition of the relay neurons. In addition, returning descending corticospinal activity will add both presynaptic and postsynaptic inhibition.

may be explained by the distribution of axons. However, a number of animals clearly had more surviving axons in the dorsal columns and did not recover SEP. Blight (1985) recently proposed that demyelination plays a major role in the functional deficits, that axonal conduction is compromised by demyelination. If confirmed, this hypothesis may explain part of the patter of SEP losses and recovery during the subacute phase of spinal cord injury. For example, demyelination peaks in the spinal cord at 2-3 days after injury and remyelination takes place at 4-12 weeks.

Further losses of axons and perhaps limited regeneration of axons may also account for some of the SEP changes in chronically injured spinal cord. In unpublished studies, Blight has also found that significant axonal losses can occur between 1-7 days after injury. Some evidence also suggests that dorsal column axons may have significant regenerative capacity, especially under the influence of applied DC electrical fields (6, 7). Although classical dogma would have us believe that regeneration does not occur in the spinal cord, most of the data that supports this dogma comes from animal studies where the spinal cord has been transected. These results apply to contusion injuries of the spinal cord where a continuous bridge of tissue remains at the injury site.

Note that regenerated axons, if they exist, labor under the same constraints as residual surviving axons. First, they are not myelinated and have to be remyelinated in order to function properly. Second, they may have unusual action potential conduction properties; i.e., rapid fatigue, slow conduction, and inability to follow rapid stimuli. Third, they may not be reconnecting to the right targets. These conditions complicate the interpretation of SEPs and raises questions whether SEP will be able to detect regenerated axons or chronic degeneration of surviving axons.

Habituation and fatigue

Somatosensory evoked responses decrease in amplitude with greater stimulus frequencies. In general, the sensitivity of SEP to stimulus frequency depends on the length of the pathway and number of synapses intervening between the stimulation and recording sites. For example, as shown in Fig. 4, peripheral nerve and spinal components of the somatosensory response in cats do not fall off as rapidly to repetitive high frequency stimulation as cortical components. Cortical responses fall off in amplitude with stimulation frequencies as low as 5Hz whereas spinal and peripheral response are usually not as affected until stimulation frequencies exceed 20Hz. Two mechanisms contribute to the sensitivity of SEPs to high frequency stimulation: Habituation and fatigue.

Habituation is characteristic of central neural responses to repetitive stimulation (9). Habituation results from inhibitory mechanisms inherent in central neuronal circuits. The most prominent way-station in the somatosensory pathway is the dorsal column nucleus. Both feedback and feed forward inhibition are present in this nucleus (22), as shown in Fig. 5. For example, ascending spinal axons inhibit dorsal column nuclei neurons both presynaptically and postsynaptically. In addition descending axons from cortex and thalamus exert presynaptic and postsynaptic inhibitory influences on the nucleus. Given these inhibitory mechanisms, the inability of cortical, thalamic, dorsal column nuclei SEP components to follow stimuli frequencies of >5-20Hz is not surprising.

Fatigue results from depletion of the membrane ionic gradients, synaptic transmitters, and refractory period membrane events. Unlike habituation, fatigue does require feedback or feed forward inhibitory circuits. Fatigue consequently is a likely cause of the decreased peripheral nerve and spinal response amplitudes to stimulation frequencies of 10-200Hz. Normally, fatigue does not play a major role in stimulation frequencies of <10Hz since most synapses and axons should respond to and recover from repetitive stimulation spaced >100mSec apart. However, fatigue is likely to be much more prominent in demyelinated or dysmyelinated axons that have prolonged refractory periods and decreased safety factors of saltatory conduction. It is likely to be the primary cause of conduction blockade and the inability of the axons to carry frequency coded sensory information. Also, in acute spinal cord injury, the deranged extracellular ionic activities (31, 32, 36, 39), cellular ionic levels (35), posttraumatic blood flow changes (27, 34), and other factors may contribute to a reversible changes in axonal conduction across the lesion site (Young, 1980).

The extent to which the amplitude frequency or latency frequency curves change in lesioned nervous systems has not been systematically investigated. Dimitrijevic et al., (1977, 1983, 1984) have shown that functional or subclinical conduction in chronic spinal cord injured patients are often mediated by residual axons that impart tonic activity. Tests like the SEP which rely primarily on phasic activation of the nervous system are not suitable for evaluating such residual function. We have observed spinal injured patients are often unable to follow stimulus rates of even 0.1Hz (26). Changes in habituation or fatigue characteristics of the somatosensory responses are likely to play a major role in determining SEP recovery in injured nervous systems.

Future directions

A clear understanding of the morphological and physiological bases of SEP changes in injured nervous systems is essential for further progress in spinal cord monitoring. The traditional assumption that loss of evoked potentials can be equated to loss of axons is being challenged by studies demonstrating that demyelination, dysmyelination, and reorganization of neural structures proximal and distal to the lesion site are critical factors determining the recovery of evoked potentials. Workers in the field are beginning to recognize that injured nervous systems do not operate by the same rules as the normal nervous system. The findings that $< 10\%$ of the axons in the spinal cord can support recovery of SEP and motor function reinforce this point. Thus, morphology and function may not have the same relationship in injured spinal cords as they do in normal spinal cords.

Many important questions remain uninvestigated. To what extent is the recovery of function a result of reorganization of proximal and distal spinal cord, as opposed to an improvement in axonal conduction through the lesion site? What is the relationship of function and morphology in injured spinal cords? What are the necessary and sufficient conditions for functional recovery? Answers to these questions will require animal studies where morphology can be directly correlated with physiology.

Clinical studies, however, will provide some answers that cannot be as easily obtained in animal studies. Chronic spinal injury animal experiments are exceedingly difficult and expensive to carry out. There is no shortage of human cases to study. The relationship of specific sensory and motor recovery to neurophysiological responses can be better assessed in humans. Finally, the mechanisms by which injured human nervous systems achieve recovery of function and neurophysiological responses may be very different from animals.

Acknowledgements

We thank our colleagues in the Neurosurgery, Orthopedics, and Neuroradiology Departments for their help in gathering the data. This work was supported by NIH NINCDS grant NS15590, NS17627, and NS10164.

References

1. Berenstein, A.; Young, W.; Ransohoff, J.; Benjamin, V.; Merkin, H.: Somatosensory evoked potentials (SEP) during spinal angiography and therapeutic transvascular embolization. J. Neurosurg., 60: 777-785, 1983.
2. Blight, A.R.: Cellular morphology of chronic spinal cord injury in the cat: Analysis of myelinated axons by line sampling. Neuroscience, 10: 521-543, 1983a.
3. Blight, A.R: Axonal physiology of chronic spinal cord injury in the cat: Intracellular recording in vitro. Neuroscience, 10: 1471-1486, 1983b.
4. Blight, A.R.: Delayed demyelination and macrophage invasion: A candidate for "secondary" cell damage in spinal cord injury. CNS Trauma, 2: 299-315, 1985.
5. Blight, A.R.; DeCrescito, V.: Morphometric analysis of experimental spinal cord injury in the cat: The relationship of injury intensity to survival of myelinated axons. Neuroscience, 19: 321-341, 1986.
6. Borgens, R.B.; Blight, A.R.; Murphy, D.J.: Axonal regeneration in spinal cord injury: A perspective and new technique. J. Comp. Neurol., 250: 156-167, 1986.
7. Borgens, R.B.; Blight, A.R.; Murphy, D.J.; Stewart, L.: Transected dorsal column axons within the guinea pig spinal cord regenerate in the presence of an applied electrical field. J. Comp. Neurol., 250: 168-180, 1986.
8. Dimitrijevic, M.R.; Dimitrijevic, M.M.; Faganel, J.; Sherwood, A.M.: Suprasegmentally induced motor unit activity in paralyzed muscles of patients with established spinal cord injury. Ann. Neurol., 16: 216-221, 1984.
9. Dimitrijevic, M.R.; Faganel, J.; Gregoric, P.; Nathan, P.W.; Trontelj, J.K.: Habituation: Effects of regular and stochastic stimulation. J. Neurol. Neurosurg. Psychiat., 35: 234-242, 1972.
10. Dimitrijevic, M.R.; Faganel, J.; Lehmkuhl, D.; Sherwood, A.M.: Motor control in man after partial or complete spinal cord injury. In: J.E. Desmedt (ed): Motor Control Mechanisms in Health and Disease. Raven Press, New York, pp. 915-925, 1983.

11. Dimitrijevic, M.R.; Spencer, W.A.; Trontelj, J.V.; Dimitrijevic, M.M.: Reflex effects of vibration in patients with spinal cord lesions. Neurology, 27: 1078-1086, 1977.

12. Flamm, E.S.; Young, W.; Collins, W.F.; Piepmeier, J.; Clifton, G.L.; Fischer, B.: A phase I trial of naloxone treatment in acute spinal cord injury. J. Neurosurg., 63: 390-397, 1985.

13. Gardner, E.P.; Hamalainen, H.A.; Warren, S.; Davis, J.; Young, W.: Somatosensory evoked potentials (SEP) and cortical single unit responses elicited by mechanical tactile stimuli in awake monkeys. Electroencephalogr. Clin. Neurophysiol., 58: 537-552, 1985.

14. Ginsburg, H.H.; Shetter, A.G.; Raudzens, P.A.: Postoperative paraplegia with preserved intraoperative somatosensory evoked potentials. Case report. J. Neurosurg., 62: 296-300, 1985.

15. Gruner, J.A.; Young, W.; DeCrescito, V.: The vestibulo spinal free fall response: A test of descending function in spinal injured cats. CNS Trauma, 1: 139-159, 1984.

16. Hahn, J.F.; Latchaw, J.P.: Evoked potentials in the operating room. Clin. Neurosurg., 31: 389-403, 1984.

17. Homma, S.; Tamaki, T.: Fundamental and Clinical Application of Spinal Cord Monitoring. Saikon Publishing Co., Tokyo, 1984.

18. Lesser, R.P.; Raudzens, P.; Luders, H. et al.: Postoperative neurological deficits may occur despite unchanged intraoperative somatosensory evoked potentials. Ann. Neurol., 19: 22-25, 1986.

19. McGarry, J.; Friedgood, D.L.; Woolsey, R. et al.: Somatosensory evoked potentials in spinal cord injuries. Surg. Neurol., 22: 341-343, 1984.

20. Nash, C.L.; Brown, R.H.: Spinal Cord Monitoring Workshop: Data Acquisition and Analysis. Case Western Univ., Cleveland, Ohio, 1979.

21. Schramm, J.; Jones, S.J.: Spinal Cord Monitoring. Springer-Verlag, Berlin, Heidelberg, 1985.

22. Willis, W.D.; Coggeshall, R.E.: Sensory Mechanisms of the Spinal Cord. Plenum Press, New York, London, pp. 230-238, 1978.

23. York, D.H.; Watts, C.; Raffensberger, M. et al.: Utilization of somatosensory evoked cortical potentials in spinal cord injury. Prognostic limitations. Spine, 8: 832-839, 1983.

24. Young, W.: The interpretation of surface recorded evoked potentials. Trends in Neurosciences, 14: 277-280 (issue 11), 1981.

25. Young, W.: Correlation of somatosensory evoked potentials and neurological findings in spinal cord injury. In: C.H. Tator (ed): Early Management in Acute Spinal Cord Injury. Raven Press, New York, pp. 153-165, 1982.

26. Young, W.: Somatosensory evoked potentials (SEP) in spinal cord injury. In: J. Schramm; S.J. Jones (eds.): Spinal Cord Monitoring. Springer-Verlag, Berlin, Heidelberg, pp. 127-142, 1985.

27. Young, W.: Blood flow, metabolic and neurophysiological mechanisms in spinal cord injury. In: D.P. Becker; J.T. Povlishock (eds.): Central Nervous System Trauma Status Report - 1986. NIH-NINCDS, Bethesda, Maryland, pp. 463-473, 1986a.

28. Young, W.: Letter. J. Neurosurg., 64: 987-988, 1986b.

29. Young, W.; Berenstein, A.: Somatosensory evoked potential monitoring of intraoperative procedures. In: J. Schramm; S. Jones (eds.): Spinal Cord Monitoring. Springer-Verlag, Berlin, Heidelberg, New York, Tokyo, pp. 197-203, 1985.

30. Young, W.; Cohen, A.; Merkin, H.; Fisher, B.; Berenstein, A.; Ransohoff, J.: Somatosensory evoked potential changes in spinal injury and during intraoperative manipulation. J. American Paraplegia Soc., 5: 44-48, 1982a.

31. Young, W.; DeCrescito, V.: Sodium ionic changes in injured spinal cords: Mechanisms of edema. Proc. Soc. Neuroscience, 16: 267, 1986.

32. Young, W.; Flamm, E.S.: Effect of high dose corticosteriod therapy on blood flow, evoked potentials, and extracellular calcium in experimental spinal injury. J. Neurosurg., 57: 667-673, 1982.

33. Young, W.; Flamm, E.S.; Blight, A.R.; Gruner, J.A.; DeCrescito, V. (in press): Pharmacological therapy of acute spinal cord injury: Studies of high dose methylprednisolone and naloxone. Clin. Neurosurg.

34. Young, W.; Flamm, E.S.; Ransohoff, J.; Demopoulos, H.B.; DeCrescito, V.; Tomasula, J.J.: Effect of naloxone on posttraumatic ischemia in experimental spinal contusion. J. Neurosurg., 55: 209-219, 1981.

35. Young, W.; Koreh, I.: Potassium and calcium changes in injured spinal cords. Brain Res., 365: 42-53, 1986.

36. Young, W.; Koreh, I.; Yen, V.; Lindsay, A.: Effects of sympathectomy on extracellular potassium ionic activity and blood flow in experimental spinal cord contusion. Brain. Res., 253: 115-124, 1982b.

37. Young, W.; Mollin, D. (in press): Intraoperative monitoring of somatosensory evoked potentials. In: J. Desmedt (ed): Advances in Neurology. Raven Press.

38. Young, W.; Tomasula, J.J.; DeCrescito, V.; Flamm, E.S.; Ransohoff, J.: Vestibulo spinal monitoring in experimental spinal trauma. J. Neurosurg, 52: 64-72, 1980.

39. Young, W.; Yen, V.; Blight, A.R.: Extracellular calcium ionic activity in experimental spinal cord contusion. Brain Res., 253: 105-113, 1982c.

Continuous Somatosensory Evoked Potential Monitoring in the Neurointensive Care Unit

R.B. Hansen;[*] C.A. Vaz; C. Borel; N.V. Thakor; D.F. Hanley

Summary

We describe a technique of continuous median nerve somatosensory evoked potential testing used as a monitoring device in our neurointensive care unit. Early experience suggests that such testing can be successfully done but demands a great deal of attention from technical and nursing personnel. Appropriate application of the technique may be in selected cases of therapeutic barbiturate anesthetic coma and in delayed ischemia from subarachnoid hemorrhage but in these and other potential situations clinical utility remains to be proven. The effects of commonly used drugs in the neurointensive care unit, such as lidocaine, as well as factors such as patient positioning, also need to be more completely evaluated.

Introduction

The intensive care unit management of patients with severe disease of the nervous system depends on the ability to monitor changes in patient state. At present, the mainstays of neurologic monitoring are serial bedside neuroexaminations and measurement of the intracranial pressure. However, both reflect the effects of pathophysiologic processes and do not directly measure cellular response to various forms of injury. In addition, certain therapies, such as induced barbiturate coma, interfere with the ability to follow the patient clinically. Under some circumstances, electrophysiologic monitoring may be useful in detecting changes in patient condition, especially when deteriorations may be expected to occur, such as during carotid and spinal surgery (1, 2). Situations in which deteriorations might be expected to occur are also encountered in the neurologic intensive care unit and include delayed ischemia (vasospasm) in aneurysmal subarachnoid hemorrhage and progressive brain injury during barbiturate coma. In order to evaluate the usefulness of SEP monitoring in such patients we devised a system to continuously record the median nerve SEP and display waveform, amplitude and latency determinations. We have made preliminary recordings of control subjects and selected patients to assess the feasibility of such testing in the neuro-ICU.

Methods

An IBM PC was interfaced to a TECA Neurolab evoked potential instrument in such a way as to control the stimulator unit and receive analog data from the amplifiers. This was passed through an A/D converter and placed in averaging buffers in the IBM. After one thousand repetitions were accumulated, the buffer would accept a new

* Departments of Neurology, Biomedical Engineering, and Anesthesiology/Critical Care, Johns Hopkins Medical Institutions, 600 North Wolfe Street, Baltimore, MD 21205

33 y/o ♂ Control

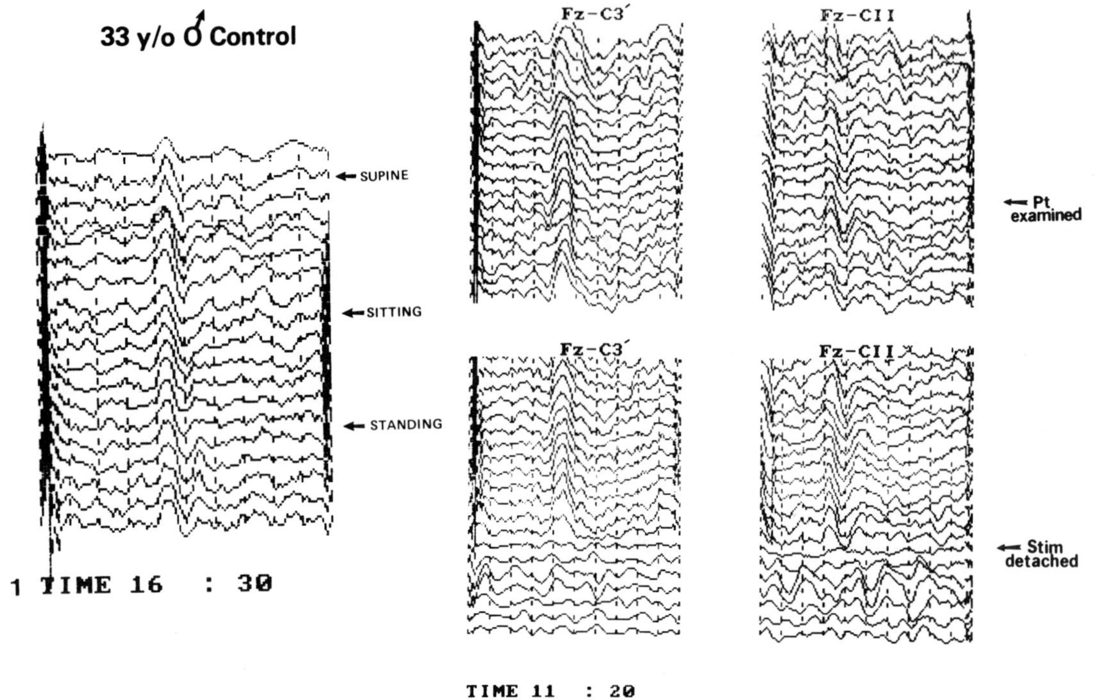

72 y/o ♂ c̄ SAH

17 y/o ♂ c̄ Closed head injury

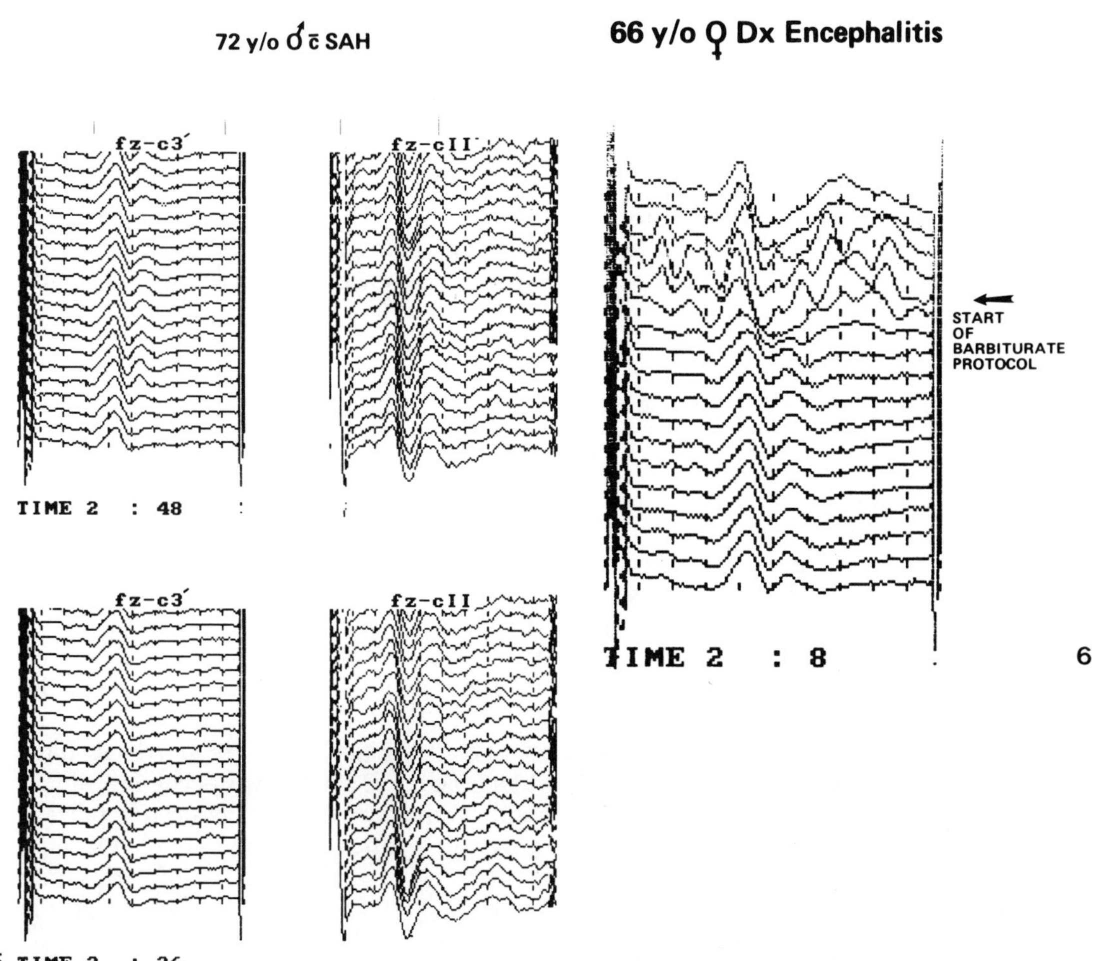

Figs. 1-6 (Facing page, and above) Single channel illustrations are those of the Fz-C3 derivation. Two channel illustrations show Fz-C3 and Fz-CII. Each column of hatched lines denotes 4ms. After obtaining a baseline study, the sensitivity on the continuous display was adjusted to produce visually satisfactory tracings. Each block represents approximately 40 minutes of recording.

stimulus while discarding the one most remote, thus creating a "moving average" of only the one thousand most recent stimuli.

Before embarking upon continuous monitoring, a standard set of bilateral median nerve SEPs were obtained using superimposed trials of at least one thousand stimuli each. Four channels were employed with electrode locations at Erb's point, C11, C3 and C4 (2cm posterior to C3 and C4 respectively) all referenced to FZ. The bandpass was 100 to 3000hz and the amplification 500,000. Square wave stimlui of 0.1ms were given to either median nerve at the wrist by means of silver electrodes embedded in a plastic bar which was secured with an elastoplast dressing. The stimuli was adjusted to 10-15% above the motor threshold, and the stimulus rate was 5.9/sec. Electrode impedances were kept below 3000ohms by periodically using ECG gel to refill tin electrodes which were held in place by collodion soaked gauze squares.

After the baseline SEPs were obtained, continuous monitoring of the hemisphere judged most likely to show changes with disease progression was begun. Recordings were made on volunteer control subjects and on patients with a variety of disorders including subarachnoid hemorrhage, closed head injury, and increased intracranial pressure from viral encephalitis. The duration of the recording ranged from a few hours to more than 24 hours. No ill-effects from prolonged stimulation was seen either in control subjects or in patients.

Every 30 seconds a printer would graphically record both the N20 latency and amplitude (baseline to peak) to produce horizontal trend plots of these parameters. Wave forms from two channels (displaying the N20 and N13 waves) were stacked vertically on a color monitor, with an updated trace appearing every 2 minutes. When the screen was full, it would be "dumped" onto the hard copy paper as well as placed into disc storage.

Results

A large number of factors work against successful continuous SEP monitoring in the neuro-ICU. The stimulating and recording electrodes are in constant jeopardy while the patient is being attended to by nurses, physicians, and technicians. A great deal of patient repositioning necessarily takes place in addition to spontaneous patient movement. These changes in position had no apparent effect on the SEP waveforms. Examples of this can be seen during changes of posture in a control subject (Fig. 1), with moving a patient during the course of a consultant examination (Fig. 2), and with a repositioning of the patient by the nurse (Fig. 3).

In patients in whom no clinical changes were occurring, the waveforms remained stable during both daytime recording, as well as during recordings done throughout the night (Fig. 4).

Preliminary studies of patients with aneurysmal subarachnoid hemorrhage have shown no changes in the continuous SEP when no clinical change was occurring, and in one patient, no changes during mild fluctuating hemiparesis during induced hypertensive treatment for vasospasm (Fig. 5). However, in a patient with persistently elevated ICP (greater than 25mmHg) during the acute phase of viral encephalitis, barbiturate anesthesia produced a distinct latency shift of N20 followed by stability of the waveforms. The patient eventually made a full recovery (Fig. 6).

The parallel display of the channel displaying N13 served as a check on the integrity of the recording system. Situations occurred in which a sudden loss of the cortical component was seen but simultaneous loss of the medullary potential suggested that the change was not due to sudden deterioration of CNS function. An example is seen in Fig. 7 in which the stimulator became detached.

Discussion

The role of evoked potential testing in the neuro-ICU is still being defined. A number of reports have reported their usefulness in prognostication of head injury and anoxia (3-9). Recent interest has focused on the application of SEPs in the assessment of cerebral ischemia (10). Some reports suggest that changes in SEP latencies and amplitudes may be seen with graded ischemia (11, 12), while others hold that EP changes depend upon a CBF "threshold effect" similar to that seen with EEG (13, 14). Clearly, SEP monitoring would be most helpful if it can detect changes in CBF before critical "pre-ischemic" levels are reached. Application of SEP testing in subarachnoid hemorrhage has been reported (15) with differences in the so-called central conduction time seen between patients with good clinical grades and those with marked deficits (grade 4). Our patient with fluctuating hemiparesis did not show changes in the SEP during the period of instability. More study of the relationship of CBF, clinical changes, and SEP changes in this setting is needed.

Several reports have detailed the usefulness of SEP monitoring in therapeutic barbiturate coma, although the magnitude of the anesthetic effect on latency is disputed

(16-19). We have seen latency shifts and subsequent stability in patients who have been successfully treated, suggesting that continuous SEP monitoring is useful in this situation.

Effects on the SEP have been reported with changes in position of the neck (20) and with movement of the stimulated limb (21). This has not proved to be a problem over the range of patient movement seen in our unit. We have also been impressed with the stability of the waveforms over long periods of recording, a finding which has been seen by others (22). The relative insensitivity of the N13 through N20 interpeak latency (central conduction time) to temperature changes (23) also adds to the utility of SEP monitoring. However, the varying effects of different anesthetics on the SEP (23) makes further quantification of the effects of drugs commonly used in the neuro-ICU, such as lidocaine, necessary.

In summary, our early experience with continuous SEP monitoring leads us to believe that this technique can be feasibly used in an intensive care setting. However, more experience with the effects of drugs, temperature, circadian rhythms and body positions is needed. Additionally, further delineation of specific clinical conditions in which such a technique could provide useful information is also necessary before this type of monitoring can be generally recommended.

References

1. Hacke, W.: Neuromonitoring. J. Neurol., 232: 125-133, 1985.
2. Markand, O.N.; Dilley, R.S.; Moorthy, S.S.; Warren, C.: Arch. Neurol., 41: 375-378, 1984.
3. Newlon, P.G.; Greenberg, R.P.; Enas, G.G.; Becker, D.P.: Effects of therapeutic pentobarbital coma on multimodality evoked potentials recorded from severely head-injured patients. Neurosurg., 12: 613-619, 1983.
4. Narayan, R.K.; Greenberg, R.P.; Miller, J.D.; Enas, G.G.; Choi, S.C.; Kishore, P.R.S.; Selhorst, J.B.; Lutz, H.A.; Becker, D.P.: Improved confidence of outcome prediction in severe head injury. J. Neurosurg., 54: 751-762, 1981.
5. Hume, A.L.; Cant, B.R.: Central somatosensory conduction after head injury. Ann. Neurol., 10: 411-419, 1981.
6. Anderson, D.C.; Bundlie, S.; Rockswold, G.L.: Multimodality evoked potentials in closed head trauma. Arch. Neurol., 41: 369-374, 1984.
7. Newlon, P.S.; Greenberg, R.P.; Hyatt, M.S.; Enas, G.G.; Becker, D.P.: The dynamics of neuronal dysfunction and recovery following severe head injury assessed with serial multimodality evoked potentials. J. Neurosurg., 57: 168-177, 1982.
8. Hume, A.L.; Cant, B.R.; Shaw, N.A.: Central somatosensory conduction time in comatose patients. Ann. Neurol., 5: 379-384, 1979.
9. Rumpl, E.; Prugger, M.; Gerstenbrand, F.; Hackl, J.M.; Pallua, A.: Central somatosensory conduction time and short latency somatosensory evoked potentials in post-traumatic coma. Electroenceph. and Clin. Neurophysiol., 56: 583-596, 1983.
10. Prior, P.F.: EEG monitoring and evoked potentials in brain ischemia. Br. J. Anaesth., 57: 63-81, 1985.
11. Sato, M.; Pawlik, G.; Umbach, C.; Heiss, W.D.: Comparative studies of regional CNS blood flow and evoked potentials in the cat. Stroke, 15: 97-101, 1984.
12. Lesnzck, J.E.; Michele, J.J.; Simeone, F.A.; DeFeo, S.; Welsh, F.A.: Alteration of somatosensory evoked potentials in response to global ischemia. J. Neurosurg., 60: 490-494, 1984.
13. Astrup, J.; Symon, L.; Branston, N.M.; et al.: Cortical evoked potential and extracellular K + and H + at critical levels of brain ischemia. Stroke, 8: 51-57, 1977.
14. Hargadine, J.R.; Branston, N.M.; Symon, L.: Central conduction time in primate brain ischemia -- A study in baboons. Stroke, 11: 494-498, 1980.
15. Rosenstein, J.; Wang, A.D.J.; Symon, L.; Suzuki, M.: Relationship between hemispheric cerebral blood flow, central conduction time and clinical grade in aneurysmal subarachnoid hemorrhage. J. Neurosurg., 62: 25-30, 1985.
16. Reisecker, F.; Witzmann, A.; Loffler, W.; Deisenhammer, E.; Valencak, E.: Somatosensory and brain stem acoustic evoked potentials during barbiturate induced comatose states. Electroenceph. and Clin. Neurophysiol., 61: 539, 1985.
17. Ganes, T.; Lundar, T.: The effect of thiopentone on somatosensory evoked responses and EEGs in comatose patients. J. Neur. Neurosurg. Psych., 46: 509-514, 1983.

18. Sutton, L.N.; Frewen, T.; Marsh, R.; Jaggi, J.; Bruce, D.A.: The effects of deep barbiturate coma on multimodality evoked potentials. J. Neurosurg., 57: 178-185, 1982.
19. Lundar, T.; Ganes, T.; Lindegaard, K.F.: Induced barbiturate coma: Methods for evaluation of patients. Crit. Care Med., 11: 559-562, 1983..
20. McPherson, R.W.; Szymanski, J.; Rogers, M.C.: Somatosensory evoked potential changes in position-related brain stem ischemia. Anesthesiology, 61: 88-90, 1984.
21. Angel, R.W.; Weinrich, M.; Rodnitzky, R.: Recovery of somatosensory evoked potential amplitude after movement. Ann. Neurol., 19: 344-348, 1986.
22. Myklebust, J.B.; Sances, A.; Cusick, J.F.; Friedman, R.H.; Larson, S.J.; Cohen, B.A.: Stationarity of the somatosensory evoked potential. Med. & Biol. Eng. & Comput., 22: 558-563, 1984.
23. Hume, A.L.; Durkin, M.A.: Central and spinal somatosensory conduction times during hypothermic cardiopulmonary bypass and some observations on the effects of fentanyl and isoflurane anesthesia. Electrophysiol. Clin. Neurophysiol., 65: 46-58, 1986.

A Comparison of Dermatomal and Major Nerve Evoked Responses with Clinical Diagnosis in Acute Spinal Injury

J. R. Toleikis;[*] T. B. Sloan

Introduction

The present status of evoked potential technology has made it relatively commonplace to monitor activity along the sensory neural pathways in human peripheral nerves, spinal cord and the brain. It has been widely applied for monitoring of these pathways during surgical procedures where undesired trauma may occur. Similarly, it can be utilized to evaluate the functional integrity of an already traumatized spinal cord and the process of recovery from that injury. Several such studies using somatosensory evoked potentials (SEPs) have been published regarding the evaluation of patients who have sustained spinal cord injuries (1-6). The patients in these studies were generally evaluated with lower extremity (e.g., posterior tibial or peroneal nerve) SEPs because the signals obtained from such stimulation in normal individuals are relatively large in amplitude and have characteristic morphologies. It has been accepted that the presence of a SEP resulting from stimulation of a nerve originating below the level of injury in the period of 24 hours to one week post injury is a favorable prognostic sign. This is thought to indicate an incomplete lesion where return of some degree of neural function is possible. On the other hand, the consistent absence of such a signal is an unfavorable prognostic indicator and generally correlates with a complete lesion.

Although useful, these major nerve SEPs test only a selected region of the spinal cord and fail to evaluate large regions where function may be lost or spared. The clinical exam is potentially far more comprehensive than evoked potential testing (both in the regions of the body that are testable and the modalities that can be tested), but lacks value in patients in coma and is subject to variability between examiners.

Dermatomally elicited evoked potentials possess the objectivity and potential to expand the SEP sensory testing to wide areas of the body for a more complete evaluation of the patient with a spinal cord injury. Techniques for eliciting segmental or dermatomal evoked potentials have been detailed in both the European and American literature (7-21) in patients with nerve root or spinal cord abnormalities. Its diagnostic value has allowed delineation of subtle disc disease or other conditions involving nerve root compression (7-14). It has also been shown to be useful in studies of patients with various spinal cord abnormalities from both space occupying and non-space occupying lesions (15-20) and for a small group of patients with chronic spinal cord injuries (22).

[*] Department of Anesthesia, Northwestern University Medical School, 303 East Superior Street, Passavant Pavillion - Room 360, Chicago, IL 60611

This study was undertaken to compare the clinical diagnosis of acute spinal injured patients with information obtained from SEP and dermatomal testing.

Methods

The techniques for acquiring dermatomal evoked potentials have been described elsewhere (14, 18, 20, 23, 24). All testing was done at the Spinal Cord Injury Unit of Northwestern Memorial Hospital. Over the past two and one half years, 515 patients have been tested using the following protocol. This protocol accommodates the patient's clinical management, the patient's comfort (usually a function of how long they could tolerate being in one position), and the time necessary to acquire enough information to make a meaningful electrophysiological assessment of the patient. A testing session of approximately two hours has appeared to be optimal.

The following assessment of a patient's sensory function (on a scale of 0 to 2) was made of those areas which receive their innervation from around the level of injury. A zero was given to those dermatomes where sensation appeared to be absent. A one was given to those dermatomes where sensation appeared to be present but the sensory threshold was obviously elevated and a two was given to those areas where sensation appeared to be normal. Traditional SEPs were then acquired from each of the extremities using median or ulnar nerve stimulation for the upper extremities and posterior tibial or peroneal nerve stimulation for the lower extremities. Duplicate responses were acquired with stimulation sufficient to elicit a motor response. Occasionally, more than one nerve per extremity was tested when it was felt that the additional information might be helpful in evaluating the patient's condition.

Finally, dermatomal responses were acquired. Repeatable responses can be obtained from normals at dermatomal levels as high as the fourth cervical (C4) dermatome through the fourth sacral (S4) area (20). Dermatomal testing focused primarily on the level of injury. Responses were normally obtained from four levels; the level of injury, a level above, and from each of the two levels below. If lower extremity major nerve responses were abnormal or absent, sacral sparing was evaluated by testing for sacral dermatomal responses. In addition, dermatomal responses were obtained from areas of questionable sensory deficit or areas of unexpected deficit.

Standardized stimulation locations were utilized for all dermatome levels. Electrodes were placed approximately 3.0cm apart with the anode distal (lateral in the case of the thoracic dermatomes) in the center of each dermatome region in order to avoid stimulating more than one dermatome at a time. When dermatomes were situated very close to each other, as in the case of the thoracic (T2 through T12) dermatomes, only even numbered dermatomes were tested. Stimulation consisted of 300 millisecond square wave constant current pulses delivered at 5.7 to 8.7Hz. These rates minimized testing time and did not appear to cause degradation of the responses. The stimulation intensity was applied at a level which the patient described as feeling strong but not painful. This approach optimized the amplitude of the response without causing excessive EMG artifact. Generally, stimulation at 1-2 milliamps above sensory threshold was sufficient to elicit small but repeatable responses in normals. Stimulation was always below that which produced muscle twitching (most troublesome in the thoracic region), as this may produce stimulation of adjacent dermatomes. Considerable care was necessary to minimize EMG contamination, while limiting the duration of the test session.

Testing was conducted on either Stryker frames or Roto-Rest beds with spinal traction intact. As a consequence, multi-channel recording of spinal cord potentials was not feasible. Single channel recordings were therefore acquired from only scalp locations using standard 10mm gold EEG cup electrodes. Electrode impedances were kept below 5k ohms. Based on previous studies (20), recording electrodes were placed at locations 2cm behind the standard EEG placements for CZ, C1, C2, C3 and C4. A reference electrode was placed at Fz and a ground electrode was placed on the patient's shoulder. A timebase of 50 milliseconds was used for acquiring upper extremity and C4 through T8 dermatomal responses and 100 milliseconds was used to acquire lower extremity and dermatomal responses from below T8. Several averages with excessive artifact were discarded and at least two grand averages were then created by summing two or more 250 sample averages. The filter bandwidth of 5 to 250Hz was always used with no 60 hertz notch filtration. The low pass setting of 250Hz appeared to cause minimal degradation in the cortical responses while limiting the amount of high frequency artifact that was commonplace in the spinal cord intensive care unit.

On admission, patients were clinically diagnosed as being neurologically intact or having a complete or incomplete lesion. An intact patient was defined as having no clinically apparent neurological deficit. Those with complete lesions had no apparent neurological function below the level of injury, whereas a patient with an incomplete lesion was determined as having some preservation of motor and/or sensory function below the level of injury. Each of the diagnoses based on the major nerve SEPs and the dermatomal SEPs were compared to the admitting clinical diagnosis.

Results

Five hundred and fifteen acute spinal cord injured patients have been tested. Of these, 25.5% were tested within 24 hours of injury and an additional 50.4% were tested in the five days following injury. Of these, 73.4% were male. The population was relatively young with 47.7% of the patients under the age of 30 and 72.5% under the age of 50. The largest proportion of spinal cord injuries occurred in the cervical region (62.9%) with over half of these occurring at C5 and C6. Of the remaining injuries, 15.7% occurred in the thoracic region and 21.4% occurred in the lumbar and sacral areas.

In 88 patients, only major nerve SEP testing was conducted because of time constraints, lack of apparent anatomic injury, high cervical injury beyond the range of testing (C1 or C2) or an occasional uncooperative patient. In the remaining 427 patients, the clinical diagnosis at the time of testing was compared to the results of dermatomal and major nerve SEP testing. These results are show in Table 1. Chi-squared analysis of these data reveals $X^2 = 45.1$ (p < 0.000001).

In 252 patients, the results of both major nerve and dermatomal testing agreed with the admitting clinical diagnosis. However, in many of these cases, dermatomal testing defined abnormalities not evident clinically or with major nerve SEP testing.

Of particular interest are the 112 cases where major nerve SEP testing supports the admitting clinical diagnosis but dermatomal responses suggest a different diagnosis. In 33 of these cases, patients were clinically diagnosed as being neurologically intact, but at least one dermatomal level had no repeatable response, thus suggesting some degree of neural deficit. Another 47 of these patients were admitted as having incomplete spinal cord injuries with diagnoses such as Brown-Sequard, central cord and anterior cord lesions. In these cases, the level-by-level dermatomal testing at the level of injury suggested that the patients were neurologically intact rather than incomplete. In many cases, however, these dermatomal responses were present and repeatable but abnormal around the level of injury. There were 21 patients who were diagnosed as having a complete lesion at a specific level and were found by dermatomal testing to have sensory responses from spinal regions below this level. Therefore, dermatomal testing suggested that they had an incomplete rather than a complete lesion at the specific level. In another 11 patients, dermatomal responses suggested a complete lesion at an even higher level than clinically suspected.

In the 15 cases where dermatomal responses agreed with the clinical diagnosis but major nerve SEP responses disagreed, the differences resulted from the way the presence and absence of major nerve responses were interpreted. When upper extremity SEPs were present and lower extremity SEP responses were absent, major nerve testing suggested a complete lesion between cervical level 7 and lumbar level 4. The presence of signals at intermediate levels or sacral signals suggested that the patient had an incomplete rather than a complete lesion.

The presence of both upper and lower extremity major nerve SEPs suggested that patients were neurologically intact. However, signal deficits at individual spinal levels would suggest an incomplete lesion. These deficits could only be detected with dermatomal responses. These two types of situations accounted for the differences in the major nerve and dermatomal SEP results for this group of 15 patients.

In the 48 cases where both the dermatomal and major nerve SEP responses disagreed with the clinical diagnosis, there were only eight patients in which some neural deficit was clinically thought to be present, but none was apparent with either major nerve or dermatomal testing. In the remaining cases, the types of neural deficits formed the basis for discrepancies between admitting diagnosis and test results. Patients thought to have incomplete lesions on clinical exams were indicated as having complete lesions when tested with both types of evoked potentials. The remaining patients thought to have clinically complete lesions were suggested as having incomplete lesions by evoked potential testing.

Discussion

At most hospitals, the admitting clinical diagnosis for spinal cord injured patients is made in the emergency room by the neurosurgical and/or orthopedic staff. This diagnosis provides a guideline for the subsequent management of these patients. However, this evaluation is subjective and therefore prone to variability between examiners. Traditionally, major nerve SEPs have played a role in the evaluation of these patients because of their objective nature. However, except for providing normally large, well defined, easily elicited waveforms, major nerve SEPs are a rather crude measure of spinal cord integrity. Many areas cannot be tested directly with conventional SEP techniques and abnormalities when found cannot be delineated to specific spinal levels.

Further, we have previously demonstrated that major nerve SEP responses can be normal when component dermatomal responses are absent (23). Dermatomal evoked potential testing appears to be a means for overcoming some of these shortcomings and providing information to the clinical staff that might not be apparent and obtainable by any other means. There are indications that this information might act as a prognostic indicator as well (24).

At Northwestern Memorial Hospital, evoked potential testing is utilized as a key assessment of the acutely injured patient. The results affect the subsequent diagnostic procedures, surgical management and rehabilitation of these patients. The results of these comparisons indicate that in the majority of patients (59%), both the results of major nerve and dermatomal testing support the clinical diagnosis. In these cases, the dermatomal results were often as informative as the major nerve SEP testing in supporting the clinical diagnosis. In 11.2% of the patients, the major nerve and dermatomal SEP results were equally effective in defining a discrepancy with the admitting diagnosis.

In those cases in which dermatomal results differed from the major nerve results and the clinical admitting diagnosis, the basis for this discrepancy appeared to arise in the shortcomings of the traditional major nerve SEP technique. Namely, it appeared to be associated with the inability of this technique to detect sacral sparing and to delineate and define central cord and other incomplete spinal cord lesions.

Discrepancies between the admitting clinical diagnosis and the SEP results are to be expected. Although the time between cord insult and testing was in most cases less than five days, this is certainly enough time for the patient's condition to change. Function that may have been present early may have disappeared and in other cases, function may have returned as the level of neural function recedes and advances with edema, secondary insults and medical management. Finally, the clinical diagnosis is based on modalities such as motor function not testable by sensory evoked responses. It is not surprising that a certain percentage of the patients should then have a significant difference between their admitting and SEP diagnoses.

Inherent in this type of analysis are problems associated with the patients being tested and the conditions under which testing is taking place. Clearly, nearly the entire cord (C4 through S4) could be tested. However, restlessness and pain inherent to the patient's injuries ultimately cause patients to introduce EMG artifact. This is one of the primary reasons for limiting a testing period to two hours. In addition, many of the patients that arrive on the spinal cord injury unit may be predisposed to poor quality tracings because of their general physical condition. Both these factors can contribute to poor responses; particularly in those instances in which the signals are expected to be small in amplitude in the best of cases. This is generally true for those areas which are minimally represented in the sensory cortex in terms of cortical surface area such as the thoracic and sacral dermatomes.

Several investigators have published papers regarding the utility of dermatomal or segmental SEPs (7-20, 22). These papers and our experience indicate that repeatable responses can be obtained using these techniques in normal as well as injured patients. Some authors have expressed reservations about the clinical utility of these studies (7, 8). However, others have found them reliable and useful (9-20, 22).

Unlike the applications to which the dermatomal or segmental techniques have been applied where statistical variability plays a large role in determining the utility of the testing procedure, spinal cord assessment is most concerned with identifying intact neural function. The primary reason for utilizing the dermatomal technique to evaluate spinal cord injured patients is that it might be a means to detect information that is subclinical or clinically ambiguous and not detectable using standard SEP techniques. In a population of patients who had experienced severe trauma to the spinal cord, this information would be expected to be abnormal and usually was. Nevertheless, the identification of spared neural function beyond that seen clinically, even if it was abnormal, dictated that the patient be managed more aggressively in an effort to spare this function. The implications of sparing even one level in a patient with a spinal cord lesion can be significant in terms of functioning in a productive and self-reliant manner. The results of this study, clearly suggest that dermatomal responses provide significantly detailed, objective information that better defines the lesions than major nerve SEPs and clinical testing. Of particular note is the identification of function not thought present (i.e., function below an otherwise complete lesion) and the identification of unexpected areas of injury.

These findings are comparable to those seen in many patients with low back pain (12, 14). In these cases, lower extremity major nerve responses were oftentimes normal and there were significant abnormalities present in the lumbar and/or sacral dermatomal responses. As shown in a previous study, it was not uncommon to find median nerve responses which appeared normal with only one normal dermatomal component; the others being either abnormal or absent (23). Based on these findings, dermatomal testing appears to be an important adjunct to traditional SEP testing and the clinical examination for monitoring neural function in spinal cord injured patients. Its ability to delineate specific dermatomally related cord injuries makes it a valuable monitor for delineating cord damage and detecting functional changes during the post injury period. Further studies are needed to evaluate the prognostic value of these results. However, specific instances have demonstrated a favorable outcome not otherwise expected (24).

Table 1 - Comparison of evoked response testing with clinical diagnosis.

| | | Major Nerve SEP | |
		Agree	Disagree
Dermatomal			
Evoked	Agree	252 (59%)	15 (4%)
Responses	Disagree	112 (26%)	48 (11%)

Total patients = 427

References

1. Perot, P.L.; Vera, C.L.: Scalp-recorded somatosensory evoked potentials to stimulation of nerves in the lower extremities and evaluation of patients with spinal cord trauma. Ann. NY Acad. Sci., 1982, 388: 359-368.
2. Perot, P.L.: The clinical Use of evoked potentials in spinal cord injury. Clin. Neurosurg., 1973, 20: 367-381.
3. Rowed, D.W.; McLean, J.A.G.; Tator, C.H.: Somatosensory evoked potentials in acute spinal cord injury: Prognostic value. Surg. Neurol., 1978, 9: 203-210.
4. Rowed, D.W.: Value of somatosensory evoked potentials for prognosis in partial cord injuries. In: C.H. Tator (ed.), Early Management of Acute Spinal Cord Injury. New York, Raven Press, 1982, pp. 167-180.
5. Chabot, R.; York, D.H.; Watts, C.; Waugh, W.A.: Somatosensory evoked potentials evaluated in normal subjects and spinal cord-injured patients. J. Neurosurg., 1985, 63: 544-551.
6. Spielholz, N.I.; Benjamin, M.V.; Engler, G.; Ransohoff, J.: Somatosensory evoked potentials and clinical outcome in spinal cord injury. In: A.J. Popp et al. (eds), Neural Trauma, New York: Raven Press, 1979, pp. 217-222.
7. Aminoff, M.J.; Goodin, D.S.; Barbaro, N.M. et al.: Dermatomal somatosensory evoked potentials in unilateral lumbosacral radiculopathy. Ann. Neurol., 1985, 17: 171-176.
8. Aminoff, M.J.; Goodin, D.S.; Parry, G.J. et al.: Electrophysiologic evaluation of lumbosacral radiculopathies: Electromyography, late responses, and somatosensory evoked potentials. Neurology, 1985, 35: 1514-1518.
9. Dvonch, V.; Scarff, T.; Bunch, W.H. et al.: Dermatomal somatosensory evoked potentials: Their use in lumbar radiculopathy. Spine, 1984, 9: 291-293.
10. Eisen, A.: The somatosensory evoked potentials. Canadian J. Neurol. Sci., 1982, 9: 65-77.
11. Eisen, A.; Hoirch, M.; Moll, A.: Evaluation of radiculopathies by segmental stimulation and somatosensory evoked potentials. Canadian J. Neuro. Sci., 1983, 10: 178-182.
12. Green, J.; Gildemeister, R.; Hazelwood, C.: Dermatomally stimulated somatosensory cerebral evoked potentials in the clinical diagnosis of lumbar disc disease. Clin. Electroenceph., 1983, 14: 152-160.
13. LaJoie, W.J.; Melvin, J.L.: Somatosensory evoked potentials elicited from individual cervical dermatomes represented by different fingers. Electromyogr. Clin. Neurophysiol., 1983, 23: 403-411.
14. Scarff, T.B.; Dallmann, D.E.; Toleikis, J.R.; Bunch, W.H.: Dermatomal somatosensory evoked potentials in the diagnosis of lumbar root entrapment. Surg. Forum, 1981, 32: 489-491.
15. Baust, W.; Ilsen, H.W.; Jorg, W.; Wambach, G.: A neurophysiological method for the localization of transverse lesions of the spinal cord. Acta. Neurochir., 1972, 26: 352-353.
16. Jorg, J.: Die Electrosensible Diagnostik In Der Neurologic. Berlin: Springer, 1977.
17. Jorg, J.; Dullberg, W.; Koeppen, S.: Diagnostic value of segmental somatosensory evoked potentials in cases with chronic progressive para- or teraspastic syndromes. In: J. Courjon et al. (eds), Clinical Applications of Evoked Potentials in Neurology. New York: Raven Press, 1982, pp. 347-358.
18. Scarff, T.B.; Toleikis, J.R.; Bunch, W.H.: Dermatomal somatosensory evoked potentials in children with myelomeningocele. Z. Kinderchir., 1979, 28: 384-387.
19. Schramm, J.; Oettle, G.J.; Pichert, T.: Clinical applications of segmental somatosensory evoked potentials (SEP) - Experience in patients with non-space occupying lesions. In: C. Barber (ed.), Evoked Potentials. Leicester: MTP-Press, 1980, pp. 455-465.
20. Toleikis, J.R.; Sloan, T.B.; Schrader, S.; Koht, A.: Scalp distribution of dermatomal evoked potentials. In: J. Schramm; S.J. Jones (eds), Spinal Cord Monitoring. Berlin-Heidelberg: Springer-Verlag, 1985, pp. 59-63.

21. Katifi, H.A.; Sedgwick, E.M.: Somatosensory evoked potentials from posterior tibial nerve and lumbo-sacral dermatomes. Electroenceph. Clin. Neurophy., 1986, 65: 249-259.
22. Louis, A.A.; Gupta, P.; Perkash, I.: Localization of sensory levels in traumatic quadriplegia by segmental somatosensory evoked potentials. Electroencephalgr. Clin. Neurophysiol., 1985, 62: 313-316.
23. Toleikis, J.R.; Sloan, T.B.: A comparison study of major nerve and dermatomal somatosensory evoked potentials in the evaluation of spinal cord injured patients. In: C. Barber; T. Blum (eds), Evoked Potentials III: The Third International Evoked Potentials Symposium. Stoneham, Massachusetts: Butterworth (In Press).
24. Schrader, S.C.; Sloan, T.B.; Toleikis, J.R.: Detection of sacral sparing in acute spinal cord injury. Spine (in press).

Evoked Spinal Cord Action Potential in Syringomyelia Level Diagnosis and Spinal Cord Monitoring

K. Satomi;[*] T. Okuma; K. Kenmotsu; Y. Nakamura; T. Hayakawa; K. Hirabayashi

Summary

Evoked spinal cord action potentials due to both median nerve and spinal cord stimulation were recorded in ten patients with cervical and dorsal syringomyelia, and the evoked spinal cord action potentials (ESCAPs) which had positive potentials were graded from normal to grade 3. The N2 deflection of MN-ESCAP, which may originate from the posterior horn of the spinal cord, was not recorded in 6 out of 8 recordings. Both N1 and N2 deflections of the SC-AESCAP, which may originate from the white matter of the spinal cord, were recorded in 6 out of 10 recordings. Therefore the syrinx could be electrophysiological- ly located in the gray matter of the spinal cord in these six cases. In three patients, an ESCAP graded 3 was observed in the cervical region, and the positive potential recorded at the spinal level was attributed to a main lesion among the cervical cord with syrinx. yringomyelia

Spinal cord monitoring using ESCAP was performed in seven cases during the myelotomy and shunting. Slight changes of the amplitude of ESCAP were observed after myelotomy in three cases, but no remarkable neurological deficiency except position sense appeared after the operation.

Introduction

Syringomyelia can be easily diagnosed by Magnetic Resonance Imaging, and there has been a subsequent increase in the number of cases treated operatively. However, it is generally difficult to predict accurately the level of the main lesion in the spinal cord, because a syrinx can develop across a wide area of the spinal cord.

To diagnose the main lesion, ESCAPs were recorded in patients with cervical and dorsal syringomyelia. Spinal cord monitoring using ESCAPs was performed to prevent the spinal cord from damage during myelotomy and tubing for treatment of the syringomyelia.

Clinical materials

Ten patients with syringomyelia were operated on in our clinic between 1978 and 1986 with ESCAP recording. There were 6 males and 4 females between the ages of 20 and 51 years, with an average of 38.9 years. Neurological signs and symptoms on admis-

---

* Department of Orthopaedics, School of Medicine, Keio University, 35, Shinano-Machi, Shinjuku-Ku, Tokyo 160, Japan

Table 1. Summary of clinical data in ten patients with syringomyelia.
* C = cervical spine; T = thoracic spine; L = lumbar spine.

<table>
<tr><td colspan="2"></td><td>1</td><td>2</td><td>3</td><td>4</td><td>5</td><td colspan="2">6</td><td>7</td><td colspan="2">8</td><td>9</td><td>10</td></tr>
<tr><td colspan="2">sex</td><td>M</td><td>F</td><td>F</td><td>M</td><td>F</td><td colspan="2">M</td><td>F</td><td colspan="2">M</td><td>M</td><td>M</td></tr>
<tr><td colspan="2">age</td><td>20</td><td>49</td><td>28</td><td>48</td><td>51</td><td>37</td><td>38</td><td>35</td><td colspan="2">43</td><td>36</td><td>42</td></tr>
<tr><td colspan="2">dissociated sensory loss</td><td>+</td><td>−</td><td>+</td><td>−</td><td>+</td><td>−</td><td>−</td><td>−</td><td colspan="2">+</td><td>+</td><td>+</td></tr>
<tr><td colspan="2">posterior column sensory loss</td><td>−</td><td>+</td><td>+</td><td>−</td><td>+</td><td>−</td><td>−</td><td>+</td><td colspan="2">+</td><td>−</td><td>+</td></tr>
<tr><td colspan="2">spastic gait</td><td>−</td><td>+</td><td>−</td><td>−</td><td>+</td><td>+</td><td>+</td><td>+</td><td colspan="2">+</td><td>−</td><td>+</td></tr>
<tr><td colspan="2">urinary disturbance</td><td>−</td><td>+</td><td>−</td><td>−</td><td>+</td><td>−</td><td>+</td><td>−</td><td colspan="2">+</td><td>−</td><td>+</td></tr>
<tr><td colspan="2">claw hand</td><td>+</td><td>+</td><td>+</td><td>+</td><td>+</td><td>−</td><td>−</td><td>−</td><td colspan="2">+</td><td>+</td><td>+</td></tr>
<tr><td colspan="2">Horner's syndrome</td><td>−</td><td>−</td><td>−</td><td>−</td><td>+</td><td>−</td><td>−</td><td>−</td><td colspan="2">+</td><td>−</td><td>+</td></tr>
<tr><td rowspan="2">needle EMG (denervation)</td><td>r</td><td>+</td><td>+</td><td>−</td><td>+</td><td>−</td><td colspan="2">╱</td><td>−</td><td colspan="2">+</td><td>−</td><td>+</td></tr>
<tr><td>l</td><td>−</td><td>−</td><td>+</td><td>−</td><td>−</td><td colspan="2">╱</td><td>−</td><td colspan="2">+</td><td>+</td><td>+</td></tr>
<tr><td colspan="2">site of syrinx (spine level) *</td><td>C</td><td>C.T</td><td>C.T</td><td>C</td><td>C.T</td><td>C.T</td><td>C.T</td><td>C.T.L</td><td>C</td><td>T</td><td>C.T</td><td>C.T</td></tr>
<tr><td colspan="2">tonsillar ectopia</td><td>−</td><td>−</td><td>−</td><td>−</td><td>−</td><td>−</td><td>−</td><td>−</td><td>−</td><td>−</td><td>+</td><td>+</td></tr>
<tr><td colspan="2">aspiration</td><td>○</td><td>○</td><td>○</td><td>○</td><td>○</td><td>○</td><td>○</td><td>○</td><td>○</td><td>○</td><td>○</td><td>○</td></tr>
<tr><td colspan="2">myelotomy</td><td></td><td>○</td><td>○</td><td>○</td><td>○</td><td>○</td><td>○</td><td>○</td><td>○</td><td>○</td><td>○</td><td>○</td></tr>
<tr><td colspan="2">syringo-subarachnoid shunt</td><td></td><td></td><td>○</td><td>○</td><td>○</td><td>○</td><td>○</td><td>○</td><td>○</td><td>○</td><td>○</td><td>○</td></tr>
</table>

sion are shown in Table 1. A dissociated type of sensory loss was demonstrable in 60%, posterior column sensory loss in 60%, spastic gait in 60%, urinary disturbance in 50%, claw hand in 80% and Horner's syndrome in 30% of the patients.

Needle EMG studies of the involved muscles were performed in nine patients preoperatively. Denervation potential of the muscle was determined as the potential having fibrillation and/or positive wave in the EMG study. This was observed in seven patients (Table 1). Those patients who were thought to have anterior horn involvement were diagnosed as to segmental location by evaluation of the muscle denervation potentials produced by the EMG studies. The ranges of syrinx were diagnosed by Amipaque myelography, delated CTM, magnetic resonance imaging (MRI) and intraoperative syringograms. These were performed at the cervical level in two cases, from cervical to thoracic in seven cases and from cervical to lumbar in one case (Table 1). Furthermore, cases 1 and 3 were diagnosed as having noncommunicating syringomyelia, while the other eight cases had communicating syringomyelia.

Methods of operation were an aspiration of the syrinx in one case, myelotomy less than 5mm in one case and myelotomy with syringo-subarachnoid shunts in eight cases. Shunting was performed in the cervical cord in seven cases, and in the dorsal cord in four cases (Table 1). Time elapsed between the onset of symptoms and the time of operation ranged from one year to 21 years, with an average of 9 years.

Table 2. Grading of the recorded ESCAPs and the change of the ESCAP in amplitude during the surgery in ten patients with syringomyelia.
* N = normal

		1	2	3	4	5	6		7	8		9	10
sex		M	F	F	M	F	M		F	M		M	M
age		20	49	28	48	51	37	38	35	43		36	42
site of syrinx		C5.6	C4 −T1	C1 −T2	C5.6	C2 −T1	C1−T5		C1 −L1	C1 −C7	T5 −T10	C2 −T10	C2 −T12
site of myelotomy			C5	C7	C5/6	C5	T4	C5	T11	C6	T3	T1	T1
MN-ESCAP (Grade)		N	2	1	1	3		2		3		2	3
SC-AESCAP (Grade)		N	N		1	3	2	2	1	3		N	N
SPINAL CORD MONITORING	method			MN ↓ C4	T11/12 ↓ C3	L3 ↓ C4/5	L3 ↓ C7	MN ↓ C4/5	L1 ↓ T9				L1 ↓ C4/5
	amplitude			→	↑	→	↓	→	↑				→

Method of ESCAP recording

After tracheal intubation, a catheter electrode was inserted into the spinal intra-arachnoidal space at the lower thoracic spine for stimulation of the spinal cord, and needle electrodes were placed around the median nerve at the elbow or wrist joint subcutaneously for stimulation of the median nerve. In both methods, 200μsec duration pulses at five times muscle threshold were used as a stimulus. For recording, a catheter electrode was inserted into the epidural space in four cases; needle electrodes were inserted into each interlaminar yellow ligament in six cases. The reference electrode was inserted into the paraspinal muscle. The bandwidths of the recording system were kept constant from 1Hz to 1KHz. Generally 32 to 64 evoked signals were amplified and averaged in a Nihonkoden MEM-3202 averaging system.

ESCAPs from both median nerve and spinal cord stimulation were

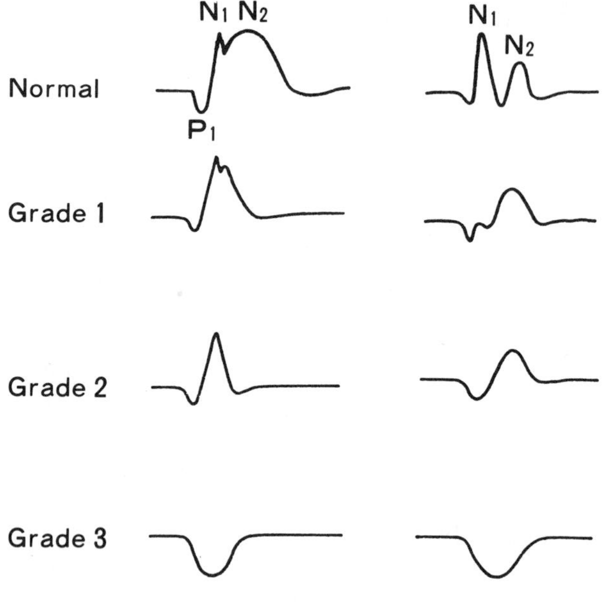

Fig. 1. Grading of the MN-ESCAP and the SC-AESCAP. ESCAPs recorded before the myelotomy were graded from normal to Grade 3 which had the positive potential. Upward deflection was defined as negative in these and in all subsequent ESCAPs.

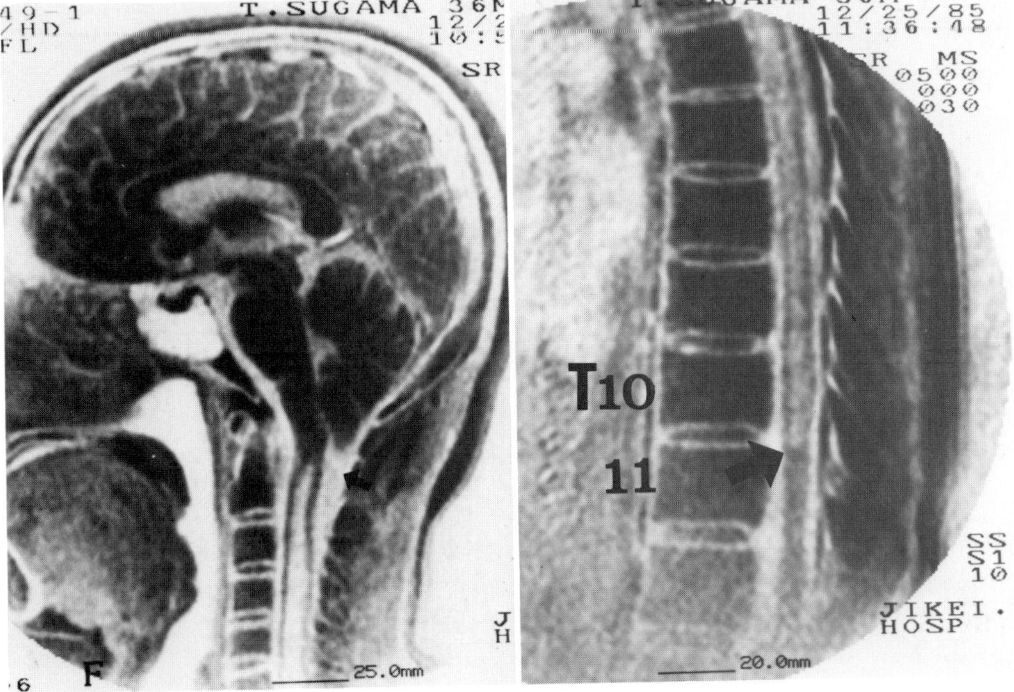

Fig. 2. MRI picture recorded in case 9. The extent of the syrinx was clearly shown in the spinal cord between C2 and T10 spinal levels. The lower end of the syrinx was indicated by large arrow. The tonsillar ectopia was also observed in this picture, which was indicated by small arrow.

recorded for syrinx level diagnosis, preoperatively in one case and intraoperatively in nine patients. Spinal cord monitoring using MN-ESCAP and/or SC-AESCAP was performed in six patients.

Analysis of the ESCAP

Normal ESCAPs were considered as potentials recorded from the patients without neurological signs and symptoms. The normal Median Nerve-ESCAP (MN-ESCAP) consisted of the P1-N1 and N2 deflections, while normal spinal cord - ascending ESCAP (SC-AESCAP) consisted of the N1 and N2 deflections (Fig. 1).

Abnormal pattern of the ESCAPs was graded for cervical myelopathy as follows: Grade 1: Slightly abnormal, the N2 decreased in amplitude in MN-ESCAP and the N1 decreased in amplitude in SC-AESCAP; Grade 2: Moderately abnormal, the N2 disappeared in MN-ESCAP and the N1 or N2 disappeared in SC-AESCAP; and Grade 3: Severely abnormal, positive potential recorded in both ESCAPs (Fig. 1).

Results

Among nine recordings of the MN-ESCAP, one was classified as normal, two were classified as Grade 1, three were classified as Grade 2 and three were classified as Grade 3 (Table 2). In patients having MN-ESCAP graded 2 and 3, the syrinx may be located in the posterior part of the gray matter of the spinal cord, because the N2 deflection of the MN-ESCAP may originate from the posterior horn of the spinal cord.

Among ten recordings of the SC-AESCAP, four were estimated as normal, two were classified as Grade 1, two were classified as Grade 2 and two were classified as Grade 3 (Table 2). In patients having SC-AESCAP graded normal, the syrinx may be located in

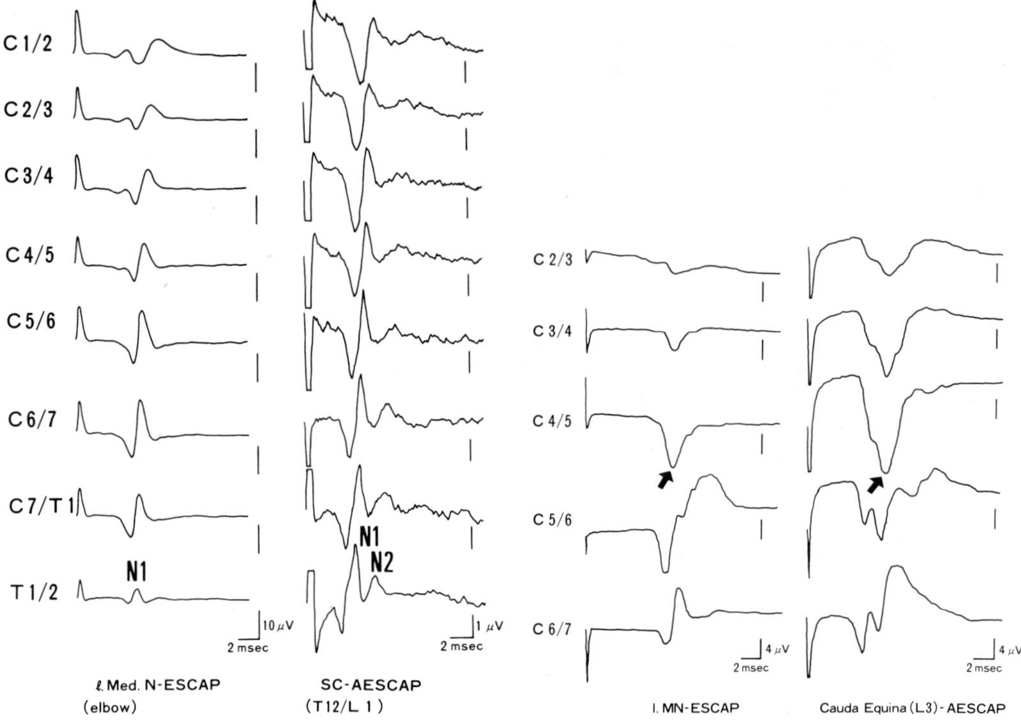

Fig. 3. ESCAPs recorded at each yellow ligament before the myelotomy in case 9. Both N1 and N2 deflections in the SC-AESCAP were recorded in every leads, but the N2 deflection of the left Med. N-ESCAP was not recorded in every leads showed in this picture.

Fig. 4. ESCAPs recorded in case 4. The positization of the potential indicated by arrows was observed in both ESCAPs recorded at C4/5 spinal level, and the main lesion in the spinal cord was supposed at this spinal level in this case.

the gray matter of the spinal cord, because the N1 and N2 deflections of the SC-ESCAP were propagated in the spinal tracts of the white matter of the spinal cord.

Therefore the patients recorded MN-ESCAP graded 2 or 3 with SC-AESCAP graded as normal or 1 were thought to have a typical syrinx located in the gray matter of the spinal cord.

Case 9.

A 36-year-old male was admitted for syringo-subarachnoid shunt in January, 1986. At the age of 33 years he had noticed numbness around the left face and left arm. He had observed the atrophy and weakness of his left hand one year before admission.

Neurologically, he had areflexia in both upper extremities, spasticity in both lower extremities, marked weakness of the grip power and bilateral hypalgesia between C2 and D5 dermatome. By needle EMG study, moderate denervation potentials were recorded in the muscles innervated by C6, 7, 8 and T1 spinal cord segments. Motor conduction velocity of the left ulnar nerve decreased to 43m/sec; however the sensory conduction velocity of the nerve was 60.8m/sec. MRI showed that the syrinx was lo-

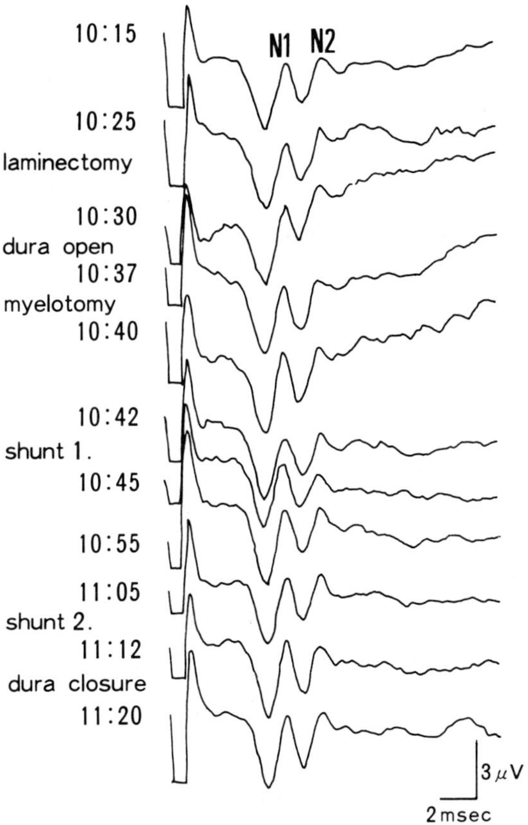

Fig. 5. Spinal cord monitoring using SC-AES-CAP (L1 qpki C4/5) in case 10. The potential decreased in amplitude during the shunting, but the recovery of the potential in amplitude was noticed at the end of the surgery.

cated in the spinal cord between the C2 and T10 spinal level with tonsillar ectopia (Fig. 2).

Operation: A syringo-subarachnoid shunt was performed at the T1 spinal level with a suboccipital craniotomy and C1 laminectomy.

ESCAP recording: The left MN-ESCAP recorded at the yellow ligaments between T1/2 and C1/2 spinal levels showed the P1-N1 complex without the N2 deflection, which might indicate pathology existing in the posterior horn of the spinal cord at the levels of the recorded ESCAPs (Fig. 3). The SC-AESCAP evoked by stimulation of the spinal cord at the T12/L1 spinal level showed almost normal N1 and N2 deflections in the leads between T1/2 and C1/2 spinal levels, which might indicate the syrinx was not expanded into the white matter in the spinal cord electrophysiologically (Fig. 3).

The positive potential graded 3 showed the conduction block in the spinal tracts corresponding to the pathway of each ESCAP in both methods. In this series, three patients had ESCAPs graded 3 in the cervical region, and from the spinal levels recorded, the positive potentials were seen as indicating a main lesion of the cervical cord with syrinx in each case (Table 2).

Case 5.

This 51-year-old housewife was admitted complaining of gait disturbance. She had discovered an inability to appreciate heat with her right trunk and leg one month before admission. She had gradually noticed a gait disturbance and bladder symptoms. Neurologically, she had spasticity in both extremities, a bilateral Babinski sign, left sided marked weakness of the muscle below C6 myotome and right sided hypalgesia and hypothermia below the C6 dermatome, while the EMG studies in the muscle of both arms were almost normal.

Myelography showed swelling of the cervical cord, and MRI showed the syrinx in the spinal cord between C2 and T1 spinal levels.

A syringo-subarachnoid shunt was performed in the spinal cord at the C6 spinal level. Before the myelotomy, ESCAPs were recorded at each yellow ligament.

Positive potentials in both left median nerve and cauda equina (L3) -AESCAPs were observed at the C4/5 spinal level, although an almost normal MN-ESCAP was recorded at the C5/6 spinal level and a decreased N1 deflection with normal N2 deflection in the CE-AESCAP was recorded at the C6/7 spinal level (Fig. 4). Accordingly, the main lesion in the spinal cord was determined electrophysiologically to be at the C4/5 spinal level in this case. Cortical evoked potentials from right and left median nerve stimulations were recorded as normal before and after operation in this case.

Spinal cord monitoring using ESCAP was performed in seven cases during myelotomy and shunting into the spinal cord. The SC-AESCAP recorded at the rostral part of the myelotomized site was used for monitoring in five cases and the MN-ESCAP was used in two cases. ESCAP amplitude did not change in four cases, decreased in one case and increased in two cases at the end of the operation (Table 2). However, the only neurological deficits to appear after the operation were a decreased sense of position of the extremities. Therefore, the pathologic changes which occurred in the posterior column longitudinally were not detected well by this monitoring system.

Case 10.

This 42-year-old male was admitted for syringo-subarachnoid shunt in April, 1986. His syringomyelia was diagnosed and the cervical cord was myelotomized nine years ago.

He had noticed increased bladder problems and disability of both arms in the preceding four months. MRI showed the syrinx in the spinal cord between the C2 and the T12 spinal levels.

A left MN-ESCAP recorded at each yellow ligament before myelotomy had positive potentials at the spinal level between C5/6 and C7/T1, although the SC-AESCAP was almost normal. Spinal cord monitoring using SC-AESCAP was performed before and after myelotomy. No remarkable change was observed in either amplitude or latency during the operation (Fig. 5).

Discussion

Spinal cord action potentials evoked by both peripheral nerve and spinal cord stimulation have been utilized to monitor function of the spinal cord during surgery and to diagnose the level or extent of lesions in the spinal cord.

In syringomyelia, the lesion in the spinal cord is diagnosed by neurological examination and EMG study, and more recently, the site of the syrinx is easily diagnosed by MRI.

In this study, we tried to diagnose the lesion by recording ESCAPs, because the origin of the ESCAP or the tract propagating the ESCAP in the spinal cord has been established by animal experiments.

The absence of N2 deflection of the MN-ESCAP might show the lesion existing in the posterior horn of the spinal cord, and the absence of N1 or N2 deflection of the SC-AESCAP might show the lesion in the white matter of the spinal cord; however the lesion in the anterior horn of the spinal cord could not be detected by these ESCAPs, but might be determined by needle EMG study.

Theoretically, the spinal level of the main lesion in syringomyelia should be identifiable by ESCAPs. In this series, the main lesion among the syrinx was diagnosed in three cases.

Spinal cord monitoring using MN-ESCAP and/or SC-AESCAP was useful to prevent the spinal cord from damage during myelotomy and shunting. However, it might be impossible to detect damage in the posterior column longitudinally less than 5mm.

References

1. Deecke, L.; Tator, C.H.: Neurophysiological assessment of afferent and efferent conduction in the injured spinal cord of monkeys. J. Neurosurg. 39: 65-74, 1973.
2. Gardner, W.J.; Bell, H.S.; Poolos, P.N.; Dohn, D.F.; Steinberg, M.: Terminal ventriculostomy for syringomyelia. J. Neurosurg. 46: 609-617, 1977.
3. Imai, T.: Human electrospinogram evoked by direct stimulation of the spinal cord through epidural space (in Japanese). J. Jpn. Orthop. Assoc. 50: 1037-1056, 1976.
4. Kaneda, A.; Yamaura, I.; Kamikozuru, M.; Shinomiya, K.; Sato, H.; Yokoyama, M.: An analysis of spinal cord potentials evoked median nerve stimulation. In: J. Schramm; S.J. Jones (eds.): Spinal cord monitoring. Springer-Verlag Berlin Heidelberg, pp. 35-42, 1985.
5. Mii, K.; Inoue, S.; Senga, K.; Itabashi, T.; Otsuka, Y.; Isobe, K.: Clinical study on 14 cases of syringomyelia (in Japanese). Rinshoseikeigeka 18: 421-428, 1983.
6. Okuma, T.; Satomi, K.; Nakamura, Y.; Kenmotsu, K.; Wakano, K.; Hirabayashi, K.: Level diagnosis using evoked spinal cord action potentials on cervical myelopathy. Rinshoseikeigeka 21: 505-512, 1986.
7. Satomi, K.; Nishimoto, G.I.: Effects of selective spinal cord transection on evoked spinal potentials in cats. In: S. Homma; T. Tamaki (eds.): Fundamentals and clinical application of spinal cord monitoring. Saikon Publishing Co., Ltd., Tokyo, pp. 87-98, 1984.
8. Satomi, K.; Nishimoto, G.I.: Comparison of evoked spinal potentials by stimulation of the sciatic nerve and the spinal cord. Spine 10: 884-890, 1985.
9. Tator, C.H.; Meguro, K.; Rowed, D.W.: Favorable results with syringosubarachnoid shunts for treatment of syringomyelia. J. Neurosurg. 56: 517-523, 1982.
10. West, R.J.; Williams, B.: Radiographic studies of the ventricles in syringomyelia. Neuroradiology 20: 5-16, 1980.
11. Williams, B.: Orthopaedic features in the presentation of syringomyelia. J. Bone Joint Surg. 61-B: 314-323, 1978.

Somatosensory Evoked Potential Recordings in Neurotrauma Patients and Value of SEPs in Diagnosing Conversion Disorders

R. McAlaster;[*] R.W. Thatcher; P. Krause; F. Geisler

Introduction

The use of evoked potentials in clinical practice has been steadily increasing for evaluating patients in the operating room and for aiding in the differential diagnosis of central nervous system disease and traumatic lesions. This chapter describes some uses of evoked potentials and the clinical differential diagnoses where they were of value. The chapter will describe recording methodology for successful bedside evoked potential recording in the acute phase of spinal cord injury in an electrically "noisy" intensive care unit environment. Ten cases are presented with the type and method of collection, digital filtering, and signal processing techniques used along with evoked potential data and their interpretation. The chapter also describes evoked potentials recorded from patients as having a "conversion disorder" due to the disparity between the claimed neurologic deficit and the physical exam. The manner in which evoked potentials clarified the anatomical nature of the lesion or lack of physiologic abnormality is described in each case and used to provide evidence for the correct clinical diagnosis.

Methods

I. Patients

The patients in this report were admitted to the Neurotrauma Service of the Maryland Institute for Emergency Medical Services Systems (MIEMSS), except for patient #6 who was tested as an outpatient. Ten cases are reported in this paper. The cause of injury in this group are: 2 by motor vehicle accidents, 2 by industrial accidents, 2 by falls, 1 by a swimming accident, 1 by assault, and 2 by gunshot wounds.

II. Somatosensory evoked responses

Evoked potentials were used to evaluate somatosensory conduction. Data were collected during median nerve stimulation as follows. Grass silver-silver-chloride discs were applied to P3, P4, and Fz of the International 10/20 system (Jasper, 1958). In some cases, tin disk electrodes were applied to these scalp sites using an elasticized cap manufactured by Electro-Cap, Int. A transorbital eye channel (electro-oculogram) was used to measure eye movements. Impedance measures for all channels were kept less

[*] Applied Neuroscience Laboratory, Maryland Institute for Emergency Medical Services Systems, 22 South Greene Street, Baltimore, MD 21201

than 3,000 ohms. An on-line artifact rejection routine was used which excluded segments of input if the voltage in any channel exceeded a pre-set limit, usually 12μV. Stimuli were constant current square wave pulses of .5 milliseconds duration, presented at a rate of 3.7 per second. The stimulus and trigger to the computer averager were synchronized by a Grass S44 waveform generator. Silver-chloride electrodes were applied to the skin over the median nerve at the wrist and were referenced to Fz. Active sites were P3 and P4 for right and left median nerves respectively. A forehead ground was used. Outputs from the electrodes were fed into amplifiers with frequency bandpasses of 20-3KHz. Stimulus current intensity was gradually increased to produce mild twitching of the thumb.

For tibial nerve tests, active sites Cz, C3, and C4 were used, referenced to Fz. Stimulating electrodes were applied to the skin over the posterior tibial nerve at the ankle. Current intensity was gradually increased to produce mild twitching of the abductor hallucis. Amplifier bandpasses were 10-300Hz. Inputs exceeding 22μV were rejected.

Digitized data were collected for a set of 10 'blocks,' with each block consisting of 50 evoked potential trials using 200mSec epochs and a sampling interval of 200μSec. Spectral analysis was performed on each trial of 50 responses (as described by Fridman et al., 1984). The final 'grand average' SEP was the sum of 10 digitally filtered blocks. A peak detection algorithm, based upon the zero crossing of the first derivative, was applied to yield estimates of mean peak latencies and amplitudes, and their standard deviations.

For long latency somatosensory recordings, as reported in one patient in whom peroneal nerve stimulation at the knee was done, all 19 sites of the International 10/20 system were used. A transorbital eye channel was used to measure eye movements and all scalp recordings were referenced to linked ear lobes. Impedance measures for all channels were less than 3,000 ohms. Amplifier bandwidths for EEG and cortical evoked potentials were nominally 1.5 to 30Hz, the outputs by a PDP 11/03 data acquisition system manufactured by Neurometrics, Inc., a subsidiary of Cordis, Inc. An on-line artifact rejection routine was used, which excluded trials if the voltage in any channel exceeded a present limit. Fifty stimuli were presented to each leg and were averaged over an 800 milliseconds epoch. The digitization rate was 100Hz, and the data were analyzed off-line by a PDP 11/70 computer and plotted by a Versatec printer/plotter.

III. Brainstem auditory evoked potentials

Brainstem auditory evoked potentials were acquired from two patients with the clinical diagnosis of 'conversion disorder.' Auditory click stimuli were delivered through air conducting tubes from piezoelectric transducers (Motorola #KSM20004A) at 10 clicks/sec. The stimulus intensity was 90dB SPL to each ear. The vertex signal was fed to an amplifier with a gain of 100,000, a noise level of 4μV peak-to-peak, a common mode rejection ratio of 200dB, and a bandwidth from 100Hz to 2.9KHz at -4dB points. The prefiltered analog signal was digitized at a rate of 10,000 points/seconds with 16 bit resolution. Digital data were collected for a set of 10 "trials" or sub-averages. Each such trial consisted of the average of 100 evoked potentials using an analysis epoch of 9mSec and a sampling interval of 200μSec. Averaged brainstem evoked potentials were thus based upon 1,000 responses.

Brainstem auditory evoked response (BAEP) peak detection involved first spectral analyzing each trial using a 422 point Fast Fourier Transform (FFT) and computing the mean amplitude of each spectral component across all 10 trials. In order to enhance the signal-to-noise ratio the BAEPs were digitally filtered with frequency components outside of the 440-2400Hz band set to zero (Fridman et al., 1984). Data were collected

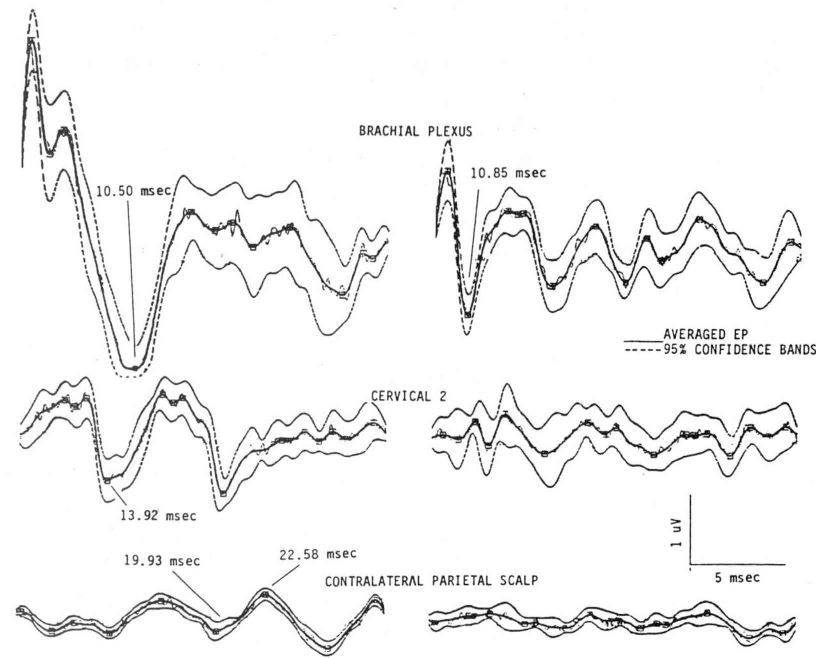

Fig. 1. Left and right median nerve stimulation -- near field and far field recordings. The three waveforms on the left are from ERBs point, cervical 2, and contralateral parietal scalp, for left median nerve stimulation at the wrist. Corresponding recordings for right median nerve stimulation are on the right. Data indicates better conduction from left periphery than from the right. The recording epoch is from 9 to 29mSec in all traces. Positive is up.

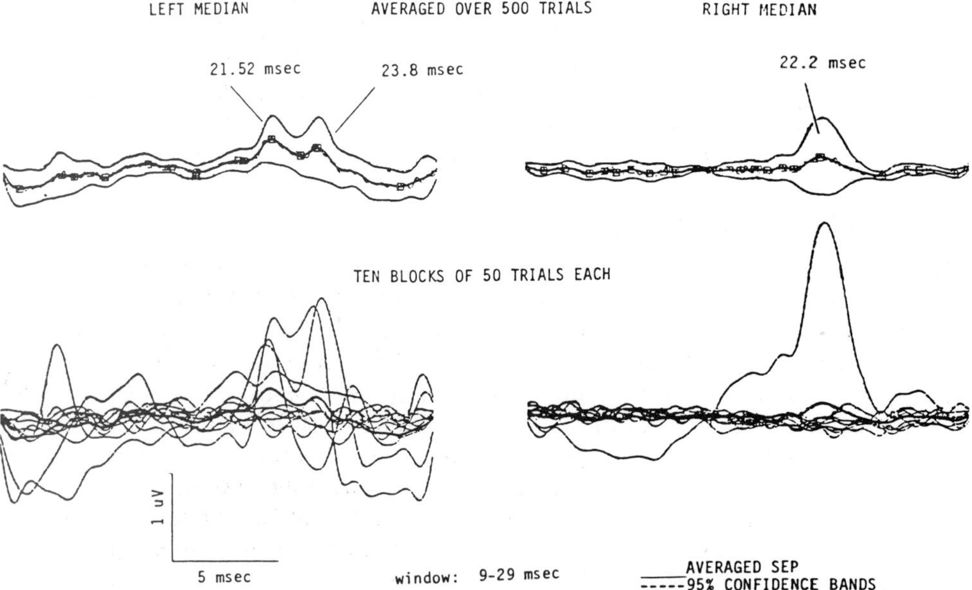

Fig. 2. Median nerve stimulation -- far field recording. The top waveforms are left and right median nerve stimulation, contralateral parietal recording sites. Data is averaged over 500 trials and 95% confidence intervals are outlined around the average. The bottom waveforms are subsets from the 500 trials, plotted as separate blocks of 50 trials each. A few large responses within each block will therefore have a large effect on one subset. Positive is up.

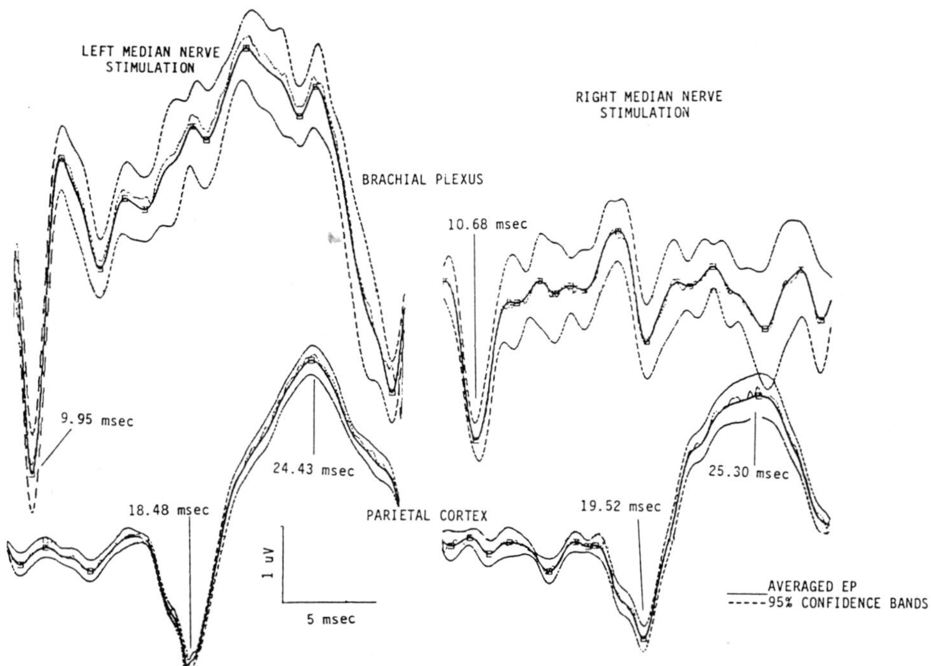

Fig. 3. Left and right median nerve stimulation -- near field and far field recordings. The top waveforms are recordings from ERBs point, indicating a small difference of about 0.75mSec between the left and right arrival times. The bottom two waveforms were from contralateral parietal regions, indicating normal arrival times at the cortex. The recording epoch is from 7 to 27mSec for the top recordings, and 9 to 29mSec for the bottom recordings. Positive is up.

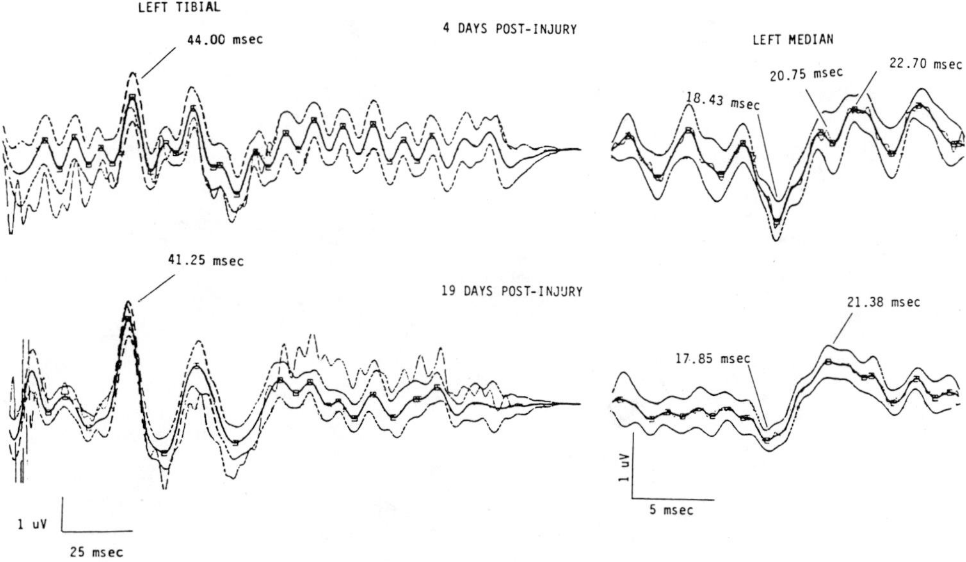

Fig. 4. Left tibial and left median nerve stimulation -- far field recordings. Tibial nerve recordings are of 500 stimuli from Cz recording site. Left median nerve recordings are from right parietal P4. The unfiltered mean is visible on the tibial nerve recording as the light waveform. The overall average is plotted with 95% confidence bands for tibial and median nerve. Recording epochs are 0 to 200mSec for tibial nerve, and 9 to 29mSec for median nerve plots. Positive is up in all recordings.

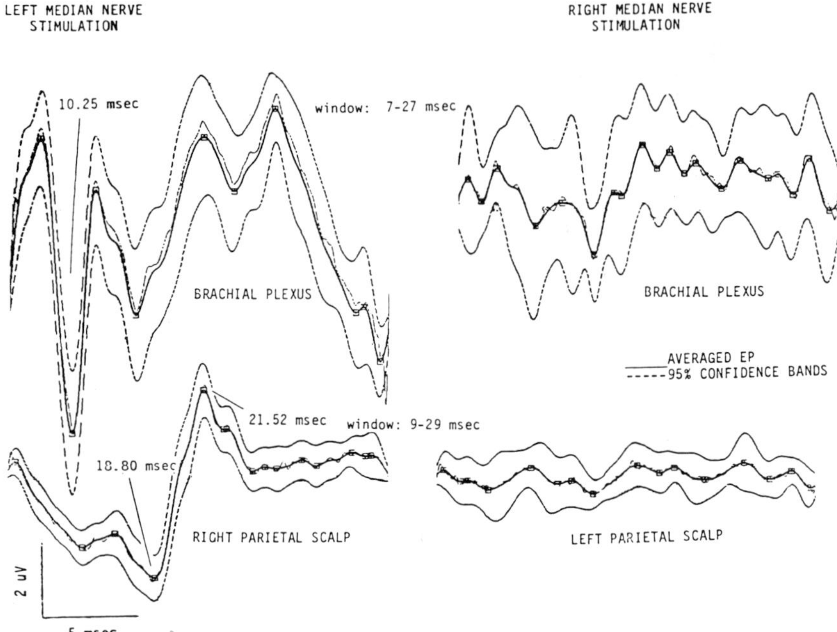

Fig. 5. Left and right median nerve stimulation -- near and far field recordings. The top waveforms are recordings from ERBs point over a recording epoch of 7 to 27mSec. Bottom waveforms are from contralateral parietal scalp. These data indicate a lack of conduction at ERBs point and at the cortex for right median nerve, but normal conduction for left median nerve stimulation to the brachial plexus and to the cortex. Positive is up.

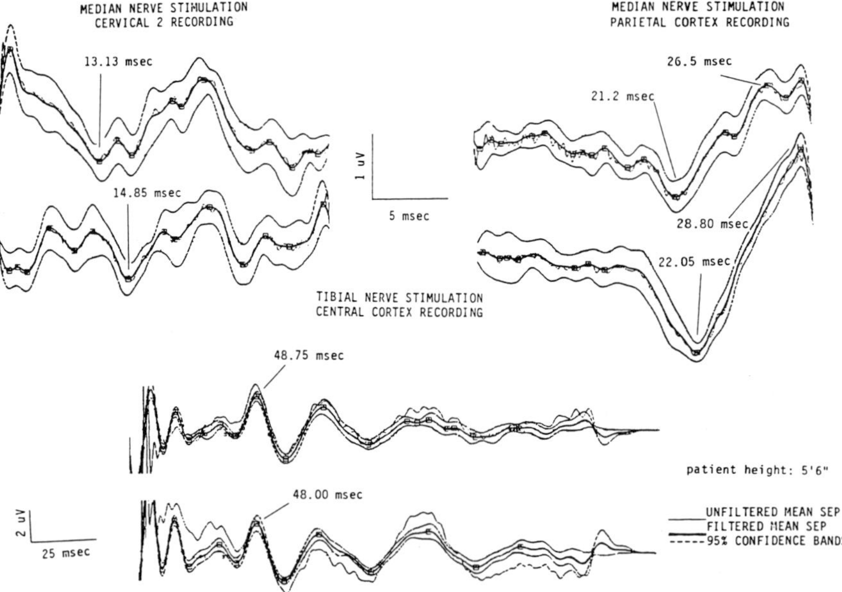

Fig. 6. Left and right median nerve and left and right tibial nerve stimulation, and near field, far field, and intermediate recordings. The two top left waveforms are left and right median nerve stimulation with recordings from C2 with a 7 to 27mSec recording epoch. The top right two recordings are of left and right median nerve stimulation, recorded from contralateral parietal cortex. The bottom two waveforms are left and right tibial nerve stimulation, recorded from Cz, over an epoch from 0 to 200mSec. The unfiltered tibial nerve average is visible as the light, unsmoothed waveform. These recordings show delayed conduction from the periphery to the cortex, for all extremities.

from a lead placed at Cz, referenced to the ipsilateral ear. Normative data used were means and standard deviations of peak latencies and interpeak intervals derived from a normative study (Starr and Achor, 1975).

Results

Diagnostic Utility

The first case involves that of a 24 year old female injured in a swimming accident, which resulted in a C5 through C6 fracture dislocation and incomplete C5 cervical spinal cord lesion myelopathy. In addition to the spinal cord injury the patient had a brief loss of consciousness at the scene of the accident, and an admission Glascow coma score of 14/15. She had movement only of the deltoid and bicep muscles bilaterally on admission. The patient had a surgical stabilization of the C5 though C6 fracture dislocation with a Casper cervical plate and eventually walked with a walker 9 months later. The first recordings were obtained two days post injury and are shown in Fig. 1. The patient at the time of the recording was in Gardner-Wells tongs. This patient presented difficulties in obtaining optimum cervical signals as a result of a fat neck (patient was 5'5", weighing 167 lbs.). The three evoked potential plots for left side stimulation (on the left side of the figure), indicate not only a response at 10.5mSec for ERBs point, but also a peak recorded at cervical 2 at 13.92mSec, confirming transmission from the left median nerve to this point, and finally, the recording from contralateral scalp indicates an early cortical response at 22.58mSec. However, the three evoked potential plots for right side stimulation (the three plots on the right side of the figure) indicate that although there is transmission at the brachial plexus, there is no clear response either at cervical 2, or at the scalp recording, indicating a lack of transmission to the cortex. This indicated an interruption of transmission somewhere between the brachial plexus and cervical 2 on the patient's right side. At 23 days post injury, this patient demonstrated greater strength on the left than on the right (graded 5/5 and 4/5 respectively according to the nurses' observations). This correlates with the difference noticed in the first recordings. The patient also showed recovery of an early cortical response for stimulation of the right median nerve at that time, although it was very poor and highly variable as compared with the left. This is not shown here due to space considerations. This improvement in the evoked potentials may show that in this patient the EPs help to predict clinical recovery.

The second patient was a 66 year old female who sustained a transient loss of consciousness and a C4 complete quadriplegia after a fall down a flight of stairs. She was originally admitted to another hospital and was then transferred to MIEMSS for definitive care of her spinal cord injury. The patient was alert and oriented on admission to MIEMSS. Her past medical history was positive for a possible seizure disorder and early organic brain syndrome. The cervical x-ray disclosed degenerative spine pathology with a C3 through C4 osteophyte which was presumed to cause the spinal cord injury in an extension injury. Evoked potentials during stimulation of the left and right median nerves, averaged over 500 trials, show little responsivity, particularly for the right median nerve (Fig. 2). However, an examination of the 95% confidence bands indicates that there have clearly been some events at about 21 to 23mSec on either side, which were so infrequent they had little effect on the overall average. More information can be gleaned from the recordings of individual averages of each of the 10 separate "blocks" of trials of 50 each (which were combined to make the overall average of 500). An examination of these recordings reveals the reason for the perturbation in the 95% confidence bands. For the left median nerve, three of the blocks of averages show respectable amplitudes at 21 to 23mSec. This indicates that conduction occurs only sporadically, and probably infrequently during each block of 50 trials, but suggests

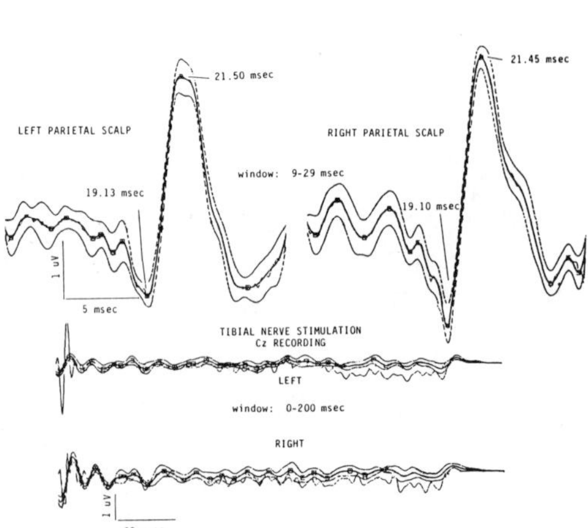

Fig. 7. Left and right median nerve stimulation -- far field recordings from contralateral parietal scalp (top two waveforms) and left and right tibial nerve stimulation intermediate latency recording from Cz. The unfiltered mean SEP on the tibial nerve recordings can be seen as a faint line. All recordings include an average over 500 trials and 95% confidence bands. These data show normal conduction from each median nerve, but a lack of conduction from each tibial nerve.

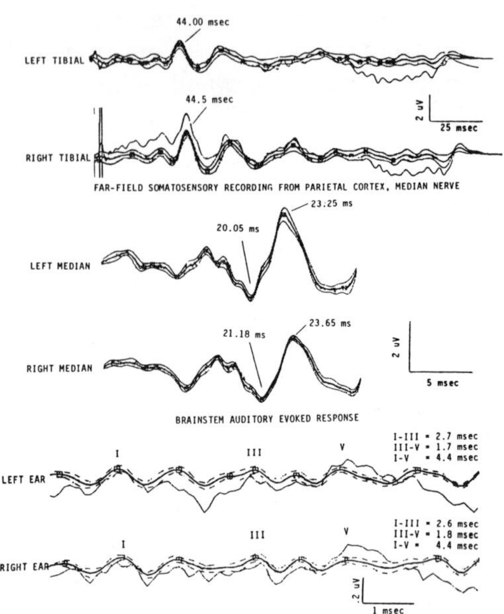

Fig. 8. Conversion disorder. The patient is right hemiparetic. The top two waveforms are left and right tibial nerve stimulation with the recordings from Cz. The unfiltered SEP is seen as a background unsmoothed waveform. The overall average of 500 stimuli and 95% confidence bands are plotted over a 200mSec epoch. The center two recordings are left and right median nerve stimulation, recorded from contralateral parietal cortex over an epoch from 9 to 29mSec. The bottom two waveforms are brainstem auditory evoked potentials, left and right ear stimulation averaged over 1000 trials and recorded from Cz. The unfiltered mean BAER is visible as a faint line. The overall average is smoothed with 95% confidence band intervals plotted around it. Although all somatosensory evoked potentials for this patient were normal, the auditory brainstem response is abnormal and exhibits an excessive interpeak latency for waves I to III.

that when conduction does occur, that the overall amplitude of the response is probably very high, since it has a considerable effect on the amplitude of the average of 50. An examination of the response for the right median nerve suggests a similar phenomenon: That conduction occurred only very rarely, but when it occurred even occasionally in one block of 50 trials, it had the effect of markedly increasing the average amplitude in that set of trials. Of course, the presence of this rare phenomenon on either side helps to corroborate the data, since the interpretation would be less certain if this occurred only on one side. This low mode of responding may explain the vague and intermittent sensations that some patients report. The clinical relevance of sporadic conduction is difficult to specify. These data may indicate 'de-enervation supersensitivity' in which a relatively small number of afferent fibers conduct to the cortex. However, these fibers elicit a de-enervation supersensitivity in their target organs. This would explain the high variance and low reliability of the response.

The third patient is a young male injured in a work-related accident in which a heavy object struck his neck and shoulder causing a right brachial plexus injury, as evidenced by weakness in the right extremity and diminished peripheral reflexes in that extremity, and normal reflexes in both lower extremities. The data show that conduction occurs on either side, but that it is only slightly slower on the right than the left (Fig. 3).

The fourth patient is a 25 year old woman injured in an automobile accident. She sustained a closed head injury and was admitted with a Glascow coma score of 14, a C4 through C5 anterior subluxation, and an L2 transverse fracture. Gardner-Wells tongs were in

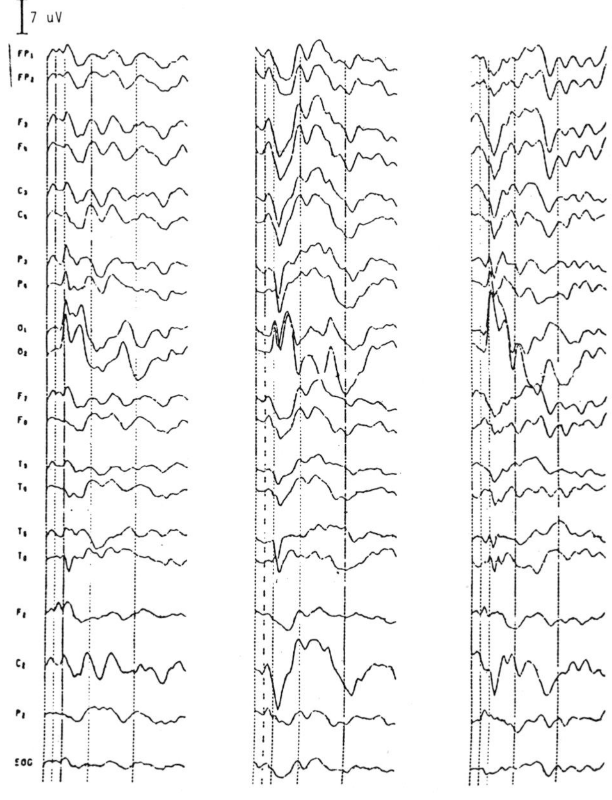

Fig. 9. Visual evoked response from a patient with a temporary conversion reaction. Recording is over the full electrode array of the 10/20 system. The recording is averaged over 50 flash stimuli over an 800mSec epoch. The first column of waveforms is left eye only; the second column is right eye only; and the final column is binocular stimulation. The data indicates that conduction to the occipital cortex has occurred (see waveforms labelled 01 and 02 which are recorded from occipital cortex).

place at the time of the test. This individual presents a response that is outside the norms for her height on the initial test, which was done 4 days following her injury. The early cortical response was at 44.00mSec (Fig. 4). However, a recording taken 19 days following her injury shows that the latency of the early cortical response has decreased to 41.25mSec, which is within 2 standard deviations of normal (Fig. 4, bottom plot). This patient showed significant changes in her clinical motor exam in the first month, remaining a mild central cord injury. This shows the value of longitudinal repeated testing to evaluate recovery of function.

The fifth patient sustained a gunshot wound to the right arm and a brachial artery disruption. Fig. 5 shows the left Erbs and right cortical response to left side stimulation, indicating the integrity of the pathway from the left side. However, the absence of

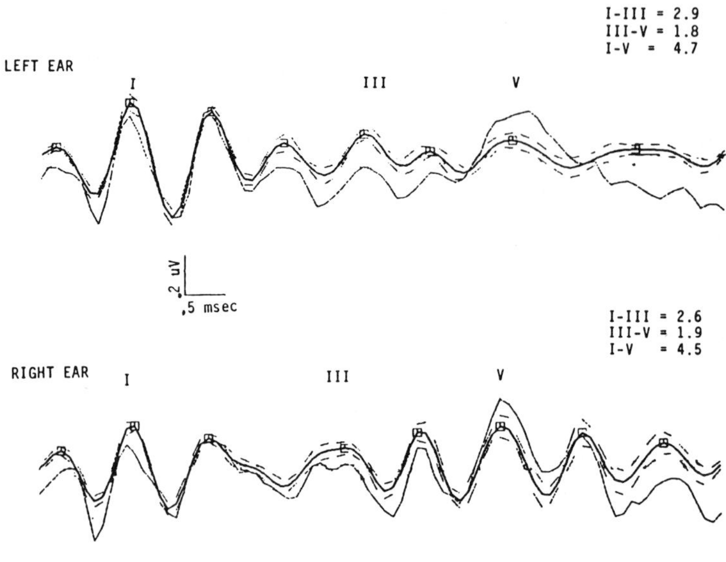

Fig. 10. Brainstem auditory evoked response from a patient with a loss of vision (apparent 'conversion reaction'). This is the same patient whose VEPs are displayed in Fig. 9. The unfiltered BAER is the faint line. The filtered average (over 1000 click stimuli) is the heavy dark line. The 95% confidence bands (dotted lines) are plotted around the filtered average. These data indicate an excessive I to III interpeak latency for either ear.

a response at the right Erbs and left parietal regions to right side stimulation supports the clinical interpretation of disrupted conduction. This verified the clinical impression.

The sixth patient had a progressive myelopathy exacerbated by a fall before his test as an outpatient. The evoked potential data show severe delays in cortical responses to stimulation of the median and tibial nerves (Fig. 6).

The results of the somatosensory evoked potential tests are consistent with a reduced axonal conduction velocity in the peripheral nervous system. The tibial nerve conduction time is very markedly lengthened: The early cortical responses (P48.75 for left and P48.00 for right posterior tibial nerve) are greater than 3 standard deviations from normal for a patient of this height (5 feet, 6 inches).

Finally, the seventh patient is a 22 year old male who received a gunshot wound to C7, completely severing the spinal cord. The evoked potential data show a normal response to left and right median nerve stimulation (Fig. 7). However, there is no response to left and right tibial nerve stimulation, confirming the absence of conduction to the cortex. Since the median nerve has dominant fibers in C6, C7, and C8, the normal evoked potential response indicates the signal is being transferred into at least one but not necessarily all of the roots. In this case, the entire median nerve response is derived from the median nerve inputs to C6.

Conversion disorders

The diagnosis of "conversion disorder" is difficult and important. The eighth patient is a man injured in a fall on his head at work. Upon admission to the trauma unit, he had a Glascow coma score of 12/15. He reported right-sided weakness of both upper and lower extremities. At 4 months post injury, he still reported right hemiparesis, and was unable to see the right visual field from his right eye, and was unable to hear from his right ear. An examination of his tibial and median nerve responses at two days post injury indicated that they were within normal limits for latencies (Fig. 8).

Thus the somatosensory evoked potentials indicated no evidence of abnormality. However, upon further testing, the visual evoked potentials and the spectral EEG were found to be mildly abnormal, and in addition, the patient presented an abnormal auditory brainstem evoked response (Fig. 8, the bottom 2 BAER plots). The auditory brainstem response showed a considerably prolonged interpeak latency for waves I to III (2.7 for left ear, and 2.6mSec for right ear stimulation). The abnormal BAER is consistent with trauma at the level of the lower pontine structures. In summary, the patient's initial presentation of 'conversion syndrome' appeared to be resolved by further testing in which a number of physiologic deficits were observed.

We have seen another trauma patient (case 9) initially diagnosed as "hysteric blindness" in which the brainstem auditory evoked response also was abnormal. In this instance, the patient was the victim of an assault and was hit on the head with a baseball bat, and was ad-

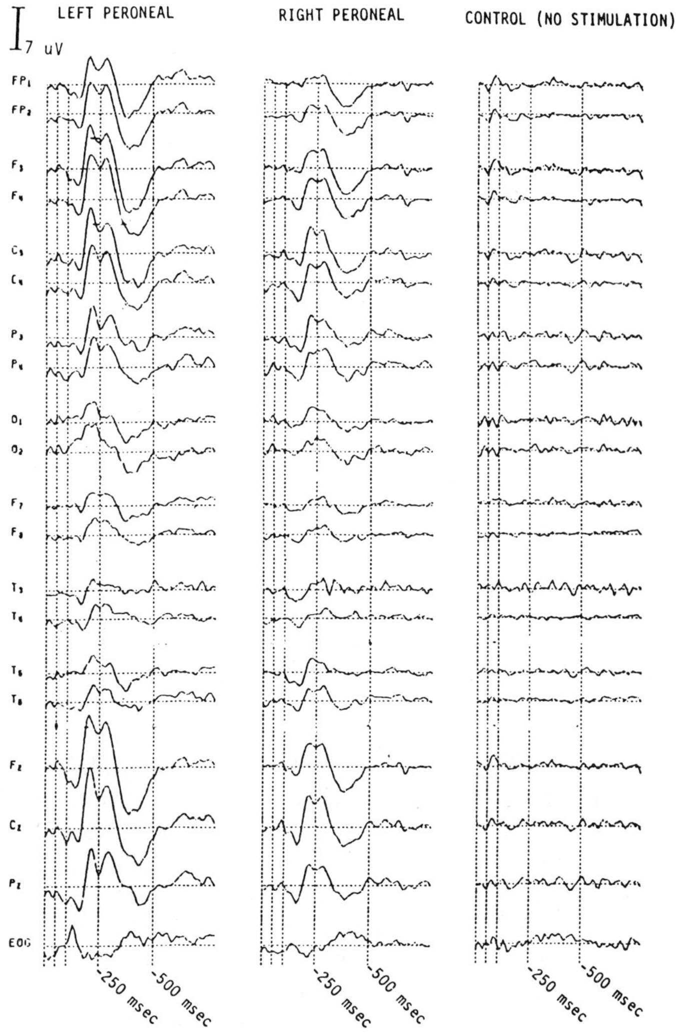

Fig. 11. Long latency somatosensory evoked response recorded over the full 10/20 system array. Recordings are averaged over 50 stimuli, and an 800mSec time epoch. Left peroneal, right peroneal, and a control condition are successively recorded. Data show a clear response over the cortex which appears to be normally distributed, indicating conduction has occurred. The patient is a 'conversion disorder' patient who reports paralysis and lack of sensation from the waist down.

mitted to MIEMSS with Glascow coma score of 14. He reported a complete loss of vision, including the ability even to detect light. The visual evoked response recorded from this patient was normal and indicated that conduction had occurred to the occipital regions (Fig. 9).

The spectral EEG for this patient was abnormal, and indicated widespread elevations of delta and theta. In addition, the auditory brainstem evoked response was abnormal (Fig. 10). The abnormality was between waves I and III.

These data, as with the previous patient, suggest that although conduction pathways to the cortex seem to be normal and the cortical response to the input has occurred as expected, there is nevertheless a cortical and a brainstem abnormality that may be at the basis of the symptoms as reported by the patient. These cases emphasize the importance of not limiting conduction studies just to the pathways that initially appear to be implicated. The next case emphasizes this point.

The tenth patient, a 39 year old female, was brought to MIEMSS as an apparent trauma patient. She was a pedestrian found by the side of the road, who said she had been hit by a car. She reported she had no sensation below the waist, and said she was unable to move her legs. Peroneal stimulation at each knee resulted in normal cortical responsivity, indicated normal integrity of the pathways. However, since a BAER and spectral EEG were not administered, there is no way to resolve the case, as was done in the previous two examples.

Discussion

In this paper we presented several examples of the application of SEPs for the diagnosis of trauma patients. Because of constraints of time, acute patient care necessities, and patient inability to tolerate long test sessions due to short attention span or other effects of head injuries, the number of stimuli delivered in most sessions was fewer than would be given in a laboratory where there is typically greater cooperation from the patient. For instance, instead of the standard repetition of 2,000 stimuli delivered to each extremity (frequently repeated, for a total of 4,000 stimuli), only 500 stimuli are delivered. Nevertheless, data quality was usually excellent. This was due to a variety of enhancement techniques. These included the following: 1) high quality differential amplification, 2) 60Hz notch filtering, 3) digital filtering, and 4) averaging.

The diagnostic utility of our methods for trauma patients was clearly demonstrated. For instance, de-enervation supersensitivity, noted in the second case described, would not have been detectable in a technique that displays only one overall average. But in a method that displays individual groups of stimuli, this occurrence can be detected.

Some limitations to giving a single SEP test were also demonstrated in cases involving questionable trauma. In a trauma ward, there are a few cases in which the symptoms are not easily explained because of the lack of evidence for structural damage, and because the integrity of the pathways in question seems to be established. Yet the patient exhibits symptoms which trouble him and for which explanations cannot be found. It was noted that for two out of three such patients exhibiting a confusion symptomatology, both an abnormal spectral EEG and an abnormal brainstem auditory evoked response was found. In the third patient, this test was not administered so the case remained unresolved.

References

1. Fridman, J.; John, E.R.; Bergelson, M.; Kaiser, J.B.; Baird, H.W.: Application of digital filtering and automatic peak detection to brainstem auditory evoked potentials. Electroenceph. Clin. Neurophysiol., 53: 405-416, 1984.
2. Jasper, H.H.: The ten-twenty electrode system of the international federation. Electroenceph. Clin. Neurophysiol., 10: 371-375, 1958.
3. Starr, A.; Achor, L.J.: Auditory brainstem responses in neurological disease. Arch. Neurol. (Chic.), 32: 761-768, 1975.

Somatosensory Evoked Potentials in the Diagnosis and Prognosis of Multiple Spinal Meningiomas

M.M. Vaghari;[*] L.J. Streletz; B.E. Northrup; R.G. Fariello; S. Duckett; A.R. Babaria

Abstract

Somatosensory evoked potentials (SEP) were recorded in a 15 year old patient with multiple thoracic cord meningiomas. Lack of improvement of the quantitative SEP parameters added objective measurement to postoperative assessment of the patient and confirmed the need for further surgery. Clinical follow-up and correlation with SEP parameters of the patient during 18 months of longitudinal study indicates the value of SEP as an electrodiagnostic and prognostic tool.

Introduction

Somatosensory evoked potentials (SEP) to stimulation of nerves in the lower extremities have been used to evaluate the integrity of the spinal cord pathways for over a decade (1, 2, 3). Many investigators have reported lower extremity SEP correlations with spinal cord pathology. Few reports have documented the effects of spinal cord decompression for tumors on SEP parameters and only one of these reports has been in children (4, 5, 6, 7). The spinal cord meningioma is a rare tumor in the pediatric age group and multiple spinal meningiomas are exceptional, being associated with von Recklinghausen's disease in most cases (6, 9). Multiple spinal cord meningiomas present significant difficulties in diagnosis, surgical management, and prognosis. The unusual spinal cord pathology and SEP correlation obtained pre- and postoperatively in this patient prompted us to describe this case.

Methods and procedure

Somatosensory evoked potentials were evoked with 200μSec duration square wave pulse presented as a constant current at 5.2 per second. The stimulus was applied to the skin over the tibial nerve at the ankle by a bipolar stimulator with a cathode placed 3cm to the anode. The current intensity was increased to produce a mild to moderate twitching of the abductor hallucism. The stimulus trigger output was synchronized to the computer averager (Nicolet CA-1000).

Four EEG electrodes were applied at the popliteal fossa, iliac crest (IC), and over the spinous process of the third lumbar (CE) and second cervical vertebral (SC2) using collodion technique. Scalp electrodes were placed at Cz (2cm behind Cz international 10/20 electrode system) and Fz. The IC electrode contralateral to the side stimulated was reference for CE. The Fz electrode was reference for SC2 and Cz. An electrode

[*] Department of Neurology, Jefferson Medical College, Thomas Jefferson University, Philadelphia, PA 19107

on the mastoid was used as a ground. The output from each lead was lead into amplifiers (Nicolet HGA-200A) with a frequency bandpass of 30-3000Hz (-3dB point with slope of 12dB/octave) and amplification of 40,000X. Three or four inputs with 512 addresses per channel, resolution of 30-50 μSec per address and a sweep time of 100 μSec were used. One thousand to two thousand responses were averaged and replicated at least once to check the reliability of the data. All channels were monitored on one oscilloscope and computer averaging was discontinued when prominent movement artifacts appeared. The data was plotted on an x-y plotter for subsequent analysis.

The absolute latency was measured at the highest peak of each component of the SEP using a single cursor on the oscilloscope display and labeled N20, N30, P38, and N47 based on the surface polarity and mean peak latency of the component derived from a group of 18 control subjects (Fig. 1, A). The P38 amplitude was measured baseline to peak distance and was obtained in the lateral decubitus position between the CE electrode side and Cz. The N20 through P38 interpeak interval was calculated as a measure of central conduction time. The distance between CE and scalp divided by the N20 through P38 interval was used to calculate central conduction velocity based on current knowledge of the neural generators of the SEP components (10, 11).

Case Report

On January 3, 1985, a 15 year old boy was admitted to TJUH with a two month history of left leg pain and swelling of the knee which worsened with walking. Initially, he was seen by his pediatrician and plain x-ray of the knee was normal. The swelling subsided with rest, but within a few weeks the pain had become progressively worse and weakness developed. Upon admission he reported that his left leg gave away causing repeated falls. He denied any numbness, paresthesias, or changes in bowel and bladder control. Trauma was specifically denied. Family history was not available since he was an orphan. There was no other relevant past medical or social history.

Upon examination this well developed, black male adolescent was noted to have one cafe au lait spot (3.0cm) on the anterior chest wall. There were no cutaneous tumors. The remainder of findings were confined to the neurologic examination. His mental state was unimpaired. Cranial nerve examination revealed a congenital isotropia of the right eye and a sensory neural hearing loss of the left ear. There was a wide based ataxic gait and tandem walking was impossible. Examination of the motor system revealed normal bulk, tone, and strength (5/5, L/R) in both arms and right leg. The left leg showed increased tone with decreased strength as follows: Quadriceps fem. m., Hamstrings m.; Abductor magn. M.; and Gluteus max. m. (4/5) and Tibialis ant. m. and Peroneus lg. m. (3/5). A positive Queen's Square test was present bilaterally. Deep tendon reflexes were hyperactive in the left lower extremity, with the plantar response extensor on the left but flexor on the right. Sensory examination revealed decreased position, vibration, and light touch on the left with preserved functions on the right. Pain and temperature sensation was diminished at the T5 level on the right but preserved on the left side, confirming a classic Brown-Sequard syndrome.

Laboratory procedures

Evoked potential studies: Bilateral SEPs were recorded to stimulation of the left and right posterior tibial nerve at the ankle. The preoperative SEP study showed an absolute delay of the P38 latency and reduced amplitude on the left side. The subcortical cervical component (N30) was absent and the lumbar potential (N20) was normal on that side (Fig. 1, A). The central conduction time and conduction velocity were slowed (Table 1). All of the cortical and subcortical components of the SEPs on the right side were within 2.5 standard deviations of mean values for the laboratory.

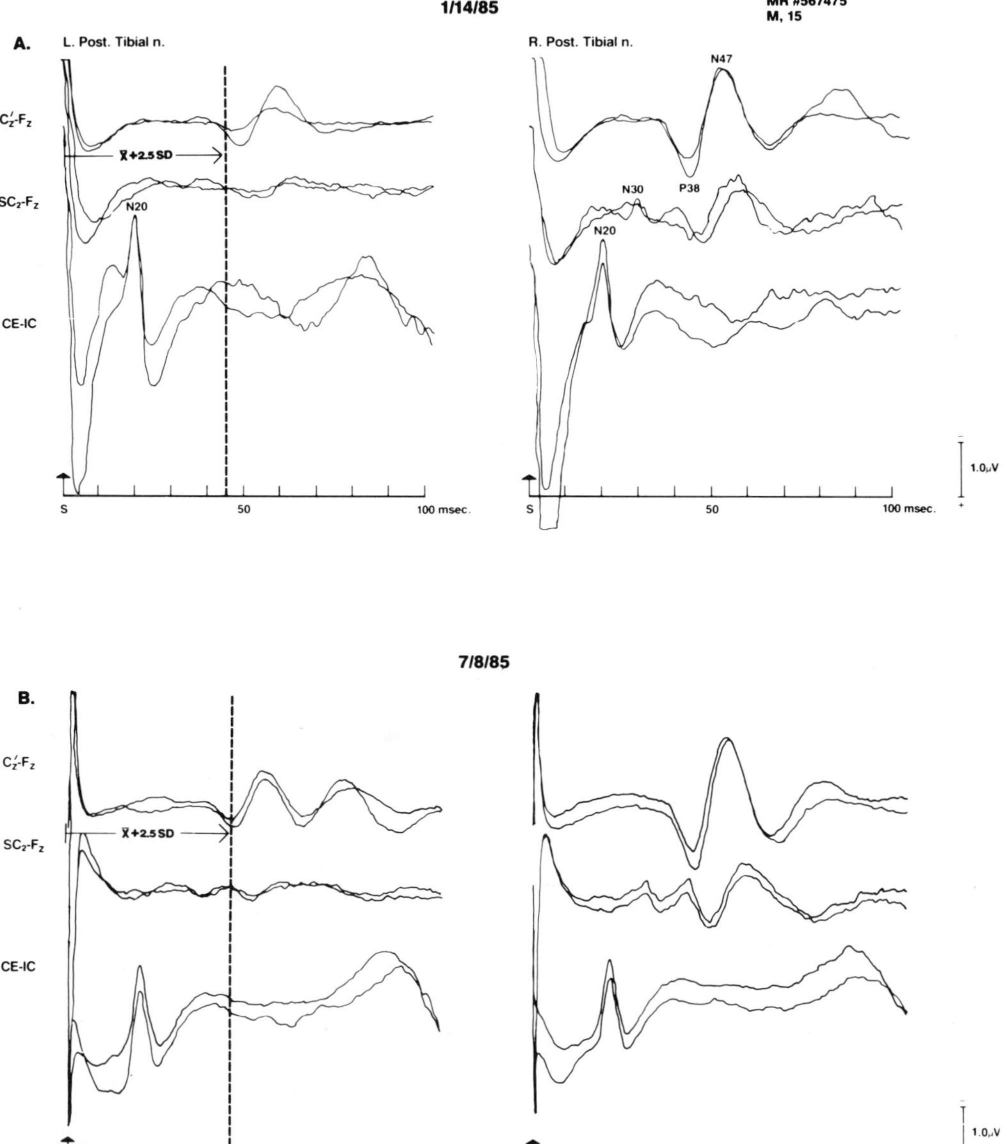

Fig. 1. The SEP to tibial nerve stimulation at the ankle recorded at various sites simultaneously. Labels were based on surface polarity and mean peak latency observed in 18 normal subjects (age = 16-36). Electrodes were placed at third lumbar (CE) and cervical spinous process (SC2) and Cz (2cm behind Cz) and Fz (10/20 electrode system). The contralateral iliac crest (IC) was the reference for CE and Fz was reference for SC2 and Cz. The analysis time was 100µSec and 1024 responses were averaged. Two replications were superimposed to indicate the consistency of components. The calibration was 1.0 microvolts.

A. Preoperative SEP reveals significant delay of P38 component with reduced amplitude on the L side. The N30 spinal cord component is absent. The R side SEP is normal.

B. Following T5 through T6 and T3 through T4 laminectomies (6 months), the P38 component has returned to normal range (less than mean plus 2.5SD) but remains of low amplitude compared top opposite.

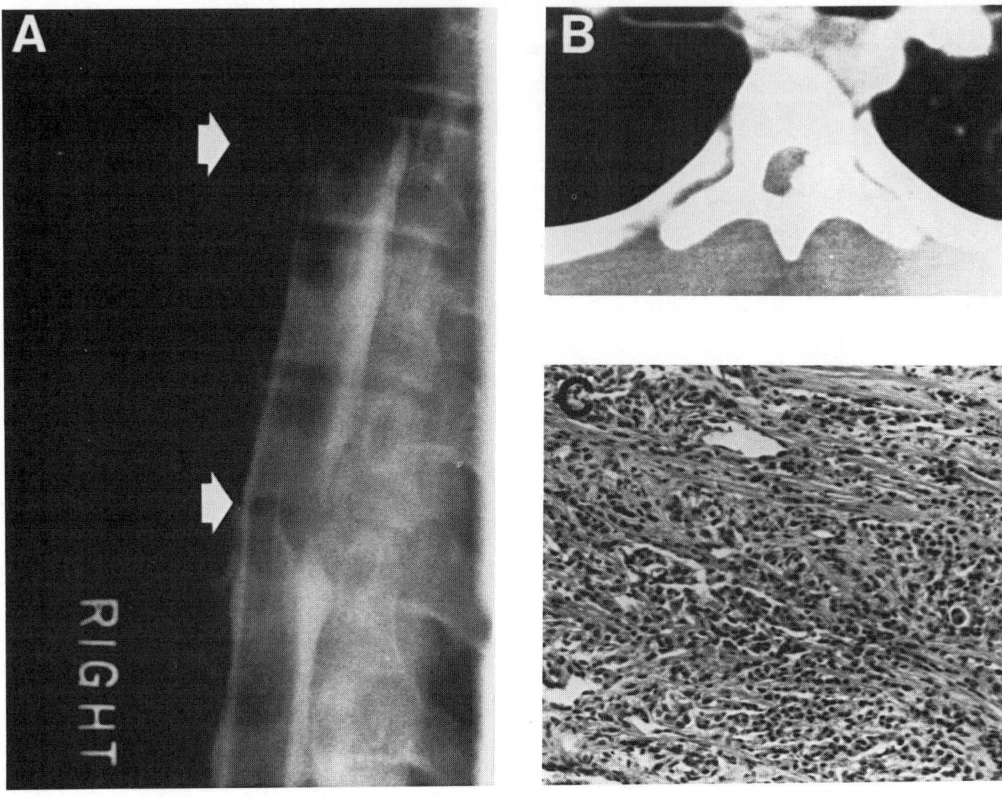

Fig. 2. This figure summarizes the radiologic and pathologic findings in the case report.

A. A myelographic study performed with water soluble contrast shows two intradural extramedullary lesions at the T3 through T4 disc level and at the T6 vertebral body (upper and lower arrows, respectively). The lesions cause compression and displacement of the spinal cord toward right anterolateral aspect. There is also slight flattening of the pedicle at the level of T6 (lower arrow) on the left side.

B. A computerized axial tomographic image of thoracic spine obtained after myelogram shows intradural extramedullary lesion at the level of T6, causing compression and displacement of the spinal cord toward right anterolateral aspect. There are no obvious bony changes seen on this study.

C. A microscopic surgical specimen showing densely packed fibrous tissue with psammoma bodies and calcification consistent with psammoma body meningioma, H and E x 95.

Radiological studies: A complete spine survey revealed an abnormality. A total myelogram revealed two intradural extramedullary masses at the T6 through T7 and T3 through T4 levels (Fig. 2, A). A contrast Computed Tomogram (CT) of the thoracic spine showed the calcified lesion at the T6 level displacing the spinal cord anterolaterally (Fig. 2, B). A CT of the brain revealed lesions with increased density in the area of the left Foramen of Monroe, right tentorial incisura, and left cerebellopontine angle. CT studies of the kidneys, liver, spleen, and pancreas were normal.

Neurosurgical procedures and follow-up

On January 17, 1985, the patient underwent a T5 through T6 thoracic laminectomy and exploration with removal of an intradural tumor and meningeal attachment. The microscopic neuropathology consisted of densely packed fibrous tissues with calcification and psommoma bodies consistent with a meningioma (Fig. 2, C).

Following three months, there was no clinical improvement in the motor and sensory function, nor any favorable changes in the SEP parameters (Table 1). Repeat myelogram and CT scan showed a complete block at T3 through T4 level. The neuropathology was similar to the first tumor.

Three months after the second operation and six months following initial studies, SEP parameters from the left leg improved significantly with the P38 peak latency entering the normal range; N20 through P38 conduction time reduced from 29.6μSec to 23.6μSec; and L3 through scalp conduction velocity increased from 19.6μSec to 24.6μSec (Table 1). The latter two SEP parameters although appreciably improved, remained outside 2.5SD of mean value for our laboratory.

In the 15 months following the second operation the patient's clinical examination gradually improved. This was associated with changes in SEP parameters with: Further reduction of P38 peak latency from 43.6μSec to 42.0μSec; decrease of the N20 through P38 conduction time from 23.6μSec to 21.6μSec; and increased L3 through scalp conduction velocity from 24.5μSec to 26.9μSec.

The amplitude of P38 did not improve until 15 months after the second operation (Table 1). This clearly followed improvement of conduction times, conduction velocities, and P38 peak latency of the affected side. At the time of our last study (07/09/86) the left leg amplitude for the P38 did not differ significantly from the right side.

Discussion and conclusion

The present investigation demonstrates that the cortical and subcortical SEP components may be used to provide clinically meaningful information about spinal somatosensory function in a patient with multiple cerebral and spinal lesions. In addition, the quantitative SEP parameters added objective measurement to the postopera-

Table I

SSEP PARAMETERS (Mean ± 2.5 SD)	PRE-OP 1/14/85	POST-OP T5-T6 Laminectomy 3/21/85 (3 Mos.)	POST-OP T3-T4 Laminectomy 7/8/85 (6 Mos.)	POST-OP 7/9/85 (18 Mos.)
P_{38} Peak Latency (43.6 m/sec.)	L 48.4*	50.8*	43.6	42.0
	R 40.4	42.4	41.8	40.4
P_{38} Peak Latency (0.2 μV)	L 0.6	0.4	0.5	1.5
	R 1.0	0.7	1.0	1.3
N_{20} → P_{38} Conduction Time (19.9 m/sec.)	L 28.4*	29.6*	23.6*	21.6*
	R 20.4	21.2*	22.8*	20.0
L_3 → Scalp Conduction Velocity (26.5 m/sec.)	L 20.7*	19.6*	24.6*	26.9
	R 28.4	27.3	25.4*	29.0

*Significant difference from normal values.

tive assessment of the patient; confirmed the need for further surgery; and heralded recovery of neurologic function following removal of the second spinal tumor.

This data confirms observations in other laboratories that indicate the latency and amplitude of the cortical components (P38) show parallel changes as a result of spinal cord compression (4). Though detailed clinical correlations were not attempted in this case, the finding of a Brown-Sequard syndrome ipsilateral to the side of the P38 prolongation and amplitude reduction argues strongly for dorsal column and spinocerebellar tract mediation of the SEP through the spinal cord on the side of stimulation (10, 11, 12, 13).

Our preoperative and longitudinal data, at 3, 6, and 18 months postoperatively revealed significant changes in latency and amplitude in the interval following the removal of the upper thoracic tumors. These findings generally support the value of SEPs in monitoring of the effects of surgery as found in the literature (4, 5, 6, 7, 15). As Shiff and coworkers indicated in their studies, the SEP parameters of conduction time and conduction velocities were sensitive and useful for evaluation of spinal cord pathology.

Riffel et al., reported that latency of the P38 was essentially normal and amplitude criteria were more reliable for diagnosis of spinal cord tumors (type and location unspecified) (16). The significant delay of the P38 latency greater than 3SD) and reduced amplitude in our case indicate that with respect to extramedullary intradural thoracic tumors latency abnormalities must be considered.

In our longitudinal studies, the restoration of the amplitude of P38 appears much later than improvement of conduction parameters with spinal cord decompression. This suggests that the measures of conduction time and velocity may be an early indicator of recovery of somatosensory pathway functions. Further, we speculate that demyelination, in part, was responsible for the preoperative SEP abnormalities. Conversely, remyelination following decompression probably underlies the restoration of latency, conduction velocity, and amplitude. Such changes suggest that for the purpose of diagnosis and prognosis, latency measurements with spinal conduction times, velocities, and amplitude measurement are necessary to optimize SEP recordings as a clinical neurophysiological tool in the evaluation of patients with spinal cord pathology.

References

1. Desmedt, J.E.; Neol, P.: Average cerebral evoked potentials in evaluation of lesion of sensory nerves and of the central somatosensory pathway. In: J.E. Desmedt (ed): New Developments in Electromyography and Clinical Neurophysiology, Vol. 2, Karger, Basel, pp. 352-71, 1973.
2. Cracco, J.B.; Cracco, R.Q.; Graziani, L.J.: The spinal evoked response in infants and children. Neurology, 25: 31-36, 1975.
3. Cracco, J.B.; Bosch, V.V.; Stolove, R.: Spinal evoked potential in man: A maturational study. Electroencephalogr. Clin. Neurophysiol., 46: 58-64, 1979.
4. Dorfman, L.J.; Perkash, I.; Bosley, T.M.; Cummins, K.L.: Use of cerebral evoked potentials to evaluate spinal somatosensory function in patients with traumatic and surgical myelopathies. J. Neurosurg., 52: 654-660, 1980.
5. Schiff, J.A.; Cracco, R.Q.; Rossini, P.M.; Cracco, J.B.: Spinal and scalp somatosensory evoked potentials in normal subjects and patients with spinal cord disease: Evaluation of afferent transmission. Electroencephalogr. Clin. Neurophysiol., 59: 374-387, 1984.
6. Baron, E.; Grover, W.; Brown, L.; Foley, C.: Somatosensory evoked potential abnormalities with spinal neoplasms. Electroenceph. Clin. Neurophysiol., 56: 542, 1982.
7. Schramm, J.: Somotosensorisch Evozierte Potentials (SEP) in der differentialdiagnose spinaler Erkranhangen. In: J. Schromm (ed): Evozierte Potentiale in der Praxis. Springer, Verlag, Berlin, Heidelberg, pp. 97-32, 1985.
8. Porras, C.L.: Meningioma in the foramen magnum in a boy aged 8 years. J. Neurosurg., 20: 167-168, 1963.
9. DiRocco, C.; Caldarelli (no initial), Puca, A.; Colosimo, J.R.C.: Multiple spinal meningiomas in children. Neurochirurgia, 27: 25-27, 1984.

10. Kakigi, R.; Shihasaki, H.; Hashipume, A.; Kurolma, Y.: Short latency somatosensory evoked spinal and scalp recorded potentials following posterior tibial nerve stimulation in man. Electroencephalogr. Clin. Neurophysiol., 53: 602-611, 1982.

11. Desmedt, S.E.; Cheron, G.: Central somatosensory conduction in man: Neural generators and interpeak latencies of the far field components recorded from neck and right or left scalp and earlobes. Electroencephalogr. Clin. Neurophysiol., 50: 382-403, 1980.

12. Halliday, A.M.; Wakefield, G.S.: Cerebral evoked potentials in patients with dissociated sensory loss. J. Neurol. Neurosurg. Psychiat., 26: 211-219, 1963.

13. Cusick, J.F.; Myklehust, J.B.; Larson, S.J.; Sanos, J.A.: Spinal cord evaluation by cortical evoked responses. Arch. Neurol., 36: 140-143, 1979.

14. Feldman, M.H.; Cracco, R.Q.; Farmer, P.; Mount, F.: Spinal evoked potentials in monkeys. Ann. Neurol., 7: 238-244, 1980.

15. Cracco, R.Q.; Evans, B.: Spinal evoked potentials in the cat: Effects of asphyxia, strychnine, cord section and compression. Electroencephalogr. Clin. Neurophysiol., 44: 187-201, 1978.

16. Riffel, B.; Stohr, M.; Korner, H.: Spinal and cortical evoked potentials following stimulation of the posterior tibial nerve in the diagnosis and localization of spinal cord diseases. Electroencephalogr. Clin. Neurophysiol., 58: 400-407, 1984.

Use of Somatosensory Evoked Potentials to Monitor Spinal Cord Ischemia during Surgery on the Thoracic and Thoraco-Abdominal Aorta

J. N. Cunningham, Jr.;[*] K. H. Lim.; D. M. Rose

Introduction

Significant controversy remains concerning the etiology of paraplegia and the usefulness of adjunctive techniques to prevent this complication during surgery on the thoraco-abdominal aorta. The disconcerting prevalence of this complication has remained between 2-10% despite many alterations in surgical and anesthetic techniques (2, 6, 7, 8, 9, 10, 13, 14, 15, 16, 17, 20, 26, 29, 30, 31, 36, 37, 38). Most of the data concerning the incidence and etiology of paraplegia for these types of surgical procedures has been obtained from retrospective studies (1, 8, 9, 11, 14, 31, 33.

Previous studies from our laboratory have demonstrated the importance of adequate distal perfusion and reimplantation of critical intercostal arteries following aortic cross clamping in order to prevent paraplegia, both clinically and experimentally (12). The present report summarizes our clinical experience using somatosensory evoked potentials in a prospective group of patients undergoing surgical procedures on the descending thoracic and thoraco-abdominal aorta. The intent of this study is to define what intraoperative events result in spinal cord ischemia; which techniques, adjuncts, or pharmacological agents may be useful in prevention of ischemia and which events result in the development of permanent paraplegia.

Methods and patient population

Patient population (Table 1)

Thirty-three consecutive patients undergoing operations for lesions of the descending thoracic or thoraco-abdominal aorta between October, 1981, and October, 1984, at N.Y.U. Medical Center, New York, N.Y. and Maimonides Medical Center, Brooklyn, N.Y. form the basis for this report. Etiologies of the aortic lesions included atherosclerotic, posttraumatic, mycotic, chronic dissection, congenital coarctation, and an aneurysm at the site of a previously ligated patent ductus arteriosis. The origin of the left carotid and/or left subclavian arteries was involved in 20 patients, and extensive thoraco-abdominal disease requiring reimplantation of mesenteric and renal vessels was encountered in 2 patients. The age ranged from 15-83, and patients were predominantly male (23 male, 10 female).

---

* Department of Surgery, Maimonides Medical Center; 4802 10th Avenue, Brooklyn, NY 11219

Cortical Response

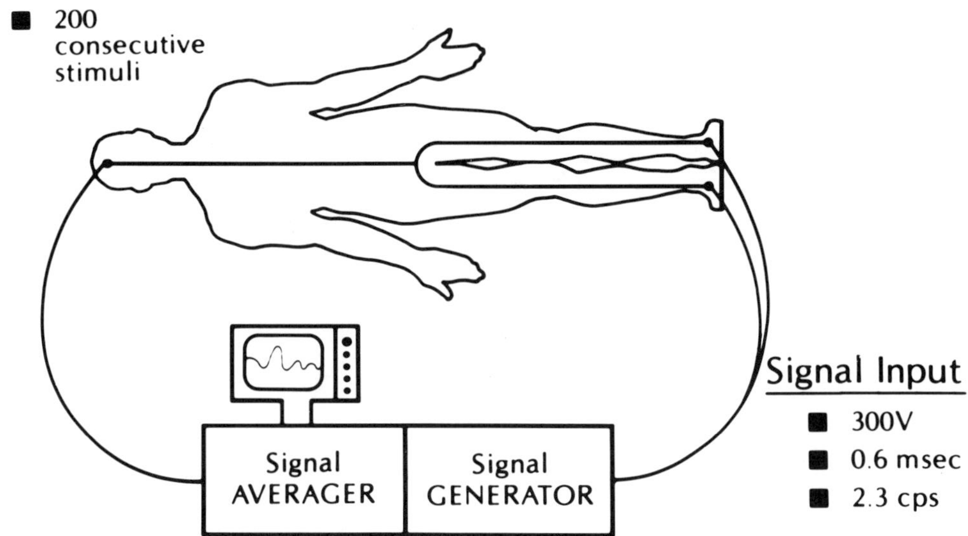

Fig. 1. Schematic representation of clinical technique of SEP monitoring.

Operative data and techniques

Decisions regarding operative technique and utilization of shunt or bypass devices were made according to individual preference of the several surgeons participating in this study and were not altered by intraoperative SEP findings.

Dacron tube grafts were inserted in 25 patients using the reimplantation technique of Crawford (8, 9). Rooftop dacron patches were employed after excision of saccular aneurysms in 4 patients, and primary end to end aortic anastomoses were performed in 3 patients with congenital coarctations. One patient underwent an extra-anatomic axillo-femoral bypass prior to excision of a T9 through T10 mycotic aneurysm.

Technique of spinal cord protection during aortic repair varied. A "clamp/repair" technique with no distal perfusion was utilized in 8 patients, femoral-femoral partial cardiopulmonary bypass was employed in 19 patients, a TDMAC heparin shunt (proximal aorta to femoral artery) was inserted in 3 patients, and there was no attempt made at distal perfusion in 3 patients with congenital coarctations since there was evidence of adequate distal perfusion via intrathoracic collateral blood flow (as indicated by distal pressure > 60mmHg and maintenance of SEP). Proximal and distal perfusion pressures were monitored in all patients. When heparinized shunts were employed, intraoperative shunt flow was assessed with an on-line electromagnetic flow probe (Gould-Statham Medical Products, Oxnard, CA).

Evaluation of neurologic status and spinal cord conduction

Complete neurologic examinations were performed in all patients preoperatively, 24 hours postoperatively, and prior to hospital discharge. All neurologic examinations were performed on the same patient by the same neurologist.

A clinical evoked potential system (TN-3000, Tracor Analytic Inc., Oak Grove Village, IL; Nicolet Biomedical Instruments, Cranford, N.J.) (Fig. 1) (12). SEP traces were generated by bilateral stimulation of posterior tibial nerves with two bipolar input

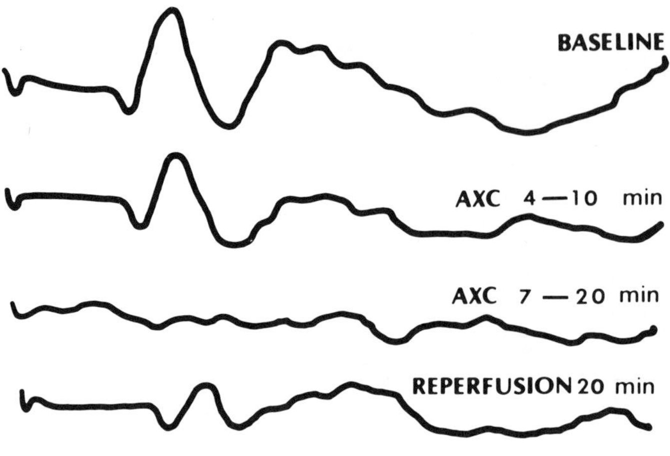

Fig. 2. Type 1 SEP response - loss of SEP secondary to spinal cord ischemia after proximal aortic cross clamping without distal aortic perfusion.

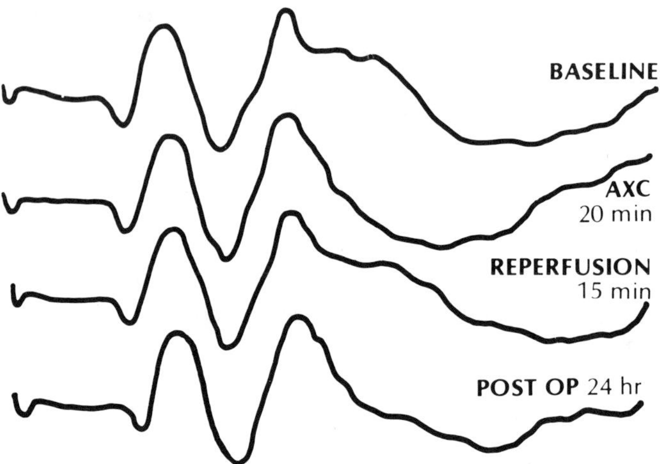

Fig. 3. Type 2 SEP response - maintenance of normal SEP by adequate distal perfusion after proximal aortic cross clamping.

channels. After conduction of impulses via the dorsal spinal columns, the cortical response to 200 consecutive stimuli was recorded from needle electrodes inserted at the nasion and 55% of the distance from the nasion to the inion in the midline of the scalp. The potentials were amplified 10,000 times and processed with a 10-H LoPass and 250-H Hipass filter. To improve the signal to noise ratio of these small potentials, 200 consecutive responses activated by supramaximal stimuli to the nerves (4 times the motor twitch threshold, 10-20 milliamps, 0.6mSec, duration pulses, 2.3 per second) were averaged for each SEP trace. A separate grounding electrode was placed in the upper thorax of each patient, and in addition, a no stimulus control trace was recorded in each patient for establishment of the background noise level. SEP responses were graded as Type 1-4 based on previous studies (12).

- Type 1 (Fig. 2). This response is characterized by the rapid deteriorations of spinal cord conduction with increased latency and diminution of SEP amplitude which progresses to total absence of conduction 7-30 minutes after proximal aortic cross clamping. This response was observed when the "clamp/repair" technique was utilized.

- Type 2 (Fig. 3). The SEP remains normal following aortic cross clamping as a result of distal perfusion >60mgHg with mechanical devices (Gott shunt, left atrial femoral artery bypass, femoral-femoral artery bypass) or extensive collateralization (coarctation with satisfactory intrathoracic collateral blood flow).

- Type 3 (Fig. 4). In the presence of adequate distal aortic perfusion following proximal aortic cross clamping (with shunt or bypass), sudden loss of evoked potentials after placement of the distal aortic cross clamp indicates spinal cord ischemia secondary to exclusion of critical intercostal arteries. Removal of the distal cross clamp will result in the rapid return of evoked potentials. This response alerts the surgeon for the need for rapid reimplantation of the intercostal arteries. Arteries, intercostal

- Type 4 (Fig. 5). In this response there is a slow insidious loss of SEP gradually disappearing 30-50 minutes following aortic cross clamping. This response is seen in patients with suboptimal distal aortic perfusion (usually < 60mmHg) as a result of technical and/or anatomic factors (profound vasodilitation, extensive aneurysmal disease, or aorto-iliac disease).

All results are expressed as mean $\pm$ standard deviation of the mean (SEM), and statistical analysis was performed by a Fisher's exact test unless otherwise indicated.

Results

Results are summarized in Table 2. As illustrated in Fig. 6, 11 patients underwent surgery without the use of shunt or bypass techniques. Three of the 11 patients had aortic coarctation and had evidence of "self-shunting" via intrathoracic collaterals with mean distal aortic pressures remaining greater than 60mmHg throughout the cross clamp interval. A temporary shunt or distal bypass technique was employed in 22 patients. Incidence of paraplegia in the entire group of 33 patients was 15.1% (5/33 patients). Paraplegia was not observed in any of the 19 patients in whom adequate distal aortic perfusion pressure (> 60mmHg) was maintained by distal bypass or collateral circulation following aortic cross clamping (p = 0.01). In contrast, the incidence of paraplegia was 35.7% (5/14 patients) when adequate distal aortic perfusion was not achieved.

Disappearance of evoked potentials with clamp/repair techniques (Type 1 response)

Aortic cross clamping without attempt at distal aortic perfusion in 8 patients resulted in loss of SEP within 7-30 minutes (mean 17.0 $\pm$ 8 minutes). In the absence of distal shunting or adequate collaterals, mean distal aortic perfusion pressure was < 20mmHg in this group. The increase in latency and diminution of amplitude noted soon after proximal cross clamping rapidly progressed, indicating significant decrease in spinal cord impulse conduction via the dorsal columns. Conduction slowly returned over a 30-50 minute period following declamping in 5 patients. These patients subsequently demonstrated normal neurologic function postoperatively. In the 3 remaining patients, the SEP failed to reappear and these patients were paraplegic. Overall incidence of paraplegia in the "clamp/repair" group was 37.5% (3/8 patients).

Maintenance of normal evoked potentials associated with adequate distal aortic perfusion (Type 2 response)

Distal aortic perfusion pressure was maintained at 60-80mmHg throughout the entire cross clamp interval in 17 patients. Femoral vein-femoral artery oxygenator bypass was employed in 11 patients, heparinized shunts (ascending aorta-femoral) in 3 patients, and 3 patients with congenital coarctations exhibited "self-shunting" through intrathoracic collaterals.

Because normal evoked potentials were maintained throughout the entire cross clamp interval (range 23-105 minutes), intercostal reimplantation was not carried out in

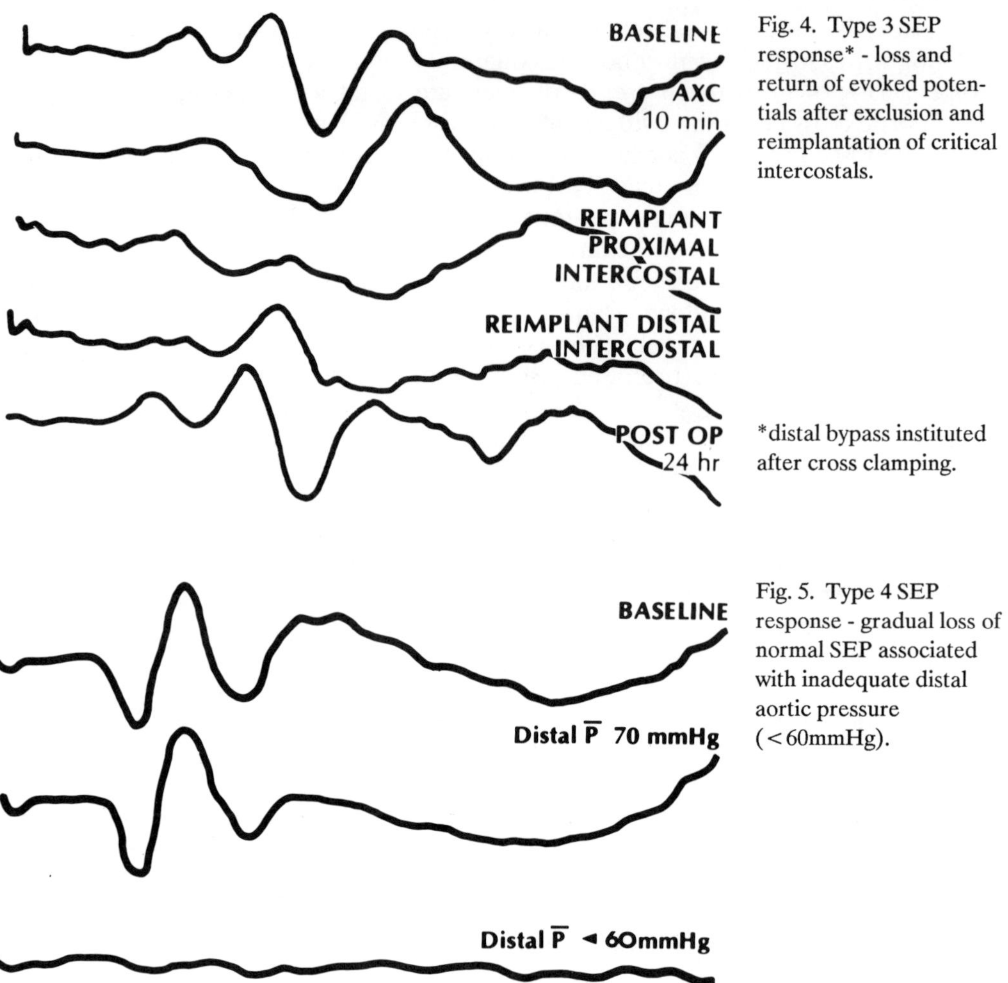

Fig. 4. Type 3 SEP response* - loss and return of evoked potentials after exclusion and reimplantation of critical intercostals.

*distal bypass instituted after cross clamping.

Fig. 5. Type 4 SEP response - gradual loss of normal SEP associated with inadequate distal aortic pressure ($<$60mmHg).

any of these patients. No paraplegia occurred in these patients although 9/17 of these patients had extensive aneurysms extending below the diaphragm.

Use of evoked potentials to identify critical intercostals (Type 3 response)

Two of the patients in whom evoked potentials were normal following placement of the proximal cross clamp and who had adequate distal aortic perfusion were noted to rapidly lose the SEP after placement of the distal aortic cross clamp. Since this phenomenon implied that there were critical intercostal vessels within the region of the proximal and distal cross clamps, appropriate measures were taken to expeditiously reimplant and reperfuse these important intercostal vessels. Evoked potentials rapidly disappeared in both patients (total duration of SEP loss was 13 and 30 minutes respectively) but quickly returned following intercostal reperfusion. Both patients had normal neurologic exams postoperatively.

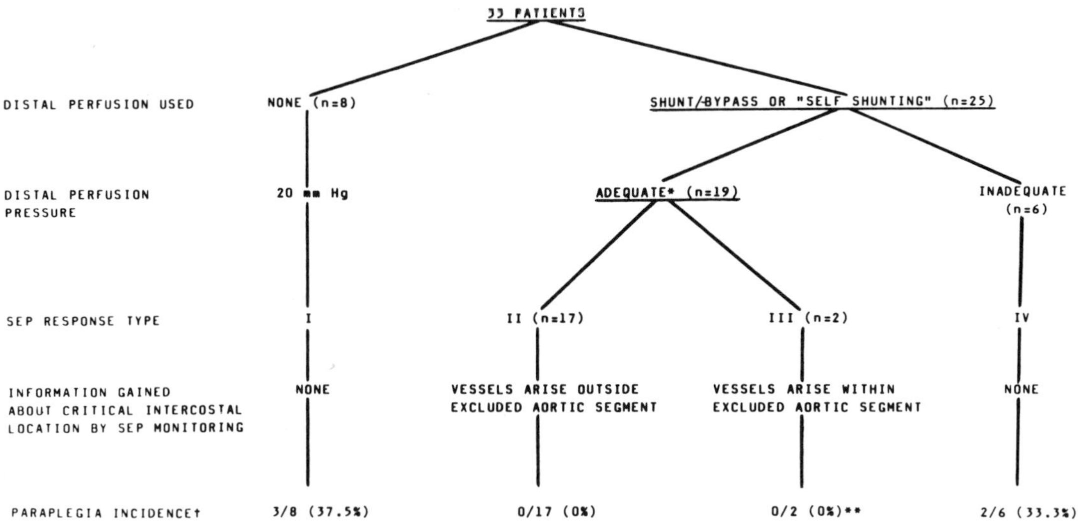

Fig. 6. Patterns of SEP response (Types 1, 2, 3, 4) and the incidence of paraplegia in 33 patients undergoing procedures on the thoraco-abdominal aorta.

"Fade-out" of evoked potentials associated with inadequate distal perfusion (Type 4 response)

In 6 patients distal shunt/bypass techniques were employed but distal aortic perfusion pressure was <60mmHg as a result of technical problems. Distal extent of the aneurysms varied from T6 to T12 and duration of aortic cross clamp time ranged from 30-85 minutes. In these patients shunt or bypass techniques were impeded by the presence of extensive distal aorto-iliac disease or profound peripheral vasodilation.

A salient observation in this group of patients was that SEP tracings gradually decayed of "faded-out" over periods as long as 51 minutes (mean 33.3 ± 13 minutes). This time course of evoked potential disappearance was significantly more prolonged than the rapid disappearance of evoked potentials seen following sudden interruption of identifiable critical intercostals (Type 3) or interruption of flow from a simple clamp/repair approach (Type 1).

Restoration of distal aortic perfusion with a mean pressure greater than 60-70mmHg following declamping resulted in prompt return of SEP and a normal postoperative neurologic status in 3 patients. The duration of total SEP loss following the "fade-out phenomena" in these patients was relatively brief (5, 11, and 15 minutes). Duration of evoked potential loss was more prolonged (43, 49, and 51 minutes) in 3 other patients, and two of these had permanent postoperative paraplegia.

Relationship of variables to incidence of postoperative paraplegia (Table 3)

Five variables were analyzed to determine their association with postoperative paraplegia (extent of aneurysm, use of shunt/bypass, distal perfusion pressure, SEP loss,

Table I: Summary of Preoperative Patient Data

Number	Patient	Age	Sex	Etiology	Extent of Lesion
1	SD	73	M	Atherosclerotic	L Subclavian origin – T 6
2	MS	83	M	Traumatic (old)	L Carotid origin – T 8
3	EG	65	F	Atherosclerotic	L Subclavian origin – bifurcation
4	AN	58	M	Atherosclerotic	L Subclavian origin – T 10
5	ES	76	F	Atherosclerotic	Distal to L Subclavian – T 9
6	ES	58	M	Atherosclerotic	L Carotid origin – T 6
7	EM	73	F	Atherosclerotic	Distal to L Subclavian – bifurcation
8	JA	60	M	Dissection (chronic)	L Subclavian origin – T 10
9	MP	52	F	Atherosclerotic	L Subclavian origin – T 12
10	JS	75	M	Atherosclerotic	T 7 – T 9
11	EA	76	M	Atherosclerotic	L Subclavian origin – T 12
12	AG	52	M	Atherosclerotic	L Subclavian origin – T 8
13	PC	15	M	Congenital Coarctation	Post Ductal – T 6
14	JG	35	F	Congenital Coarctation	Post Ductal – T 6
15	DD	2	F	Congenital Coarctation	Post Ductal – T 6
16	MR	67	M	Atherosclerotic	L Subclavian origin – T 11
17	CC	42	F	Dissection (chronic)	L Subclavian origin – T 12
18	DV	74	M	Traumatic (old)	L Subclavian origin – T 7
19	VS	62	M	Atherosclerotic	L Subclavian origin – T 10
20	HF	68	F	Atherosclerotic	L Subclavian origin – T 11
21	DH	60	M	Atherosclerotic	Distal to L Subclavian – T 12
22	MH	62	M	Atherosclerotic	T 6 – T 11
23	VH	65	M	Dissection (chronic)	L Carotid origin – T 8
24	HC	68	F	Atherosclerotic	L Subclavian origin – T 10
25	RD	46	M	Traumatic (old)	T 6 – T 9
26	EB	60	M	Mycotic	T 9 – T 10
27	HV	70	M	Atherosclerotic	T 8 – T 12
28	JO	69	M	Atherosclerotic	L Subclavian origin – T 5
29	CJ	59	M	Atherosclerotic	L Subclavian origin – T 10
30	JU	58	M	Atherosclerotic	Distal to L Subclavian – T 10
31	NH	67	M	Atherosclerotic	T 6 – T 9
32	BK	19	M	Istrogenic	L Subclavian origin – T 6
33	MP	66	F	Atherosclerotic	L Subclavian origin – T 11

and duration of spinal cord ischemia). The relationship of distal extent of aneurysm was closely scrutinized as a contributing factor to postoperative neurologic injury. Of the 5 cases of paraplegia occurring, 2 involved patients with extensive aneurysms extending to the aorto-iliac bifurcation while the remaining 3 occurred at or above the level of the diaphragm. No statistically significant difference in incidence of postoperative paraplegia was found when upper thoracic aneurysms (above T8) were compared to more extensive thoraco-abdominal aneurysms (T9 to bifurcation).

The failure to use a shunt or bypass device per se was not found to be a statistical predictor of subsequent neurologic injury. While the 37.5% occurrence of paraplegia seen without use of such devices did not reach statistical significance, there was a strong suggestion that utilization of these techniques was quite important. Maintenance of distal perfusion pressure $>60\text{mmHg}$ following cross clamping by either "self-shunting" or shunt/bypass techniques was associated with a 0% (0/19) incidence of paraplegia, a statistically significant observation when compared to the higher incidence of paraplegia in inadequately perfused patients (35.7%, 5/14; $p = .01$).

The loss of evoked potentials during the cross clamp interval was a significant predictor of subsequent paraplegia. Analysis of this data revealed that the duration of ischemia during the cross clamp interval was the most sensitive predictor of postoperative neurologic injury. Paraplegia did not occur in any patient in whom SEP loss was limited to 30 minutes or less. In contrast, permanent neurologic injury occurred in 71% of patients (5/7) when evoked potential disappearance exceeded 30 minutes following aortic cross clamping ($p < .001$).

Discussion

Evoked potential monitoring - a measure of spinal cord viability

Adams, in 1956, first analyzed paraplegia following surgical attempts to repair lesions of the thoraco-abdominal aorta. This important work identified both duration of aortic cross clamping and variability of spinal cord blood supply as the primary determinants of injury (1). Numerous authors have suggested abdominal causes of neurologic injury, but unfortunately these reports are limited by their retrospective nature and the absence of technology to accurately determine the intraoperative status of spinal cord conduction (3, 4, 5, 8, 9, 14, 31, 32, 34).

The development and clinical use of somatosensory evoked potential monitoring during surgical procedures on the descending thoracic and thoraco-abdominal aorta has provided a sensitive means for intraoperative detection of spinal cord ischemia (5, 12, 21, 23, 24, 25). Important observations concerning the temporal occurrence of neurologic injury are now possible as a result of SEP monitoring. For example, the Type I response (no distal perfusion after cross clamping) demonstrated loss of SEP in 17 ± 8 minutes after cross clamping, and this was associated with a 71% incidence of paraplegia if these changes were not reversed within 30 minutes. This correlates well with previous empiric clinical observations that the incidence of paraplegia is very rare if aortic cross clamp time is less than 30 minutes but increases with longer ischemic intervals (1, 8, 9, 11, 14, 16, 26, 31, 38).

What has been wrong with shunts in the past?

Historical evidence from previous clinical studies suggests that the mere use of shunt or bypass techniques following proximal aortic cross clamping has not necessarily prevented paraplegia. However, the absence of paraplegia occurring in this study when patients underwent adequate shunting or distal aortic perfusion (distal aortic pressure

Table II: Summary of Intraoperative and Postoperative Patient Data

Number	Patient	Shunt Used	Aortic Crossclamp Placement		Duration (min)	Distal Aortic Pressure (mmHg)	SEP Response Type	Time to SEP Loss Post AXC (min)	Duration of SEP Loss
			Proximal	Distal					
1	SD	None	Pre L subclavian	T8	29	20	1	8	21
2	MS	None	Pre L carotid	T9	37	20	1	9	28
3	EG	None	Pre L subclavian	bifur.	58	20	1**	22	36
4	AN	None	Post L subclavian	T10	23	20	1	20	3
5	ES	None	Post L subclavian	T10	23	20	1	19	4
6	ES	None	Pre L carotid	T6	84	20	1**	22	62
7	EM	None	Post L subclavian	bifur.	124	20	1**	29	95
8	JA	None	Pre L subclavian	T11	16	20	1	7	110
9	MP	Fem/Fem Bypass	Pre L subclavian	T12	75	65	2	-	--
10	JS	Fem/Fem Bypass	T6	T9	35	60	2	-	--
11	EA	TDMAC-heparin	Pre L subclavian	T12	65	65	2	-	--
12	AG	TDMAC-heparin	Pre L subclavian	T9	65	65	2	-	--
13	PC	"self shunt"	Post L subclavian	T7	27	70	2	-	--
14	JG	"self shunt"	Post L subclavian	T7	23	60	2		
15	DD	"self shunt"	Post L subclavian	T7	26	60	2	-	--
16	MR	Fem/Fem Bypass	Pre L subclavian	T12	62	65	2	-	--
17	CC	Fem/Fem Bypass	Pre L subclavian	T12	68	60	2	-	--
18	DV	Fem/Fem Bypass	Pre L subclavian	T7	35	60	2	-	--
19	VS	Fem/Fem Bypass	Pre L subclavian	T12	55	60	2	-	--
20	HF	Fem/Fem Bypass	Pre L subclavian	T12	53	70	2	-	--
21	DH	Fem/Fem Bypass	Post L subclavian	T12	47	75	2	-	--
22	MH	Fem/Fem Bypass	T5	T12	26	60	2	-	--
23	VH	Fem/Fem Bypass	Pre L carotid	T8	105	70	2	-	--
24	HC	Fem/Fem Bypass	Pre L subclavian	T11	61	60	2	-	--
25	RD	Fem/Fem Bypass	T6	T9		60	2	-	--
26	EB	Extra-anatomic (Ax/Fem)	T5	L1	40	65	3	10	30
27	HV	Fem/Fem Bypass	T7	T12	37	70	3	20	13
28	JO	Fem/Fem Bypass	Pre L subclavian	T6	30	37	4	25	5
29	CJ	TDMAC-heparin	Pre L subclavian	T11	85	20	4 **	36	49
30	JU	Fem/Fem Bypass	Post L subclavian	T10	68	30	4 **	25	43
31	NH	Fem/Fem Bypass	T6	T9	62	45	4	51	11
32	BK	Fem/Fem Bypass	Pre L subclavian	T7	67	20	4	45	15
33	MP	Fem/Fem Bypass	Pre L subclavian	T12	69	30	4	18	51

> 60mmHg, normal evoked potentials) suggests that regulation of these devices based on SEP monitoring techniques is effective in preventing neurologic injury.

The most likely explanation for failure of shunts or distal bypass to eliminate postoperative paraplegia following aortic surgery is related to two factors: Basic inadequacy of the shunt or perfusion device and inappropriate identification of critical intercostal vessels. Small shunts may provide inadequate distal aortic perfusion resulting in spinal cord ischemia and neurologic injury. These heparinized aorto-femoral shunts may not provide adequate distal flow as a result of small tubing size, low proximal aortic pressure, kinking, etc. Regardless of the circumstances of distal aortic perfusion, appropriate flow and pressure must be achieved to prevent ischemia. Our previous experience clearly indicates that requirements for "optimal" distal perfusion vary greatly from patient to patient and the level of appropriate retrograde perfusion can only be determined by on-line monitoring of evoked potentials and distal aortic pressure (12). Interruption or exclusion of critical intercostal segments is then an identifiable event signaled by sudden loss of evoked potentials following placement of a distal aortic cross clamp.

Single artery of Adamkiewicz - myth or fact?

The long standing theory that a single critical artery or pair of intercostals exists and must be preserved or reimplanted to prevent paraplegia will likely be more thoroughly evaluated in the future. It is unclear from historical and present data exactly where critical pairs of intercostal vessels are anatomically located and it is unlikely that a single artery is critical to overall spinal cord blood supply. It is generally recognized that the incidence of paraplegia is highest following thoraco-abdominal aneurysm repair and interruption of intercostals between the diaphragm and renal arteries. Nonetheless, neurologic injury may occur in 3-8% of patients with repair of isolated upper thoracic aneurysms (1, 8, 9, 11, 14, 20, 31). Therefore, the variable incidence of paraplegia relative to location and extensiveness of aneurysm resection may be a function of a combination of factors including anatomic location of single or multiple critical intercostals as well as individual surgical technique.

It is unlikely that paraplegia following aortic surgery occurs more frequently as increasing numbers of collaterals are excluded. This situation might result in a cumulative increase in resistance to spinal cord blood flow and diminished collateral supply. In this situation direct reimplantation of any large intercostal or cluster of intercostals as suggested by Crawford might be sufficient to improve regional blood supply (9). Paradoxically, prolongation of ischemia time associated with endeavors to reimplant large numbers of intercostals might increase the frequency of paraplegia. It seems more reasonable during repair of extensive aneurysm that the surgeon simply plan on re-establishing flow to one or more major groups of intercostals within 30 to 40 minutes of the time the aorta is cross clamped rather than postponing re-establishment of flow until all anastomoses are done. This concept will be tested with further experience, but can only be evaluated appropriately if evoked potential monitoring is utilized.

Ongoing research is exploring pharmacologic methods to prolong the safe ischemic time interval. In some animal studies there have been favorable results with the use of corticosteroids (22), calcium channel blocking agents (18, 35), and free radical scavengers (27, 28, 35) (DMSO and superoxide dismutase). Also, the use of intrathecal papaverine appears to have some experimental efficacy (35). Certainly further investigative work is necessary to elucidate more precisely the cellular mechanisms of injury and other possible methods to prevent this serious injury.

Table III: Relationship of Variables to Incidence of Postoperative Paraplegia

Variable	Incidence of Postoperative Paraplegia	Significance of Difference*
I. Extent of Aneurysm		
a. Above T8	10.0% (1/ 10)	NS
b. T9 to Bifurcation	17.4 (4/ 23)	
II. Use of Shunt/Bypass		
a. Yes	8.0% (2/ 25)	NS
b. No	37.5% (3/ 8)	
III. Distal Perfusion Pressure Following AXC		
a. >60 mm Hg	0% (0/ 19)	p= .01
b. <60 mm Hg	35.7% (5/ 14)	
IV. Spinal Cord Ischemia (SEP loss) After AXC		
a. No	0% (0/ 17)	p= .02
b. Yes	31.2% (5/ 16)	
V. Duration of Spinal Cord Ischemia		
a. 30 min or <	0% (0/ 26)	p= .001
b. >30 min	71.2% (5/7)	

* p value for incidence of paraplegia a vs. b for each variable.

Significance of differences calculated by Fishers Exact Test.

AXC = Aortic crossclamp

SEP = Somatosensory Evoked Potentials

Conclusions

The first common pathway to postoperative paraplegia is spinal cord ischemia, an event recognized by SEP monitoring. Once ischemia occurs secondary to any cause, the patient is at risk for postoperative paraplegia. The duration of this ischemic state is the single most important factor affecting the subsequent incidence of paraplegia. Ischemic intervals associated with less than 30 minutes of SEP loss were not observed to result in injury in this study, whereas greater periods were associated with a 71% incidence of paraplegia. Use of devices to maintain adequate distal flow and perfusion pressure prevents spinal cord ischemia and paraplegia unless critical intercostal arteries originate from within the excluded aortic segment. Exclusion of critical groups of intercostals can be appropriately identified by sudden loss of evoked potential responses following distal aortic cross clamping when adequate distal perfusion is maintained. In such situations, spinal cord ischemia time can be minimized by early reimplantation of clusters of intercostals located in a "patch" of aorta. Perhaps other modalities of protection such as direct perfusion of critical intercostals during cross clamping or the use of various pharmacologic adjuncts may be important in preventing neurologic injury. It should be expected that unfavorable anatomy (e.g., aorto-iliac disease, extensive thoraco-abdominal aneurysm, distal aortic dissection, etc.) often results in an inability to provide adequate distal perfusion following proximal aortic cross clamping. In this situation SEP monitoring may not be helpful, and the incidence of paraplegia will vary with the length of total spinal cord ischemia occurring before reperfusion of blindly reimplanted intercostal arteries is accomplished. Certainly, large prospective studies during surgical procedures on the thoraco-abdominal aorta are needed to elucidate completely the many unknown factors in production and prevention of paraplegia.

References

1. Adams, H.D.; Van Geertruyden, H.H.: Neurologic complications of aortic surgery. Ann. Surg., 144: 574-610, 1956.
2. Bahson, H.T.: Definitive treatment of saccular aneurysms of the aorta with excision of the sac and aortic suture. Surg. Gynecol. Obstet., 96: 383-402, 1953.
3. Blaisdell, F.W.; Cooley, D.A.: The mechanism of paraplegia after temporary thoracic aorta occlusion and its relationship to spinal fluid pressure. Surgery, 51: 351-355, 1962.
4. Brewer, L.A.; Fosburg, R.G.; Muldar, G.A.; Verska, J.J.: Spinal cord complications following surgery for coarctation on the aorta. J. Thorac. Cardiovasc. Surg., 64: 368-381, 1966.
5. Coles, J.G.; Wilson, G.J.; Sima, A.F.; Klement, P.; Tait, G.A.; Williams, W.G.; Baird, R.J.: Intraoperative management of thoracic aortic aneurysm. J. Thorac. Cardiovasc. Surg., 85: 292-299, 1983.
6. Connolly, J.E.; Wakabayashi, A.; German, J.C.; Stemmer, E.A.; Serres, E.J.: Clinical experience with pulsatile left heart bypass without anticoagulation for thoracic aneurysms. J. Thorac. Cardiovasc. Surg., 62: 568-576, 1971.
7. Connors, J.P.; Ferguson, T.B.; Roper, C.L.; Weldon, C.S.: The use of the TDMAC-heparin shunt in replacement of the descending thoracic aorta. Ann. Surg., 181: 735-741, 1975.
8. Crawford, E.S.; Fenstermacher, J.M.; Richardson, W.; Sandiford, F.: Reappraisal of adjuncts to avoid ischemia in the treatment of thoracic aneurysms. Surgery, 67: 182-196, 1970.
9. Crawford, E.S.; Waler, H.S.; Saleh, S.A.; Normann, N.A.: Graft replacement of aneurysm in descending thoracic aorta: Results without bypass or shunting. Surgery, 89: 73-85, 1981.
10. Cukingham, R.A.; Carey, J.S.: Repair of lesions of the thoracic aorta with the TDMAC-heparin shunt. J. Thorac. Cardiovasc. Surg., 75: 227-231, 1978.
11. Culliford, A.T.; Ayvaliotis, B.; Shemin, R.; Colvin, S.B.; Isom, O.W.; Spencer, F.C.: Aneurysms of the descending aorta. J. Thorac. Cardiovasc. Surg., 85: 98-104, 1983.
12. Cunningham, J.N., Jr.; Laschinger, J.C.; Merkin, H.A.; Nathan, I.M.; Colvin, S.B.; Ransohoff, J.; Spencer, F.C.: Measurement of spinal cord ischemia during operations upon the thoracic aorta. Ann. Surg., 196: 285-296, 1982.
13. DeBakey, M.; Cooley, D.A.: Successful resection of aneurysms of the thoracic aorta and replacement by graft. JAMA, 152: 673-676, 1953.
14. DeBakey, M.; McCollum, C.H.; Graham, J.M.: Surgical treatment of aneurysms of the descending thoracic aorta. Long-term results in 500 patients. J. Cardiovasc. Surg., 19: 571-576, 1978.

15. DeMeester, T.R.; Cameron, J.L.; Gott, V.L.: Repair of a through-and-through gunshot wound of the aortic arch using a heparinized shunt. Ann Thorac. Surg., 16: 193-198, 1973.

16. Donahoo, J.S.; Brawley, R.K.; Gott, V.L.: The heparin-coated vascular shunt for thoracic aortic and great vessel procedures; a ten year experience. Ann. Thorac. Surg., 23: 507-513, 1977.

17. Frantz, P.T.; Murray, G.F.; Shallal, J.A.; Lucas, C.L.: Clinical and experimental evaluation of left ventriculoiliac shunt bypass during repair of lesions of the descending thoracic aorta. Ann. Thorac. Surg., 31: 551-557, 1981.

18. Gelbfish, J.S.; Phillips, T.; Lim, K.H.; Wait, R.B.; Rose, D.M.; Connolly, M.W.; Acinapura, A.J.; Cunningham, J.N., Jr.: Acute spinal cord ischemia: Prevention of paraplegia with verapamil. Circulation, 2: 1-5, 1986.

19. Gelman, S.; Reves, J.G.; Fowler, K.; Samuelson, P.N.; Lell, W.A.; Smith, L.R.: Regional blood flow during cross clamping of the thoracic aorta and infusion of sodium nitroprusside. J. Thorac. Cardiovasc. Surg., 85: 287-291, 1983.

20. Hilgenberg, A.D.; Rainer, W.G.; Sadler, T.R.: Aneurysm of the descending thoracic aorta. Replacement with the use of shunt or bypass. J. Thorac. Cardiovasc. Surg., 81: 818-824, 1981.

21. Laschinger, J.C.; Cunningham, J.N., Jr.; Catinella, F.C.; Nathan, I.M.; Knopp, E.A.; Spencer, F.C.: Detection and prevention of intraoperative spinal cord ischemia after cross clamping of the thoracic aorta: Use of somatosensory evoked potentials. Surgery, 92: 1109-1117, 1982.

22. Laschinger, J.C.; Cunningham, J.N., Jr.; Cooper, M.; Krieger, K.; Nathan, I.M.; Spencer, F.C.: Prevention of ischemic spinal cord injury following aortic cross clamping: Use of corticosteriods. Ann. Thorac. Surg., 38: 500, 1984.

23. Laschinger, J.C.; Cunningham, J.N., Jr.; Isom, O.W.; Nathan, I.M.; Spencer, F.C.: Definition of the safe lower limits of aortic resection during surgical procedures on the thoraco-abdominal aorta. J. Amer. Coll. Cardiol., 2: 959-965, 1983.

24. Laschinger, J.C.; Cunningham, J.N., Jr.; Nathan, I.M.; Knopp, E.A.; Cooper, M.M.; Spencer, F.C.: Experimental and clinical assessment of the adequacy of partial bypass in the maintenance of spinal cord blood flow during operations on the thoracic aorta. Ann. Thorac. Surg., 36: 417-426, 1983.

25. Laschinger, J.C.; Cunningham, J.N., Jr.; Nathan, I.M.; Krieger, K.; Isom, O.W.; Spencer, F.C.: Intraoperative identification of vessels critical to spinal cord blood supply - use of somatosensory evoked potentials. Current Surg., 41: 107-109, 1984.

26. Lawrence, G.H.; Hessel, E.A.; Sauvage, L.R.; Krause, A.H.: Results of use of the TDMAC-heparin shunt in surgery of aneurysms of the descending thoracic aorta. J. Thorac. Cardiovasc. Surg., 73: 393-398, 1977.

27. Lim, K.H.; Connolly, M.; Rose, D.M.; Siegman, F.; Jacobowitz, I.J.; Acinapura, A.J.; Cunningham, J.N., Jr.: Prevention of reperfusion injury of the ischemic spinal cord - use of recombinant superoxide dismutase. Ann. Thorac. Surg., 42: 282, 1986.

28. Lim, K.H.; Weiss, M.; Connolly, M.W.; Rose, D.M.; Jacobowitz, I.J.; Cunningham, J.N., Jr.: Can diamethyl sulfoxide (DMSO) prolong the ischemic tolerance of the spinal cord? J. Am. Coll. Cardiol., 7: 160, 1986.

29. May, I.A.; Ecker, R.R.; Inversion, L.I.G.: Heparinless femoral venoarterial bypass without an oxygenator for surgery on the descending thoracic aorta. J. Thorac. Cardiovasc. Surg., 73: 387-392, 1977.

30. Murray, G.F.; Young, W.G.: Thoracic aneurysmectomy utilizing direct left ventriculo femoral shunt (TDMAC-heparin) bypass. Ann. Thorac. Surg., 21: 26-29, 1976.

31. Najafi, H.; Javid, H.; Hunter, J.; Serry, C.; Monson, D.: Descending aortic aneurysmectomy without adjuncts to avoid ischemia. Ann. Thorac. Surg., 30: 326-335, 1980.

32. Nylander, W.A., Jr.; Plunkett, R.J.; Hammon, J.W., Jr.; Oldfield, E.H.; Menacham, W.F.: Thiopental modifications of ischemic spinal cord injury in the dog. Ann. Thorac. Surg., 33: 64-68, 1982.

33. Reul, G.J.; Cooley, D.A.; Hallman, G.L.; Reddy, S.B.; Kuger, E.R., III; Wukasch, D.C.: Dissecting aneurysms of the descending aorta. Improved surgical results in 91 patients. Arch. Surg., 110: 632-640, 1975.

34. Spencer, F.C.; Zimmerman, J.M.: The influence of ligation of intercostal arteries on paraplegia in dogs. Surg. Forum, 9: 340-342, 1958.

35. Svensson, L.G.; Von Ritter, C.M.; Groeneveld, H.T.; Rickards, E.S.; Hunter, S.J.S.; Robinson, M.F.; Hinder, R.A.: Cross clamping of the thoracic aorta. Influence of aortic shunts, laminectomy, papaverine, calcium channel blocker, allopurinol, and superoxide dismutase on spinal cord blood flow and paraplegia in baboons. Ann. Surg., 204: 38, 1986.

36. Wakabayashi, A.; Connolly, J.E.: Heparinless left heart bypass for resection of thoracic aortic aneurysms. Am. J. Surg., 130: 212-218, 1975.

37. Wakabayashi, A.; Connolly, J.E.: Prevention of paraplegia associated with resection of extensive thoracic aneurysms. Arch. Surg., 111: 1186-1189, 1976.

38. Wolfe, W.G.; Kleimman, L.H.; Weschler, A.S.; Sabiston, D.C., Jr.: Heparin coated shunts for lesions of the descending thoracic aorta. Arch. Surg., 112: 1481-1487, 1977.

The Effect of Spinal Cord Blood Flow on Evoked Potentials

J.V. Lieponis;[*] K. Jacobs; W.H. Bunch; C. Robinson; E.J. Neafsey

Introduction

The treatment of spinal deformity has undergone considerable evolutionary change (1, 2, 3, 11, 12, 13, 14, 15, 22). Initially prevention of progression, not correction, was emphasized. Although successful, this technique did nothing to resolve the existing deformity. The next logical step was to correct the position of the spine and then to fuse it in a more anatomic alignment. Various techniques were developed: Localizer cast, traction, rods, or a combination of some or all of the above. Regardless of the technique used, two factors limited the amount of obtainable correction: Inherent stability of the spine and the potential for neurologic injury (19).

Initially the window of vulnerability was small. The amount of maximal obtainable correction was limited equally by mechanical and neurophysiological constraints. The development of improved surgical techniques has not only increased the amount of obtainable correction, but it has also widened the window of vulnerability. It is now possible to mechanically correct the spine far beyond the physiological limits of the spinal cord. With the advent of such powerful devices and techniques the amount of correction is limited primarily by the risks of neurologic injury. Now, more than ever, it is imperative that the surgeon have an active means of monitoring spinal cord function.

To be a clinically useful tool, spinal cord monitoring must fulfill certain criteria. It needs to be:

1. Accurate and sensitive. The number of false positive and false negative occurrences must be close to nil.

2. It should ideally indicate a pre-injury state or at least an injury to the cord which is reversible.

3. It should not increase the operative time or operative risks.

Several monitoring techniques have been developed. These can be grouped into two large categories: The wake-up test (10, 28) and electrophysiologic monitoring (4, 5, 16, 20, 21, 23).

Variations and modifications of each of these techniques are many, but the relative advantages and disadvantages are similar within each group. Disadvantages of the wake-up test include:

1. Limited number of tests.

2. This technique tests motor function only.

3. Results can be equivocal.

4. It is not easily applied to patients with existing neurological defects.

* Department of Orthopaedics and Rehabilitation, Yale University School of Medicine, New Haven, CT 06510

TABLE I
SPINAL CORD BLOOD FLOW

	Control	Hypocarbia	Recontrol	Distraction	Distraction and Hypocarbia
C1-C7	14.57	9.35	10.05	15.94	7.0
T1-T6	13.17	7.85	9.34	13.77	5.74
T7-T12	12.01	8.26	9.61	12.23	6.17
L1-L5	9.41	6.46	7.70	9.17	7.33

TABLE II
BRAIN BLOOD FLOW

	Control	Hypo-carbia	Re-control	Distract-tion	Distraction and Hypocarbia
Pre CCor LL	109.7	65.8	65.7	118.8	57.9
Post CCor LL	119.8	61.4	65.8	136.6	60.4
Pre CCor RL	85.9	48.6	48.5	109.8	37.9
Post CCor RL	57.2	43.7	41.6	98.4	56.0
Pre CCor LA	134.7	69.3	73.4	139.7	51.7
Post CCor LA	113.5	70.2	62.1	133.1	47.2
Pre CCor RA	96.8	63.7	73.0	132.4	66.1
Post CCor RA	78.7	61.1	71.3	120.6	70.4

ml/min/100 gr tissue

Fig. 1   **BRAIN BLOOD FLOW**
(Microsphere Determination)

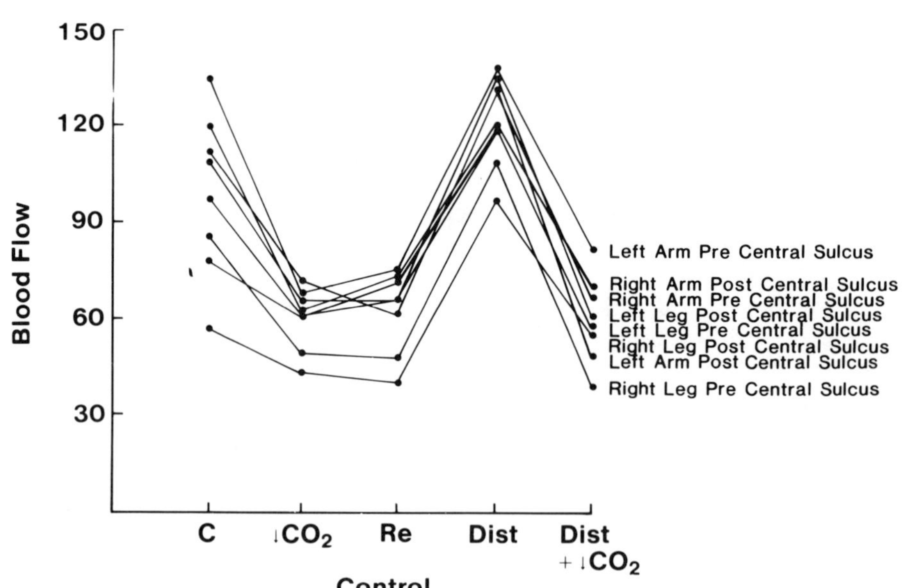

5. It is indicative of an injury to the cord after the fact, and therefore is not necessarily reversible.

These limitations have prompted the development of more sophisticated electrophysiological monitoring techniques. Electrophysiological monitoring can be performed repeatedly during the entire surgical procedure and it is applicable to patients with neurological deficits. Most importantly, however, is that it appears to be an indicator of impending cord damage, a reversible state of neurologic compromise. Although attractive, this system of evoked potentials is not perfect. There are shortcomings. First, evoked potentials monitor primarily the dorsal columns of the cord (sensory paths). Status of the motor tracts is inferred (5). Secondly, evoked potential signals demonstrate considerable normal variation and can be affected by a number of variables in the operating room setting: Anesthetic agents, depth of anesthesia, physiologic parameters, electrical interference, and technical problems (5, 9, 20, 27). Third, the mechanisms responsible for changes in evoked potentials have not been fully elucidated (8, 27).

Evoked potentials have two measurable parameters, amplitude and latency. Changes in each of these parameters have been used as indicators of spinal cord injury. Unfortunately, the pathophysiology responsible for these changes has not been clearly defined. Neurologic deficits during spinal surgery can result from the mechanical deformation of the neural elements. Interruption of spinal cord blood flow during correction of spinal deformities can also result in identical sequelae.

Materials and methods

The present study was designed to define the interrelationships between spinal cord blood flow, spinal distraction, and changes in evoked potentials. Previous studies employ models with normal blood supplies to the cord. This does not necessarily correspond with the clinical setting. The blood supply to the cord can be compromised by the configuration of the spinal canal or as a result of surgical intervention; i.e., loss of segmental anterior arteries secondary to anterior approaches to the spine. To more closely approximate a clinical situation, the following animal model was created.

Five skeletally immature baboons (papio, anubis), ages 3-6 months, were subjected to anterior transthoracic exposure of the lower thoracic spine (T7 through T11). Two or three segmental vessels in each animal were sacrificed during the exposure. Two adjacent disks and the intervening vertebral body were then excised. The defect was filled with morselized rib graft to stimulate anterior fusion of these motion segments.

Postoperatively, the animals were followed with periodic radiographic examination to document the development of deformity. Each animal was also observed for the development of any neurologic defects. The follow-up period lasted 2-3 years. None of the animals developed significant kyphosis and none demonstrated any appreciable neurologic deficits.

Each animal was then subjected to the following experimental protocol.

Sedation for endotracheal intubation was obtained by the administration of intramuscular ketamine. The animals were intubated. Ventilatory support was provided by an Ohio pediatric ventilator and adequate levels of anesthesia were maintained by the administration of intravenous morphine drip. Venous and arterial access was obtained thru external jugular, femoral, and radial artery cutdowns. The left atrial appendage was cannulated through a left thoracotomy for the injection of microspheres. Blood pressure, heart rate, and EKG were continuously monitored and recorded on a Grass strip chart recorder. Body temperature was monitored with a rectal probe and maintained constantly with a heating blanket. The animal was placed prone in a stereotaxic frame.

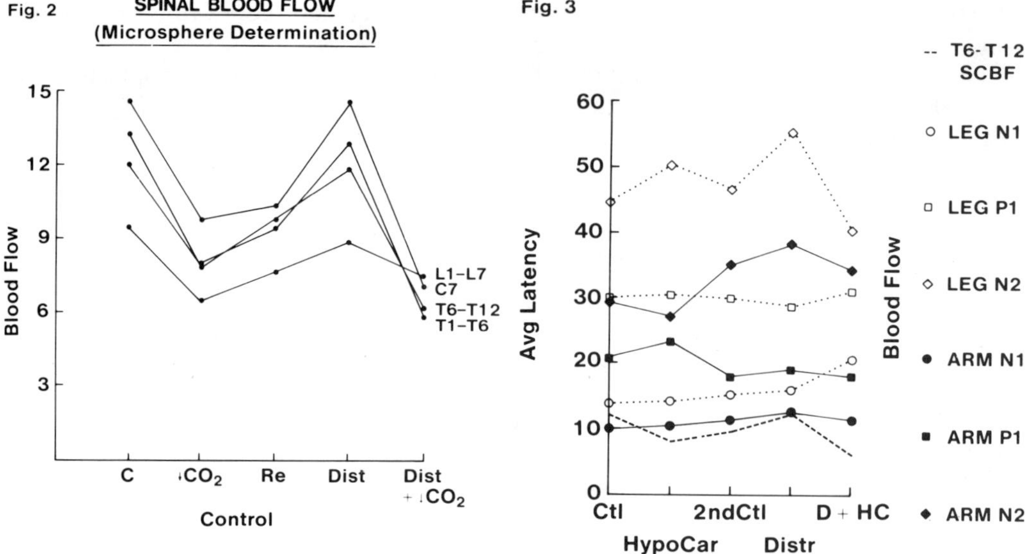

Fig. 2
SPINAL BLOOD FLOW
(Microsphere Determination)
Blood Flow
15
12
9
6
3
L1–L7
C7
T6–T12
T1–T6
C
↓CO₂
Re
Dist
Dist + ↓CO₂
Control
Fig. 3
T6-T12 SCBF
LEG N1
LEG P1
LEG N2
ARM N1
ARM P1
ARM N2
Avg Latency
Blood Flow
60
50
40
30
20
10
0
Ctl
2ndCtl
D + HC
HypoCar
Distr

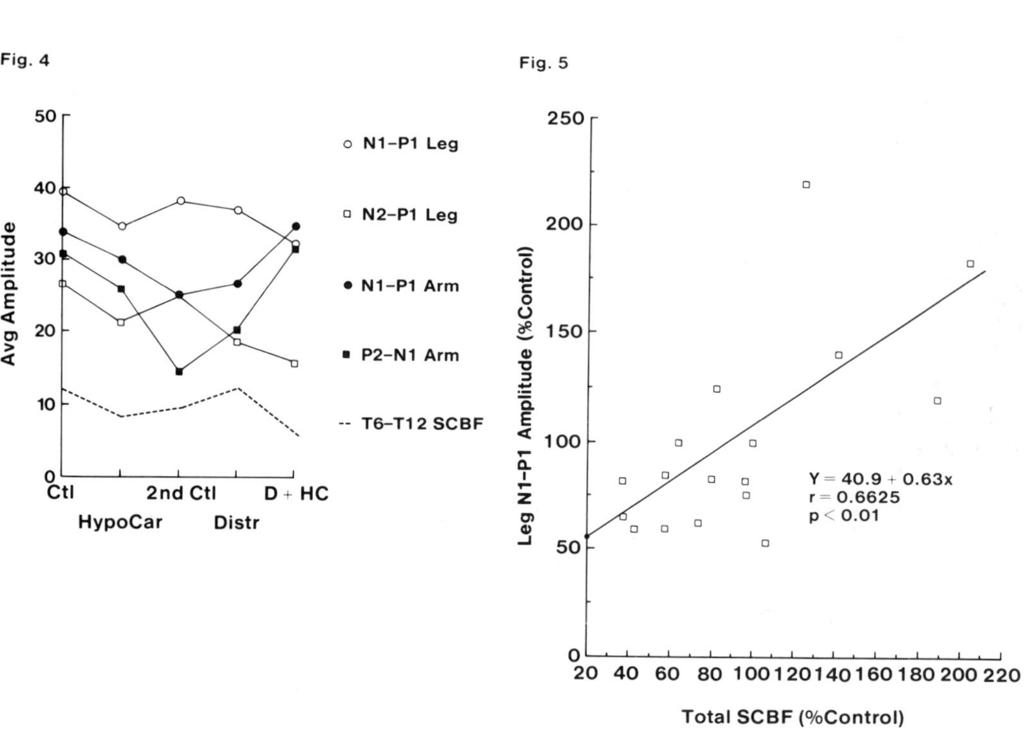

Fig. 4
N1–P1 Leg
N2–P1 Leg
N1–P1 Arm
P2–N1 Arm
T6–T12 SCBF
Avg Amplitude
50
40
30
20
10
0
Ctl
2nd Ctl
D + HC
HypoCar
Distr
Fig. 5
250
200
150
100
50
0
Leg N1–P1 Amplitude (%Control)
Y = 40.9 + 0.63x
r = 0.6625
p < 0.01
20 40 60 80 100 120 140 160 180 200 220
Total SCBF (%Control)

Fig. 6

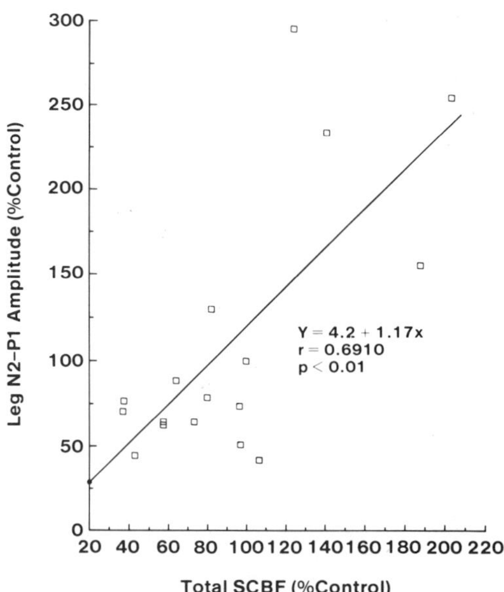

Fig. 7

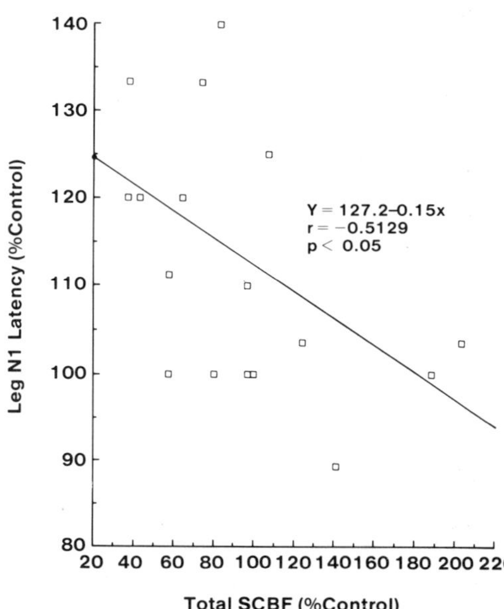

Fig. 8

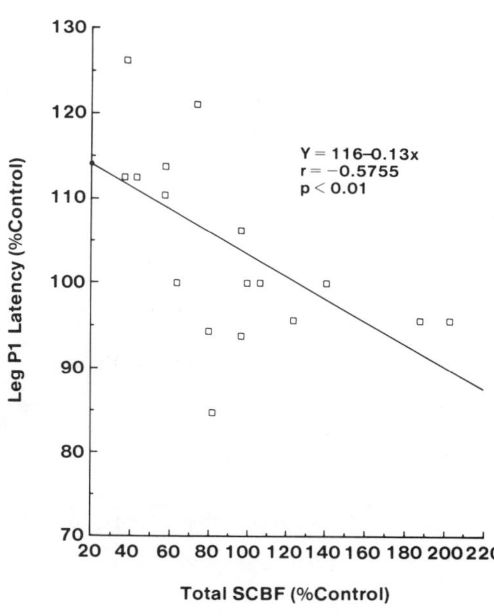

Fig. 9

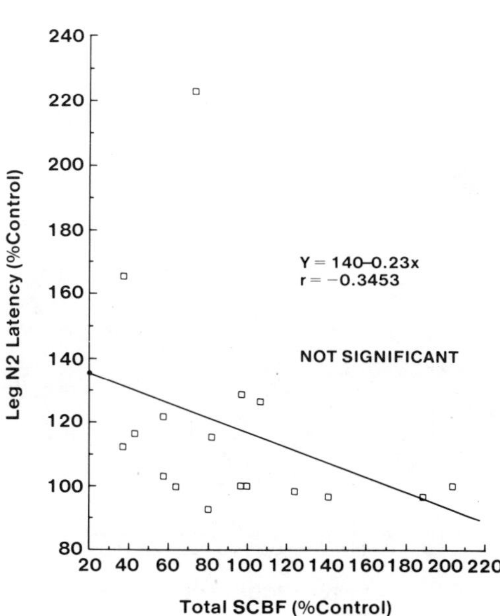

Fig. 10

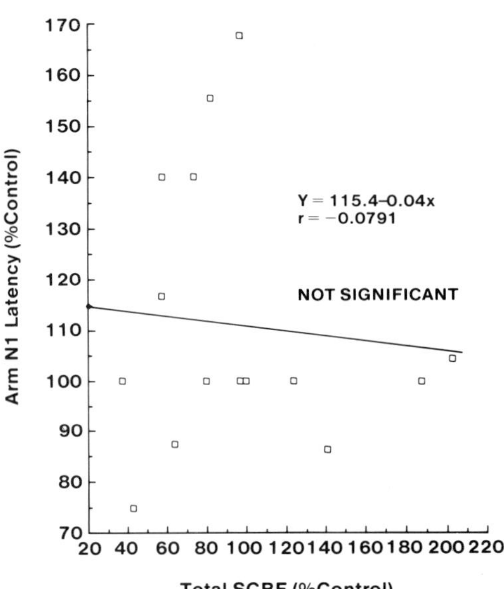

Fig. 11

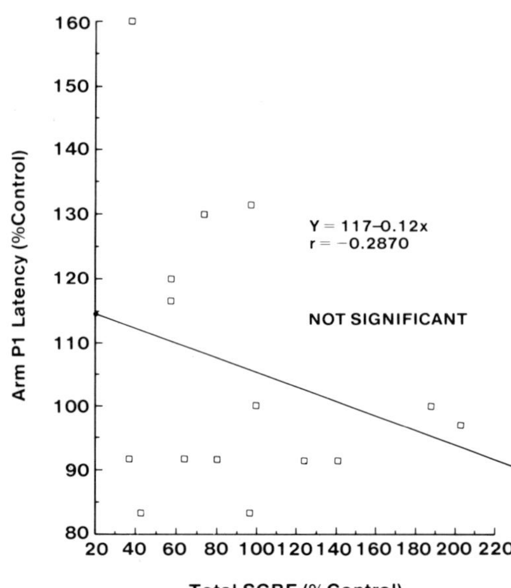

Fig. 12

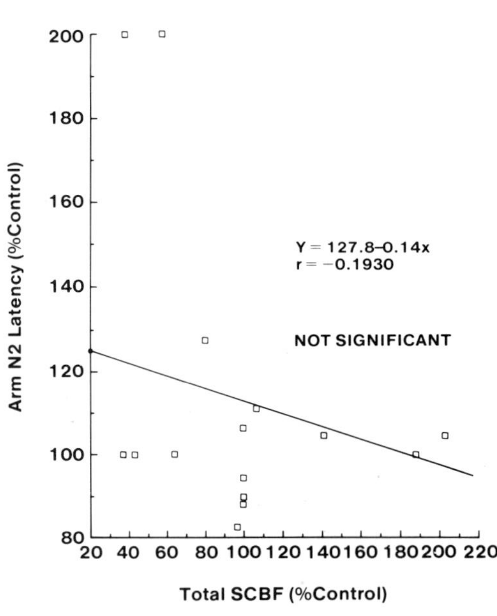

Fig. 13

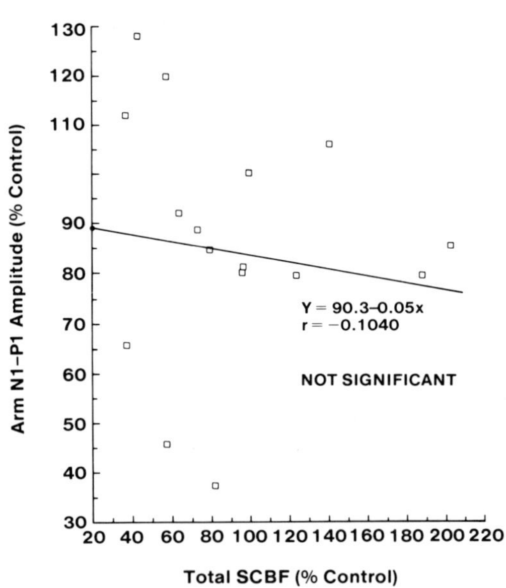

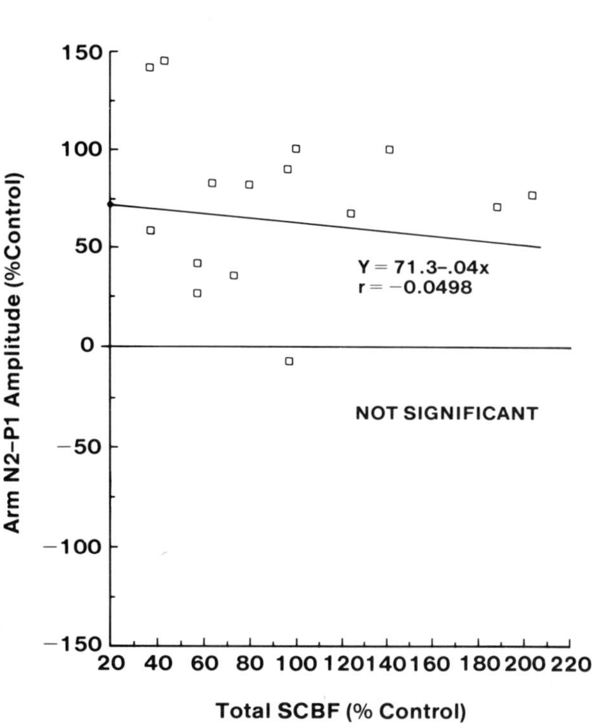

The left somatosensory cortex was exposed by removing an appropriate amount of the skull. The dura was incised and retracted over the edges of the exposed bone to prevent the accumulation of blood on the brain.

Neurophysiologic monitoring was performed with six electrodes. Two ball tipped silver electrodes were positioned on the somatosensory cortex. A stimulus applied to one electrode elicited a reproducible twitch of the ankle dorsiflexors. Similar stimulation of the second electrode produced wrist extension. Each of these responses were also recorded by a peripheral EMG electrode and stored on an oscilloscope.

Two relay switches were integrated in the circuit to permit easy reversal of the monitoring process. A stimulus applied to the peripheral electrode (either upper or lower extremity) could now be used to elicit an evoked potential which was recorded by the cortical electrodes, amplified and stored on an oscilloscope. Permanent records of the tracing were obtained by photographing the oscilloscope screen on Polaroid film.

Fig. 14

The above arrangement permitted independent documentation of the integrity of both ascending and descending motor tracts under each experimental condition.

Spinal cord blood flow was measured using a radioactive labelled isotope technique. A different isotope was used for each experimental condition.

A bolus of isotope was injected into the left atrium while simultaneous withdrawal of a reference sample from the femoral artery was performed. Blood flow to a particular vascular bed including the brain and spinal cord could then be calculated by measuring the amount of emitted gamma radiation per milligram of tissue.

The final preparatory step consisted of exposing the posterior thoracic and lumbar spine. Small spinal hooks were placed beneath the lamina of T7 and L5. A modified rod allowed controlled distraction in 1mm increments. The purchase of the superior hook was re-enforced with methyl methacrylate.

The study consisted of five experiment conditions:

1. Control - control condition consisted of physiologic respiratory rate 12-15 cycles/minute, normal pCO_2 (36-43mmHg) and pO_2 as determined by arterial blood gases, and no distraction.

2. Hypocarbia - each animal was then hyperventilated 25-40 cycles/minute, producing a significant decrease in pCO_2 (20-30mmHg). Electrophysiologic data were collected and the second microsphere was injected.

3. Re-control - ventilatory rate was decreased allowing a return to control conditions. Spinal monitoring and spinal cord blood flow determinations were repeated.

4. Distraction - the spine was then slowly distracted by advancing the distraction device. Evoked potentials and EMG recordings were performed following each 5mm of distraction. Distraction was continued until evidence of eminent failure at the superior hook was noted. Blood flow to the spinal cord was recorded at this point by injection of the next microsphere.

5. Distraction and hypocarbia - the final condition consisted of a worse case scenario. After maximum distraction was obtained, the ventilatory rate was increased. Arterial blood gases documented the level of hypocarbia. Electrophysiologic data was gathered and the final microsphere was injected.

During each experimental condition spinal evoked responses were recorded for both the upper and the lower extremities. The polarity of the electrodes was then reversed permitting stimulation of the motor cortex and recording of distal EMG tracings as well as visually observing movement of first the upper and then lower extremities.

The experiment was terminated. X-rays of the spine in the distracted position were obtained prior to the harvesting of the spinal cord, the brain, and other reference organs.

Results

Data was divided into two major categories; blood flow and evoked potentials. Blood flows to the spinal cord, cerebral cortex, and heart were analyzed. Spinal cord flow was calculated separately for the cervical, upper thoracic (T1 through T6), lower thoracic (T7 through T12), and lumbar cord. Actual blood flow values for each experimental condition are depicted in Table 1. Cerebral flows are similarly shown in Table 2.

An excellent correlation between cerebral and spinal cord blood flow was found (Fig. 1 and Fig. 2). There was a significant decrease in blood flow with hypocarbia in all anatomic sections. Re-control resulted in an increase in blood flow but not to previous control values. Distraction surprisingly produced an increase in both spinal cord and cerebral blood flow although this did not prove to be statistically significant. The addition of hypocarbia to maximal distraction caused the blood flow to again decrease.

Evoked potentials for both the arm and the leg were recorded for each experimental condition. The first upward deflection was labelled N1, the first downward deflection labelled P1. By convention the second upward deflection was called N2. The amplitude consisted of the distance in millimeters between major peaks. Amplitude of N1 through P1 and P1 through N2 was calculated for each experimental condition.

The time from stimulus artifact to corresponding peak represented the latency for that peak in milliseconds. There were considerable differences in signal configuration and, therefore, amplitude and latencies among individual animals. A recognizable wave form could, however, be identified for each animal and appeared to persist throughout the experiment. Analysis of the data did not demonstrate any consistent correlation between experiment conditions and either evoked potential latencies or amplitudes (Fig. 3 and Fig. 4).

There was, however, a statistically significant correlation between total spinal cord blood flow and evoked potentials to the leg. Fig. 5 and Fig. 6 demonstrate a positive correlation between a SCBF and leg amplitudes. An inverse relationship between SCBF and leg latencies is seen in Fig. 7 and Fig. 8. Fig. 9 shows a similar relationship for N2 although this did not prove to be statistically significant. Interestingly, there was no such correlation found between arm evoked potential latencies or amplitudes and spinal cord blood flow (Fig. 10, 11, 12, 13, 14) in the upper extremity.

Conclusions

The limitations of this type of study are clear. The authors are acutely aware of the hazards of drawing conclusions from such a small study group. Nonetheless, we feel that several trends were identified and have clinical relevance.

1. The study clearly demonstrated the similarity in the auto-regulatory capacity of the primate brain and spinal cord when subjected to conditions mimicking an operating room setting.

2. Distraction did not result in a significant change in spinal cord blood flow. The premise that catastrophic neurologic compromise during distraction is the result of vascular alterations might not prove accurate (17).

3. Spinal evoked potentials are affected by spinal cord blood flow. Statistically significant changes in both amplitude and latency in leg evoked potentials were noted with changes in spinal cord blood flow. Similar changes were not seen in arm evoked potentials. The mechanisms responsible for this discrepancy were not identified by this study. The possibility that membrane dysfunction could be related to axonal length is suggested and merits further investigation (29).

4. The study design permitted simultaneous testing of ascending and descending tracts. Changes in evoked potentials were identified without evidence that descending motor tracts were impaired, thus, supporting the contention that changes in evoked potentials are indicators of a pre-injury state and, therefore, a clinically useful tool.

References

1. Albee, F.H.: A report of bone transplantation and osteoplasty in the treatment of Pott's disease of spine. N.Y. Med. J., 95: 469, 1912.
2. Blount, W.P.: Bracing for scoliosis. Orthotics, 1966.
3. Blount, W.P.; Schmidt, A.C.; Bidwell, R.G.: Making the Milwaukee brace. J.B.J.S., 40A: 511, 1958.
4. Britt, R.; Ryan, T.: Use of a flexible epidural stimulating electrode for intraoperative monitoring of spinal somatosensory evoked potentials. Spine, 11: 4, 348-351, 1986.
5. Brown, R.H.; Nash, C.L.: Current status of spinal cord monitoring. Spine, 4: 6, 1979.
6. Cheng, M.; Robertson, C.; Grossman, R.; Foltz, R.; Williams, V.: Neurologic outcome correlated with spinal evoked potentials in a spinal cord ischemia model. J. Neurosurg., 60: 786-795, 1984.
7. Dolan, E.; Transfelt, E.; Tator, C.; Simmons, E.; Hughes, K.: The effect of spinal distraction on regional spinal cord blood flow in cats. J. Neurosurg., 53: 756-764, 1980.
8. Gaines, R.; York, C.; Watts, C.: Identification of spinal cord pathways responsible for the peroneal-evoked response in the dog. Spine, 9: 8, 810-814, 1984.
9. Gelfan, S.; Tarlov, I.M.: Physiology of spinal cord, nerve root and peripheral nerve compression. Am. J. Physiol., 185: 217-229, 1956.
10. Hall, J.; Levine, C.; Sudhir, K.: Intraoperative awaking to monitor spinal cord function during Harrington instrumentation and spine fusion: Description of procedure and report of three cases. J.B.J.S. (Am), 60A: 533-536, 1978.
11. Hare, S.: Practical Observations on the Prevention, Causes, and Treatment; of Curvatures of the Spine. London, Churchill, 1849.
12. Harrington, P.R.: Treatment of scoliosis-correction and internal fixation by spine instrumentation. J.B.J.S., 44A: 591, 1962.
13. Hibbs, R.A.: An operation for progressive spinal deformities. N.Y. Med. J., 93: 1013, 1911.
14. Hodgson, A.R.; Stack, F.E.: Anterior spine fusion. Br. J. Surg., 44: 266, 1956.
15. Hoffa, A.: Lehrbuch der orthopadischen chireurigie. Stuttgart, Verlag von Ferdinand Emke, 1898..
16. Jones, S.; Edgar, M.; Ransford, A.; Thomas, N.: A system for the electrophysiologic monitoring of the spinal cord during operations for scoliosis. J.B.J.B., 65B: 134-139, 1983.
17. Kling, T.; Ferguson, N.; Zeach, A.; Hensinger, R.; Lane, G.; Knight, P.: The influence of induced hypotension and spine distraction on canine spinal cord blood flow. Spine, 10: 10, 878-883, 1985.
18. Kling, T.; Wilton, N.; Hensinger, R.; Knight, P.: The influence of trimethapon (arfonad) induced hypotension with and without spinal cord distraction on canine spinal cord blood flow. Spine, 11: 3, 219-224, 1986.
19. MacEwen, G.D.; Bunnell, W.P.; Sriran, K.: Acute neurological complications in the treatment of scoliosis. J.B.J.S., 57: 404-408, 1975.
20. Machida, M.; Weinstein, S.L.; Yamada, T.; Kimura, J.: Spinal cord monitoring electrophysiologic measures of sensory and motor function during spinal surgery. Spine, 10: 5, 407-413, 1985.

21. Nordwall, A.; Axelgaard, J.; Harada, Y.; Valencia, P.; McNeal, D.; Brown, J.: Spinal cord monitoring using evoked potentials. Spine, 4: 6, 486-494, 1979.
22. Pare, A.: Opera ambrossi parie. Paris apud jocabum du-puys, 1582.
23. Ryan, T.; Britt, R.: Spinal and cortical somatosensory evoked potential monitoring during corrective spinal surgery with 108 patients. Spine, 11: 4, 352-361, 1986..
24. Sato, M.; Pawlik, G.; Umbach, C.; Heiss, W.: Comparative studies of regional CNS blood flow and evoked potentials in the cat. Stroke, 15: 1, 1984.
25. Satomi, K.; Nishimoto, G.: Comparison of evoked spinal potentials by stimulation of the sciatic nerve and the spinal cord. Spine, 10: 10, 884-890, 1985.
26. Schramm, J.; Shigeno, T.; Broch, M.: Clinical signs and evoked response alterations associated with chronic experimental cord compression. J. Neurosurg., 58: 734-741, 1983.
27. Shimoyi, K.; Kano, T.: Evoked electrospinogram: Interpretation of origin and effects of anesthesia. Int. Anesthesiol. Clin., 13: 171-189, 1975.
28. Stagnara, V.; Journroux, P.: Functional monitoring of spinal cord activity during spinal surgery. Clin. Orthop., 93: 173, 1973.
29. Waxman, S.; Brill, M.; Geschwind, N.; Sabin, T.; Lettvin, J.: Probability of conduction deficit as related to fiber length in random-distribution models of peripheral neuropathies. J. Neurol. Sci., 29: 39-53, 1976.

Continuous, Chronic Changes in Evoked Potentials: Summary

M. R. Dimitrijevic[*]

Whenever we record or monitor dynamic physiological functions, regardless of whether they represent the condition of the vascular supply or the functional electrical properties of the posterior columns of the spinal cord, we always seek to describe short and long-term changes. Short-term changes in recorded cortical somatosensory evoked potentials (cortical SEP) in healthy humans will mainly reflect fluctuations in the excitability of the brainstem and brain structures involved in the processing of volleys from the posterior columns of the spinal cord. It is important, therefore, to incorporate these changes in the methodology only then shall we be able to differentiate between methodological and physiological changes in the shape, latency, and amplitude of the cortical evoked potentials which reflect the tested malfunctioning portion of the nervous system. In order to continuously test the reliability of obtained responses, we have found that it is very informative to monitor not only the electrical parameters of the stimulus applied over the peripheral nerve trunk but also the effects of depolarization on the nerve. This can be done by recording sensory nerve action potentials of distal cutaneous branches, or electromyographic responses of stimulated muscles. Moreover, while studying changes in cortical somatosensory responses, it is possible to examine the changes that the stimulus applied to peripheral nerves was also causing at the spinal cord level by recording lumbosacral or cervical somatosensory evoked responses. This protocol to record cortical somatosensory potentials evoked by the stimulation of the tibial and median nerves is shown in Fig. 1.

In recent years, professionals working in neuro-intensive care units have continuously monitored cortical somatosensory evoked potentials in patients suffering from impaired consciousness. They have shown that this procedure can be of value to recognize the deteriorating effects of the impaired brain metabolism on brain functions, and that it also makes it possible to differentiate between metabolic brain dysfunctions and the primary region of neurogenic changes.

By applying the protocols for cortical somatosensory evoked potentials shown in Fig. 1, we have learned that, when applied in head injured patients, they can be very useful to detect early the clinically unrecognized or subclinical traumatic lesions of peripheral nerves and the spinal cord.

In practice, there is a need to expand peripheral input to cortical somatosensory evoked potentials from the most conveniently situated median, ulnar, tibial, or peroneal nerves to many other sites of the body which can be done by stimulating different dermatomes (F44), either electrically or mechanically. Another possibility lies in

[*] Section of Restorative Neurology and Clinical Neurophysiology, Department of Rehabilitation, Baylor College of Medicine, Houston, TX 77030

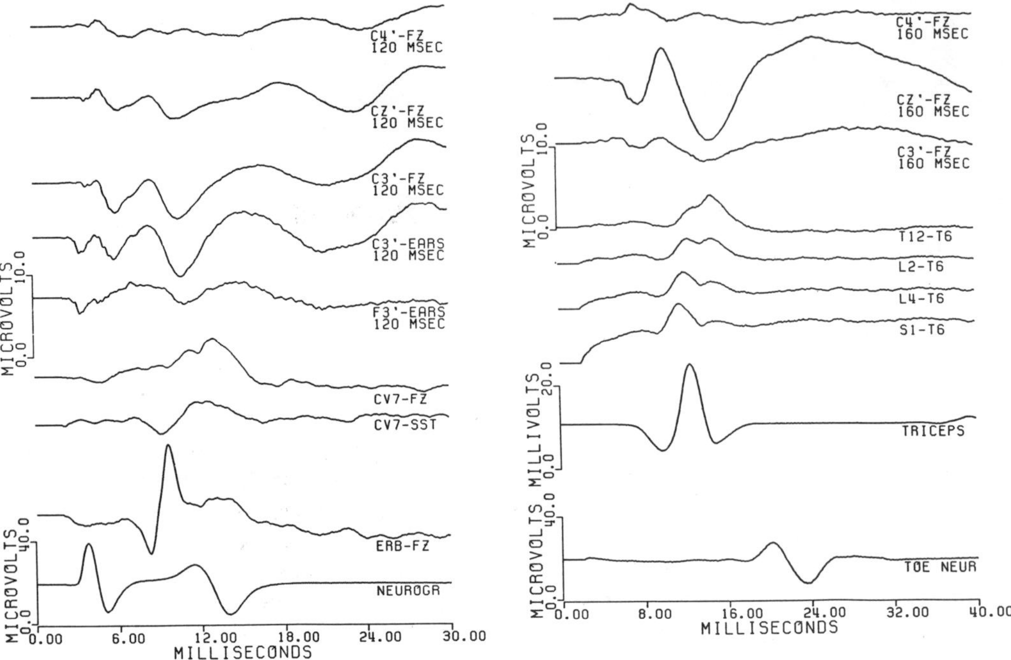

Fig. 1A. Simultaneous recording of SNAP, spinal SEP, brainstem, and cortical SEP after median nerve stimulation.

Fig. 1B. Simultaneous recording of SNAP, spinal SEP, brainstem, and cortical SEP after tibial nerve stimulation.

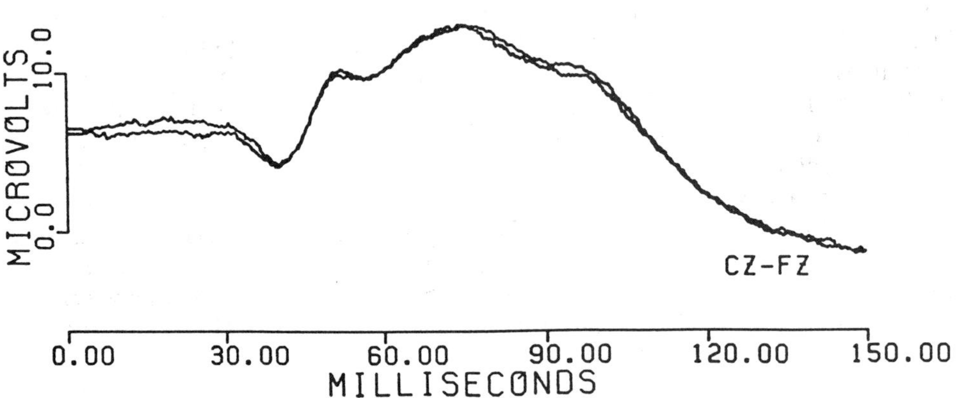

Fig. 2. Recording of cortical SEP after pudendal nerve stimulation.

the use of pudendal penile cutaneous stimulation to elicit cortical somatosensory potentials so that large, fast conducting fibers of the posterior columns are examined from the most caudal part. An example of such cortical somatosensory potentials elicited by pudendal nerve stimulation is shown in Fig. 2.

Neurotrauma services deal with a wide variety of patients and their posttraumatic condition. It is not surprising to find out that cortical somatosensory potentials also have value in such an environment for the diagnosis of conversion disorders (F47). Hysterical anesthesia can also be tested with objective methods applying recording of contingent negative variation (CNV). After documentation of the integrity of posterior column functions by cortical somatosensory evoked potentials, CNV methodology can also add objective evidence of the subject's ability to perceive touch stimulus.

Multiple lesions of the spinal cord, multiple meningiomas (F48), multiple trauma and other pathologies are clinically common and rarely recognized since more rostral lesions of the spinal cord mask all evidence and make it impossible to delineate other more caudal lesions. In such circumstances, cortical somatosensory evoked potentials, spinal somatosensory potentials (lumbosacral and cervical), dermatomal cortical somatosensory evoked potentials can be of significant diagnostic and follow-up assistance. In the past 15 years, the examination of the functions of the posterior column in different neurological disorders has led to the acknowledgement of the diagnostic value of cortical somatosensory evoked responses in several demyelinating neurological conditions. Moreover, this new knowledge has contributed to the understanding that it is not rare to find abnormalities of posterior column functions also in patients with minimal abnormality or even without clinical findings. This possibilitiy, if not recognized before surgical intervention, can cause difficulties when monitoring spinal cord functions. Therefore, it is useful to be informed about the functions of the patient's posterior column before monitoring cortical somatosensory evoked responses and other somatosensory evoked potentials in surgery.

Important improvements in the monitoring of spinal cord functions during intraspinal surgery can be achieved by using somatosensory cortical evoked responses together with segmental, stationary spinal cord posterior horn generated negative potentials as has been illustrated (F46).

Experimental models for studies of the effect of spinal cord blood flow, spinal cord ischemia, influence of mechanical injury on spinal and cortical evoked potentials demonstrate that electrical events are dependent upon vascular functional conditions (F50, F51, F52). This work shows that there is no simple and direct relationship between "blood volume - potential volume." Evoked responses are based on compound potentials and have their own dynamics - electrolytic, metabolic - and they also depend on released influences resulting from temporary dysfunctions of some elementary portion of neuronal integrated circuits.

Spinal vascular malformation and neuroradiological procedures are opening new avenues to the study of the relation between the vascular conditions, and spinal and cortical electrical events (F41). This is an opportunity, while radiologists perform interventions on spinal vascular malformations, to learn about the human spinal cord neurophysiology.

In conclusion, the work which is reviewed in this chapter on Continuous/Chronic Changes in Evoked Potentials illustrates how clinicians and researchers are continuously improving monitoring functions of the spinal cord, not only by studying the functions of the healthy spinal cord and results of experimental models but also by studying the results of newly acquired knowledge of neurological conditions.

Conclusions

Orthopaedic Review and Summary

G. L. Engler[*]

Review

Since the inception of intraoperative spinal cord monitoring in the early 1970's, much has been written regarding the various techniques, methodology, experimentation, and results of this monitoring system.

In 1981, Maccabee and his co-workers presented their work on "Monitoring of Spinal and Subcortical Somatosensory Evoked Potentials During Harrington Rod Instrumentation," (9). Simultaneous spinal and scalp SEP's from bilateral peroneal nerve stimulation were recorded. Cauda equina potentials were occasionally transmitted by the standard Harrington outrigger apparatus. These were eliminated and improvement in signal-to-noise ratio was accomplished with a custom made non-conducting outrigger. They concluded that relative amplitude change was not considered a valid criterion with their technique.

By utilizing spinal recording electrodes rather than scalp electrodes, the ability to analyze the early latency waves becomes possible. A careful analysis of these early potentials, usually less than 40ms, will provide the investigator and spinal surgeon with the ability to detect changes in the conduction pathway and possible alterations of neurologic function.

It has been recognized by others that the outrigger and, indeed, the insertion of the Harrington distraction rod itself, may cause alteration of the pathway transmission by acting as a conduction element. Additionally, it may produce artifacts which interfere with the analysis of the early latency potentials. This situation is complicated by the utilization of spinous process recording electrodes but may in some small way influence the recordings obtained even from scalp electrodes.

One further development might be the application of the recording electrodes to the hook holders or the hooks themselves, once a method for artifact rejection because of the outrigger and the rod has been developed. This would eliminate the necessity of additional electrodes being placed in the spinous processes.

The concept that the relative changes in amplitude are not to be considered as criteria for impending cord damage is extremely important. It may be that only the latency changes can be used for impending neurologic deficiencies. Nevertheless, once the SEP has been totally obliterated, the neurologic dysfunction has already run its course and possible permanent neurologic deficit has ensued.

A major contribution to the field of evoked potentials is indicated in the theory that postsynaptic spinal potentials are generated by distal sensory stimulation. These postsynaptic signals occur in a phase reversal configuration presumably indicating that

[*] Department of Neurological Surgery, New York University Medical Center, 530 First Avenue, New York, NY 10016

they occur after entrance of the dorsal ganglion to the cord. If this is true, this technique of spinal cord monitoring is not relegated to the dorsal columns alone for transmission. Indeed, injury occurring in other areas of the cord may be reflected in the SEP obtained. It is further interesting to note that the stimulation rate affects the development of the spinal generated potentials. Spinal injury may be mechanical distortion of tissue, vascular ischemia or occlusion, or neurochemical interruption of transmitters in ascending and descending fibers. Any of these factors or combinations may influence abnormal conduction or lack of conduction in a spinal cord when a peripheral nerve is stimulated.

The more sophisticated aspects which have been accomplished during the development of Maccabee's work include the relation of levels of sleep to the latency of the SEP wave form, especially the later latency potentials. Furthermore, the montage of the scalp placement does, indeed, alter the recordings when utilization of the scalp electrodes is employed. By changing various filters on the electronic devices used, the band width can be altered and a more critical analysis of the finely dispersed potentials can be made, thus arriving at a more critical analysis of the early wave latency forms. Hidden spikes may thus become more obvious.

Although it has been thought that anesthesia affects the entire SEP recording, certain changes in the early latency spikes are also realized upon alteration of anesthetic agents.

This work presents material which forms a background for many investigative projects and may indeed form the basis for a more critical analysis of the wave forms and patterns which are varied in smaller windows of latency periods.

Dr. Donald McNeal (10) and his co-workers at Rancho Los Amigos Hospital in California have used epidural electrodes for spinal cord monitoring, similar to the work of Tamaki. They attempted to compare this technique with cortical evoked potentials. They found that evoked potentials acquired with epidural electrodes were relatively noise free and remarkably stable.

The obvious disadvantage of epidural electrodes might influence some clinicians to shy away from its utilization. It is the most invasive technique of monitoring spinal cord function but perhaps gives the best critical recording by its close approximation to the cord itself. The location of the electrode can presumably be identified by either a caliper or x-ray technique. It is only used in the intraoperative situation when posterior surgical approaches are employed.

It is interesting to note their work also alludes to the enhanced spinal recordings obtained from spinous processes between the T9 to L1 levels. Other work by Maccabee (9) also suggested this electronic watershed phenomenon with reduction of the amplitude when spinous recordings are made above the T9 level. It is uncertain whether this is strictly a distance phenomenon with dissipation of the potentials or some other factor is implicated.

Further corroboration of the fact that amplitude change has little significance, whereas latency change may be of extreme importance is supplied by this technique.

Comparison of the epidural and bipolar bone recordings was carried out by Bunch and associates (1) to ascertain whether or not it is truly necessary to invade the canal for appropriate recordings. It is their contention that the more invasive method is not necessary. They found no apparent discrepancy between the recording from bone or epidural electrode. Additionally, no difference of significant importance appears to be obtained from unilateral or bilateral stimulation sites.

A very important question now comes to the fore: How critically shall we analyze the wave patterns obtained and what is the significance of these minute differences? For the individual interested in "overall" neurologic function of the spinal cord, a simplistic approach might be sufficient. In more complex neurologic states or deformities, a more critical analysis of different windows and latencies might be necessary. Finally, if one is interested in monitoring anesthetic levels or agents, blood flow, blood pressure, or other transient phenomena of the spinal cord during surgery, it might be necessary to use the more invasive techniques.

The role of the more invasive techniques may become relegated to the realm of investigation and data accumulation rather than the orthopedist interested in surgical safety during spinal manipulation.

In 1981, Satomi, Nishimoto, and Passoff described their work which was designed to analyze the properties of SEP's due to both sciatic nerve and spinal cord stimulation and also to quantify the contributions of each column of the spinal cord to the SEP (15). The spinal cord stimulation involved both afferent and efferent stimulation.

They found that each SEP had a specific wave pattern. Further, sciatic nerve SEP's had more components derived from synaptic and postsynaptic activities then afferent SEP's. In asphyxia studies, a temporal augmentation of the SEP's was observed. They concluded that for precise clinical evaluation, both methods should be employed.

By selective transection of the spinal cord and analysis of the SEP and histologic sections, it was shown that about 30% of the early component SEP arises from the lateral columns with sciatic nerve stimulation. With the spinal cord evoked SEP, about 60% of the N-1 deflection has its origin in the lateral columns and partially from the anterior columns. 70% of the N-2 deflection was relegated to the dorsal columns. It is left for the N-3 deflection to be committed to dorsal column of the spinal cord.

These findings again lend credibility to the fact that the SEP, whether it be from sciatic nerve stimulation or spinal cord, does indeed reflect transmission of the intact spinal cord in areas other that the dorsal columns.

This work, however, evinces the dilemma of technique versus analysis of the wave forms. These areas of investigation bring to light the principle that much information can be gleaned from discrete analysis of the SEP and basic research into the various techniques used to obtain these patterns.

This work utilizes experimentation with cats. There has been some discussion in previous years about the efficacy of transmitting the data obtained from that laboratory animal to the human. A direct correlation has never been established.

By moving the electrodes and recording from different areas, emphasis is placed on the importance of the placement of the electrode in the spinal canal. Slight variations can result in dramatic changes in the configuration of the SEP, but may not influence the resultant patterns obtained when neurologic dysfunction ensues.

Threshold and asphyxia studies again point out the necessity to standardize the method being utilized so that information can be acquired on a broad base by many investigators.

After spinal cord transection studies, it was shown that there was an increase in amplitude both across and below the section site. This enhancement may be due to the abolished postsynaptic cord potentials or the severance of the inhibitor descending pathways.

Good evidence is presented that the first positive peak might be generated in the dorsal root ganglia and subsequent negative potentials might arise in the dorsal roots, their intramedullary continuation.

This extensive study deserves critical analysis, but may provide certain insights into the pathway transmission of the SEP recording. If selective spinal cord monitoring can be perfected, isolation and analysis of spinal cord lesions may be possible. Furthermore, selectivity of the technique required would be possible.

Much has been written about the effects of anesthesia on the recordable SEP. Halothane and sodium nitroprusside (SNP) are frequently used singly or in combination to deliberately induce hypotension when spinal surgery is performed. Schroeder and his co-workers found that this combination of drugs will allow savings in total dose and thereby decrease the major potential complication of metabolic acidosis which results from the release of cyanide as SNP is metabolized (16). They concluded that combination of these two agents has no statistically significant effect on the SEP as recorded from canine vertebral bone. Cortical recordings do not accord this advantage.

The obliteration of the cortical evoked potential with halothane and other anesthetic agents has been well documented. Some indication that no alteration in the spinal evoked potential occurs with halothane may make it's utilization preferable when the type of anesthesia is required. Furthermore, little diminution in the latency was observed with decrease in blood pressure and cardiac output. The amplitude, however, did show significant reduction.

As stated earlier, the critical analysis of the evoked potential recording may reflect neurophysiologic changes occurring elsewhere in the body. Homeostasis and correction of blood flow to the spinal cord maintains the integrity of the neurologic function during alteration of blood pressure and temperature.

If the techniques of hypothermia and/or induced hypotension are required during surgery, it is likely that the technique of recording from the spinous processes would be an essential method.

The fact that the amplitude which has been shown in previous works to be an inappropriate indicator of neurologic function may indeed represent a more sensitive indication of spinal cord blood flow. The fear of hypotension in correction of spinal deformities is based on the belief that reduction in blood flow to the spinal cord in an attenuated vessel will result in ischemia to the cord itself. If, indeed, the amplitude is a reflection of the reduction in blood flow, it may be possible to prevent irreversible damage due to ischemia in that type of situation.

At present, studies at New York University Medical Center in the Department of Neuroradiology provide further proof that alteration of blood flow to the spinal cord can be monitored effectively with the SEP (2). During selective angiography and embolization for spinal cord arteriovenous malformations, alteration in the SEP can provide the investigator with a safe pathway. Again, the amplitude change appears to be the early indicator of the altered blood flow to the cord.

Shaffer and his associates reported their findings during segmental sub-laminar wiring of the spine for spinal deformities (17). The inherent risks of epidural hemorrhage, dural tear, and spinal cord contusion by direct passage of the segmental wires on many fusion levels have led many surgeons to avoid this method. Nevertheless, in some cases, it is still employed.

Their study showed transient changes in the form of decreased amplitude during passage of the sub-laminar wires. These changes were similar to those reported in Harrington rod distraction and generally resolved within 15 minutes. One case of neurologic impairment was encountered but it was not due directly to the passage of the sub-laminar wires. It was, however, reflected in alterations of the SEP tracing.

The neurologic catastrophe reiterates and supports the notion that the alteration of amplitude is a reflection of blood pressure and blood flow to the spinal cord. It was not a complication of the sub-laminar wiring, but rather due to a hypotensive episode occurring prior to the wire passage.

The second point which is implied by their work relates to the decision of the surgeon in the face of an altered SEP recording. It is obvious that the SEP must be interpreted properly. Nevertheless, it is but another variable to which the operating surgeon must harken. It does not, nor will it ever, take the place of appropriate surgical judgement and technique.

Donaldson, Odom, and Brown have detailed their program to establish a spinal cord monitoring facility in their local non-university hospital (4). This acknowledges the area of practicality and the utilization of the technique in the everyday practice. It is true that most of the hardware can be found in the usual EMG laboratory in most hospitals today. With slight alteration, primarily the purchase of an averager, this EMG machine can be transformed into a spinal cord monitor.

In truth, it is unlikely that spinal cord monitoring would be required for the normal healthy idiopathic scoliotic patient undergoing surgery with a 50 degree right thoracic curve. On the other hand, open reduction and internal fixation of spinal fractures in the face of partial neurologic deficit or where no neurologic deficit exists, might constitute an absolute indication for monitoring spinal cord function. The individual case and the decision of the surgeon play a significant role in determination of the need for a spinal cord monitor.

An absolute diminution of the recording showed a positive correlated with the wake-up test. Once confidence in the SEP is established, the wake-up test might not be required. In the face of a pre-existent partial neurologic deficit, the wake-up test might indeed by impossible to perform.

Amplitude alterations were noted with hypotension and hypothermia while the latency stability was demonstrated. Demonstration of a decrease in the latency after anterior decompression correlated with an improvement in motor strength.

Donaldson cautions that the SEP must not provide the surgeon with a "false sense of security." Nevertheless, with appropriate investigation and analysis, variability of technique, and further elucidation of spinal cord pathways, it can provide necessary information to allow the spinal surgeon greater latitude in his ability to assist those afflicted with spinal deformity and disease.

Nickodem and associates have done work to establish spinal cord conduction velocity intra-operatively (12). The average spinal cord conduction velocity in ten patients with idiopathic scoliosis undergoing surgery was 48ms. Their technique involved stimulation of the posterior tibial nerve while recording from spinous processes.

This is a direct rather than an indirect method to determine spinal cord conduction velocity. This method, being closer to the spinal cord itself, avoids the extraneous noise and interference patterns obtained when too much tissue is interposed between the stimulus and recording sites.

The question to be asked is: Does it produce more reliable information than ordinary, less invasive techniques with respect to spinal cord injury? Further, what are the effects of blood pressure and anesthesia?

There is also report of an increase in the recorded latency and a change in the shape and size of the action potentials as depicting "problems with spinal cord function." Increases in latency are usually related to abolition of large caliber fiber transmission tracts while decreases in amplitude are the result of fewer fiber tracts firing.

Finally, what are the responsible "neural generators?" Are there specific tracts or nuclei responsible for the peaks recorded? If so, are they only posterior column tracts or might they be dorsal cord potentials? The possibility that they represent a reflex phenomenon must also be considered.

Nainzadeh and associates have described their series of surgical spinal procedures while utilizing 1) peripheral nerve stimulation with epidural recording, 2) epidural stimulation with epidural recording, 3) peripheral nerve stimulation with spinous process recording, and 4) peripheral nerve stimulation with cortical recording (13). When utilizing scalp electrodes for recording somatosensory evoked potentials, one must be cautious in ascribing changes to spinal cord dysfunction rather than alteration in cerebral action potentials. These are well known to be affected by anesthesia, blood pressure, and blood gas changes. Only by comparing cortical evoked potentials with spinal evoked potentials might one dissociate the cerebral changes with such techniques.

In a review of spinal surgery with segmental wiring performed at Case Western Reserve University, Wilber and associates presented a retrospective study which concluded that there was an increased risk factor for neurologic injury with these methods of segmental wiring (21). Spinal cord monitoring was found to be of benefit in recognizing impending neurologic deficits but was inconsistent with transient sensory neurologic changes that appear to be unique to segmental spinal wiring techniques.

The observation of false negative results reiterates the notion that somatosensory evoked potentials do not monitor the entire spinal cord, but will be sensitive enough to pick up minimal deficits which eventually clear or, if persistent and worsening, indicate impending neurologic disaster.

Unfortunately, even with proper techniques and precautions, false negative recordings may be noted with neurologic disaster. Ginsburg, Shetter, and Raudzens have reported a case of postoperative paraplegia with preserved intraoperative somatosensory evoked potentials (6). An anterior spinal release was followed by a posterior spinal fusion with instrumentation in a patient with achondroplasia.

As a possible causation of the paralysis, the literature and transection of the intersegmental vessels during the anterior approach might be incriminated. By altering the anterior blood supply, a diminution in the evoked potential was indeed realized, but the second stage procedure was carried out. The change in the preliminary SEP of 20-21ms did indeed change to 25ms at the time of the second operation. Complicating this reduction in anterior blood supply by posterior surgery may have caused an attenuation of the remaining vessels. The result was an anterior cord syndrome.

Further, it must be realized that in congenital lesions of the spine, abnormal intersegmental vessels may be present. When anterior and posterior surgery is contemplated on these cases, angiography might be essential to define these vessels. A noted high incidence of paraplegia with resection of hemivertebra is well known.

The increased high stimulus frequency of 8.1 cycles per second for 360ms in their technique is far above that normally used on other centers. This higher frequency may be producing some other neural generators while masking the sensitivity to subtle impending neurologic abnormalities.

The fact that halothane was used and had no specific influence on the resultant recording may support the contention that the frequency rate is too high, thus introducing other variables. It has been suggested that standard methods employ 2 cycles per second with 100-200 repetitions. This has resulted in little if any attenuation of the recording. The values of 20-21ms latency found in their patient is slightly abnormal for cortical recordings, and may be related to the dwarfism.

Highland, Goshgarian, and LaMont presented their work which involved recording evoked potentials after dorsal rhizotomy (8). This brings to the fore the question whether or not the ventral roots contain afferent fibers. Because of the methodology employed, the findings reported could be explained by antidromic conduction in motor fibers (which constitute the major part of the ventral root) and not orthodromic conduction in hypothesized afferent fibers in the ventral roots. The figures shown in their work represent recordings taken from the T-10 spinous process. If this vertebral level is close to the spinal level the authors identified as the territory of the posterior tibial nerve (e.g., L-6, 7, and S-1), then the potentials recorded after dorsal rhizotomy were probably antidromic conduction in the unaffected motor (ventral) roots. The afferent fibers that some investigators believe to be in the ventral roots are unmyelinated fibers. If these investigators were indeed seeing conduction transmitted in these fibers, the latencies recorded should have been much longer (i.e., slower) than those actually recorded. However, the latencies which were recorded were in the range of myelinated fibers, which is consistent instead with antidromic conduction in motor axons. It should be remembered that the conduction of antidromic impulses in motor fibers refers to the "F-Response."

Recently, Cohen and Huizenga described their experience utilizing dermatomal monitoring to track changes in conduction characteristics of specific nerves and root entry levels (3). Their utilization of this technique was prompted by the finding that routine somatosensory monitoring of evoked potentials did not give the same specificity to nerve roots as may be desirable to accurately detect changes during surgical correction of spinal deformities, specifically severe spondylolisthesis.

The dermatomal analysis of evoked potentials is more specific than the multi-rooted mixed nerve. This does not appear to be too important when dealing with manipulation and possible spinal cord injury but may be of great importance when nerve root decompression is performed. Furthermore, the diagnosis of radiculopathies may be accurately determined by this method according to the work of Stolov and Slimp (19), as cited by Dr. Cohen. It is yet to be determined what type of injury the nerve root can tolerate before SEP changes are noted and irreversible damage occurs.

Sunderland, in his book "Nerve and Nerve Injuries," states that nerve fibers conduct almost to the point of rupture (18). Nerve roots fail before peripheral nerves and suffer the effects of compression earlier than peripheral nerves. This results in a more severe injury to the nerve root than the comparable injury to the peripheral nerve. The reason purported by Sunderland is the lack of a protective epineurium on the nerve root.

While previous studies have assumed a quantitative convention of 50% reduction in amplitude and 10% increase in latency as the basis for impending neural injury with this method, Dr. Cohen's reasoning that a 20% amplitude decrease and 4% latency increase may indeed be more accurate as the harbinger of injury. Caution must be exercised, however, since the work of Transveldt which demonstrated that an increase in amplitude and diminished latency occurs just prior to neurologic disaster (20). Abnormal responses must be confirmed by multiple trails, and fluctuations within this range may occur.

Continuing his extensive work with spinal cord monitoring techniques, Machida with his co-workers recorded the response from surface electrodes placed over the soleus muscle following stimulation of the spinal cord with electrodes placed in the epidural space (11). They found that the wave form was consistent, irrespective of the level of the stimulation along the spine. The conduction velocity within the spinal cord was unexpectedly slow, 10-30m/sec. He concluded that the potential recorded by this method was indeed the muscle action potential mediated through the motor pathway. He suggested that simultaneous recordings of both conventional and motor stimulation may provide enhanced information during spinal surgery.

Dr. Machida alleges that there is a significant correlation between the two methods. However, changes were not noted in the spinal evoked potentials with alteration in blood pressure as usually seen in cortical evoked potentials. The vascular studies included in his investigation demonstrate the dissociation between motor evoked and spinal evoked potentials when the circulation of the anterior spinal artery is interrupted. Anterior cord syndromes may be recognized while they may produce a false negative with the technique of sensory evoked potentials alone.

Transient change can be observed in the motor evoked potential which resolves without apparent evasive action taken by the surgeon. With somatosensory cortical evoked potentials, although these tests may be sensitive, they are not specific.

The recordings obtained in Dr. Machida's animal experiments were the result of much greater stimulation magnitudes than those utilized in the human subjects. This results in fewer descending fibers being stimulated in the humans. If the locus of injury corresponds to these unstimulated fibers, they would not be portrayed or demonstrate an abnormal response. It appears necessary to invoke a greater number of fibers in order to elicit changes occurring throughout the fiber spectrum when this technique is used in the clinical setting in humans.

Recent work by Owen and associates proposed to develop a noninvasive technique for eliciting a neurogenic motor evoked potential (NMEP) (14). They attempted to determine the relative sensitivity and specificity of SEP's and NMEP's to spinal cord compression, ischemia, and distraction. NMEP data was elicited from patients undergoing surgery for spinal deformities, aortic aneurysms, and intra-medullary tumors.

Although preliminary results are reported, it appeared that the NMEP's were more sensitive and specific to the effects of spinal cord compression, ischemia, and distraction than SEP's in animal experimentation. They too utilize both SEP's and NMEP's as spinal cord monitors during surgery where spinal cord function might be jeopardized.

When stimulating spinous processes and recording from peripheral nerves, although some prodromic motor fibers are stimulated, perhaps some antidromic stimulation of posterior column sensory fibers does occur. The clinical finding of paralysis supports the conclusion that it is primarily motor pathway stimulation. Perhaps a posterior column lesion can be produced experimentally to isolate the motor tracts to confirm this theory. Cortical stimulation circumvents this problem but may invoke the theoretical production of seizures, e.g., "kindling."

Simultaneous stimulation of sensory evoked and motor evoked potentials has resulted in "modulation" of the motor evoked response, according to Levy. This appears to enhance the sensitivity and specificity of the resultant recording.

The utilization of Pavulon mandates the recording of motor evoked response from peripheral nerve rather than muscle since the transmission is blocked at the motor end plate. Other agents may be used to permit EMG analysis of the response.

Although many centers have embarked on a preliminary study in the use of motor evoked potentials, Gianutsos and his co-workers have found that the spinal cord of patients with a lesion considered to be "complete" still contains fibers which descend through the lesion and are capable of conveying impulses leading to muscle contraction (7).

The definition of "monitor" according to Webster is "something that reminds or warns." In this regard, the "wake-up test" which in the past has been the only device to ascertain the function of the anterior cord must be considered a "determinator" since it, again according to Webster, establishes that which is "settled or decided and conclusive." The works summarized and the years of investigation they represent establish the foundation on which our goal to find a complete anterior and posterior monitor of the spinal cord will be forged.

Requirements of Monitoring Equipment

In order for a monitoring device to be utilized in a safe environment, the equipment used must have proven safety to the patient. This would involve proper electric grounding as well as the assurance that no untoward side effects would ensue from the proper utilization of the particular system employed. Makeshift devices or devices not normally used for proper monitoring should not be adapted. Proper equipment already proven and tested is mandatory to assure that no ill effects will result from the testing session.

If the equipment is to be used in the operating room environment, certain requirements mandated by the hospital should have appropriate "check-out" prior to the employment of the device within the operating room confines. It is prudent to use a dedicated unit in the operating room. Devices used in animal laboratories are usually inappropriate for utilization in the operating room field. Exhaust fans are known to be a source of contamination when such a protocol is followed. Furthermore, storage of the unit should be in a relatively germ free environment with easy access to and from the operating room theater. Proper shielding and cleanliness of the apparatus must be maintained. The internal fire codes and electric maintenance must have proper safety precautions assured.

Before any device is used for patient evaluation, it should have full evaluation in appropriate laboratory testing on non-human subjects. This would allow the user of the equipment to become familiar with all aspects of the testing device and the idiosyncrasies of the testing situation. It is inappropriate to begin with a new piece of equipment in a critical situation. Proper training might include sojourns to other institutions and individuals who are familiar with the particular piece of equipment in order to prepare the user for proper applications.

False negatives and false positives should be kept to a minimum. Although all testing devices do have some false results, by proper application and interpretation, these can be minimized. Whether or not the change in a record occurs during a critical time is extremely important in determining the overall reliability of the signal.

With the increasing problems of cost effective analysis, some impediment has been applied to the utilization of the monitoring equipment both in the treatment of disease and research activities. By sharing the device as an interdepartmental function, this cost effective situation can be alleviated. At present, the cardiovascular surgeons, the Departments of Neuroradiology and Neurosurgery and the Departments of Pediatric and Pediatric Neurosurgery are all involved in utilization of spinal cord monitoring equipment for various disease states and procedures. In addition, it has been proven to be of great assistance in the clinical evaluation of spinal cord injuries both in the initial evaluation and the prognosis of the nature of the injury. Finally, in the field of research, it has had vast applications in all of the above areas.

Finally, the monitor equipment must be appropriate for the intended use and current case. The nature of the field of spinal cord monitoring now permits us to monitor a variety of situations and spinal cord tracts. Various devices can monitor auditory, visual, and somatosensory evoked potentials. It is the nature of the field to select the device and methodology which best serves the particular patient in his clinical course. This necessitates a complete evaluation of exactly what is demanded prior to the utilization of any particular device or method. In this way, the occurrence of false negatives and positives can also be kept to a minimum.

Interpretation of Published Articles

It is extremely important when analyzing recently published articles on the subject of spinal cord monitoring to review with great care the methodology and results in order to arrive at the efficacy of the reports. The interpretation and recommendations based upon these papers has led to false assumptions. By careful scrutiny of the techniques used and the recordings obtained, erroneous results are often the result of inappropriate utilization of a particular monitor system which has no place in the cases presented. For example, one would not monitor a median nerve when one is interested in a spinal cord injury below the T6 level. Appropriate windows of latency must also be selected in order to isolate and amplify changes occurring during the monitoring session.

It is also important to select appropriate stimulation sights to avoid false negatives. One example would be monitoring bilateral extremities to avoid false negatives in cases of a Brown-Sequard deficit.

Utilization of Monitor by Surgeon

When the monitor is used in the operating room to avoid neurologic complications, the final decision must be made by the surgeon as to the appropriate course of action based upon the monitor findings. To this end, the surgeon must know the changes which may occur during the surgical procedure as well as have some knowledge of the function of the monitor machine. In this way, he can interpret properly the changes which occur at a specific time during the surgical procedure in order to take appropriate action. If the surgical procedure cannot be altered in spite of an ominous alteration on the monitor, it is often the surgeon's decision to disregard the recording and proceed with the surgical procedure. In these cases, the safety to the life of the patient becomes relatively more important than the potential hazard to the neurologic system. Nevertheless, the presence of a spinal cord monitor in the operating room may indeed act as a deterrent for taking unnecessary chances during the surgical procedures.

Scoliosis

The utilization of the somatosensory evoked potential as a neurologic determinant as to the cause of idiopathic scoliosis has been reported in the past. There is a theory that a neurologic lesion does exist in idiopathic scoliosis. This was reported by a Japanese study in Kyoto at the SRS meeting in 1977 by Dr. Hiroshi Yamamoto (21). Alterations seen in cases of "normal" idiopathic scoliosis do occur after corrective surgery on the spinal deformity. These somatosensory evoked potential changes may be due to an anatomic change in the configuration of the spinal cord or to an alteration (enhancement?) of the vascular supply to the cord. These changes often occur within the two hour period of surgery and may be due to either recruitment of greater numbers of fibers or larger caliber fibers, thereby resulting in decreased latency and greater amplitude. These changes occur in approximately 60% of all surgically corrected scoliosis cases (5).

Perhaps some optimization of correction of spinal deformities can be monitored by evoked potentials in order to obtain the "best" neurophysiologic response. This work is difficult to assess due to unavailable methods, at present, of determining the neurologic status of the normal idiopathic scoliotic patient.

Spinal Cord Trauma

Based on the new methods of testing with both motor and somatosensory evoked potentials, perhaps the time has come to redefine the term, "complete lesion." It is now necessary to qualify this statement inasmuch as some degree of neurologic function may be electrically detectable but clinically imperceptible.

Summary

In general it can be seen that the utilization of the spinal cord monitor has indeed progressed over the past several years. Newer techniques and fuller mapping of the spinal cord are progressing but have not yet achieved the state of absolute certainty to avoid neurologic complication. It appears that even today the monitor can be only used to determine the catastrophic result of spinal cord injury rather than provide the premonition of impending disaster to the surgeon during manipulation of the spine and spinal cord. It is left to the future to provide us with the tools and methodology which will allow us to have a safe time frame of anticipation of impending disaster in order that we may take appropriate steps and have the ability to alter our proposed plan of surgical treatment. This then will allow us to avoid the epoch of neurologic catastrophe.

Fortunately, when certain neurologic deficits do become manifest and are recorded by our monitors, certain steps can be taken more rapidly with resultant reversal of the neurologic deficit. Such a case would be the immediate removal of the Harrington distraction instrumentation during a case of corrective scoliosis. This has been shown to result in reversal of the neurologic deficit with return of the monitor response. At present, this is the most prudent and available course of treatment for these identifiable cord lesions.

Bibliography

1. Bunch, W.H.; Dallman, D.: A comparison of the epidural and bipolar bone recordings of a spinal evoked potential. Ortho. Transactions, Spring, 1981; Vol. 6: 1, p. 19.
2. Berenstein, A.; Young, W.; Ransohoff, J.; Benjamin, V.; Merkin, H.: Somatosensory evoked potentials during spinal angiography and therapeutic transvascular embolization. J. Neurosurg., April, 1984; 60: 4, pp. 777-785.
3. Cohen, B.A.; Huizenga, B.: Dermatomal monitoring for surgical correction of spondylolisthesis; A case report. Presented at Annual Meeting Scoliosis Research Society, Vancouver, BC; Sept., 1987.
4. Donaldson, D.H.; Odom, J.A.; Brown, C.W.: Establishing a spinal cord monitoring system in a community hospital - pointers and pitfalls. Ortho. Transactions, Spring, 1982; Vol. 6: 1, p. 20.
5. Engler, G.L.; Spielholz, N.I.; Bernhard, W.N.; Danziger, F.; Merkin, H.; Wolff, T.: Somatosensory sensory evoked potentials during Harrington instrumentation for scoliosis. J. Bone Jt. Surg., June, 1978; Vol. 60-A: 4, pp. 528-532.
6. Ginsburg, H.H.; Shetter, A.G.; Raudzens, P.A.: Postoperative paraplegia with preserved intraoperative somatosensory evoked potentials. J. Neurosurg., Aug., 1985; Vol. 63, pp. 296-300.
7. Gianutsos, J.; Eberstein, A.; Ma, D.; Holland, T.; Goodgold, J.: A noninvasive technique to assess completeness of spinal cord lesions in humans. Experimental Neurol., 1987; 98: pp. 34-40.
8. Highland, T.R.; Goshgarian, H.G.; LaMont, R.L.: The effect of root transection on somatosensory evoked potential in spinal cord monitoring - where does the signal travel? Ortho. Transactions, Spring, 1986; Vol. 10: 1, p. 15.
9. Maccabee, P.; Levine, D.B.; Kahanovitz, N.; Pinhasov, E.: Monitoring of spinal and subcortical somatosensory evoked potentials during Harrington rod instrumentation. Ortho. Transactions, Spring, 1982; Vol. 6: 1, p. 19.
10. McNeal, D.R.; Passoff, T.; Swank, S.; Satomi, K.: Spinal Cord monitoring using epidural electrodes for stimulation and recording. Ortho. Transactions, Spring, 1982; Vol. 6: 1, p. 19.
11. Machida, M.; Weinstein, S.L.; Yamada, T.; Kimura, J.: Dissociation of muscle action potentials and spinal somatosensory evoked potentials after ischemic damage of spinal cord. Ortho. Transactions, Spring, 1986; Vol. 10: 1, p. 17.
12. Nickodem, R.J.; Bunch, W.H.; Scarff, T.B.; Dvonch, V.M.; Ibrahim, K.: Intraoperative determination of spinal cord conduction velocity. Ortho. Transactions, Spring, 1984; Vol. 8: 1, p. 160.
13. Nainzadeh, N.; Graham, J.; Lane, M.; Neuwirth, M.; Bernstein, R.: Somatosensory evoked potentials as intraoperative indicators of spinal cord function. Ortho. Transactions, Spring, 1984; Vol. 8: 1, p. 160.
14. Owen, J.H.; Laschinger, J.C.; Bridwell, K.H.; Shimon, S.M.; Nielsen, C.H.: Sensitivity and specificity of somatosensory and neurogenic-motor evoked potentials in animals and humans. Presented at Annual Meeting Scoliosis Research Society, Vancouver, BC, Sept., 1987.
15. Satomi, K.; Nishimoto, G.I.; Passoff, T.L.; Axlegaard, J.; McNeal, D.R.: Effects of selective spinal cord transections on spinal evoked potentials. Ortho. Transactions, Spring, 1982; Vol. 6: 1, p. 20.
16. Schroeder, F.W.; Hyman, A.A.; Powell, S.H.; Boettner, R.B.: Spinal evoked potentials - response alterations to anesthesia, hypotension, and hypothermia. Ortho. Transactions, Spring, 1982; Vol. 6: 1, p. 20.
17. Shaffer, J.W.; Brown, R.H.; Nash, C.L.; Kim, W.C.: Spinal cord monitoring in segmental spine wiring and fusion for idiopathic scoliosis. Ortho. Transactions, 1982; Vol. 6: 1, p. 20.
18. Sunderland, S.: The mechanical properties of peripheral nerve trunks. In Nerve And Nerve Injuries, 2nd Ed., pp. 62-66. Churchill Livingstone, 1978.
19. Turella, N.; Stolov, W.C.; Slimp, J.C.: Dermatomal somatosensory evoked potentials in spinal stenosis. Presented at Eighth International Congress of Electromyography and Related Clinical Neurophysiology, Sorrento, Italy, May, 1987.
20. Transfeldt, E.E.: Spinal evoked potentials in experimental spinal cord distraction. Ortho. Transactions, Nov., 1978; Vol. 3: 1.
21. Wilber, R.G.; Thompson, G.H.; Shaffer, J.W.; Brown, R.H.: Spinal cord monitoring and neurologic deficits in segmental spinal instrumentation. Ortho. Transactions, Spring, 1984; Vol. 8: 1, p. 161.
22. Yamamoto, H.; Tani, T.; MacEwen, G.D.; Herman, R.: An evaluation of brainstem function as a prognostication of early idiopathic scoliosis. J. Pediatr. Ortho., 1982; 2: 5, pp. 521-528.

Overview of Fundamental and Clinical Aspects of Monitoring the Spinal Cord during Spinal Cord Surgery

C.H. Tator;[*] R.D. Linden; M.G. Fehlings; C.M. Benedict; I. Bell

Introduction

There is an urgent need to monitor the physiological integrity of the spinal cord during surgery on the cord. In many centers somatosensory evoked potentials (SSEP) are routinely recorded to monitor spinal cord function (5, 6, 22, 31), and there are several reports which suggest that a change in the intraoperative SSEP is predictive of a postoperative change in the neurologic status of the patient (19, 38). However, it has been reported that technically satisfactory responses are recordable in only 57% (1) to 81% (33) of cases monitored. In contrast, we have developed a protocol using SSEPs and spinal evoked potentials (SEP), which has allowed successful monitoring of all patients with a variety of spinal and spinal cord lesions including scoliosis, syringomyelia, and various extradural and intradural neoplasms. Our protocol is described in detail in this paper.

Despite this success with SSEPs, there are still serious disadvantages to limiting monitoring to this modality. For example, it was hoped SSEPs would serve as an index of the integrity of both the sensory and motor function of the cord, but unfortunately this has not been possible due to the differing anatomical location and blood supply of the sensory and motor tracts in the cord. The SSEPs are assumed to be generated primarily in the dorsal columns (8, 28) and/or the dorsal spinocerebellar tracts (23, 40), whereas the motor tracts of the spinal cord have a different anatomical location and blood supply (25, 32, 39). The ventral cord is composed of both motor and sensory fibers, and its predominant blood supply is by the anterior spinal artery, whereas the dorsal columns are mainly supplied by the posterior spinal arteries. Accordingly, the motor tracts may be injured while the sensory components remain undamaged. Indeed, recently there have been "false-negative" results reported (15, 26) wherein postoperative paraplegia occurred despite preserved intraoperative SSEPs. In addition, we have observed a "false-positive" result detailed below in a patient in whom SSEPs were used to monitor cord function: Intraoperatively, the SSEPs disappeared, yet postoperatively there was preserved sensory and motor function.

Motor evoked potentials (MEPs) have been advocated as a solution to the problem of monitoring the motor tracts intraoperatively. Levy et al., (27) recorded MEPs intraoperatively in response to transcranial stimulation of the motor cortex and Boyd et

* Division of Neurosurgery, Toronto Western Hospital, 399 Bayhurst Street, Toronto, Ontario, Canada M4N 3M5

al., (4) stimulated the motor cortex intraoperatively with a low output resistance stimulator.

In order to facilitate the use of MEPs in the operating room, we have performed a series of fundamental studies of the MEP in an animal model of spinal cord injury. In this paper we will also review our experience in the combined recording of MEPs and SSEPs from normal and spinal cord injured rats. Finally, the present paper summarizes the important developments in this field and makes recommendations for the future.

Methods

Somatosensory evoked potential intraoperative monitoring in patients

SSEPs were recorded preoperatively, intraoperatively, and postoperatively in 44 patients described in Table 1. Both posterior tibial nerves (PTB) were stimulated at the ankles by stimulating cathodes taped to the skin between the medial malleolus and the medial border of the achilles tendon (Fig. 1). The anodes were attached to the skin 3cm distal to the cathodes. The intensity was set at 10% greater than the threshold intensity required to elicit a twitch resulting in plantar flexion of the toes. The stimulation sites used preoperatively were marked with indelible ink, and the same sites were used intraoperatively. Each leg was stimulated separately and then bilaterally with constant current electrical stimuli at a pulse duration of 250μs and a presentation rate of 4.7/sec.

Popliteal fossa (PF) potentials were recorded from active electrodes (Ag/AgCl) taped to the skin of each leg 4cm above the popliteal crease. Reference electrodes were placed on the medial aspect of each knee. SEPs were recorded from L3 and T12 with reference electrodes placed on the iliac crest. In addition, SEPs were recorded from bipolar epidural electrodes placed rostral and caudal to the operative site after completion of the laminectomy. Scalp potentials were recorded between Cz'-Fpz', Cz''-Fpz'' and C3'-C4' with electrodes attached to the scalp by collodion impregnated gauze. The prime marking indicates a position 2cm behind the location designated by the international 10-20 system. Negativity at the vertex was plotted as an upward deflection. A large belt ground electrode was placed around the calf, 8cm below the popliteal crease, and close to the recording PF electrode to decrease stimulus artifact (29). Inter-electrode impedances were below 3k. The low and high frequency filters were set at 30Hz and 2000Hz. Two replications of 1000 responses each were recorded. After a 4ms delay, the responses were recorded for 50ms after stimulus presentation. If there was no cortical response the sweep time was increased to 100ms. These stimulation and recording parameters are summarized in Table 2.

Induction of anesthesia was carried out with fentanyl citrate 5μg/kg and sodium thiopental 3-5mg/kg, and neuromuscular blockade instituted with vercuronium 0.1mg/kg. Following endotracheal intubation, controlled mechanical ventilation was set to produce pCO$_2$ 35-40mmHg. Anesthesia was maintained with 60% nitrous oxide and 0.25-0.60% isoflurane. An effort was made to establish an optimal level of isoflurane as quickly as possible, and make no further changes in the inspired concentration. Several patients received a continuous intravenous infusion of fentanyl.

Systemic blood pressure was recorded continuously from an intra-arterial catheter and was maintained at ±20% of preoperative valves. Esophageal temperature was recorded, and the mean drop was 1.8°C over the duration of surgery. Acid base and ventilatory variables were monitored with a pulse oximeter, frequent arterial blood gas analyses, and a mass spectrometer. The latter was also used to record continuous inspired and expired levels of all anesthetic gases. Neuromuscular blockade was maintained with pancuronium bromide. This technique permits the use of lower doses of

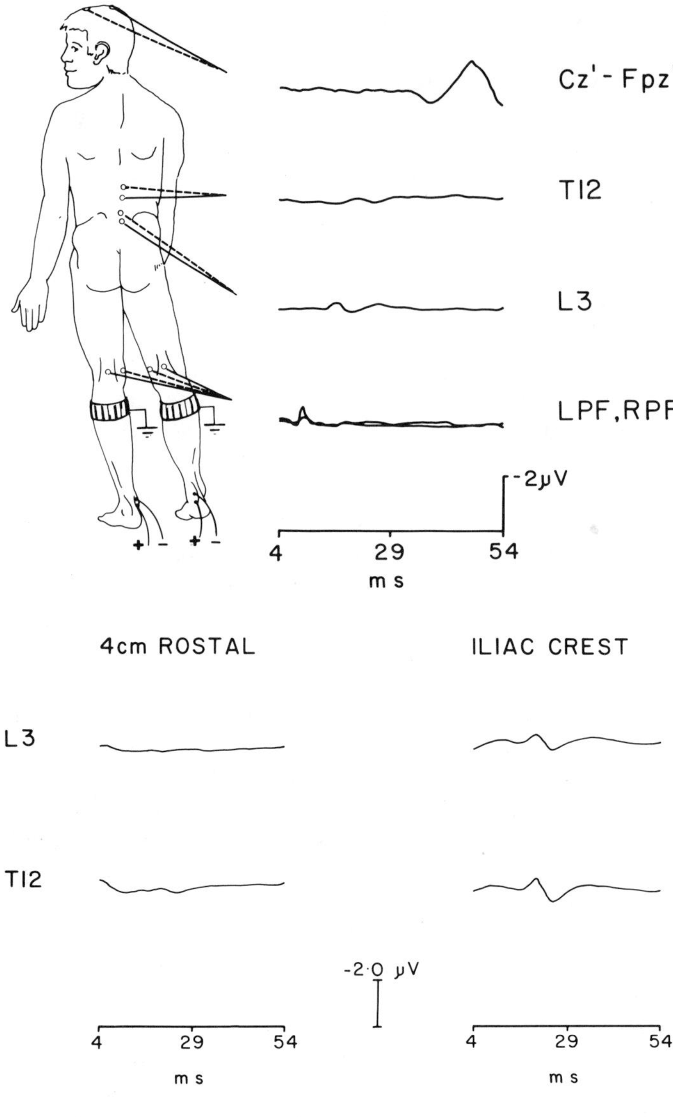

Fig. 1. Intraoperative monitoring protocol for somatosensory evoked potentials. The posterior tibial nerve at the ankle is stimulated bilaterally. The solid lines indicate the active electrodes, and the dotted lines indicate the reference electrodes. The responses are recorded at the left (LPF) and right (RPF) popliteal fossae. Spinal evoked potentials are recorded at L3 and T12. The scalp recorded potential is recording using a Cz' to Fpz' derivation. Each wave form was elicited by averaging 1000 constant current stimuli. Negativity at the vertex is plotted as an upward deflection.

Fig. 2. Grand mean spinal evoked potentials recorded from ten normal young adult subjects. The wave forms are the averaged response of 4000 constant current stimuli delivered bilaterally to the posterior tibial nerves at the ankle. The responses on the left (4cm rostral) were recorded with the reference electrode placed 4cm rostral to the active electrodes at L3 and T12, whereas the responses on the right (iliac crest) were recorded with the reference electrode placed on the iliac crest. The latter results in wave forms that are larger and easier to identify.

anesthetic agents, with the advantages of earlier awakening postoperatively, and less derangement of recorded SSEPs.

Unavoidably deviations in measured variables or changes in anesthetic drugs or doses were recorded and an attempt was made to correlate any changes in the evoked response with changes in these variables.

SSEPs and MEPs recorded from normal and spinal cord injured rats

A detailed description of the methodology to record MEPs and SSEPs from normal and spinal cord injured rats has been described elsewhere (11, 12, 13). Briefly, adult Wistar rats were anesthetized with chlorolose and urethane, and pancuronium bromide was administered for neuromuscular blockade. A tracheostomy was performed, and arterial and venous catheters inserted into the femoral artery. The rats were mechanically ventilated and maintained on a mixture of nitrous oxide and oxygen. Blood pressure, rectal temperature, and arterial blood gases were monitored.

To elicit an MEP, a 2mm platinum ball electrode was placed on the dura over the exposed sensorimotor cortex with a reference electrode placed between the tongue and the hard palate. The MEP was recorded from the cord at T9 through T10 after a laminectomy had been performed and two platinum/iridium microelectrodes were inserted into the cord. The MEP was also recorded from bipolar electrode placed on the exposed sciatic nerve. Constant current anodal stimuli were used to elicit the MEPs.

To elicit a SSEP, the exposed sciatic nerve was stimulated and the response recorded from the microelectrodes in the cord at T9 through T10 and from the electrode overlying the exposed sensorimotor cortex. The effects of varying stimulus and recording parameters on both the MEP and SSEP were determined.

To examine the effect of spinal cord injury on the MEPs and SSEPs the rat model of spinal cord injury developed in our laboratory was used (34). A C7 through T1 laminectomy was performed, and a one minute clip compression injury of the cord was inflicted with a modified aneurysm clip. Three levels of injury were examined: 56g, 20g, and 1.5g. MEPs and SSEPs were collected 30 minutes post injury. Spinal cord blood flow was measured by the hydrogen clearance technique (10A).

Results

Protocol development

Fig. 1 shows the SEP and SSEP waveforms recorded from a normal 29 year old male. Each waveform represents the averaged response to 2000 stimuli. The PF potentials were large and easy to elicit. We prefer to elicit the lower limb SSEP by stimulating the posterior tibial nerve at the ankle rather than the common peroneal nerve at the knee because the PF potentials may then be used to monitor stimulus integrity, in place of visual confirmation of twitch, thereby permitting the intraoperative administration of muscle relaxants.

The SEP waveform recorded at L3 and T12 consists of one or two negative peaks. We compared the SEPs with the reference electrodes placed 4cm rostral to the active electrode, to SEPs recorded with the reference electrode placed on the iliac crest. The distal (iliac crest) reference is superior, providing SEPs of twice the amplitude compared to SEPs recorded with the reference electrode placed 4cm rostral to the active electrode. Fig. 2 illustrates the grand mean SEP data collected from 10 normal young adults using the two reference sites.

It is advantageous to the monitoring team to have multiple back-up recording channels. Fig. 3 illustrates the grand mean data collected from 10 normal young adults with three different scalp montages. The Cz' and Fpz' montage is the conventional electrode placement for recording lower limb SSEPs. The back-up montage of Cz''-Fpz'' provides a scalp potential smaller in amplitude. The C3' through C4' montage has been recommended as a potential auxiliary electrode combination (7). Fig. 3 illustrates that care must be taken when recordings are made with the C3' through C4' montage because the P37 and N45 potential are small and variable.

Intraoperative results

To date we have monitored spinal cord function intraoperatively in 44 patients, and our protocol produced technically satisfactory responses in all cases. Table 2 lists the patients monitored according to the conditions for which surgery was required.

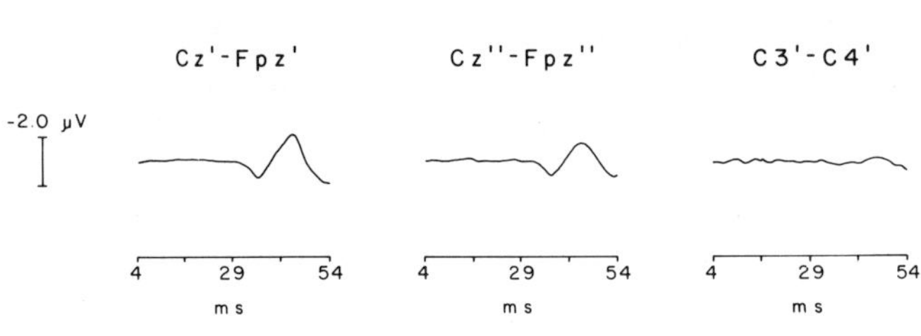

Fig. 3. The grand mean (n = 10) cortical somatosensory evoked potentials recorded in response to bilateral stimulation of the posterior tibial nerve at the ankles. The Cz'-Fpz' derivation wave form consists of P37 and N45 potentials. The amplitude of the response recorded with the Cz''-Fpz'' derivation is smaller than that recorded with the Cz'-Fpz' derivations. Often there is no observed response from a C3'-C4' derivation.

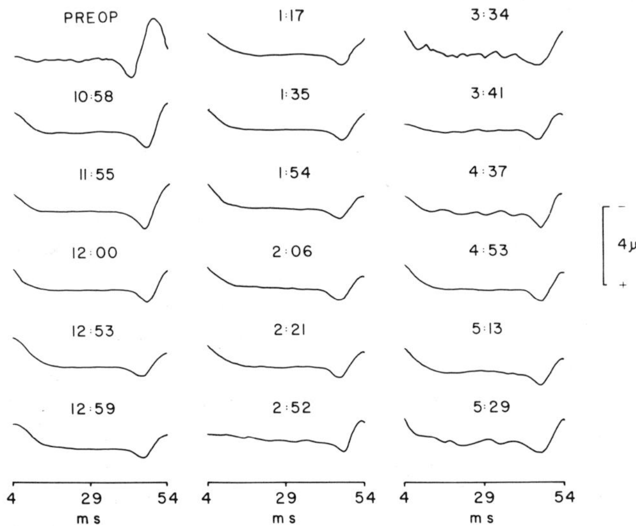

Fig. 4. Intraoperative monitoring of cortical somatosensory evoked potentials during the removal of a spinal meningioma. Each tracing is the averaged response to 1000 stimuli delivered bilaterally to the posterior tibial nerve at the ankle. The illustrated response was recorded from Cz' referenced to Fpz'. The preoperative (preop) tracing, recorded before the operation is larger and has a shorter latency than the intraoperative recording.

Fig. 4 illustrates the intraoperative SSEP in a patient during removal of a meningioma that extended from C4 to C6. The tumor arose from the dura anterolaterally on the right side, and displaced the spinal cord posteriorly and to the left. Preoperatively, the cortical P37 response to right PTB nerve stimulation was 4ms later than the response to left PTB nerve stimulation. The scalp recorded wave forms in Fig. 4 were recorded in response to bilateral PTB stimulation. Preoperatively, as illustrated in the upper left corner of the figure, the amplitude was larger and the latency earlier than the intraoperative potentials. These changes are considered due to the effects of anesthesia and are routinely observed.

Intraoperative improvement in the recorded wave forms is frequently observed. For example, after decompression an increase in the response amplitude may be observed. Fig. 5 illustrates the changes observed during decompression of the L5 root: The cortical SSEP recorded from Cz' to Fpz' increased in amplitude. Occasionally, after a surgical decompression there is both an increase in the response amplitude and a decrease

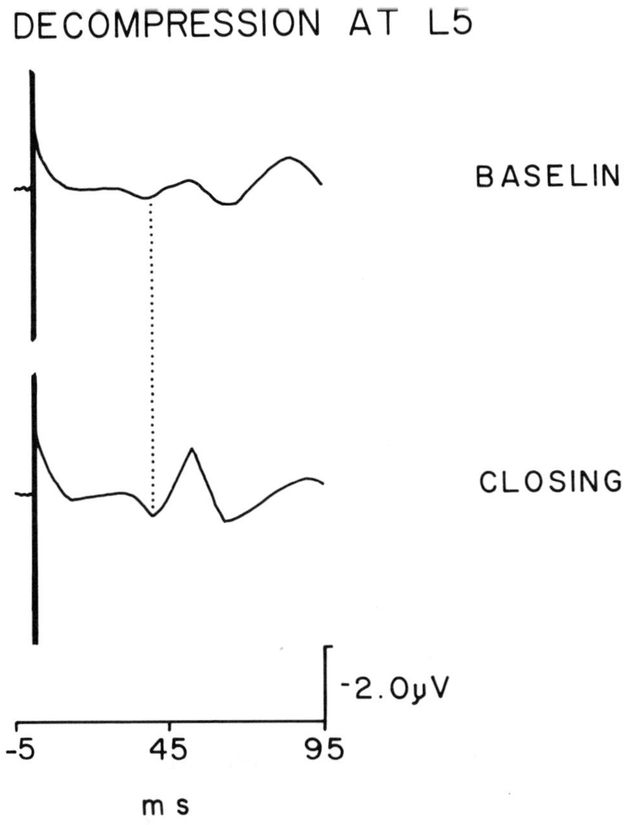

Fig. 5. Cortical somatosensory evoked potentials recorded during decompression of the L5 root. After induction of anesthesia, but before the surgery commenced a baseline tracing was obtained (baseline). After the decompression the amplitude of the response increased (closing).

in the response latency (Fig. 6). The positivity at 37ms is the peak used for calculation of the latency, and amplitude measurement is based upon the peak-to-peak amplitude between the positivity at 37ms and the negativity at 45ms.

When the operative field permits epidural placement of electrodes, we routinely record SEPs from epidural bipolar electrodes, and Fig. 7 illustrates epidural SEPs recorded intraoperatively. The recorded wave forms are large, polyphasic, and easy to elicit. The onset of the responses is earlier in the electrode placed caudal to the operative site compared to the rostral electrode. However, these potentials may vary when recorded from patients with myelopathy. The figure illustrates epidural SEPs recorded during surgery for insertion of shunt for syringomyelia and for removal of an ependymoma.

Report of a false-positive response

The patient was a 23 year old male who suffered a neck injury while playing hockey in April, 1985. Two months later he developed weakness in his left arm and had difficulty running. His left arm and leg strength continued to deteriorate. On admission to the neurosurgical service at the Toronto Western Hospital, he had a Brown-Sequard syndrome with a level at about C4. The myelogram demonstrated a complete block at C5 through C6, and the delayed metrizamide CT scan showed a possible syrinx. The MRI demonstrated a syrinx at C2 through C3, and gave some evidence of an adjacent

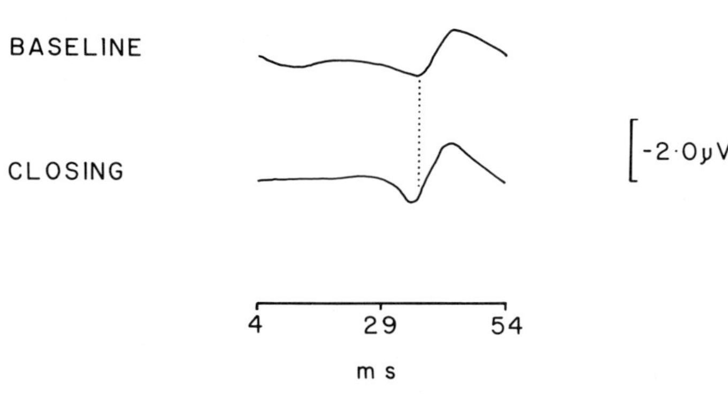

Fig. 6. Intraoperative cortical somatosensory evoked potentials recorded during an anterior cervical decompression procedure. After the decompression the amplitude of the response increased and the latency decreased. The improved response was present for the remainder of the operation.

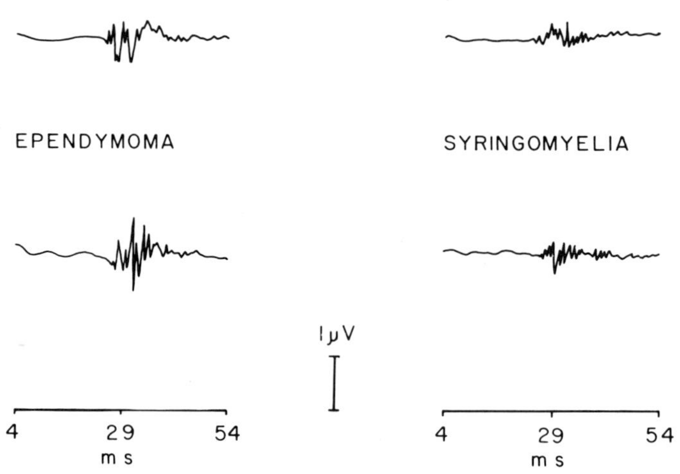

Fig. 7. Spinal evoked potentials recorded with epidural bipolar electrodes. Electrodes are placed caudal and rostral to the operative site. The wave forms on the left were recorded from a patient who had an ependymoma removed from the spinal cord. The wave forms on the right were recorded from a patient who had a shunt inserted for syringomyelia. Note the variability of the responses due to the varying types of myelopathy.

intramedullary tumor from C4 to T2, although the findings were not definite. After a laminectomy at C5, the dura was found to be under increased tension. A 2mm midline opening was made in the dura, after which there was slight herniation of cord tissue through the opening. A 1.5mm midline myelotomy was performed, the dorsal columns were gently retracted, and a small biopsy was made with cup forceps at a depth of 2mm. The quick section was reported as "suggestive of astrocytoma." A complete laminectomy was then performed from C3 to T2 inclusive, and the dura was opened in the midline and retracted. The cord was markedly expanded over the entire distance, with the left side more expanded than the right. Through a midline myelotomy, the tumor with adjacent syringomyelic cavities was gradually separated from the cord circumferentially, and completely removed. The final pathological diagnosis was ependymoma. Postoperatively, the patient's left arm and leg were slightly weaker than preoperatively, and there was a patchy diminution of pin prick sensation, although he had normal position sense in his arms and legs. Vibration sense was decreased bilaterally in the legs, more so on the left, but it was still present at the toes. These sensory findings were only

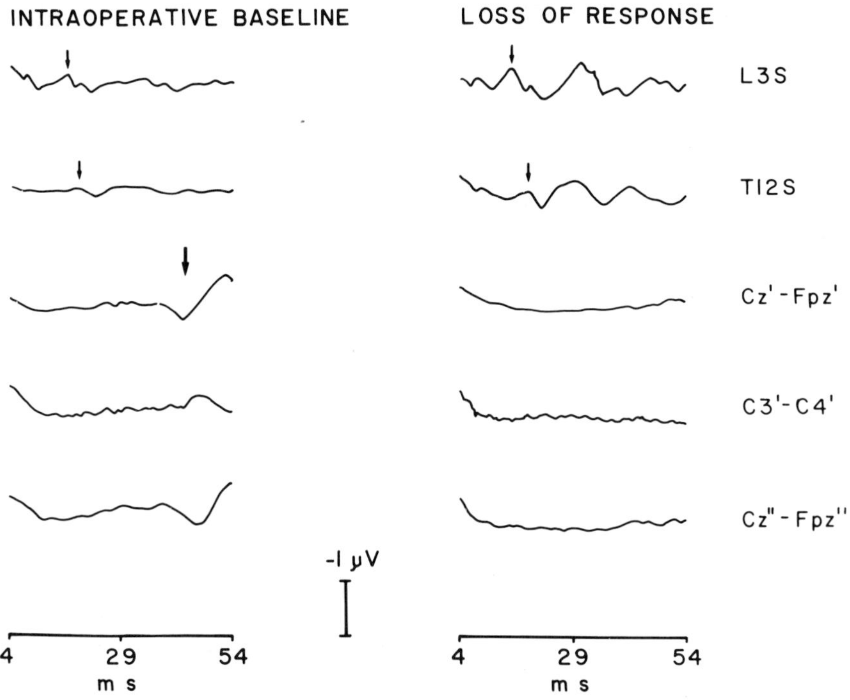

Fig. 8. The case of a false-positive result recorded with intraoperative somatosensory evoked potentials. The tracings on the left are the baseline responses recorded after a stable level of anesthesia was achieved, and on the right are the responses recorded immediately after tumor biopsy. The arrows mark the monitored peaks. Caudal to the operative site, the signal was still observed, but rostrally no response was present in any of the scalp montages.

slightly worse than preoperatively. During the next 12 months there was progressive neurological improvement, and one year after the operation the neurological examination showed no weakness or sensory loss.

The preoperative evoked potential recordings consisted of a negative potential at PF at approximately 9ms in both legs. The L3 potential consisted of a small negativity at 17ms and the T12 negativity was observed at 20ms. The cortical P37 potential was observed in all the scalp recordings. There was no N45 response recorded with the C3' through C4' electrodes in response to bilateral PTB stimulation. When each leg was stimulated separately, an N45 potential was recorded in the C3' through C4' channel. The cortical response to left stimulation was 2ms later than the response to right leg stimulation. Fig. 8 illustrates the intraoperative monitoring baseline on the left side of the figure and the loss of the SSEP, on the right side of the figure. Each tracing represents the averaged response to 2000 stimuli presented bilaterally. Immediately after the tumor was biopsied, the cortical response disappeared, and remained absent for the remainder of the operation. The sweep time was changed to 100ms post stimulus, but there was still no response.

Fig. 9 illustrates the SEPs recorded with bipolar epidural electrodes placed rostral and caudal to the operative site in this case. Before the biopsy, a multiphasic potential was observed caudally, and rostrally a similar potential was recorded with onset latency approximately 2ms later than the caudal response. After biopsy of the tumor, there was still a caudal response, but no rostral response. The epidural electrodes were removed and reversed as a check on electrode malfunction, but again only the caudal response

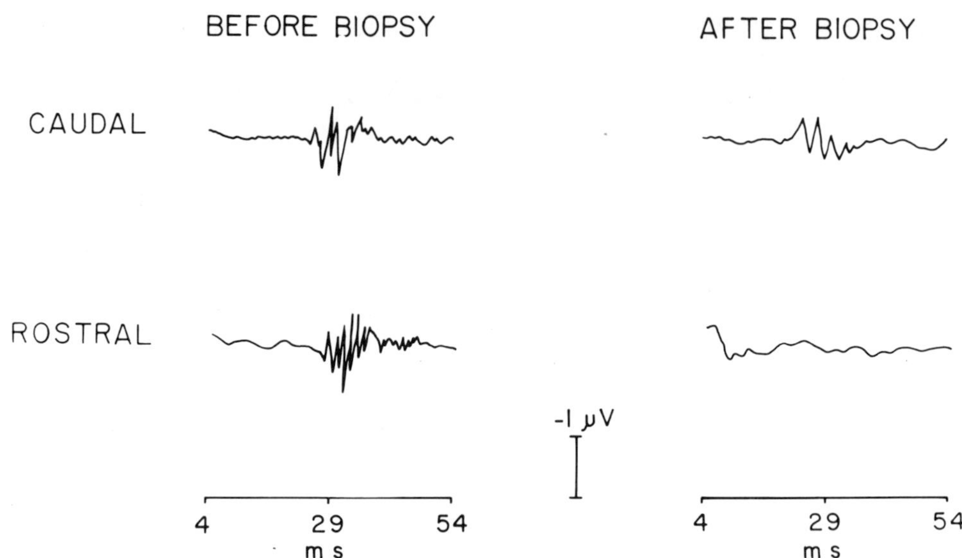

Fig. 9. The spinal evoked potentials in the same case as in Fig. 8 recorded with epidural bipolar electrodes placed rostral and caudal to the operative site. The tracings on the left were recorded before tumor biopsy, while those on the right were recorded after tumor biopsy, and showed absence of the rostral response.

was present. Throughout the remainder of the operation during which the tumor was completely removed, the rostral response remained absent.

The SSEPs were recorded at one week, and again at two months postoperatively. Fig. 10 illustrates the preoperative and postoperative recordings of the responses to left (LPTB), right (RPTB) and bilateral posterior tibial nerve (BPTB) stimulation recorded from Cz'-Fpz'. The latencies of the BPTB and RPTB response were the same preoperatively, and postoperatively, but the amplitude of the preoperative BPTB response was twice .the amplitude of the postoperative BPTB response, and there was no LPTB response recorded postoperatively.

Animal model of spinal cord injury

Motor and sensory evoked potentials recorded from normal rats

The MEP recorded from the spinal cord of normal rats consists of five waves: An initial D wave, thought to reflect direct activity in the pyramidal tracts; and then four i waves, thought to be generated by cortical interneuronal loops, although wave i4 is variable. As the intensity of the stimulus is increased, the latencies of all 5 peaks decreases. Peaks D through i4 increase in amplitude as the stimulus intensity is increased, and the effect of increasing the stimulus intensity on these peaks reaches a plateau at approximately 9mA. When MEPs are recorded from a peripheral nerve, higher stimulus intensities are required to elicit a response.

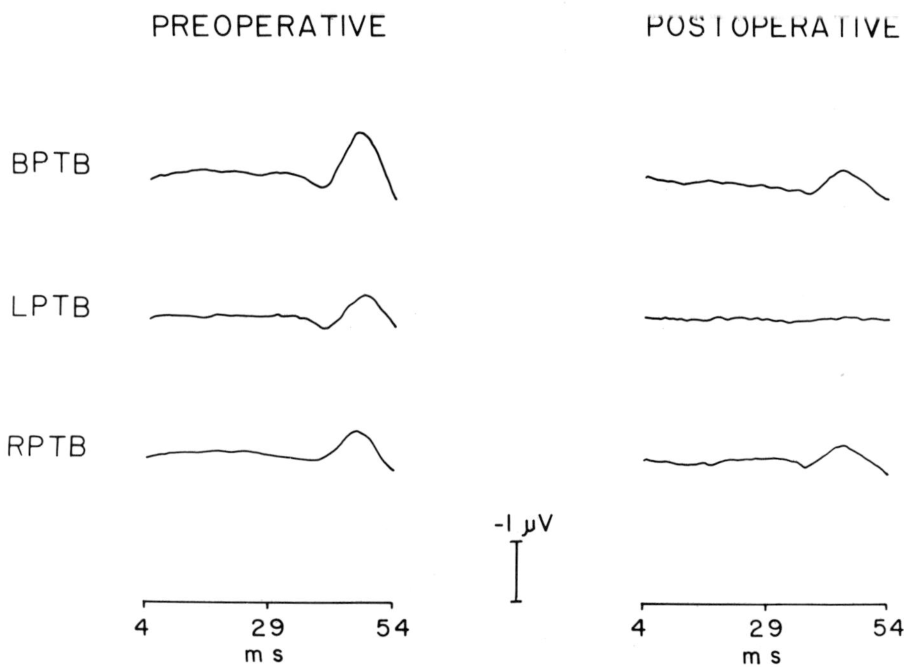

Fig. 10. The cortical SSEPs recorded preoperatively and postoperatively in the same case as Fig. 8. On the left are the preoperative recordings, and on the right are the tracings recorded two months postoperatively. The responses to left (LPTB), right (RPTB), and bilateral (BPTB) stimulation of the posterior tibial nerve at the ankle were recorded. Preoperatively, the P37 and N45 responses had a longer latency when the left leg was stimulated as compared to right leg stimulation. Postoperatively, the amplitude of the SSEPs was smaller when recorded in response to either BPTB or RPTB stimulation, but there was no response to LPTB stimulation.

The SSEP recorded from normal rats in response to direct stimulation of the sciatic nerve consists of three negative and four positive peaks. Unlike the MEP, the sciatic SSEP is relatively insensitive to alteration in the intensity of the stimulus parameters, in duration and repetition rate.

Recently, we have examined the relationship between severity of spinal cord injury, MEPs and SSEPs (14). During mild spinal cord injury the SSEP remains unchanged whereas the MEP recorded from the cord and peripheral nerve is decreased in amplitude and increased in latency.

Motor and sensory evoked potentials from rats after spinal cord trauma and ischemia

Following spinal cord injury the spinal cord blood flow (SCBF) has been shown to decrease (20, 34), and this posttraumatic ischemia is thought to promote further tissue destruction following the initial mechanical injury (20). SSEPs have been shown to be sensitive to alterations in spinal cord blood flow (SCBF), that therefore the recording of these potentials after injury may provide a non-invasive method of assessing SCBF. Indeed, Bennett (3) recommended that the recording of both MEPs and SSEPs may provide an assessment of SCBF in both the dorsal and ventral portions of the cord which have differing blood supply.

Fig. 11 illustrates the effect of spinal cord ischemia on the MEP and SSEP recorded from the rat spinal cord, with spinal cord blood flow measured by the hydrogen

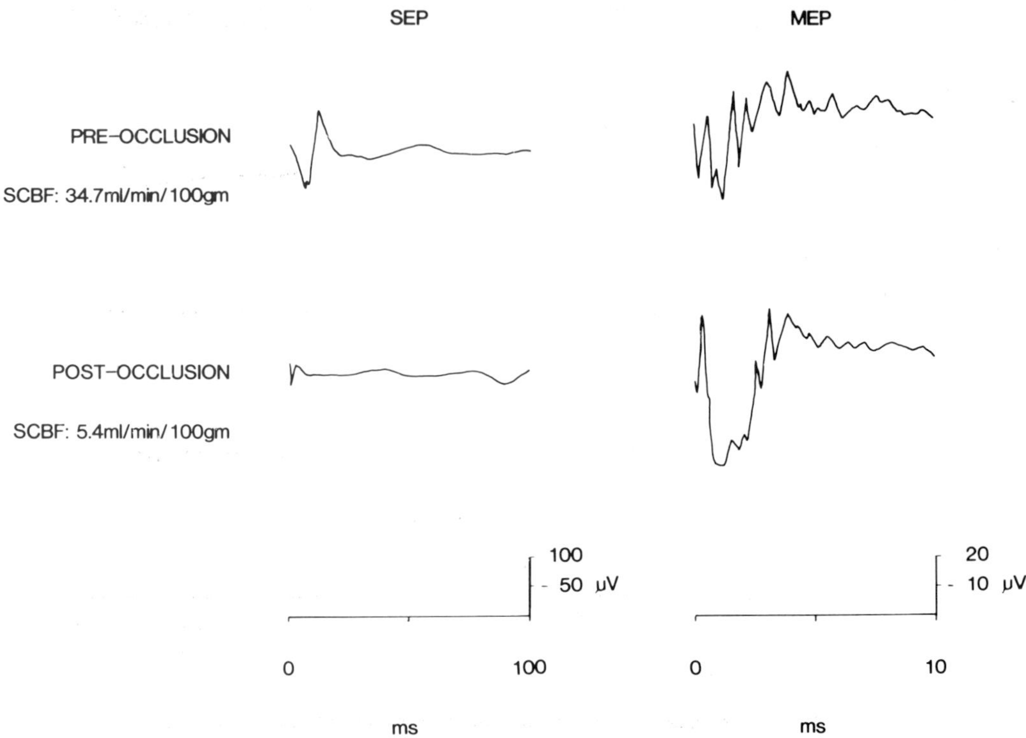

Fig. 11. The effect of spinal cord ischemia on the motor evoked potentials and spinal evoked potentials recorded from the rat spinal cord. Spinal cord blood flow (SCBF) was measured with the hydrogen clearance technique. The somatosensory evoked potential was lost at 5 minutes post-occlusion. An abnormal MEP response was present at 10 minutes post-occlusion.

clearance technique. A transabdominal, retroperitoneal approach to the abdominal aorta was performed, the renal arteries identified, and a ligature placed around the abdominal aorta 1cm proximal to the renal artery. After occlusion, the femoral artery pressure was 0mmHg, and the SEP response was completely lost at 5 minutes post-occlusion. The MEP response was still present 10 minutes post-occlusion, although the latency was delayed and a deep injury potential was present (8a). We have evaluated the MEP and SEP changes following trauma to the cord and posttraumatic ischemia (14). The SCBF threshold for loss of axonal function is 28.9 ± 4.0ml/100g/min for the MEP and 21.5 ± 3.5ml/100g/min for the SSEP. Therefore, posttraumatically, the MEP is a more sensitive index of ischemia.

Discussion

Current techniques allow the intraoperative recording of technically satisfactory evoked potentials, and the methods we have described are easy to use and have several advantages. By stimulating the posterior tibial nerve at the ankle, as recommended by Chiappa (7) for intraoperative monitoring, large responses are obtained. Furthermore, the use of this stimulus site is helpful to the anesthesiologist because a muscle twitch is no longer required to monitor stimulus integrity. Thus the popliteal fossa potentials provide an index of stimulus integrity. The optimal reference site for the recording of SEPs is unclear: The American Electroencephalograph Society (2) recommended that the SEP reference electrode be placed 4cm rostral to the active electrode for clinical evoked potential studies, whereas Chiappa (7) recommended the iliac crest as a reference site. Grundy (9) suggested that either position could be used. We compared the SEPs recorded from both reference sites from 10 normal young adults: The iliac crest reference provided potentials of twice the amplitude of those recorded with the reference electrode placed 4cm rostral to the active electrode, and therefore we recommend the iliac crest as the reference site. Bipolar epidural electrodes rostral and caudal to the operative site provide the best method of monitoring: The responses are large and may be recorded with a small number of averages (50-100).

The use of multiple back-up channels for recording is important. The three derivations we routinely record are easy to use. The Cz"-Fpz" combination provides responses of slightly smaller amplitude than the Cz'-Fpz' derivation. The C3' through C4' is the least helpful, because the cortical field orientation of these responses often cancel, resulting in a wave form that indicates an absent response.

Despite reports of the ability of SSEPs to accurately monitor spinal cord function intraoperatively, the technique has serious limitations. With respect to motor function, there have been false-negative and false-positive results. By definition, a false-negative result denotes a preserved intraoperative SSEP in the presence of a postoperative deficit in motor function, and a false-positive result denotes a lost intraoperative SSEP with no postoperative deficit in motor function. Ginsburg et al., (15) and Lesser et al., (26) have documented several false-negative results, and the current paper demonstrates a false-positive result, in which both SSEPs and SEPs were monitored, and both showed the loss of the responses above the operative site in a patient without a significant postoperative neurological deficit.

It is important that the patient's physiologic status be maintained as stable as possible during the period of intraoperative monitoring of spinal cord function with SSEPs, and that deviations from normal be recorded. Changes in the body temperature effect the scalp recorded SSEPS: Both hypothermia (17, 24) and hyperthermia (9) have been reported to alter the responses, although the exact effect of changes in body temperature on the SSEP is unclear. Van Rheinech Leyssius et al., (41) reported that only the latency of the response was affected by changes in temperature while the amplitude remained unchanged. SSEPs are also affected by hypoxia (16, 18) and hypotension. Furthermore, McPherson et al., (30) reported that intracranial subdural gas close to the recording electrode may have false-positive results, and advocated the use of plain skull roentgenograms intraoperatively to demonstrate when subdural air is in close proximity to the SSEP electrode. In the example of a false-positive result described above, there was no change in body temperature, and therefore it is unlikely that this factor caused the loss of the response. Furthermore, because the response was lost from all three scalp channels, and from the rostral epidural bipolar electrode, it is unlikely that intracranial or spinal subdural gas caused this false-positive result.

It is important that intraoperative pharmacological parameters be kept as stable as possible because of the known effects of certain anesthetic agents as evoked potentials. Not all anesthetic agents, or clinically useful combinations of agents have been studied for their effects on evoked potential monitoring, but the available literature suggests that drugs used to induce and maintain general anesthesia, as well as some preoperative sedatives, can be expected to have an effect on SSEPs. Of particular relevance to our false-positive recording are the effects of nitrous oxide and isoflurane on the SSEPs. Nitrous oxide causes a dose-dependent increase in the latency of the SSEP but, apparently, does not effect the amplitude of the response (35, 37), while isoflurane has a dose-dependent effect on both the amplitude and latency of the SSEPs (21, 36): As the concentration of inspired isoflurane increases, the amplitude decreases, and the latency of the response increases. The effect of fentanyl on the SSEP is controversial. Hume and Durkin (21) reported that anesthetic doses of fentanyl do not affect the response, while Grundy et al., (10) reported that fentanyl diminishes the SSEP response. In the described false-positive case, intraoperative baseline recordings were made after a stable level of anesthesia was achieved, and there were no subsequent changes in the end-tidal concentration of isoflurane or nitrous oxide, nor any further fentanyl administered. Thus, it is unlikely that this false-positive result was due to pharmacological factors.

There are two possible explanations for the loss of the response described in this case. First, the slight herniation of cord tissue at the time of dural opening and myelotomy may have caused damage; and second, the retraction of the dorsal columns during the biopsy or the biopsy itself may have damaged cord tissue. It should be noted that pathological analysis of the biopsied tissue demonstrated that it contained only tumor and no cord tissue. Indeed, when the tumor was ultimately completely exposed, the biopsy site was confined to the tumor.

During the past two years we have rigorously evaluated the usefulness of recording MEPs and SSEPs in the rat model of acute compression injury of the spinal cord. In this model both the MEP and SSEP are easy to elicit and highly reproducible. In patients and in experimental studies we recommend that both MEPs and SSEPs be recorded. Previously, we have demonstrated that these potentials are transmitted through separate tracts, and therefore this methodology provides a comprehensive, neurophysiological method of monitoring both the sensory and motor function of the spinal cord.

Over the last decade we have been interested in evaluating the effect of posttraumatic ischemia on spinal cord function (20), and have developed a method based on the hydrogen clearance technique to monitor spinal cord blood flow. Our rate model of spinal cord injury provides a unique opportunity to evaluate the effect of alterations in spinal cord blood flow on the MEPs and SSEPs. Our data indicate that because of the differential sensitivity of the MEP and SSEP both should be measured as a means of obtaining information non-invasively which may reflect spinal cord blood flow.

Conclusions

Although there has been considerable progress over the past few years, we still have a long way to go to develop a safe, reliable system of physiological monitoring of the spinal cord both in and out of the operating room. The following is a brief summary of what we have learned at this conference on intraoperative monitoring.

1. Anesthetics: The basic principle is to use the lowest case compatible with adequate anesthesia. This will vary from patient to patient. Once established, the anesthetic state should be maintained with constant concentrations of inhalation agents and/or continuous infusion of intravenous agents. Many different combinations are in current use.

2. Baseline studies: Preoperative measurements should always be performed, and intraoperative measurements should always be made prior to surgical approach to the spine or cord.

3. Spinal evoked potentials: Epidural recording techniques should be refined. Our colleagues from Japan have had extensive experience with SEPs from which we may benefit. However, we should be aware of the problems of electrode placement and migration. If they are too close to the lesion, precision will be lost. The exact spatial, resolving power of these electrodes needs to be determined. For example, our early work on monkeys demonstrated a positive injury potential at a considerable distance from the injury site (9). Epidural recordings are less sensitive to anesthetic changes and multiple back-up electrodes should be used.

4. Dermatomal SSEPs: The potential value of dermatomal SSEPs requires further study. This technique may play an important role in determining the level and completeness of spinal cord lesions.

5. Nomenclature: We must standardize and improve the nomenclature used for intraoperative monitoring, and criteria for abnormal intraoperative responses must be established. These criteria should include amplitude, latency, wave form morphology, conduction velocity and response reproducibility.

6. Bilaterality: Both sides of the cord should be monitored in both ascending and descending directions. Further research is required, to improve existing, and to develop new methods of monitoring as many tracts of the spinal cord as possible such as vestibulo spinal and reticulo spinal tracts.

7. Physiologic stability: Blood pressure, core temperature, and partial pressures of oxygen and carbon dioxide should be kept as constant as possible.

8. Intensive care unit: Postoperative continuous monitoring techniques should be established for use in the intensive care unit for early detection of postoperative complications.

9. Indications for intraoperative monitoring: Virtually all cases of spinal cord surgery should be monitored.

10. Training of technologists: There should be established guidelines for the training of monitoring technologists.

Table 1: Stimulation and recording parameters SSEP and SEP recordings in patients

A. Stimulus
Type: Constant current
Electrode placement: Cathode behind the medial malleolus; anode 3cm distal
Intensity: 1.1 x threshold for twitch causing plantar toe flexion
Duration: 250μs
Repetition rate: 4.7/s

B. Recording
Electrode placement and montage:
1. Left popliteal fossa referenced to medial surface of the knee
2. Right popliteal fossa referenced to medial surface of the knee
3. L3S - iliac crest
4. T12S - iliac crest
5. Cz'-Fpz'
6. C3' through C4'
7. Cz"-Fpz"
8. Bipolar epidural electrode below the operative site
9. Bipolar epidural electrode above the operative site

Number of responses: 1000
Sweep time: 50 or 100ms
Locut filter setting: 30Hz
Hicut filter setting: 2kHz

Table 2: Summary of surgical procedures performed with monitoring of somatosensory and spinal evoked potentials

Decompression/discotomy	22
Spinal fusion	9
Shunting procedures for syringomyelia	9
Removal of spinal cord tumor	4
Total	44

Acknowledgements

Dr. M.G. Fehlings is a fellow of the Medical Research Council of Canada. Dr. I. Bell is a fellow of the Natural Sciences and Engineering Research Council of Canada. This research was funded by the Medical Research Council of Canada and the Canadian Paraplegic Association. Ms. M. Vraz, Mr. J. Loukides, and Ms. D. Wilken provided technical assistance.

References

1. Allen, A.; Starr, A.; Nudleman, K.: Assessment of sensory function in the operating room utilizing cerebral cortex potentials. A study of fifty-six surgically anesthetized patients. Clin. Neurosurg., 28: 457-481, 1981.
2. American Electroencephalic Society: Guidelines for clinical evoked potential studies. J. Clin. Neurophysiol., 1: 3-53, 1984.
3. Bennett, M.A.: Effects of compression and ischemia on spinal cord evoked potentials. Exp. Neurol., 80: 508-519, 1983.
4. Boyd, S.G.; Rothwell, J.C.; Cowan, J.M.A.; Webb, P.J.; Morley, T.; Asselman, P.; Marsden, C.D.: A method of monitoring function in corticospinal pathways during scoliosis surgery with a note on motor conductions velocities. J. Neurol. Neurosurg. Psychiat., 49: 251-257, 1986.
5. Brown, R.H.; Nash, C.L., Jr.: Current status of spinal cord monitoring. Spine, 4: 466-470, 1979.
6. Bunch, W.H.; Scarff, T.B.; Trimble, J.: Spinal cord monitoring. J. Bone Joint Surg. (AM), 65: 707-710, 1983.
7. Chiappa, K.H.: Evoked Potentials in Clinical Medicine. Raven Press, New York, 1985.
8. Cohen, A.R.; Young, W.; Ransohoff, J.: Intraspinal localization of the somatosensory evoked potential. Neurosurg., 9: 157-162, 1981.
8a. Deeke, L.; Tator, C.H.: Neurophysiological assessment of afferent and efferent conduction in the injured spinal cord of monkeys. J. Neurosurg., 39: 65-74, 1973.
9. Grundy, B.L.: Intraoperative monitoring of sensory evoked potentials. Anesthesiology, 58: 72-87, 1983.
10. Grundy, B.L.; Brown, R.H.; Berilla, J.A.: Fentanyl alters somatosensory cortical evoked potentials. Anesth. Anal., 59: 544-545, 1980.
10a. Guha, A.; Tator, C.H.; Piper, I.: Effect of a calcium channel blocker on posttraumatic spinal cord blood flow. J. Neurosurg., 66: 423-430, 1987.
11. Fehlings, M.G.; Tator, C.H.; Linden, R.D.: The combined recording of motor and somatosensory evoked potentials to assess spinal cord function. In: C. Barber; T. Blum (eds.): Evoked Potentials III. The Third International Evoked Potentials Symposium. In press.
12. Fehlings, M.G.; Tator, C.H.; Linden, R.D.; Piper, I.R.: The combined recording of motor and somatosensory evoked potentials from the rat. EEG and Clin. Neurophysiol. (in press).
13. Fehlings, M.G.; Tator, C.H.; Linden, R.D.; Piper, I.R.: Motor and somatosensory evoked potentials recorded from normal and spinal cord injured rats. Neurosurg., 20: 125-130, 1987.
14. Fehlings, M.G.; Tator, C.H.; Linden, R.D.; Piper, I.R.: The relationship between severity of spinal cord injury, motor and somatosensory evoked potentials and spinal cord blood flow. Presented at the 1987 meeting of the American Association of Neurological Surgeons, Dallas, Texas.
15. Ginsburg, H.H.; Shetter, A.G.; Raudzens, P.A.: Postoperative paraplegia with preserved intraoperative somatosensory evoked potentials. J. Neurosurg., 63: 296-300, 1985.
16. Grundy, B.L.; Heros, R.C.; Tung, A.S.; Doyle, E.: Intraoperative hypoxia detected by evoked potential monitoring. Anesth. Analg. (Clev.), 60: 437-439, 1981.
17. Grundy, B.L.; McPhail, J.; Bottoms, C.; Jolly, L.; Cullivan, J.: Effect of hypothermia on somatosensory evoked potentials during cardiopulmonary bypass. Electroenceph. Clin. Neurophysiol., 58: 41, 1984.
18. Grundy, B.L.; Nash, C.L.; Brown, R.H.: Deliberate hypotension for spinal fusion: Prospective randomized study with evoked potential monitoring. Can. Anesth. Soc. J., 29: 452-461, 1982.

19. Grundy, B.L.; Nelson, P.B.; Doyle, E.; Procopio, P.T.: Intraoperative loss of somatosensory evoked potentials predicts loss of spinal cord function. Anesth., 57: 321-322, 1982.
20. Guha, A.; Tator, C.H.; Piper, I.: Increase in rat spinal cord blood flow with the calcium channel blocker nimodipine. J. Neurosurg., 3: 250-259, 1985.
21. Hume, A.L.; Durkin, M.A.: Central and spinal somatosensory conduction times during cardiopulmonary bypass and some observations on the effects of fentanyl and isoflurane anesthesia. Electroenceph. Clin. Neurophysiol., 65: 46-58, 1986.
22. Jones, S.J.; Edgar, M.A.; Ransford, A.O.; Thomas, N.P.: A system for the electrophysiological monitoring of the spinal cord during operations for scoliosis. J. Bone Joint Surg. (Br.), 65: 134-139, 1983.
23. Jones, S.J.; Edgar, M.A.; Ransford, A.O.: Sensory nerve conduction in the human spinal cord: Epidural recordings made during scoliosis surgery. J. Neurol. Neurosurg. Psychiat., 45: 446-451, 1982.
24. Lam, A.M.; Contreras, J.; Keane, J.F.; Manninen, P.H.; Brown, S.: Effects of mild hypothermia on brainstem auditory and somatosensory evoked responses. Anesth. Analyg., 64: 242, 1985.
25. Lazorthes, G.; Gouaze, A.; Zadeh, J.O.; Santin, J.J.; Lazorthes, Y.; Burdin, P.: Anterior vascularization of the spinal cord. J. Neurosurg., 35: 253-262, 1971.
26. Lesser, R.P.; Raudzens, P.; Luders, H.; Newer, M.R.; Goldie, W.D.; Dinner, D.S.; Klem, G.; Hahn, J.F.; Shetter, A.G.; Ginsburg, H.H.; Gurd, A.R.: Postoperative neurological deficits may occur despite unchanged intraoperative somatosensory evoked potentials. Ann. Neurol., 19: 22-25, 1986.
27. Levy, W.J.; York, D.H.; McCaffrey, M.; Tanzer, F.: Motor evoked potentials from transcranial stimulation of the motor cortex in humans. Neurology, 15: 287-302, 1984.
28. Macon, J.B.; Poletti, C.E.; Sweet, W.H.; Ojemann, R.G.; Zervas, N.T.: Conducted somatosensory evoked potentials during spinal surgery. Part 1: Clinical applications. J. Neurosurg., 57: 354-359, 1982.
29. McGill, K.C.; Cummins, K.L.; Dorfman, L.J.; Berlizot, B.B.; Luetkemeyer, K.; Nishimura, D.G.; Widrow, B.: On the nature and elimination of stimulus artifact in nerve signals evoked and recorded using surface electrodes. IEEE Transactions on Biomed. Engineer., 29: 129-137, 1982.
30. McPherson, R.W.; Toung, T.J.K.; Johnson, R.M.; Rosenbaum, A.E.; Wang, H.: Intracranial subdural gas: A cause of false-positive change of intraoperative somatosensory evoked potentials. Anesth., 62: 816-819, 1985.
31. Nash, C.L., Jr.; Brown, R.H.: The intraoperative monitoring of spinal cord function: Its growth and current status. Orthop. Clin. North Am., 10: 919-926, 1979.
32. Powers, S.K.; Bolger, C.A.; Edwards, M.S.B.: Spinal cord pathways mediating somatosensory evoked potentials. J. Neurosurg., 57: 472-482, 1982.
33. Raudzens, P.A.: Intraoperative monitoring of evoked potentials. Ann. N.Y. Acad. Sci., 388: 308-326, 1982.
34. Rivlin, A.S.; Tator, C.H.: Regional spinal cord blood flow in rats after severe cord trauma. J. Neurosurg., 49: 844-853, 1978.
35. Sebel, P.S.; Flynn, P.J.; Ingram, D.A.: Effect of nitrous oxide on visual, auditory, and somatosensory evoked potentials. Br. J. Anesth., 56: 1403-1410, 1984.
36. Sebel, P.S.; Ingram, D.A.; Flynn, P.J.; Rutherford, C.F.; Rogers, H.: Evoked potentials during isoflurane anesthesia. Br. J. Anesth., 58: 580-585, 1986.
37. Sloan, T.B.; Koht, A.: Depression of cortical somatosensory evoked potentials by nitrous oxide. Br. J. Anesth., 57: 849-852, 1985.
38. Sloan, T.; Koht, A.; Ronai, A.; Toleikis, J.R.: Events associated with intraoperative evoked potential changes correlation with postoperative neurolgocial status. Anesth. Analg., 64: 285, 1985.
38a. Tator, C.H.: Seminars in Neurological Surgery: Early Management of Acute Spinal Cord Injury. Raven Press, New York, 1982.
39. Turnbull, I.M.; Briey, A.; Hassler, O.: Blood supply of cervical spinal cord in man: A microangiographic cadaver study. J. Neurosurg., 24: 951-965, 1966.
40. York, D.H.: Somatosensory evoked potentials in man: Differentiation of spinal pathways responsible for conduction from the forelimb vs. hindlimb. Prog. in Neurobiol., 25: 1-25, 1985.
41. Van Rheineck Leyssius, A.J.; Kalkman, C.J.; Bovill, J.G.: Influence in moderate hypothermia on posterior tibial nerve somatosensory evoked potentials. Anesth. Analg., 65: 475-480, 1986.

Index

R

S

T

V

W